Lecture Notes in Bioinformatics 9096

Subseries of Lecture Notes in Computer Science

More information about this series at http://www.springer.com/series/5381

Robert Harrison · Yaohang Li
Ion Măndoiu (Eds.)

Bioinformatics Research and Applications

11th International Symposium, ISBRA 2015
Norfolk, USA, June 7–10, 2015
Proceedings

 Springer

Editors
Robert Harrison
Georgia State University
Atlanta, Georgia
USA

Ion Măndoiu
University of Connecticut
Storrs, Connecticut
USA

Yaohang Li
Old Dominion University
Norfolk, Virginia
USA

ISSN 0302-9743   ISSN 1611-3349   (electronic)
Lecture Notes in Bioinformatics
ISBN 978-3-319-19047-1   ISBN 978-3-319-19048-8   (eBook)
DOI 10.1007/978-3-319-19048-8

Library of Congress Control Number: 2015939350

LNCS Sublibrary: SL8 – Bioinformatics

Springer International Publishing AG Switzerland is part of Springer Science+Business Media
(www.springer.com)

Preface

The 11th edition of the International Symposium on Bioinformatics Research and Applications (ISBRA 2015) was held during June 7–10, 2015 in Norfolk, Virginia. The symposium provided a forum for the exchange of ideas and results among researchers, developers, and practitioners working on all aspects of bioinformatics and computational biology and their applications.

There were 98 submissions received in response to the call for papers. The Program Committee decided to accept 48 of them for publication in the proceedings and oral presentation at the symposium: 34 for Track 1 publication (up to 12 pages) and 14 for Track 2 (up to 2 pages). The technical program also featured invited keynote talks by four distinguished speakers: Prof. Michael Brudno from University of Toronto spoke on (computationally) solving rare disorders, Prof. Benny Chor from Tel-Aviv University spoke on what every biologist should know about computer science, Prof. Aidong Zhang from State University of New York at Buffalo spoke on dynamic tracking of functional modules in massive biological data sets, and Prof. Yang Zhang from University of Michigan spoke on protein structure prediction and protein design. Additionally, the technical program of the symposium included tutorials, poster sessions, and invited talks presented at the 4th Workshop on Computational Advances in Molecular Epidemiology (CAME 2015).

We would like to thank the Program Committee members and external reviewers for volunteering their time to review and discuss symposium papers. We would also like to thank the Chairs of CAME 2015 for enriching the technical program of the symposium with a workshop on an important and active area of bioinformatics research. We would like to extend special thanks to the Steering and General Chairs of the symposium for their leadership, and to the Finance, Publicity, Workshops, Local Organization, and Publications Chairs for their hard work in making ISBRA 2015 a successful event. Last but not least we would like to thank all authors for presenting their work at the symposium.

June 2015

Robert Harrison
Yaohang Li
Ion Măndoiu

Symposium Organization

Steering Chairs

Dan Gusfield   University of California, Davis, USA
Ion Măndoiu   University of Connecticut, USA
Yi Pan   Georgia State University, USA
Marie-France Sagot   Inria, France
Alexander Zelikovsky   Georgia State University, USA

General Chairs

Mitra Basu   National Science Foundation, USA
Desh Ranjan   Old Dominion University, USA

Program Chairs

Robert Harrison   Georgia State University, USA
Yaohang Li   Old Dominion University, USA
Ion Măndoiu   University of Connecticut, USA

Finance Chairs

Raj Sunderraman   Georgia State University, USA
Anu Bourgeois   Georgia State University, USA

Publications Chair

Ashraf Yaseen   Texas A&M University-Kingsville, USA

Local Organization Chair

Jing He   Old Dominion University, USA

Publicity Chairs

Steven Pascal   Old Dominion University, USA
Yanqing Zhang   Georgia State University, USA

Workshops Chair

Alexander Zelikovsky   Georgia State University, USA

Webmaster

Igor Mandric   Georgia State University, USA

Program Committee

Sahar Al Seesi	University of Connecticut, USA
Srinivas Aluru	Georgia Institute of Technology, USA
Mukul S. Bansal	University of Connecticut, USA
Robert Beiko	Dalhousie University, Canada
Paola Bonizzoni	Università degli Studi di Milano-Bicocca, Italy
Zhipeng Cai	Georgia State University, USA
Doina Caragea	Kansas State University, USA
Patrick Chan	South China University of Technology, China
Tien-Hao Chang	National Cheng Kung University, Taiwan
Ovidiu Daescu	University of Texas at Dallas, USA
Bhaskar Dasgupta	University of Illinois at Chicago, USA
Amitava Datta	University of Western Australia, Australia
Jorge Duitama	International Center for Tropical Agriculture, Colombia
Oliver Eulenstein	Iowa State University, USA
Lin Gao	Xidian University, China
Lesley Greene	Old Dominion University, USA
Katia Guimaraes	Universidade Federal de Pernambuco, Brazil
Jiong Guo	Universität des Saarlandes, Germany
Steven Hallam	University of British Columbia, Canada
Jieyue He	Southeast University, China
Jing He	Old Dominion University, USA
Matthew He	Nova Southeastern University, USA
Steffen Heber	North Carolina State University, USA
Wei Hu	Houghton College, USA
Xiaohua Tony Hu	Drexel University, USA
Jinling Huang	East Carolina University, USA
Ming-Yang Kao	Northwestern University, USA
Wooyoung Kim	University of Washington Bothell, USA
Danny Krizanc	Wesleyan University, USA
Guojun Li	Shandong University, China
Jing Li	Case Western Reserve University, USA
Min Li	Central South University, China
Shuai Cheng Li	City University of Hong Kong, China
Fenglou Mao	University of Georgia, USA
Osamu Maruyama	Kyushu University, Japan
Giri Narasimhan	Florida International University, USA

Bogdan Pasaniuc | University of California at Los Angeles, USA
Steven Pascal | Old Dominion University, USA
Andrei Paun | University of Bucharest, Romania
Nadia Pisantı | Università di Pisa, Italy
Teresa Przytycka | NCBI, USA
Sven Rahmann | University of Duisburg-Essen, Germany
David Sankoff | University of Ottawa, Canada
Daniel Schwartz | University of Connecticut, USA
Russell Schwartz | Carnegie Mellon University, USA
Joao Setubal | University of São Paulo, Brazil
Xinghua Shi | University of North Carolina at Charlotte, USA
Ileana Streinu | Smith College, USA
Zhengchang Su | University of North Carolina at Charlotte, USA
Wing-Kin Sung | National University of Singapore, Singapore
Sing-Hoi Sze | Texas A&M University, USA
Gabriel Valiente | Technical University of Catalonia, Spain
Jianxin Wang | Central South University, China
Li-San Wang | University of Pennsylvania, USA
Lusheng Wang | City University of Hong Kong, China
Peng Wang | Chinese Academy of Science, China
Seth Weinberg | Old Dominion University, USA
Fangxiang Wu | University of Saskatchewan, Canada
Yufeng Wu | University of Connecticut, USA
Minzhu Xie | Hunan Normal University, China
Dechang Xu | Harbin Institute of Technology, China
Zhenyu Xuan | University of Texas at Dallas, USA
Ashraf Yaseen | Texas A&M University-Kingsville, USA
Noah Zaitlen | University of California, San Francisco, USA
Alex Zelikovsky | Georgia State University, USA
Fa Zhang | Institute of Computing Technology, China
Yanqing Zhang | Georgia State University, USA
Fengfeng Zhou | Chinese Academy of Sciences, China
Leming Zhou | University of Pittsburgh, USA

Additional Reviewers

Abdelaal, Maha
Aguiar-Pulido, Vanessa
Armaselu, Bogdan
Artyomenko, Alexander
Arunchalam, Harish
Bedford, John
Benshoof, Brendan
Biton, Anne
Brown, Brielin

Brown, Robert
Chen, Yong
Cho, Dongyeon
Chu, Chong
Collins, Jason
Cowman, Tyler
Dondi, Riccardo
Dong, Ya Fei
Durno, William

Elhefnawy, Wessam
Ganegoda, Upeksha
Hahn, Aria
Hanson, Niels W.
Hoinka, Jan
Hu, Jialu
Hu, Ke
Hu, Qiwen
Jain, Chirag
Kichaev, Gleb
Kim, Dongjae
Kim, Yoo-Ah
Konwar, Kishori
Kurdia, Anastasia
Li, Yang
Liang, Jimin
Liu, Bingqiang
Liu, Juntao
Liu, Yongchao
Luo, Junwei
Mancuso, Nicholas
Mandric, Igor
Mateescu, Cristina
Mefford, Joel
Moloney, Janet
Moret, Bernard
Morgan-Lang, Connor
Nenastyeva, Ekaterina

Nihalani, Rahul
Park, Danny
Patané, José
Perea, Claudia
Pirola, Yuri
Purcaru, Victor Gabriel
Qi, Enfeng
Rizzi, Raffaella
Roman, Theodore
Roytman, Megan
Scarlet, Emanuel
Shi, Huwenbo
Sun, Penggang
Sun, Sunah
Tabari, Ehsan
Tang, Yu
Thankachan, Sharma
Tudor, Ana Maria Mihaela
Valdes, Camilo
Vasconcelos, Elton
Walker, Karl
Wang, Yan
Weir, Michael
Wójtowicz, Damian
Xue, Yu
Zhong, Jiancheng
Zhou, Chan

Invited Keynote Talks

(Computationally) Solving Rare Disorders

Michael Brudno[1,2,3]

[1]Department of Computer Science, University of Toronto, Toronto M5S 2E4
[2]Centre for Computational Medicine,
[3]Genetics and Genome Biology, Hospital for Sick Children, Toronto M5G 1L7, Canada
brudno@cs.toronto.edu

Abstract. Gene mutations cause not only well-recognized rare diseases such as muscular dystrophy and cystic fibrosis, but also thousands of other rare disorders. While individually rare, these disorders are collectively common, affecting one to three percent of the population. The last several years have seen the identification of hundreds of novel genes responsible for rare disorders, and an even greater number of cases where a known gene was implicated in a new disease.

In this talk I will describe the computational approaches that are required to make this identification possible, and describe the tools that we (and others) have developed to enable clinicians to diagnose their patients by analyzing the patient genomes and sharing de-identified patient data.

What Every Biologist Should Know About Computer Science

Benny Chor

School of Computer Science, Tel-Aviv University, Tel Aviv, Israel
benny@cs.tau.ac.il

Abstract. We join the increasing call to take computational education of life science students a step further, beyond teaching mere programming and employing existing software tools. We describe a new course, focusing on enriching life science students with abstract, algorithmic and logical thinking, and exposing them to the computational culture. The design, structure and content of our course are influenced by recent efforts in this area, collaborations with life scientists, and our own instructional experience. Specifically, we suggest that an effective course of this nature should: (1) devote time to explicitly reflect upon computational thinking processes, resisting the temptation to drift to purely practical instruction, (2) focus on discrete notions, rather than on continuous ones, and (3) have basic programming as a prerequisite, so students need not be preoccupied with elementary programming issues. We strongly recommend that the mere use of existing bioinformatics tools and packages should not replace hands-on programming. Yet, we suggest that programming will mostly serve as a means to practice computational thinking processes. This talk deals with the challenges and considerations of such computational education for life science students. It also describes a concrete implementation of the course, and encourages its use by others.

Reference

1. Rubinstein, A., Chor, B.: Computational Thinking in Life Science Education. PLoS Comput. Biol. 10(11), e1003897 (2014)

This wark was published in the educational column of PLOS Computational Biology [1]. This is a joint work with Amir Rubinstein, School of Computer Science, Tel-Aviv University, Tel Aviv, Israel.

Dynamic Tracking of Functional Modules in Massive Biological Data Sets

Aidong Zhang

Department of Computer Science and Engineering
State University of New York at Buffalo
Buffalo, NY 14260
azhang@buffalo.edu

Abstract. Functional modules are an important aspect of living cells and are made up of proteins that participate in a particular cellular process while they may not be directly interacting with each other at all times. In recent years, while most researchers have focused on detecting functional modules from static protein-protein interaction (PPI) networks where the networks are treated as static graphs derived from aggregated data across all available experiments or from a single snapshot at a particular time, temporal nature of genomic and proteomic data has been realized by researchers. Recently, the analysis of dynamic networks has been a hot topic in data mining. Dynamic networks are structures with objects and links between the objects that vary in time. Temporary information in dynamic networks can be used to reveal many important phenomena such as bursts of activities in social networks and evolution of functional modules in protein interaction networks. In this talk, I will present our computational approaches to identify the roles of functional modules and to track the patterns of modules in dynamic biological networks. Significant modules which are correlated to observable biological processes can be identified, for example, those functional modules which form and progress across different stages of a cancer. Through identifying these functional modules in the progression process, we are able to detect the critical groups of proteins that are responsible for the transition of different cancer stages. Our approaches will discover how the strength of each detected modules changes over the entire observation period. I will also demonstrate the application of our approach in a variety of biomedical applications.

Keywords: Biological networks · Bioinformatics · Gene expression data

Protein Structure Prediction and Protein Design

Yang Zhang[1,2]
[1]Department of Computational Medicine and Bioinformatics
[2]Department of Biological Chemistry
University of Michigan, Ann Arbor, MI 48109
zhng@umich.edu

Abstract. Protein structure prediction aims to determine the spatial location of every atom in protein molecules from the amino acid sequence by computational simulations, while protein design is the reverse procedure of structure prediction which aims to engineer novel protein sequences that have desirable structure and function. In this presentation, we first review recent progress in computer-based protein structure prediction, and show that a new approach combining ab initio folding and profile-based fold-recognition methods can break though the barrier of physics-based protein folding, which resulted in the successful folding of proteins larger than 150 residues in the community-wide blind CASP experiments. Next, we extend the profile alignment method to protein design, and introduce an evolutionary profile based approach to design new functional XIAP (X-linked Inhibitor of Apoptosis Protein) BIR3 domains that bind Smac peptide but do not inhibit caspase-9 activity, representing a new therapeutic potential to change the caspase-9 initiated apoptosis pathway through computational protein design. The work shows that protein family-based profiling is an efficient tool to both problems of protein folding and protein design.

Contents

Deriving Protein Backbone Using Traces Extracted from Density Maps
at Medium Resolutions 1
    Kamal Al Nasr and Jing He

Binary Contingency Table Method for Analyzing Gene Mutation
in Cancer Genome 12
    Emi Ayada, Atsushi Niida, Takanori Hasegawa, Satoru Miyano,
    and Seiya Imoto

A Filter-Based Approach for Approximate Circular Pattern
Matching 24
    Md. Aashikur Rahman Azim, Costas S. Iliopoulos,
    M. Sohel Rahman, and Mohammad Samiruzzaman

Fast Algorithms for Inferring Gene-Species Associations 36
    Arkadiusz Betkier, Paweł Szczęsny, and Paweł Górecki

Couplet Supertree Based Species Tree Estimation 48
    Sourya Bhattacharyya and Jayanta Mukhopadhyay

A Novel Computational Method for Deriving Protein Secondary
Structure Topologies Using Cryo-EM Density Maps and Multiple
Secondary Structure Predictions 60
    Abhishek Biswas, Desh Ranjan, Mohammad Zubair, and Jing He

Managing Reproducible Computational Experiments with Curated
Proteins in KINARI-2 72
    John C. Bowers, Rose Tharail John, and Ileana Streinu

Protein Crystallization Screening Using Associative
Experimental Design 84
    İmren Dinç, Marc L. Pusey, and Ramazan S. Aygün

DAM: A Bayesian Method for Detecting Genome-wide Associations
on Multiple Diseases 96
    Xuan Guo, Jing Zhang, Zhipeng Cai, Ding-Zhu Du, and Yi Pan

MINED: An Efficient Mutual Information Based Epistasis Detection
Method to Improve Quantitative Genetic Trait Prediction 108
    Dan He, Zhanyong Wang, and Laxmi Parada

Domain Adaptation with Logistic Regression for the Task of Splice Site
Prediction 125
    Nic Herndon and Doina Caragea

A Stacking-Based Approach to Identify Translated Upstream Open
Reading Frames in *Arabidopsis Thaliana* 138
    Qiwen Hu, Catharina Merchante, Anna N. Stepanova,
    Jose M. Alonso, and Steffen Heber

PRESS-PLOT: An Online Server for Protein Structural Analysis
and Evaluation with Residue-level Virtual Angle Correlation Plots 150
    Yuanyuan Huang, Kejue Jia, Robert Jernigan, and Zhijun Wu

Calcium Ion Fluctuations Alter Channel Gating in a Stochastic
Luminal Calcium Release Site Model 162
    Hao Ji, Yaohang Li, and Seth H. Weinberg

Interleaving Global and Local Search for Protein
Motion Computation 175
    Kevin Molloy and Amarda Shehu

On the Complexity of Duplication-Transfer-Loss Reconciliation
with Non-binary Gene Trees...................................... 187
    Misagh Kordi and Mukul S. Bansal

On the Near-Linear Correlation of the Eigenvalues Across BLOSUM
Matrices 199
    Jin Li, Yen Kaow Ng, Xingwu Liu, and Shuai Cheng Li

Predicting RNA Secondary Structures: One-grammar-fits-all
Solution 211
    Menglu Li, Micheal Cheng, Yongtao Ye, Wk Hon, Hf Ting,
    Tw Lam, Cy Tang, Thomas Wong, and Sm Yiu

An Approach for Matching Mixture MS/MS Spectra with a Pair
of Peptide Sequences in a Protein Database 223
    Yi Liu, Weiping Sun, Gilles Lajoie, Bin Ma, and Kaizhong Zhang

Diploid Alignments and Haplotyping 235
    Veli Mäkinen and Daniel Valenzuela

Structural Comparative Analysis of Ecto- NTPDase Models
from *S. Mansoni* and *H. Sapiens* 247
    Vinicius Schmitz Nunes, Eveline Gomes Vasconcelos,
    Priscila Faria-Pinto, Carlos Cristiano H. Borges,
    and Priscila V.S.Z. Capriles

Assessing the Robustness of Parsimonious Predictions
for Gene Neighborhoods from Reconciled Phylogenies:
Supplementary Material 260
    Ashok Rajaraman, Cedric Chauve, and Yann Ponty

Sorting Signed Circular Permutations by Super Short Reversals 272
    Gustavo Rodrigues Galvão, Christian Baudet, and Zanoni Dias

Multiple Alignment of Structures Using Center of ProTeins   284
  Kaushik Roy, Satish Chandra Panigrahi, and Asish Mukhopadhyay

NRRC: A Non-referential Reads Compression Algorithm   297
  Subrata Saha and Sanguthevar Rajasekaran

New Heuristics for Clustering Large Biological Networks   309
  Md. Kishwar Shafin, Kazi Lutful Kabir, Iffatur Ridwan,
  Tasmiah Tamzid Anannya, Rashid Saadman Karim,
  Mohammad Mozammel Hoque, and M. Sohel Rahman

A Novel Algorithm for Glycan *de novo* Sequencing Using Tandem Mass
Spectrometry   320
  Weiping Sun, Gilles A. Lajoie, Bin Ma, and Kaizhong Zhang

Community Detection-Based Feature Construction for Protein
Sequence Classification   331
  Karthik Tangirala, Nic Herndon, and Doina Caragea

Curvilinear Triangular Discretization of Biomedical Images   343
  Jing Xu and Andrey N. Chernikov

The Role of miRNAs in Cisplatin-Resistant HeLa Cells   355
  Yubo Yang, Cuihong Dai, Zhipeng Cai, Aiju Hou, Dayou Cheng,
  and Dechang Xu

DNA AS X: An Information-Coding-Based Model to Improve
the Sensitivity in Comparative Gene Analysis   366
  Ning Yu, Xuan Guo, Feng Gu, and Yi Pan

A Distance-Based Method for Inferring Phylogenetic Networks
in the Presence of Incomplete Lineage Sorting   378
  Yun Yu and Luay Nakhleh

Predicting Protein Functions Based on Dynamic Protein Interaction
Networks   390
  Bihai Zhao, Jianxin Wang, Fang-Xiang Wu, and Yi Pan

An Iterative Approach for Phylogenetic Analysis of Tumor Progression
Using FISH Copy Number   402
  Jun Zhou, Yu Lin, William Hoskins, and Jijun Tang

Phenome-Based Gene Discovery Provides Information about
Parkinson's Disease Drug Targets   413
  Yang Chen and Rong Xu

Estimating Features with Missing Values and Outliers:
a Bregman-proximal Point Algorithm for Robust Non-negative
Matrix Factorization with Application to Gene Expression Analysis   415
  Stéphane Chrétien, Christophe Guyeux, Bastien Conesa,
  Régis Delage-Mouroux, Michèle Jouvenot, Philippe Huetz,
  and Françoise Descôtes

Conservation and Network Analysis of the $(4\beta+\alpha)$ Fold
of the Immunoglobulin-Binding B1 Domain of Protein G to Elucidate
the Key Determinants of Structure, Folding and Stability   417
  Jason C. Collins, John Bedford, and Lesley H. Greene

Assessment of Transcription Factor Binding Motif and Regulon
Transfer Methods   420
  Sefa Kilic and Ivan Erill

Short Tandem Repeat Number Estimation from Paired-end Sequence
Reads by Considering Unobserved Genealogy of Multiple Individuals ...   422
  Kaname Kojima, Yosuke Kawai, Naoki Nariai, Takahiro Mimori,
  Takanori Hasegawa, and Masao Nagasaki

PnpProbs: Better Multiple Sequence Alignment by Better Handling
of Guide Trees   424
  Yongtao Ye, Tak-Wah Lam, and Hing-Fung Ting

A Novel Method for Predicting Essential Proteins Based on Subcellular
Localization, Orthology and PPI Networks   427
  Gaoshi Li, Min Li, Jianxin Wang, and Yi Pan

BASE: A Practical *de novo* Assembler for Large Genomes Using Longer
NGS Reads   429
  Binghang Liu, Ruibang Luo, Chi-Man Liu, Dinghua Li, Yingrui Li,
  Hing-Fung Ting, Siu-Ming Yiu, and Tak-Wah Lam

InteGO2: A Web Tool For Measuring and Visualizing Gene Semantic
Similarities Using Gene Ontology   431
  Jiajie Peng, Hongxiang Li, Yongzhuang Liu, Liran Juan,
  Qinghua Jiang, Yadong Wang, and Jin Chen

Predicting Drug-Target Interactions for New Drugs via Strategies
for Missing Interactions   433
  Jian-Yu Shi, Jia-Xin Li, and Hui-Meng Lu

A Genome-Wide Drug Repositioning Approach toward Prostate Cancer
Drug Discovery   435
  Rong Xu and QuanQiu Wang

Clustering Analysis of Proteins from Microbial Genomes at Multiple
Levels of Resolution   438
  Leonid Zaslavsky and Tatiana Tatusova

Systematic Analyses Reveal Regulatory Mechanisms of the Flexible
Tails on Beta-catenin   440
    Bi Zhao and Bin Xue

GRASPx: Efficient Homolog-Search of Short-Peptide Metagenome
Database Through Simultaneous Alignment and Assembly   442
    Cuncong Zhong, Youngik Yang, and Shibu Yooseph

Author Index   445

Deriving Protein Backbone Using Traces Extracted from Density Maps at Medium Resolutions

Kamal Al Nasr$^{(\boxtimes)}$, and Jing He

Department of Computer Science, Tennessee State University, Nashville, TN 37209, USA
Department of Computer Science, Old Dominion University, Norfolk, VA 23529, USA
kalnasr@tnstate.edu, jhe@cs.odu.edu

Abstract. Electron cryomicroscopy is an experimental technique that is capable to produce three dimensional gray-scale images for protein molecules, called density maps. At medium resolution, the atomic details of the molecule cannot be visualized from density maps. However, some features of the molecule can be seen such as the locations of major secondary structures and the skeleton of the molecule. In addition, the order and direction of the detected secondary structure traces can be inferred. We introduce a method to construct the entire model of a protein directly for traces extracted from the density map. The initial results show that this method has good potential. A single model was built for each of the 12 proteins used in the test. The $RMSD_{100}$ of the models is slightly improved from our previous method.

Keywords: Cryo-EM · Volume image · Skeletonization · Protein modeling · Loop modeling

1   Introduction

Electron cryomicroscopy (cryo-EM) is an emerging technique that produces three-dimensional (3D) electron density maps at a wide-range of resolutions [1-4]. When the resolution of density maps is higher than 4Å, the atomic structure can often be derived [5-9]. At the medium resolutions, such as 5-10Å, the backbone and the characteristic features of amino acids are not resolved. It is still challenging to derive the atomic structure from such a density map. When a component of the protein has atomic structure available, fitting can be performed to derive the atomic structure [10-12]. When a homologous model is available, rigid or flexible fitting can be used to derive the atomic structure [13-18]. However, it is still challenging to find a suitable template for many proteins. De novo modeling is an alternative method to derive atomic structures without relying on template structures [19-24]. It relies on the detection of secondary structure positions and the connection patterns encoded in the skeleton of the density map.

A number of computational methods have been developed to detect α-helices from the density maps [25-31]. Most helices longer than two turns can be detected. Most of the major β-sheets can also be detected using various methods such as *SheetTracer*, *SSEhunter*, *SSELearner* and *SSETracer* [29-32]. By analyzing the twist of β-sheet

© Springer International Publishing Switzerland 2015
R. Harrison et al. (Eds.): ISBRA 2015, LNBI 9096, pp. 1–11, 2015.
DOI: 10.1007/978-3-319-19048-8_1

density, the position of β-strands can be predicted [31]. A detected helix/β-strand (Fig. 1C) is represented by its central axial line, and the backbone of the helix needs to be built. In addition to α-helices and β-sheets that can be detected from the density map, skeleton can also be derived from the density map [33-36]. Skeleton (red wire in Figure 1) represents possible connection patterns among helices and β-strands (yellow and green in Figure 1 C and D).

The de novo modeling combines information from the density map and the amino acid sequence of the protein to derive the topology of secondary structure traces [21, 23, 37-39]. A topology maps the secondary structure traces from the density map to the amino acid sequence, and therefore determines how the protein chain thread through the traces. We previously showed that using a dynamic programming method combined with the *K*-shortest path algorithm, it takes $O(\Delta^2 N^2 2^N)$ time to rank the top *K* topologies using *DP-TOSS* [23, 38]. Here N is the number of secondary structure traces detected in the density map and M is the number of secondary structure sequence segments, and $\Delta = M - N + 1$.

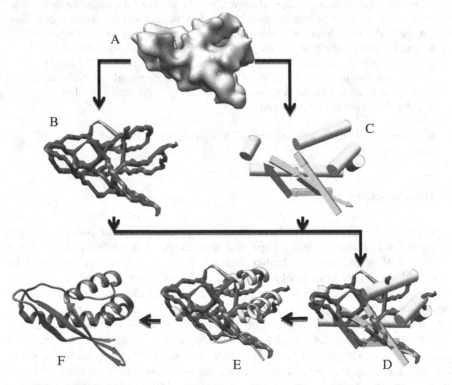

Fig. 1. De novo technique to construct a protein model. (A): Protein density map; (B): Skeleton; (C): Secondary structure   traces for helices (yellow), and β-strands (green); (D): Superposition of the skeleton and the secondary structure traces; (E): Superposition of secondary structures built and the skeleton; (F): The atomic structure.

This paper investigates the problem of constructing backbone of a protein when the topology of secondary structures is given. FM-fold uses Rosetta to construct the backbone [20, 21]. Pathwalking uses pseudo atoms derived from the density map and a constraint satisfaction solver to place Cα atoms [40]. Since the topology of secondary structures determines how the protein sequence thread through the secondary structure traces, our current approach aims to build the backbone directly from the traces and the topology information. We present a method that sequentially builds a backbone chain from the N-terminal to C-terminal though iterative fragment-based Cyclic Coordinate Descent (CCD) and Forward Backward CCD (FBCCD) method [41]. We previously proposed a fragment-based method to construct secondary structure pieces using and then connect them using loops [22]. We here report an extension of our previous method with improved capabilities. The current method constructs a chain for proteins with both helices and β-sheets. Our previous method was only applied to proteins with helices only.

2   Methods

A skeleton is a compact shape representation of a 3D image. It contains possible connection patterns, some of which are correct but most are wrong connections. A skeleton is processed in the topology determination process when the positions of secondary structures in 3D image are correlated with the sequence segments of the secondary structures. As a result, only those connections that satisfy the pattern in the amino acids sequence are selected.

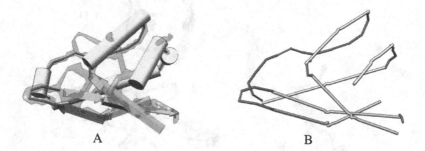

A   B

Fig. 2. Three kinds of traces in a topology. (A): Superposition of the secondary structures (α-helices yellow and β-strands in green) detected from the density map and the skeleton (red opache). (B): α-trace: yellow, β-traces: in green; loop-traces: in purple.

The first step of the method is to initialize three types of traces: α-traces, β-traces, and loop/turn traces (Figure 2).The first two may be generated from secondary structure detection methods [30, 31], and the third one is from the topology determination. The second step is to construct the backbone using the three types of traces as a guide. The idea is to start building from an end and build one segment at a time until it reaches the other end. Depending on which type of trace it is going to use, the implementation details are slightly different, but the principle is the same. The process starts by determining the number of amino acids to be constructed for the next trace. A piece of chain with random conformations is constructed using the torsion angles from Ramachandran plot at one of the three regions α-helix, β-sheet, loop/turn. Let's refer to the piece of backbone being constructed as a spline that will be forced eventually to align with the trace.

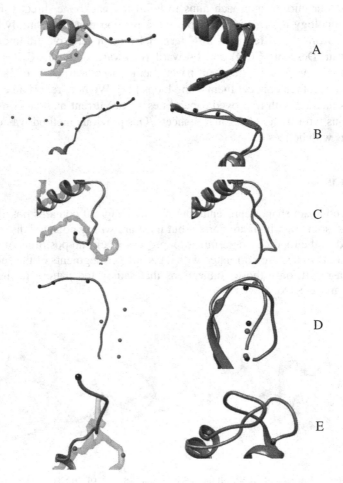

Fig. 3. Examples of backbone fragments constructed. In the left panels, a spline of random conformation with moving points (black) to be aligned with target points (red) on the trace. In the right panels, the spline is aligned with the trace after running FBCCD. (A) A short loop (2 amino acids) followed by a β-strand). The trace of the loop (if any) is added to the trace of the secondary structure. (B) A strand. (C) A loop with no trace in the skeleton. (D) A loop with a trace. (E) A loop with a trace comprising only two points.

Starting from the beginning of the trace, a point will be placed every 6Å except for the last two points near the end. 6Å distance corresponds approximately to four amino acids in a helix and two amino acids of a β-strand or a loop. Recall that an α-helix includes a rise of 1.5Å along the central axis and about 3Å for a β-sheet and a loop.

Similarly, same number of points is extracted from the spline with a random conformation. The points are placed on the central axis (spline) of the conformation (Fig. 3). The alignment process starts aligning the spline with the trace from the second point using FBCCD [41].

Our approach aligns a line segment formed by two consecutive points of the spline with a corresponding line segment from the trace. The following process is iterated for the $(n-2)$ segments, where n is the number of points to align. The amino acids corresponding to line segment $(i-1)$ on the spline are used to align the line segment i. these are the amino acids immediately before the line segment i. To preserve the structural characteristics of secondary structures, torsions used are $\phi \in [-80°, -40°]$ and $\psi \in [-60°, -20°]$ for helices; $\phi \in [-170°, -60°]$ and $\psi \in [90°, 175°]$ for β-strands; and $\phi \in [-170°, 170°]$ and $\psi \in [-170°, 170°]$ for loops. The process terminates either when the cutoff Root Mean Square Distance (RMSD) distance between the target line segment and the spline segment is reached or the maximum number of cycles is reached. In our current implementation, the cutoff distance is 0.5Å and the maximum number of cycles is 200. If, after aligning $(n-2)$ segments, the distance between the last point on the spline and the corresponding point on the trace is more than 1Å, FBCCD is applied only for these points. The amino acids involved in this process are the last half of the conformation fragment. This step is to assure that the end of the conformation fragment is close to the trace so the next fragment to be built is not misplaced from the trace.

Optimization techniques for loop closure, like FBCCD, are known to have low success rate when work with short fragments. In general, the methods fail to close loops shorter than four amino acids [41-43]. To overcome this problem, short loops less than four amino acids are always built with its successive secondary structure as one fragment (Figure 3A). Their traces are combined into one trace and the method is applied as usual. On other hand, due to the quality of the cryo-EM image and, therefore, the skeleton, the traces will be missed for some loops/turns. Our method will apply FBCCD with only one target point and one moving point. The target point will be the first point from the next secondary structure's trace and the moving point will be the centroid of the last amino acid of the fragment being constructed (Figure 3C). Similarily, if the trace has only two points, the last point is the only point targeted by FBCCD (Figure 3E).

3   Results

We tested the current method using a data set consisting of five α-proteins, one β-protein, and six α-β proteins. An α-protein / β-protein contains only α-helices / β-sheets, and a α-β protein contains both helices and β-sheets. The native structures were downloaded from the PDB database. For each native structure, a density map was simulated to 8Å resolution (except 3FIN_R [44]) using EMAN [45]. The density map for 3FIN_R was extracted from cryo-EM density map thatwas downloaded from the Electron Microscopy Data Bank (EMDB ID: 5030) [46]. For the α-proteins, we applied *SSETracer* to detect the helices [30]. For the α/β proteins and β-proteins (except 3FIN_R), the α-traces and β-traces were derived from the native structure by calculating the geometrical center of each three consecutive Cα atoms from the native structure. To imitate the challenges with short helices and strands, we only derived traces for helices longer than six amino acids and for β-strands longer than four amino acids. *DP-TOSS* was applied to produce top K ranked topologies, among which the true topology was chosen for the construction of the backbone. For each protein, one model was built using our sequential method. The model was constructed starting from one end of the sequence and was built for one trace at a time till the end of the sequence. A trace can be an α-trace, a β-trace or a loop/turn trace that was derived from the skeleton. To build the backbone for each trace, a spline of random conformation was built and then aligned with the trace quickly using fragment-based CCD. We report the backbone $RMSD_{100}$ for the model constructed for each protein in Table 1. The $RMSD_{100}$ was calculated for each constructed model against the entire native structure except for the first loop before the first secondary structure or the last loop after the last secondary structure, if any.

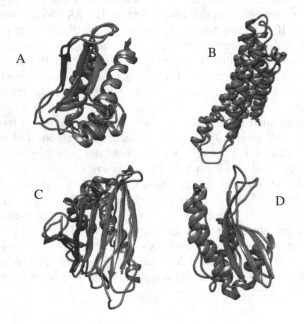

Fig. 4. Four constructed models. The constructed models (green) are superimposed with their native structures (purple) in (A) for 3FIN_R, in (B) for 1BZ4, in (C) for 1ICX, and in (D) for 4OXW. The native structures are labeled with PDB ID.

Figure 4 shows four models constructed using our method. As an example in 3FIN_R, a small protein with four helices and three β-strands, *SSETracer* detected all four helices from the density map . and produced four α-traces. The three β-traces were derived from the native structure by calculating the geometrical centers of consecutive Cα atoms. The model constructed for 3FIN_R has $RMSD_{100}$ of 3.96Å for backbone atoms (Figure 4A and Table 1). Note that the current model has improved accuracy compared to our previous model that has $RMSD_{100}$ of 5.98Å (row 3 of Table 3). A few practices might have contributed to the improved accuracy. The current method sequentially builds the backbone and the previous method first builds all secondary structures and then builds loops to connect them. Previous method does not build β-strands. β-strands were built as loops. The current approach uses β-traces to construct β-strands using torsion angles of β-strands. Current loop traces are derived from the skeleton extracted using newly developed skelEM method [36]. Previous skeleton was calculated using Gorgon [19]. We expect that the skeleton generated by skelEM has less gaps. In the case of 1ICX (Figure 4C), the model built three of the six helices that are longer than six amino acids. For helices shorter than six amino acids and strands shorter than four amino acids long, the current method builds it as a loop. The constructed model has $RMSD_{100}$ 3.47Å. Our previous method builds the structure of proteins in two steps [22]. In the first step, it builds the conformations of helices based on their central axes extracted from cryo-EM images using *SSETracer*. In the second step, it connects the helices by building the conformations of the loops/turns using curve skeletons. In contrast, the current method constructs the backbone structure of the protein in the same order of its secondary structure elements.

Current implementation of our method has encountered a number of challenges that negatively impact the accuracy of the constructed models and, therefore, the final $RMSD_{100}$ values. The first challenge is modeling a loop/turn fragment when no trace was found on the skeleton or a trace found with only two points at the beginning and the end of the trace (Fig. 3C and D). If the loop has no trace, our method set the target point to be the first point on the trace of the next secondary structure. If the loop has a trace with only two points, the target point is set to be the one at the end of the trace that is close to the next secondary structure. Consequently, our method uses FBCCD to connect the loop with the trace of the next secondary structure using only one point. This results in a structure that is not guaranteed to be aligned with the native structure. The second challenge is the length of the traces for some turns. Generally, the skeleton trace of a turn is shorter than the actual length of the native loop. This is expected since the skeleton represents the dense points in the image and the dense points of a turn are often off the backbone. The third challenge is modeling the missing secondary structures (i.e., helices) as loops. If the trace of an α-helix is not detected from the map or if it is shorter than six amino acids as in current implementation, our method constructs it as a loop. This is expected to increase the final $RMSD_{100}$ due to the conformational differences between helices and loops. One possible method to overcome this challenge is to build a helix fragment if additional information suggests a possible small helix.

The test involving 12 cases can be partitioned into two groups. The first group of proteins (row 1-5 in Table 1) was tested using two methods, the current sequential method and the previous piece-wise method. The accuracy is comparable between the two methods for the four α-proteins. The sequential method shows better accuracy for two larger proteins of the four cases and less accuracy for the two smaller proteins. As for the α-β protein 3FIN_R, the current method shows better accuracy, possibly due to the modeling of β-strands that was not available for the previous piece-wise method (see more detailed discussion in a previous paragraph). However, the sequential method builds only one conformation of the backbone directly from the traces. The previous piece-wise method builds many possible conformations, and the best conformation is listed in Table 1 (the 9th column). The second group of proteins (row 6-12) was tested using the sequential method. The $RMSD_{100}$ for these α-β proteins or the β-protein ranges from 3.30Å to 4.53Å. The α-traces and β-traces of these proteins were derived from the native structure and therefore are fairly accurate. The loop traces were derived from the skeleton of the 3D image and they face challenges from gaps and inaccuracy in some situations. Yet the backbone constructed using sequential method is around 4Å on average. This result shows the potential of the sequential method when the traces are fairly accurately derived.

Table 1. The backbone accuracy of the constructed models

No	PDB ID[a]	#AA[b]	HlxSeq[c]	StrSeq[d]	HlxMap[e]	StrMap[f]	RMSD[g]	RMSD2012[h]
1	1A7D	118	6	0	4	0	4.80	3.87
2	1BZ4	144	5	0	5	0	4.30	3.34
3	3FIN_R	117	4	3	4	3	3.96	5.98
4	1HZ4	373	21	0	19	0	3.19	3.87
5	3LTJ	201	16	0	12	0	3.32	4.07
6	4OXW	119	5	3	3	3	4.21	N/A
7	1YD0	96	5	4	3	3	4.01	N/A
8	1OZ9	150	5	5	5	4	3.61	N/A
9	1ICX	155	6	7	3	7	3.47	N/A
10	2y4z	140	6	2	6	2	4.10	N/A
11	4U3H	100	0	8	0	7	3.30	N/A
12	4YOK	204	1	16	1	15	4.53	N/A

a: Protein ID

b: The number of amino acids in the protein.

c: The number of actual helices in the protein.

d: The number of actual strands in the protein.

e: The number of helices detected from the density map.

f: The number of strands detected from the density map.

g: The backbone $RMSD_{100}$ of the constructed model with the native structure.

h: The backbone $RMSD_{100}$ of the best model built from the previous study [22]. Previous method does not apply for proteins with β-sheets, indicated as N/A.

4   Conclusions

We present a method to sequentially construct the backbone from the traces detected from density maps. It uses the topology derived from *DP-TOSS* and the α-traces, β-traces of the secondary structures and loop traces derived from skeleton of the density map. The initial results show that this method has good potential. Judging from the single model built for each of the 12 proteins, the $RMSD_{100}$ is slightly improved from our previous method [22]. The current method applies to proteins containing β-sheets, while the previous method applies to only α-proteins.

Acknowledgements. The work in this paper is partially supported by NSF DBI-1356621.

References

1. Ludtke SJ, C.D., Song JL, Chuang DT, Chiu W.: Seeing GroEL at 6 A resolution by single particle electron cryomicroscopy. Structure 12, 1129-1136 (2004)
2. Zhou, Z.H., Dougherty, M., Jakana, J., He, J., Rixon, F.J., Chiu, W.: Seeing the herpesvirus capsid at 8.5 A. Science 288, 877-880 (2000)
3. Chiu, W., Schmid, M.F.: Pushing back the limits of electron cryomicroscopy. Nature Structural Biology 4, 331-333 (1997)
4. Chiu, W., Baker, M.L., Jiang, W., Zhou, Z.H.: Deriving folds of macromolecular complexes through electron cryomicroscopy and bioinformatics approaches. Current opinion in structural biology 12, 263-269 (2002)
5. Zhang, X., Jin, L., Fang, Q., Hui, W.H., Zhou, Z.H.: 3.3 Å Cryo-EM Structure of a Nonenveloped Virus Reveals a Priming Mechanism for Cell Entry. Cell 141, 472-482 (2010)
6. Cheng, L., Sun, J., Zhang, K., Mou, Z., Huang, X., Ji, G., Sun, F., Zhang, J., Zhu, P.: Atomic model of a cypovirus built from cryo-EM structure provides insight into the mechanism of mRNA capping. Proceedings of the National Academy of Sciences 108, 1373-1378 (2011)
7. Gonen, T., Sliz, P., Kistler, J., Cheng, Y., Walz, T.: Aquaporin-0 membrane junctions reveal the structure of a closed water pore. Nature 429, 193-197 (2004)
8. Brown, A., Amunts, A., Bai, X.-c., Sugimoto, Y., Edwards, P.C., Murshudov, G., Scheres, S.H.W., Ramakrishnan, V.: Structure of the large ribosomal subunit from human mitochondria. Science 346, 718-722 (2014)
9. Hussain, T., Llácer, Jose L., Fernández, Israel S., Munoz, A., Martin-Marcos, P., Savva, Christos G., Lorsch, Jon R., Hinnebusch, Alan G., Ramakrishnan, V.: Structural Changes Enable Start Codon Recognition by the Eukaryotic Translation Initiation Complex. Cell 159, 597-607 (2014)
10. Topf, M., Lasker, K., Webb, B., Wolfson, H., Chiu, W., Sali, A.: Protein structure fitting and refinement guided by cryo-EM density. Structure 16, 295-307 (2008)
11. DiMaio, F., Tyka, M.D., Baker, M.L., Chiu, W., Baker, D.: Refinement of protein structures into low-resolution density maps using rosetta. Journal of Molecular Biology 392, 181-190 (2009)
12. Topf, M., Baker, M.L., Marti-Renom, M.A., Chiu, W., Sali, A.: Refinement of protein structures by iterative comparative modeling and CryoEM density fitting. Journal of Molecular Biology 357, 1655-1668 (2006)
13. 13. Pintilie, G.D., Zhang, J., Goddard, T.D., Chiu, W., Gossard, D.C.: Quantitative analysis of cryo-EM density map segmentation by watershed and scale-space filtering, and fitting of structures by alignment to regions. Journal of Structural Biology 170, 427-438 (2010)
14. Kovacs, J.A., Chacón, P., Cong, Y., Metwally, E., Wriggers, W.: Fast rotational matching of rigid bodies by fast Fourier transform acceleration of five degrees of freedom. Acta Crystallographica. Section D, Biological Crystallography 59, 1371-1376 (2003)
15. Wriggers, W., Chacón, P.: Modeling tricks and fitting techniques for multiresolution structures. Structure 9, 779-788 (2001)
16. Tama, F., Miyashita, O., Brooks Iii, C.L.: Flexible Multi-scale Fitting of Atomic Structures into Low-resolution Electron Density Maps with Elastic Network Normal Mode Analysis. Journal of Molecular Biology 337, 985-999 (2004)
17. Suhre, K., Navazab, J., Sanejouand, Y.-H.: NORMA: a tool for flexible fitting of high-resolution protein structures into low-resolution electron-microscopy-derived density

maps. Acta Crystallographica. Section D, Biological Crystallography 62, 1098-1100 (2006)

18. Ming, D., Kong, Y., Wakil, S.J., Brink, J., Ma, J.: Domain movements in human fatty acid synthase by quantized elastic deformational model. Proceedings of the National Academy of Sciences of the United States of America (PNAS) 99, 7835-7899 (2002)

19. Baker, M.L., Abeysinghe, S.S., Schuh, S., Coleman, R.A., Abrams, A., Marsh, M.P., Hryc, C.F., Ruths, T., Chiu, W., Ju, T.: Modeling protein structure at near atomic resolutions with Gorgon. Journal of Structural Biology 174, 360-373 (2011)

20. Lindert, S., Staritzbichler, R., Wötzel, N., Karakaş, M., Stewart, P.L., Meiler, J.: EM-Fold: De Novo Folding of α-Helical Proteins Guided by Intermediate-Resolution Electron Microscopy Density Maps. Structure 17, 990-1003 (2009)

21. Lindert, S., Alexander, N., Wötzel, N., Karaka, M., Stewart, Phoebe L., Meiler, J.: EM-Fold: De Novo Atomic-Detail Protein Structure Determination from Medium-Resolution Density Maps. Structure 20, 464-478 (2012)

22. Al Nasr, K., Chen, L., Si, D., Ranjan, D., Zubair, M., He, J.: Building the initial chain of the proteins through de novo modeling of the cryo-electron microscopy volume data at the medium resolutions.   Proceedings of the ACM Conference on Bioinformatics, Computational Biology and Biomedicine, pp. 490-497. ACM, Orlando, Florida (2012)

23. Al Nasr, K., Ranjan, D., Zubair, M., Chen, L., He, J.: Solving the Secondary Structure Matching Problem in Cryo-EM De Novo Modeling Using a Constrained K-Shortest Path Graph Algorithm. Computational Biology and Bioinformatics, IEEE/ACM Transactions on 11, 419-430 (2014)

24. Al Nasr, K., Sun, W., He, J.: Structure prediction for the helical skeletons detected from the low resolution protein density map. BMC Bioinformatics 11, S44 (2010)

25. Jiang, W., Baker, M.L., Ludtke, S.J., Chiu, W.: Bridging the information gap: computational tools for intermediate resolution structure interpretation. Journal of Molecular Biology 308, 1033-1044 (2001)

26. Del Palu, A., He, J., Pontelli, E., Lu, Y.: Identification of Alpha-Helices from Low Resolution Protein Density Maps. Proceeding of Computational Systems Bioinformatics Conference(CSB) 89-98 (2006)

27. Baker, M.L., Ju, T., Chiu, W.: Identification of secondary structure elements in intermediate-resolution density maps. Structure 15, 7-19 (2007)

28. Rusu, M., Wriggers, W.: Evolutionary bidirectional expansion for the tracing of alpha helices in cryo-electron microscopy reconstructions. Journal of Structural Biology 177, 410-419 (2012)

29. Kong, Y., Zhang, X., Baker, T.S., Ma, J.: A Structural-informatics approach for tracing beta-sheets: building pseudo-C(alpha) traces for beta-strands in intermediate-resolution density maps. Journal of Molecular Biology 339, 117-130 (2004)

30. Si, D., Ji, S., Al Nasr, K., He, J.: A machine learning approach for the identification of protein secondary structure elements from cryoEM density maps. Biopolymers 97, 698-708 (2012)

31. Si, D., He, J.: Tracing Beta Strands Using StrandTwister from Cryo-EM Density Maps at Medium Resolutions. Structure 22, 1665-1676 (2014)

32. Kong, Y., Ma, J.: A structural-informatics approach for mining beta-sheets: locating sheets in intermediate-resolution density maps. Journal of Molecular Biology 332, 399-413 (2003)

33. Ju, T., Baker, M.L., Chiu, W.: Computing a family of skeletons of volumetric models for shape description. Computer-Aided Design 39, 352-360 (2007)

34. Abeysinghe, S.S., Ju, T.: Interactive skeletonization of intensity volumes. Vis. Comput. 25, 627-635 (2009)
35. Abeysinghe, S.S., Baker, M., Wah, C., Tao, J.: Segmentation-free skeletonization of grayscale volumes for shape understanding. IEEE International Conference on Shape Modeling and Applications, SMI. , pp. 63-71, Stony Brook, NY (2008)
36. Al Nasr, K., Liu, C., Rwebangira, M., Burge, L., He, J.: Intensity-Based Skeletonization of CryoEM Gray-Scale Images Using a True Segmentation-Free Algorithm. IEEE/ACM Trans. Comput. Biol. Bioinformatics 10, 1289-1298 (2013)
37. Abeysinghe, S., Ju, T., Baker, M.L., Chiu, W.: Shape modeling and matching in identifying 3D protein structures. Computer-Aided Design 40, 708-720 (2008)
38. Al Nasr, K., Ranjan, D., Zubair, M., He, J.: Ranking Valid Topologies of the Secondary Structure elements Using a constraint Graph. Journal of Bioinformatics and Computational Biology 9, 415-430 (2011)
39. Biswas, A., Si, D., Al Nasr, K., Ranjan, D., Zubair, M., He, J.: A Constraint Dynamic Graph Approach to Identify the Secondary Structure Topology from cryoEM Density Data in Presence of Errors. Proceedings of the 2011 IEEE International Conference on Bioinformatics and Biomedicine, pp. 160-163. IEEE Computer Society (2011)
40. Baker, Mariah R., Rees, I., Ludtke, Steven J., Chiu, W., Baker, Matthew L.: Constructing and Validating Initial Cα Models from Subnanometer Resolution Density Maps with Pathwalking. Structure 20, 450-463 (2012)
41. Al Nasr, K., He, J.: An effective convergence independent loop closure method using Forward-Backward Cyclic Coordinate Descent. International Journal of Data Mining and Bioinformatics 3, 346-361 (2009)
42. Canutescu, A.A., Dunbrack, R.L.J.: Cyclic coordinate descent: A robotics algorithm for protein loop closure. Protein Science 12, 963-972 (2003)
43. Boomsma, W., Hamelryck, T.: Full cyclic coordinate descent: Solving the protein loop closure problem in Cα space. BMC Bioinformatics 6, 159 (2005)
44. Schuette, J.-C., Murphy, F.V., Kelley, A.C., Weir, J.R., Giesebrecht, J., Connell, S.R., Loerke, J., Mielke, T., Zhang, W., Penczek, P.A., Ramakrishnan, V., Spahn, C.M.T.: GTPase activation of elongation factor EF-Tu by the ribosome during decoding. EMBO J 28, 755-765 (2009)
45. Ludtke, S.J., Baldwin, P.R., Chiu, W.: EMAN: Semi-automated software for high resolution single particle reconstructions. Journal of Structural Biology 128, 82-97 (1999)
46. Lawson, C.L., Baker, M.L., Best, C., Bi, C., Dougherty, M., Feng, P., van Ginkel, G., Devkota, B., Lagerstedt, I., Ludtke, S.J., Newman, R.H., Oldfield, T.J., Rees, I., Sahni, G., Sala, R., Velankar, S., Warren, J., Westbrook, J.D., Henrick, K., Kleywegt, G.J., Berman, H.M., Chiu, W.: EMDataBank.org: unified data resource for CryoEM. Nucleic Acids Research 39, D456-D464 (2011)

Binary Contingency Table Method for Analyzing Gene Mutation in Cancer Genome

Emi Ayada[1], Atsushi Niida[1], Takanori Hasegawa[2], Satoru Miyano[1],
and Seiya Imoto[1(✉)],

[1] Human Genome Center, The Institute of Medical Science,
The University of Tokyo 4-6-1 Shirokanedai, Minato-ku, Tokyo, Japan
emi.ayada@gmail.com, {aniida,miyano,imoto}@ims.u-tokyo.ac.jp
[2] Bioinformatics Center, Institute for Chemical Research,
Kyoto University Gokasyo Uji, Kyoto, Japan
t-hasegw@kuicr.kyoto-u.ac.jp

Abstract. Gene mutations are responsible for a large proportion of genetic diseases such as cancer. Hence, a number of computational methods have been developed to find loci subject to frequent mutations in cancer cells. Since normal cells turn into cancer cells through the accumulation of gene mutations, the elucidation of interactive relationships among loci has great potential to reveal the cause of cancer progression; however, only a few methods have been proposed for measuring statistical significance of pairs of loci that are co-mutated or exclusively mutated. In this study, we proposed a novel statistical method to find such significantly interactive pairs of loci by employing the framework of binary contingency tables. Using Markov chain Monte Carlo procedure, the statistical significance is evaluated by sampling null matrices whose marginal sums are equal to those of the input matrix. We applied the proposed method to mutation data of colon cancer patients and successfully obtained significant pairs of loci.

Keywords: Cancer · Gene mutation · Binary contingency tables · Markov chain Monte Carlo

1 Introduction

Gene mutations can change the normal function of proteins, leading to genetic diseases such as cancer. For example, the mutation can result in loss of function that helps to repair damaged DNA. In cancer research, mutations observed only in tumor cells have been intensively investigated, however, these mutations are analyzed independently and combinations of mutations are not well studied. Since the accumulation of gene mutations causes cancer, it is important to detect pairs of genes, which contribute to the accumulation interactively, *e.g.*, genes with mutations that tend to occur together among a lot of samples. The elucidation of these relationships allows us to identify how genes interact with other genes. To achieve it, some studies[3,14,9] concerned the identification of

© Springer International Publishing Switzerland 2015
R. Harrison et al. (Eds.): ISBRA 2015, LNBI 9096, pp. 12–23, 2015.
DOI: 10.1007/978-3-319-19048-8_2

the relationships between genes in cancer cells based on the fact that oncogenesis is a process with multiple stages, in which normal cells transform into cancer cells via multiple genetic mutations. However, almost these researches have not measured the statistical significance of the relationships. Although Dees et al. [3] considered statistical significance for identifying the relationships based on permutation test-based method, they did not consider the varieties of the numbers of the mutations in each of samples and genes.

Therefore, to provide an accurate statistical assessment, we propose a novel method to measure the statistical significance of relationships between genes using a statistical framework of binary contingency tables (BCTs), which are defined as binary tables with fixed column and row sums. BCTs are utilized as tables that are composed of entries with binary values indicating absence (0) or presence (1) of mutations. In this study, we analyze the following two types of gene relationships: co-mutated relationships, which represent pairs of genes getting mutated together and exclusive relationships, which represent pairs of genes including both a mutated gene and a gene without a mutation. Since the numbers of mutations vary among samples and the numbers of mutated samples vary among genes, we propose random sampling of BCTs keeping the sums of mutations on each samples and genes using Markov chain Monte Carlo procedure. By using gene mutation data as a BCT, the proposed method measures the statistical significance of an observed state of each gene pair by the algorithm developed by Bezáková [1]. This algorithm samples random BCTs and non-BCTs satisfying almost marginal sums of the original data and obtains p-values for all combinations of gene pairs; however, an accurate statistical test cannot be performed when sampling non-BCTs. In order to sample BCTs only, we further proposed a novel algorithm, termed Perfect BCT(PBCT)-sampling, that samples BCTs under the restrictions of the number of mutations occurring on each sample and gene, and measures the statistical significance of the relationships between genes in the mutation data.

To show the effectiveness of the proposed method, we compared the performance of our proposed method, the proposed method using BCT-sampling and an existing method (Fisher's exact test) through a simulation study. As a result, our method outperformed other methods and we confirmed the advantage of using BCTs. Next, we analyzed gene mutation data downloaded from The Broad Institute (http://gdac.broadinstitute.org/) using our proposed method and obtained significantly co-mutated and exclusively mutated gene pairs. We confirmed that the result of the analysis by our proposed method contains pairs of genes, which have been thought as genes related to cancer.

2   Method

2.1   Binary Contingency Tables

BCTs are typically used to represent two exclusive events, such as "absence" or "presence" by 0 and 1, respectively. We apply the framework of BCTs to a $m \times n$ binary table B containing the presence of mutations for pairs of genes and samples in order to clarify the relationships between genes in cancer cells using gene mutation data. In the table B, rows and columns respectively correspond to genes and samples, and the entry at the ith row and jth column is set to 1 if the ith gene of the jth sample is mutated, and 0 otherwise. From the framework of BCTs, we analyze the two types of gene relationships, co-mutated relationships and exclusive relationships.

2.2   BCT-Sampling

From the number of mutations that occur on each column and row, we measure the statistical significance of gene pairs such as co-mutated gene pairs or gene pairs with exclusive mutations. For statistical testing, we first build the null distribution by sampling null matrices, and then measure the statistical significance in terms of the number of samples for each co-mutated gene pair or for gene pair with exclusive mutations in the original binary table B. For sampling null matrices, we adopted BCT-sampling proposed by Bezáková [1], which samples binary tables keeping the given marginal sums. Let define perfect and near-perfect tables as tables satisfying completely the marginal sums of B and tables with one row and one column sum decreased by 1, respectively. Starting from a perfect table, the above method recursively samples a table from the previously sampled table as follows:

1. Let (i, j) be a pair of indices for indicating the ith row and jth column of a table or matrix. If the current $m \times n$ table T is a perfect table, randomly select a pair of indices from $\{(i,j)|T_{i,j} = 1\}$, where $T_{i,j}$ is the ith row and jth column of T, and set $T_{i,j}$ to 0. A near-perfect table with marginal sums of rows $r_1, \ldots, r_i - 1, \ldots, r_m$ and columns $c_1, \ldots, c_j - 1, \ldots, c_n$ and a deficiency at (i, j) is obtained.
2. If the current $m \times n$ table T is a near-perfect table with a deficiency at (i, j), randomly select a pair of indices in $\{(i,j)\} \cap \{(k,l)|T_{k,l} = 1\}$. If $(k, l) = (i, j)$ holds, then set $T_{i,j}$ to 1 and a perfect table is obtained. Otherwise, select one of the following two procedures randomly:
   (a) If $T_{k,j} = 0$ holds, then set $T_{k,j}$ to 1 and $T_{k,l}$ to 0 (the deficiency moves from (i, j) to (i, l)).
   (b) If $T_{i,l} = 0$ holds, then set $T_{i,l}$ to 1 and $T_{k,l}$ to 0 (the deficiency moves from (i, l) to (k, l)).

These sampled tables from the above steps contain both near-perfect and perfect tables.

2.3   PBCT-Sampling

In order to keep precisely the condition of the original binary table B, we developed PBCT-sampling by extracting only perfect tables from all the samples tables with BCT-sampling. Sampling different perfect tables enables us to calculate precisely the frequencies of that gene pairs with the number of mutated samples, which happen together or exclusively, under the given conditions. We design PBCT-sampling to check if a newly generated perfect table is different from the previously sampled tables in order to obtain various different perfect tables. Algorithm 1 shows the detail of PBCT-sampling.

Algorithm 1. PBCT-sampling

M: the number of perfect tables to be sampled
f: the function of sampling BCTs proposed by Bezáková
X_i: ith sampled binary table from f
$\mathcal{X}$: a set of sampled binary tables
Set $\mathcal{X} = \emptyset$
Set $n = 1$
Set $count = 0$
while $count < M$ **do**
    $X_n = f(X_{n-1})$
    if X_n *is a Perfect Table and* $X_n \neq X_{count}$ **then**
        Put X_n to $\mathcal{X}$
        $count = count + 1$
    $n = n + 1$

2.4 Computation of p-value

For the pair of ith and jth genes in B, the test statistic for the detection of co-mutated gene pairs is defined by

$$\sum_{k=1}^{n_s} B_{ik} \times B_{jk}, \tag{1}$$

and that for the exclusively mutated gene pairs is defined by

$$\sum_{k=1}^{n_s} I(k), \ I(k) = \begin{cases} 1 & \text{if } B_{ik} + B_{jk} = 1 \\ 0 & \text{otherwise} \end{cases}, \tag{2}$$

where n_s is the number of samples. Here, we calculate p-value referring to the way adopted by CDCOCA [10]. Let T_{ij} be the test statistic for the ith and jth gene pair in a sampled table, O_{ij} be the test statistic for that in B, C_{ij} be the total number of the test statistics that satisfy $O_{ij} > T_{ij}$ and M be the total number of sampled matrices. The algorithm for the calculation of p-value is shown in Algorithm 2.

3 Result

3.1 Simulation Study

In the simulation study, we compared our proposed method using PBCT-sampling with the method using BCT-sampling and an alternative statistical method,

---

Algorithm 2. Computation of p-value

M: the number of perfect tables to be sampled
n_g: the number of genes
p_{ij}: p-value for gene pair, i and j
Set $count = 0$
Set $C = 0$
while $count < M$ **do**
    Sample binary tables X
    for $i < n_g$ **do**
        for $j < n_g$ **do**
            calculate T_{ij} for X
            if $T_{i,j} \geq O_{ij}$ **then**
                $C_{ij} = C_{ij} + 1$
    count = count + 1
$p_{ij} = C_{ij}/M$

Fisher's exact test [5,8]. The Fisher's exact test in this context calculates the probability of obtaining the observed state under the given conditions as follows. Let a be the number of samples without any mutations on both gene x and y, b be the number of samples with mutations on only gene y and not on gene x, c be the number of samples with mutations on only gene x and not on gene y and d be the number of samples with mutations on both genes x and y. Then the probability of occurring such event is calculated by

$$p_{xy} = \frac{\binom{a+b}{a}\binom{c+d}{c}}{\binom{n}{a+c}}, \tag{3}$$

where n is the number of all samples, which means n is identical to the sum of a, b, c and d. Table 3.1 shows the relationship between these letters and genes.

Table 3.1. The summary of the relationships between two genes, x and y. Fisher's exact test calculates the probability of obtaining the tables using a, b, c and d.

		Gene y	
		Not mutated	Mutated
Gene x	Not mutated	a	b
	Mutated	c	d

We measure the performance of the methods by area under the precision recall curve (AUC) [2]. The performance is shown by the AUC score, which is the space under the curve plotted according to the precision and recall of p-values at each threshold. Fig. 3.1 shows an example of obtained p-values and AUC on our simulation study. The precision is defined as the ratio of the number of relevant records retrieved to the total number of irrelevant and relevant records retrieved. The recall is defined as the ratio of the number of relevant records retrieved to the total number of relevant records.

To apply the method toward two problems, *i.e.*, co-mutated and exclusively mutated problems, we assume that simulated data have two pairs of co-mutated genes or exclusively mutated genes, as statistical significant gene pairs, respectively. We also suppose that the numbers of mutations vary among samples and genes. Then, we prepared simulation data as follows:

1. Generate a mutation rate for each column and row, r_i and r_j ($0 < r_i, r_j < 1$), which controls how often mutations occur on entities.
2. Set the mutation rate for each entry $r_{i,j} = r_i r_j$ based on the mutation rate of columns and rows.
3. Set noise rate that regulates the amount of mutations in the simulated mutation matrix. If the noise rate is 4, for example, we control the amount of entries with mutations to be approximately 40% of all entries. The rate is from 0 to 10.
4. Insert two pairs that have interactive relationships, which are co-mutated or exclusively mutated gene pairs, and are named as true mutation pairs. We define the number of co-mutated and exclusively mutated samples as "signal". In the co-mutated problem, when the signal is 4, two true mutation pairs have four samples, where mutations happen in both of paired genes. In the exclusively mutated problems, when the signal is 4, two true mutation pairs have four samples, where mutations happen in either of paired genes. For each true pair, we choose these samples at random.

Figs. 3.2(a)-3.2(c) show background data, simulated data with two true co-mutated pairs and simulated data with two true exclusively mutated pairs, respectively. We had a simulation using data, which is composed of 50 samples and 50 genes, and several kinds of signal and noise.

Fig. 3.3 shows the result of the co-mutated problem. Consequently, BCT-sampling has a higher AUC score than other methods in case of small noise and large signal; however, PBCT-sampling demonstrates stably superior performance in any parameter compared to other two methods. We also confirmed the advantage on employing the BCT framework to the exclusively mutated problem as concluded in Table 3.4. The result shows that our proposed method has higher AUC scores than other methods and detects true mutation pairs on both co-mutated and exclusively mutated problems.

We can consider that the simulation results of our proposed method are better than others because it takes the number of mutations on each gene and sample of the mutation data into account. In contrast, Fisher's exact test utilizes only the number of mutations happening on each gene. Also, we can see that PBCT-sampling had better performance compared with BCT-sampling since PBCT-sampling keeps completely the marginal sums and it utilizes more accurate conditions of the original mutation data.

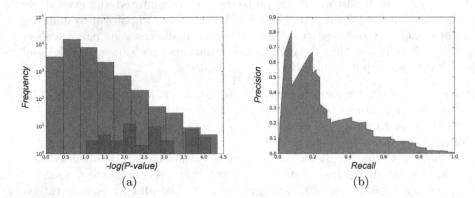

(a)   (b)

Fig. 3.1. The example for the calculation of AUC. We use a dataset of twenty simulated data to measure the performance of methods. Fig. 3.1(a) plots the number of obtained p-values through a simulation, and the red and blue bars are the number of p-values for gene pairs with true mutations and other gene pairs, respectively. Blue area on Fig. 3.1(b), which consists precision and recall axes, shows AUC of data plotted on Fig. 3.1(a).

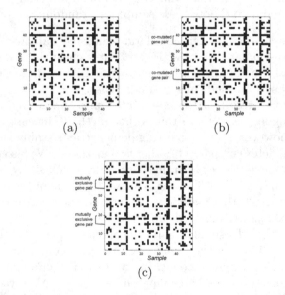

Fig. 3.2. A left figure shows background data with 20% noise and no signal. Figs. 3.2(b) and 3.2(c) are simulated data containing two true co-mutated and exclusively pairs, respectively. Black in the figure represents a gene mutation. These data are 50×50 size matrices, whose vertical and horizontal axes are genes and samples, respectively.

	BCT–sampling			PBCT–sampling		
	Fisher test	30000	300000	3000000	30000	300000	3000000
Signal15 noise3	0.349	0.048	0.148	0.223	0.178	0.250	0.471
Signal10 noise3	0.070	0.030	0.207	0.192	0.211	0.211	0.222
Signal7 noise3	0.028	0.010	0.019	0.020	0.020	0.019	0.079
Signal15 noise2	0.733	0.197	0.514	0.915	0.671	0.835	0.877
Signal10 noise2	0.537	0.288	0.376	0.553	0.460	0.545	0.561
Signal7 noise2	0.222	0.130	0.225	0.179	0.187	0.263	0.299

Fig. 3.3. The simulation results on 50 × 50 size matrices with two true co-mutated pairs. Shown values are AUC scores, and methods with high scores are thought to have high performance of assessing the relationships between genes. We tested Fisher's exact test, BCT-sampling and PBCT-sampling with several kinds of signal and noise.

	BCT–sampling			PBCT–sampling		
	Fisher test	5000	50000	500000	5000	50000	500000
Signal15 noise3	0.006	0.009	0.072	0.151	0.098	0.171	0.199
Signal10 noise3	0.002	0.008	0.033	0.041	0.042	0.055	0.084
Signal7 noise3	0.002	0.008	0.008	0.015	0.011	0.015	0.028
Signal15 noise2	0.007	0.023	0.152	0.177	0.139	0.193	0.242
Signal10 noise2	0.003	0.018	0.037	0.041	0.036	0.045	0.042
Signal7 noise2	0.002	0.006	0.011	0.012	0.020	0.013	0.013

Fig. 3.4. The simulation results on 50 × 50 size matrices with two true exclusively mutated pairs. Shown values are AUC scores, and methods with high scores are thought to have high performance of assessing the relationships between genes. We tested Fisher's exact test, BCT-sampling and PBCT-sampling with several kinds of signal and noise.

Table 3.2. Statistically significant co-mutated gene pairs obtained by PBCT-sampling

Gene	Gene	p-value	q-value
PKHD1L1	RFC1	6.6×10^{-9}	4.5×10^{-6}
LAMA3	DOCK10	6.6×10^{-9}	4.5×10^{-6}
EPB41L3	DOCK5	6.6×10^{-9}	4.5×10^{-6}
EPB41L3	TRPM2	6.6×10^{-9}	4.5×10^{-6}
EPB41L3	SLIT1	6.6×10^{-9}	4.5×10^{-6}
ADAM7	DNAH1	6.6×10^{-9}	4.5×10^{-6}
ADAM7	CACNA1C	6.6×10^{-9}	4.5×10^{-6}
ADAM7	CATSPERB	6.6×10^{-9}	4.5×10^{-6}

3.2 Real Data Experiment

We used binary gene mutation data of colorectal adenocarcinoma, which is downloaded from The Broad Institute. We sorted 631 samples according to the number of mutations and retrieved top 155 samples. The data contains 699 genes and 155 samples, whose rows and columns correspond to genes and samples, respectively. We analyzed the data with PBCT-sampling to identify statistically significant gene pairs. Consequently, we obtained eight co-mutated gene pairs with 10,000,000 samplings as concluded in Tables 3.2. For the exclusively mutated

Table 3.3. Statistically significant exclusively mutated gene pairs obtained by PBCT-sampling

Gene	Gene	p-value	q-value
TP53	XIRP2	5.0×10^{-8}	3.3×10^{-6}
BRAF	DSCAM	5.0×10^{-8}	3.3×10^{-6}
ACVR1B	POSTN	5.0×10^{-9}	3.3×10^{-6}
SPHKAP	VAV1	5.0×10^{-9}	3.3×10^{-6}
TEX15	VAV1	5.0×10^{-9}	3.3×10^{-6}
LAMA3	CELSR1	5.0×10^{-9}	3.3×10^{-6}
KCNQ3	EPB41L3	5.0×10^{-9}	3.3×10^{-6}
SEMA4D	VAV1	5.0×10^{-9}	3.3×10^{-6}
PTPRT	VAV1	5.0×10^{-9}	3.3×10^{-6}
DOCK5	DSCAM	5.0×10^{-9}	3.3×10^{-6}
TMEM132B	VAV1	5.0×10^{-9}	3.3×10^{-6}
PXDNL	DSCAM	5.0×10^{-9}	3.3×10^{-6}

problem, we sampled 2,000,000 BCTs and obtained twelve gene pairs, which are listed as significant exclusively mutated pairs shown in Table 3.3. These results of both kinds, co-mutated and exclusively mutated gene pairs, show that gene pairs of each kind have extremely small and the same p-value and q-value [13]. This is because there was no sampled BCT with greater test statistics than those of the original mutation data at the significant gene pairs.

In the obtained results, we focus on one of significant co-mutated pairs, LAMA3 and DOCK10, as illustrated in Fig. 3.5. The figure shows the frequencies that mutations appear on each gene and sample. We can observe that DOCK10 and LAMA3 have twelve and eight mutations, respectively, and the pair has 7 co-mutated samples. Some of them are recurrently mutated, but others are not likely to be mutated. Our proposed method utilizes these frequencies of mutations on each sample and gene, and detects gene pairs with co-mutated samples including infrequently mutated sample, such as LAMA3 and DOCK10.

In addition, Fig. 3.6 shows the frequencies and locations of mutations on gene-pair, TEX15 and VAV1, which is obtained as the statistically significant and exclusively mutated gene pair. We can confirm that their mutations happens in mutually exclusive way.

LAMA3 [12] and DOCK10 [6], which are obtained as a co-mutated gene pair, are considered to be involved in tumor cell invasion and progression. Additionally, among exclusively mutated gene pairs listed on Table 3.3, TP53 is widely know as tumor suppressor gene and XIRP2 was suggested as a potential driver gene in melanoma [7]. VAV1 [4] and TEX15 [11] are also identified as the candidates of genes related to cancer. VAV1 has been thought as a gene, which is associated with decreased survival and contributes to the tumorigenic properties of pancreatic cancer cells. Also, TEX15 was observed in a significant fraction of tumor samples different histological types.

Fig. 3.5. The summary of gene mutations on LAMA3 and DOCK10. A horizontal axis represents samples. Black points and blue bars along the horizontal axis in the figure show gene mutations and the total number of gene mutations occurring on each sample, respectively. Two blue bars paralleled to the horizontal axis are the total number of samples with gene mutations on each gene. We can see these two genes have gene mutations at the same samples.

Fig. 3.6. The summary of gene mutations on TEX15 and VAV1. A horizontal axis represents samples. Black points and blue bars along the horizontal axis in the figure show gene mutations and the total number of gene mutations occurring on each sample, respectively. Two blue bars paralleled to the horizontal axis are the total number of samples with gene mutations on each gene. We can see that these two genes have mutation in mutually exclusive way.

4   Conclusion and Discussion

There exist many studies focusing on mutation genes and they play a key role in the field of cancer research. In this study, we aimed to elucidate the relationships of gene pairs using binary mutation data. The developed method, PBCT-sampling, enables us to utilize the frequencies that mutations occur on each gene and sample of the mutation data and assess the statistical significance of relationships between genes.

Through the simulation study, we prepared two synthetic data representing co-mutated and exclusively mutated problems and demonstrated that PBCT-sampling outperformed BCT-sampling and Fisher's exact test, which is without the BCTs framework. These results indicated that the BCT framework is reasonable for assessing the statistical significance of gene mutation data and PBCT-sampling, which uses only BCTs, is capable of performing more accurate assessment and has superior performance compared to BCT-sampling using both of BCTs and non-BCTs, and also with Fisher's exact test. Therefore, we confirmed the advantage of the BCT-framework, which allows us to sample binary tables keeping the the varieties of the marginal sums of the mutation data. Furthermore, the analysis of real data with PBCT-sampling showed the statistically significant co-mutated and exclusively mutated gene pairs. Since they have been indicated as cancer related genes, the performance of detecting significant pairs may be suggested. In these gene pairs, we confirmed that some of obtained gene pairs as exclusively mutated gene pairs comprise of tumor suppressor gene and driver gene and some of obtained gene pairs as co-mutated gene pairs are the combination of genes, which contribute to the progression of cancer.

In this study, we focused on the detection of interactive gene pairs but we further expect that our proposed method can be applied to the problem of detecting more than three genes with interactive relationships. Also, since the BCT framework is practical for the detection of relationships in binary data, we consider that we can analyze other binary data, such as copy number data, with PBCT-sampling.

References

1. Bezáková, I.: Sampling binary contingency tables. Computing in Science & Engineering 10(2), 26–31 (2008)
2. Davis, J., Goadrich, M.: The relationship between Precision-Recall and ROC curves. In: Proceedings of the 23rd International Conference on Machine Learning, pp. 233–240. ACM (2006)
3. Dees, N.D., Zhang, Q., Kandoth, C., Wendl, M.C., Schierding, W., Koboldt, D.C., Mooney, T.B., Callaway, M.B., Dooling, D., Mardis, E.R., et al.: MuSiC: identifying mutational significance in cancer genomes. Genome Research 22(8), 1589–1598 (2012)
4. Fernandez-Zapico, M.E., Gonzalez-Paz, N.C., Weiss, E., Savoy, D.N., Molina, J.R., Fonseca, R., Smyrk, T.C., Chari, S.T., Urrutia, R., Billadeau, D.D.: Ectopic expression of VAV1 reveals an unexpected role in pancreatic cancer tumorigenesis. Cancer Cell 7, 39–49 (2005)

5. Fisher, R.A.: Statistical methods for research workers (1934)
6. Gadea, G., Sanz-Moreno, V., Self, A., Godi, A., Marshall, C.J.: DOCK10-mediated Cdc42 activation is necessary for amoeboid invasion of melanoma cells. Current Biology 18(19), 1456–1465 (2008)
7. Hayward, N.K.: Mutation spectrum of the first melanoma genome points finger firmly at ultraviolet light as the primary carcinogen. Pigment Cell & Melanoma Research 23(2), 153–154 (2010)
8. Irwin, J.: Tests of significance for differences between percentages based on small numbers. Metron 12(2), 84–94 (1935)
9. Klijn, C., Bot, J., Adams, D.J., Reinders, M., Wessels, L., Jonkers, J.: Identification of networks of co-occurring, tumor-related DNA copy number changes using a genome-wide scoring approach. PLoS Computational Biology 6(1), 1000631 (2010)
10. Kumar, N., Rehrauer, H., Ca, H., Baudis, M.: CDCOCA: a statistical method to define complexity dependent co-occurring chromosomal aberrations. Genome Biology 11(1), 23 (2010)
11. Loriot, A., Boon, T., De Smet, C.: Five new human cancer-germline genes identified among 12 genes expressed in spermatogonia. International Journal of Cancer 105(3), 371–376 (2003)
12. Sathyanarayana, U.G., Padar, A., Suzuki, M., Maruyama, R., Shigematsu, H., Hsieh, J.-T., Frenkel, E.P., Gazdar, A.F.: Aberrant promoter methylation of laminin-5–encoding genes in prostate cancers and its relationship to clinicopathological features. Clinical Cancer Research 9(17), 6395–6400 (2003)
13. Storey, J.D., Tibshirani, R.: Statistical significance for genomewide studies. Proceedings of the National Academy of Sciences 100(16), 9440–9445 (2003)
14. Vandin, F., Upfal, E., Raphael, B.J.: De novo discovery of mutated driver pathways in cancer. Genome Research 22(2), 375–385 (2012)

A Filter-Based Approach
for Approximate Circular Pattern Matching

Md. Aashikur Rahman Azim[1], Costas S. Iliopoulos[2], M. Sohel Rahman[1,2,(✉)],
and Mohammad Samiruzzaman[2]

[1] Department of CSE, AℓEDA Group, BUET, Dhaka-1215, Bangladesh
[2] Department of Informatics, King's College London,
Strand, London WC2R 2LS, UK
sohel.kcl@gmail.com

Abstract. This paper deals with the Approximate Circular Pattern
Matching (ACPM) problem, which appears as an interesting problem
in many biological contexts. Here the goal is to find all approximate occurrences of the rotations of a pattern $\mathcal{P}$ of length m in a text $\mathcal{T}$ of
length n. In this article, we present a filter-based approach to solve the
problem. We experimentally compare our approach with the state of the
art algorithms in the literature and the results are found to be excellent.

1 Introduction

The classical pattern matching problem is to find all the occurrences of a given
pattern $\mathcal{P}$ of length m in a text $\mathcal{T}$ of length n, both being sequences of characters
drawn from a finite character set Σ. This problem is interesting as a fundamental
computer science problem and is a basic requirement of many practical applications. However in most practical applications it is some sort of approximate
version of the classic patterning matching problem that is of more interest.

The circular pattern, denoted $\mathcal{C}(\mathcal{P})$, corresponding to a given pattern $\mathcal{P} =
\mathcal{P}_1 \ldots \mathcal{P}_m$, is formed by connecting $\mathcal{P}_1$ with $\mathcal{P}_m$ and forming a sort of a cycle;
this gives us the notion where the same circular pattern can be seen as m different
linear patterns, which would all be considered equivalent. In the Circular Pattern
Matching (CPM) problem, we are interested in pattern matching between the
text $\mathcal{T}$ and the circular pattern $\mathcal{C}(\mathcal{P})$ of a given pattern $\mathcal{P}$. We can view $\mathcal{C}(\mathcal{P})$
as a set of m patterns starting at positions $j \in [1 : m]$ and wrapping around the
end. In other words, in CPM, we search for all 'conjugates'[1] of a given pattern
in a given text.

Part of this research has been supported by an INSPIRE Strategic Partnership
Award, administered by the British Council, Bangladesh for the project titled
"Advances in Algorithms for Next Generation Biological Sequences".

M. Sohel Rahman – Commonwealth Academic Fellow, supported by the UK government. Currently, on a sabbatical leave from BUET.

[1] Two words x, y are conjugate if there exist words u, v such that $x = uv$ and $y = vu$.

R. Harrison et al. (Eds.): ISBRA 2015, LNBI 9096, pp. 24–35, 2015.
DOI: 10.1007/978-3-319-19048-8_3

1.1 Applications and Motivations

Along with being interesting from the pure combinatorial point view, CPM has applications in areas like, geometry, astronomy, computational biology etc. This type of circular patterns occur in the DNA of viruses [7,18], bacteria [17], eukaryotic cells [14], and archaea [3]. As a result, as has been noted in [10], algorithms on circular strings seem to be important in the analysis of organisms with such structures. Circular strings have also been studied in the context of sequence alignment. In [16], basic algorithms for pair wise and multiple circular sequence alignment have been presented. These results have later been improved in [8], where an additional preprocessing stage is added to speed up the execution time of the algorithm. In [12], the authors also have presented efficient algorithms for finding the optimal alignment and consensus sequence of circular sequences under the Hamming distance metric. For further details on the motivation and applications of this problem in computational biology and other areas the readers are kindly referred to [3, 7, 8, 10, 12, 14, 16–18] and references therein.

In this paper we focus on the Approximate Circular Pattern Matching (ACPM) problem. As has been mentioned above, the DNA sequence of many viruses has a circular structure. So if a biologist wishes to find occurrences of a particular virus in a carrier's (linear) DNA sequence, (s)he must locate all positions in $\mathcal{T}$ where at least one rotation of $\mathcal{P}$ occurs. This motivates one to study CPM. However, from practical consideration, the biologists are more interested in locating the approximate occurrences of one of the rotations of $\mathcal{P}$ in $\mathcal{T}$. This is why in this paper we are interested to solve ACPM i.e., the approximate version of the problem.

1.2 Our Contribution

The main contribution of this paper is a fast and efficient algorithm for the approximate circular pattern matching problem based on some filtering techniques. The main idea behind our approach is quite simple and intuitive. We employ a number of simple and effective filters to preprocess the given pattern and the text. After this preprocessing, we get a text of reduced length on which we can apply any existing state of the art algorithms to get the occurrences of the circular pattern.

1.3 Road Map

The rest of the paper is organized as follows. Section 2 gives a preliminary description of some terminologies and concepts related to stringology that will be used throughout this paper. Section 3 presents a brief literature review. In Section 4 we describe our filtering algorithms. Section 5 presents the experimental results. Section 6 draws conclusion mentioning some future research directions.

2   Preliminaries

Let Σ be a finite *alphabet*. An element of Σ^* is called a *string*. The length of a string w is denoted by $|w|$. The empty string ϵ is a string of length 0, that is, $|\epsilon| = 0$. Let $\Sigma^+ = \Sigma^* - \{\epsilon\}$. For a string $w = xyz$, x, y and z are called a *prefix*, *factor* (or equivalently, *substring*), and *suffix* of w, respectively. The i-th character of a string w is denoted by $w[i]$ for $1 \le i \le |w|$, and the factor of a string w that begins at position i and ends at position j is denoted by $w[i : j]$ for $1 \le i \le j \le |w|$. For convenience, we assume $w[i : j] = \epsilon$ if $j < i$. A k-factor is a factor of length k.

A circular string of length m can be viewed as a traditional linear string which has the left-most and right-most symbols wrapped around and stuck together in some way. Under this notion, the same circular string can be seen as m different linear strings, which would all be considered equivalent. Given a string $\mathcal{P}$ of length m, we denote by $\mathcal{P}^i = \mathcal{P}[i : m]\mathcal{P}[1 : i - 1]$, $0 < i < m$, the i-th *rotation* of $\mathcal{P}$ and $\mathcal{P}^0 = \mathcal{P}$.

Example 1. Suppose we have a pattern $\mathcal{P} = atcgatg$. The pattern $\mathcal{P}$ has the following rotations (i.e., conjugates): $\mathcal{P}^1 = tcgatga, \mathcal{P}^2 = cgatgat, \mathcal{P}^3 = gatgatc,$ $\mathcal{P}^4 = atgatcg, \mathcal{P}^5 = tgatcga, \mathcal{P}^6 = gatcgat.$

The *Hamming distance* between strings $\mathcal{P}$ and $\mathcal{T}$, both of length n, is the number of positions i, $0 \le i < n$, such that $\mathcal{P}[i] \ne \mathcal{T}[i]$. Given a non-negative integer k, we write $\mathcal{P} \equiv_k \mathcal{T}$ or equivalently say that $\mathcal{P}$ k-matches $\mathcal{T}$, if the Hamming distance between $\mathcal{P}$ and $\mathcal{T}$ is at most k. In biology, the *Hamming distance* is popularly referred to as the *Mutation distance*. A little mutation could be considered and in fact anticipated while finding the occurrences of a particular (circular) virus in a carrier's DNA sequence. This scenario in fact refers to *approximate circular pattern matching* (ACPM). If, $k = 0$, then we get the exact CPM, i.e., mutations are not considered. Note carefully that in this setting, ACPM also returns all the occurrences returned by CPM; it computes the occurrences allowing **up to** k mismatches/mutations.

We consider the DNA alphabet, i.e., $\Sigma = \{a, c, g, t\}$. In our approach, each character of the alphabet is associated to a numeric value as follows. Each character is assigned a unique numbers from the range $[1...|\Sigma|]$. Although this is not essential, we conveniently assign the numbers from the range $[1...|\Sigma|]$ to the characters of Σ following their inherent lexicographical order. We use $num(x), x \in \Sigma$ to denote the numeric value of the character x. So, we have $num(a) = 1, num(c) = 2, num(g) = 3$ and $num(t) = 4,$. For a string S, we use the notation S_N to denote the numeric representation of the string S; and $S_N[i]$ denotes the numeric value of the character $S[i]$. So, if $S[i] = g$ then $S_N[i] = num(g) = 3$. The concept of circular string and their rotations also apply naturally on their numeric representations as is illustrated in Example 2 below.

Example 2. Suppose we have a pattern $\mathcal{P} = atcgatg$. The numeric representation of $\mathcal{P}$ is $\mathcal{P}_N = 1423143$. And this numeric representation has the following rotations:

$\mathcal{P}_N^1 = 4231431, \mathcal{P}_N^2 = 2314314, \mathcal{P}_N^3 = 3143142, \mathcal{P}_N^4 = 1431423, \mathcal{P}_N^5 = 4314231, \mathcal{P}_N^6 = 3142314.$

The problem we handle in this article can be formally defined as follows.

Problem 1. (Approximate Circular Pattern Matching with k-mismatches (i.e mutations) (ACPM)). Given a pattern $\mathcal{P}$ of length m, a text $\mathcal{T}$ of length $n > m$, and an integer threshold $k < m$, find all factors $\mathcal{F}$ of $\mathcal{T}$ such that $\mathcal{F} \equiv_k \mathcal{P}^i$ for some $0 \leq i < m$. And when we have a factor $\mathcal{F} = \mathcal{T}[j : j + |\mathcal{F}| - 1]$ such that $\mathcal{F} \equiv_k \mathcal{P}^i$ we say that the circular pattern $\mathcal{C}(\mathcal{P})$ k-matches $\mathcal{T}$ at position j. We also say that this k-match is due to $\mathcal{P}^i$, i.e., the ith rotation of $\mathcal{P}$.

In the context of our filter based algorithm the concept of false positives and negatives is important. So, we briefly discuss this concept here. Suppose we have an algorithm $\mathcal{A}$ to solve a problem $\mathcal{B}$. Now suppose that $\mathcal{S}_{true}$ represents the set of true solutions for the problem $\mathcal{B}$. Further suppose that $\mathcal{A}$ computes the set $\mathcal{S}_\mathcal{A}$ as the set of solutions for $\mathcal{B}$. Now assume that $\mathcal{S}_{true} \neq \mathcal{S}_\mathcal{A}$. Then, the set of false positives can be computed as follows: $\mathcal{S}_\mathcal{A} \setminus \mathcal{S}_{true}$. In other words, the set computed by $\mathcal{A}$ contains some solutions that are not true solutions for problem $\mathcal{B}$. And these are the false positives, because, $\mathcal{S}_\mathcal{A}$ falsely marked these as solutions (i.e., positive). On the other hand, the set of false negatives can be computed as follows: $\mathcal{S}_{true} \setminus \mathcal{S}_\mathcal{A}$. In other words, false negatives are those members in $\mathcal{S}_{true}$ that are absent in $\mathcal{S}_\mathcal{A}$. These are false negatives because $\mathcal{S}_\mathcal{A}$ falsely marked these as non-solutions (i.e., negative).

3 Brief Literature Review

The problem of circular pattern matching has been considered in [15], where an $\mathcal{O}(n)$-time algorithm is presented. A naive solution with quadratic complexity consists in applying a classical algorithm for searching a finite set of strings after having built the *trie* of rotations of $\mathcal{P}$. The approach presented in [15] consists in preprocessing $\mathcal{P}$ by constructing a suffix automaton of the string $\mathcal{PP}$, by noting that every rotation of $\mathcal{P}$ is a factor of $\mathcal{PP}$. Then, by feeding $\mathcal{T}$ into the automaton, the lengths of the longest factors of $\mathcal{PP}$ occurring in $\mathcal{T}$ can be found by the links followed in the automaton in time $\mathcal{O}(n)$. In [9], the authors have presented an optimal average-case algorithm for CPM, by also showing that the average-case lower bound for the (linear) pattern matching of $\mathcal{O}(n \log_\sigma m/m)$ also holds for CPM, where $\sigma = |\Sigma|$. Recently, in [6], the authors have presented two fast average-case algorithms based on word-level parallelism. Very recently, we have presented a filter-based approach to solve the problem in [4]. Our approach in [4] turns out to be highly effective. In fact, as will be clear shortly, in this paper, we extend our approach in [4] to solve the approximate version of the problem.

The approximate version of the problem has also received attention in the literature very recently [5]. In [5], Barton et al. have first presented an efficient algorithm for CPM that runs in $\mathcal{O}(n)$ time on average. Based on the above, they have also devised fast average-case algorithms (ACSMF-Simple) for approximate circular string matching with k-mismatches. They have built a library for ACSMF-Simple algorithm. The library is freely available [1]. Notably, indexing circular patterns [11] and variations of approximate circular pattern matching under the edit distance model [13] have also been considered in the literature.

4   Filtering Algorithm

As has been mentioned above, our algorithm is based on some filtering techniques. Suppose we are given a pattern $\mathcal{P}$ and a text $\mathcal{T}$. We will frequently and conveniently use the expression "$\mathcal{C}(\mathcal{P})$ k-matches $\mathcal{T}$ at position i" (or equivalently, "$\mathcal{P}$ circularly k-matches $\mathcal{T}$ at position i") to indicate that one of the conjugates of $\mathcal{P}$ k-matches $\mathcal{T}$ at position i (or equivalently, $\mathcal{C}(\mathcal{P}) \equiv_k \mathcal{T}$). We start with an brief overview of our approach below.

4.1   Overview of Our Approach

Our approach follows our recent work in [4] where we have used a number of filters to solve the exact circular pattern matching problem. In particular we will extend the ideas of [4] and adapt the filters presented there so that those filters become useful and effective for the approximate version as well. We employ a number of filters to compute a set $\mathcal{N}$ of indexes of $\mathcal{T}$ such that $\mathcal{C}(\mathcal{P})$ k-matches $\mathcal{T}$ at position $i \in \mathcal{N}$ in such a way that there are no false negatives.

4.2   Our Filters

We employ a total of 4 filters. The key to our observations and the resulting filters is the fact that each function we devise results in a unique output when applied to the rotations of a circular string. For example, consider a hypothetical function $\mathcal{X}$. We will always have the relation that $\mathcal{X}(\mathcal{P}) = \mathcal{X}(\mathcal{P}^i)$ for all $1 \leq i < n$. Recall that, $\mathcal{P}^0$ actually denotes $\mathcal{P}$. For the sake of conciseness, for such functions, we will abuse the notation a bit and use $\mathcal{X}(\mathcal{C}(\mathcal{P}))$ to represent $\mathcal{X}(\mathcal{P}^i)$ for all $0 \leq i < |\mathcal{P}|$.

Filter 1. We define the function sum on a string $\mathcal{P}$ of length m as follows: $sum(\mathcal{P}) = \sum_{i=1}^{m} P_N[i]$. Our first filter, Filter 1, is based on this sum function. We have the following observation.

Observation 1. *Consider a circular string $\mathcal{P}$ and a linear string $\mathcal{T}$ both having length n. If $\mathcal{C}(\mathcal{P}) \equiv_k \mathcal{T}$, where $0 \leq k < n$, then we must have*

$$sum(\mathcal{T}) - k \times 4 + k \times 1 \leq sum(\mathcal{C}(\mathcal{P})) \leq sum(\mathcal{T}) + k \times 4 - k \times 1.$$

Example 3. Consider $\mathcal{P} = atcgatg$. We can easily calculate that $sum(\mathcal{C}(\mathcal{P})) = 18$. Now, consider $\mathcal{T}1 = aacgatg$, slightly different from $\mathcal{P}$, i.e, $\mathcal{P}[2] = t \neq \mathcal{T}1[2] = a$. As can be easily verified, here $\mathcal{P} \equiv_1 \mathcal{T}1$. According to Observation 1, in this case the lower (upper) bound is 15 (18). Indeed, we have $\mathcal{T}1_N = 1123143$ and $sum(\mathcal{T}1) = 15$, which is within the bounds. Now consider $\mathcal{T}2 = ttcgatg$, slightly different from $\mathcal{P}$, i.e, $\mathcal{P}[1] = a \neq \mathcal{T}2[1] = t$. As can be easily verified, here $\mathcal{P} \equiv_1 \mathcal{T}2$. Therefore, in this case as well, the lower and upper bound mentioned above hold. And indeed we have $\mathcal{T}2_N = 4423143$ and $sum(\mathcal{T}2) = 21$, which is within the bounds. Finally, consider another string $\mathcal{T}' = atagctg$. It can be easily verified that $\mathcal{C}(\mathcal{P}) \not\equiv_1 \mathcal{T}'$. Again, the previous bounds hold in this case and we find that $\mathcal{T}'_N = 1413243$ and $sum(\mathcal{T}') = 18$. Clearly this is within the bounds of Observation 1 and in fact it is exactly equal to $sum(\mathcal{C}(\mathcal{P}))$. This is an example of a false positive with respect to Filter 1.

Filters 2 and 3. Our second and third filters, i.e., Filters 2 and 3, depend on a notion of distance between consecutive characters of a string. The *distance* between two consecutive characters of a string $\mathcal{P}$ of length m is defined by $distance(\mathcal{P}[i], \mathcal{P}[i+1]) = \mathcal{P}_N[i] - \mathcal{P}_N[i+1]$, where $1 \leq i \leq m-1$. We define $total\_distance(P) = \sum_{i=1}^{m-1} distance(\mathcal{P}[i], \mathcal{P}[i+1])$. We also define an absolute version of it: $abs\_total\_distance(P) = \sum_{i=1}^{m-1} abs(distance(\mathcal{P}[i], \mathcal{P}[i+1]))$, where $abs(x)$ returns the magnitude of x ignoring the sign. Before we apply these two functions on our strings to get our filters, we need to do a simple pre-processing on the respective string, i.e., $\mathcal{P}$ in this case as follows. We extend the string $\mathcal{P}$ by concatenating the first character of $\mathcal{P}$ at its end. We use $ext(\mathcal{P})$ to denote the resultant string. So, we have $ext(\mathcal{P}) = \mathcal{P}\mathcal{P}[1]$. Since, $ext(\mathcal{P})$ can simply be treated as another string, we can easily extend the notation and concept of $\mathcal{C}(\mathcal{P})$ over $ext(\mathcal{P})$ and we continue to abuse the notation a bit for the sake of conciseness as mentioned at the beginning of Section 4.2 (just before Section 4.2).

Now we have the following observation which is the basis of our Filter 2.

Observation 2. *Consider a circular string $\mathcal{P}$ and a linear string $\mathcal{T}$ both having length n and assume that $\mathcal{A} = ext(\mathcal{P})$ and $\mathcal{B} = ext(\mathcal{T})$. If $\mathcal{C}(\mathcal{P}) \equiv_k \mathcal{T}$, where $0 \leq k < n$, then we must have*

$$abs\_total\_distance(\mathcal{B}) - k \times 4 + k \times 1 \leq abs\_total\_distance(\mathcal{C}(\mathcal{A}))$$

$$\leq abs\_total\_distance(\mathcal{B}) + k \times 4 - k \times 1.$$

Example 4. Consider the same strings of Example 3, i.e., $\mathcal{P} = atcgatg$, $\mathcal{T}1 = aacgatg$ and $\mathcal{T}2 = ttcgatg$. As can be easily verified, here $\mathcal{P} \equiv_1 \mathcal{T}1$ and $\mathcal{P} \equiv_1 \mathcal{T}2$. Now consider the extended strings and assume that $\mathcal{A} = ext(\mathcal{P})$, $\mathcal{B}1 = ext(\mathcal{T}1)$ and $\mathcal{B}2 = ext(\mathcal{T}2)$. It can be easily verified that $abs\_total\_distance(\mathcal{C}(\mathcal{A}))$ is 14. Recall that $\mathcal{T}1$ is slightly different from $\mathcal{P}$, i.e, $\mathcal{P}[2] = t \neq \mathcal{T}1[2] = a$. Now we have $\mathcal{T}1_N = 1123143$. Hence $\mathcal{B}1_N = 11231431$. Hence, $abs\_total\_distance(\mathcal{B}1) = 10$ which is indeed within the bounds of Observation 2. Now consider $\mathcal{T}2$, which is slightly different from $\mathcal{P}$, i.e, $\mathcal{P}[1] = a \neq \mathcal{T}2[1] = t$. Now we have $\mathcal{T}2_N =$

4423143. Hence $\mathcal{B}2_N = 44231434$. Hence, $abs\_total\_distance(\mathcal{B}2) = 10$, which is also within the bounds. Finally, consider $\mathcal{T}' = atagctg$, which is again slightly different from $\mathcal{P}$. It can be easily verified that $\mathcal{C}(\mathcal{P}) \not\equiv_1 \mathcal{T}'$. However, assuming that $\mathcal{B}' = ext(\mathcal{T}')$ we find that $abs\_total\_distance(\mathcal{B}')$ is still 14, which is in the range of Observation 2. This is an example of a false positive with respect to Filter 2.

Now we present the following related observation which is the basis of our Filter 3. Note that Observation 2 differs with Observation 3 only through using the absolute version of the function used in the latter.

Observation 3. *Consider a circular string $\mathcal{P}$ and a linear string $\mathcal{T}$ both having length n and assume that $\mathcal{A} = ext(\mathcal{P})$ and $\mathcal{B} = ext(\mathcal{T})$. If $\mathcal{C}(\mathcal{P}) \equiv_k \mathcal{T}$, where $0 \leq k < n$, then we must have*

$$total\_distance(\mathcal{B}) - k \times 4 + k \times 1 \leq total\_distance(\mathcal{C}(\mathcal{A}))$$

$$\leq total\_distance(\mathcal{B}) + k \times 4 - k \times 1.$$

Example 5. Consider the same strings of Example 3, i.e., $\mathcal{P} = atcgatg$, $\mathcal{T}1 = aacgatg$ and $\mathcal{T}2 = ttcgatg$. As can be easily verified, here $\mathcal{P} \equiv_1 \mathcal{T}1$ and $\mathcal{P} \equiv_1 \mathcal{T}2$. Now consider the extended strings and assume that $\mathcal{A} = ext(\mathcal{P})$, $\mathcal{B}1 = ext(\mathcal{T}1)$ and $\mathcal{B}2 = ext(\mathcal{T}2)$. It can be easily verified that $abs\_total\_distance(\mathcal{C}(\mathcal{A}))$ is 14. Recall that $\mathcal{T}1$ is slightly different from $\mathcal{P}$, i.e, $\mathcal{P}[2] = t \neq \mathcal{T}1[2] = a$. Now we have $\mathcal{T}1_N = 1123143$. Hence $\mathcal{B}1_N = 11231431$. Hence, $total\_distance(\mathcal{B}1) = 0$ which is indeed within the bounds of Observation 2. Now consider $\mathcal{T}2$, which is slightly different from $\mathcal{P}$, i.e, $\mathcal{P}[1] = a \neq \mathcal{T}2[1] = t$. Now we have $\mathcal{T}2_N = 4423143$. Hence $\mathcal{B}2_N = 44231434$. Hence, $total\_distance(\mathcal{B}2) = 10$, which is also within the bounds. Finally, consider $\mathcal{T}' = atagctg$, which is again slightly different from $\mathcal{P}$. It can be easily verified that $\mathcal{C}(\mathcal{P}) \not\equiv_1 \mathcal{T}'$. However, assuming that $\mathcal{B}' = ext(\mathcal{T}')$ we find that $total\_distance(\mathcal{B}')$ is still 0, which is in the range of Observation 2. This is an example of a false positive with respect to Filter 3.

Filter 4. Filter 4 uses the $sum()$ function used by Filter 1, albeit, in a slightly different way. In particular, it applies the $sum()$ function on individual characters. So, for $x \in \Sigma$ we define $sum_x(\mathcal{P}) = \sum_{1 \leq i \leq |\mathcal{P}|, \mathcal{P}[i] = x} \mathcal{P}_N[i]$. Now we have the following observation.

Observation 4. *Consider a circular string $\mathcal{P}$ and a linear string $\mathcal{T}$ both having length n. If $\mathcal{C}(\mathcal{P}) \equiv_k \mathcal{T}$, where $0 \leq k < n$, then we must have*

$$sum_x(\mathcal{T}) - k \times num(x) \leq sum_x(\mathcal{C}(\mathcal{P})) \leq sum_x(\mathcal{T}) + k \times num(x)$$

for all $x \in \Sigma$.

Algorithm 1. Approximate Circular Pattern Signature using Observations 1 : 4 in a single pass

---

1: **procedure** $ACPS\_FT(\mathcal{P}[1:m])$
2:    define three variables for observations 1, 2, 3
3:    define an array of size 4 for observation 4
4:    define an array of size 4 to keep fixed value of A, C, G, T
5:    $s \leftarrow \mathcal{P}[1:m]\mathcal{P}[1]$
6:    initialize all defined variables to zero
7:    initialize fixed array to $\{1, 2, 3, 4\}$
8:    **for** $i \leftarrow 1$ to $|s|$ **do**
9:        **if** $i \neq |s|$ **then**
10:            calculate different filtering values via observations 1 & 4 and make a running sum
11:        **end if**
12:        calculate different filtering values via observations 2, 3 and make a running sum
13:    **end for**
14:    **return** all observations values
15: **end procedure**

4.3   Reduction of Search Space in the Text

Now we present an $\mathcal{O}(n)$ runtime algorithm to reduce the search space of the text applying the four filters presented above. It takes as input the pattern $\mathcal{P}[1:m]$ of length m and the text $\mathcal{T}[1:n]$ of length n. It calls Procedure $ACPS\_FT$ with $\mathcal{P}[1:m]$ as parameter and uses the output. It then applies the same technique that is applied in Procedure $ACPS\_FT$ (*Algorithm* 1). We apply a sliding window approach with window length of m and calculate the values applying the functions according to Observations 1 : 4 on the factor of $\mathcal{T}$ captured by the window. Note that for Observations 2, and 3, we need to consider the extended string and hence the factor of $\mathcal{T}$ within the window need be extended accordingly for calculating the values. After we calculate the values for a factor of $\mathcal{T}$, we check it against the returned values of Procedure $ACPS\_FT$. If it matches, then we output the factor to a file. Note that in case of overlapping factors (e.g., when the consecutive windows need to output the factors to a file), Procedure $ACPS\_FT$ outputs only the non-overlapped characters. And Procedure $ACPS\_FT$ uses a $ marker to mark the boundaries of non-consecutive factors, where $ \notin \Sigma$.

Now note that we can compute the values of consecutive factors of $\mathcal{T}$ using the sliding window approach quite efficiently as follows. For the first factor, i.e., $\mathcal{T}[1..m]$ we exactly follow the strategy of Procedure $ACPS\_FT$. When it is done, we slide the window by one character and we only need to remove the contribution of the left most character of the previous window and add the contribution of the rightmost character of the new window. The functions are such that this can be done very easily using simple constant time operations. The only other issue that needs be taken care of is due to the use of the extended

Algorithm 2. Reduction of Search Space in a Text String using procedure ACPS_FT

```
1: procedure RSS_FT(T[1 : n], P[1 : m])
2:      CALL ACPS_FT(P[1 : m])
3:      save the return value of observations 1 : 4 for further use here
4:      define an array of size 4 to keep fixed value of A, C, G, T
5:      initialize fixed array to {1, 2, 3, 4}
6:      lastIndex ← 1
7:      for i ← 1 to m do
8:          calculate different filtering values in T[1 : m] via observations 1 : 4 and
      make a running sum
9:      end for
10:     if 1 : 4 observations values of P[1 : m] vs 1 : 4 observations values of T[1 : m]
      have a match then
11:                                                          ▷ Found a filtered match
12:         Output to file T[1 : m]
13:         lastIndex ← m
14:     end if
15:     for i ← 1 to n − m do
16:         calculate different filtering values in T[1 : m] via observations 1 : 4 by
      subtracting i-th value along with wrapped value and adding i + m-th value and
      new wrapped vale to the running sum
17:         if 1 : 4 filtering values of P[1 : m] vs 1 : 4 filtering values of T[i + 1 : i + m]
      have a match then
18:                                                          ▷ Found a filtered match
19:             if i > lastIndex then
20:                 Output an end marker $ to file
21:             end if
22:             if i + m > lastIndex then
23:                 if i < lastIndex then
24:                     j ← lastIndex + 1
25:                 else
26:                     j ← i + 1
27:                 end if
28:                 Output to file T[j : i + m]
29:                 lastIndex ← i + m
30:             end if
31:         end if
32:     end for
33: end procedure
```

---

string in two of the filters. But this too do not need more than simple constant time operations. Therefore, overall runtime of the algorithm is $\mathcal{O}(m) + \mathcal{O}(n - m) = \mathcal{O}(n)$. The algorithm is presented in the form of Procedure RSS_FT (Algorithm 2).

Table 1. Elapsed-time (in seconds) and speed-up comparisons among ACSMF-Simple [5] and our algorithm considering the first three filters only for a text of size 299MB

m	k=2			k=3			k=4			k=5		
	Elapsed Time(s) of ACSMF-simple	Elapsed Time(s) of Our Algorithm	Speed up	Elapsed Time(s) of ACSMF-simple	Elapsed Time(s) of Our Algorithm	Speed up	Elapsed Time(s) of ACSMF-simple	Elapsed Time(s) of Our Algorithm	Speed up	Elapsed Time(s) of ACSMF-simple	Elapsed Time(s) of Our Algorithm	Speed up
500	9.079	9.932	1	8.664	9.885	1	10.45	17.986	1	10.821	20.124	1
700	9.994	22.798	0	10.211	23.059	0	10.604	38.117	0	10.031	25.686	0
900	8.87	59.881	0	8.781	61.15	0	8.778	59.747	0	10.621	98.298	0
1000	9.949	32.957	0	11.214	31.264	0	10.814	50.094	0	10.849	55.664	0
1600	9.218	17.648	1	8.309	15.143	1	9.724	25.51	0	11.905	25.855	0
1800	10.889	94.992	0	11.794	92.022	0	13.708	100.813	0	14.149	135.089	0
2000	9.89	33.822	0	11.585	31.419	0	11.625	49.459	0	13.156	50.133	0
2200	10.42	4.717	2	12.689	5.983	2	13.804	7.299	2	13.974	7.76	2
2400	10.906	70.06	0	12.573	68.952	0	10.36	74.988	0	12.655	105.017	0
2600	9.928	9.31	1	8.279	8.563	1	14.268	15.141	1	13.77	15.326	1
2800	9.11	5.219	2	12.163	5.298	2	13.242	7.255	2	11.368	5.372	2
3000	8.826	8.321	1	11.466	9.058	1	10.534	9.767	1	13.535	13.286	1

Table 2. Elapsed-time (in seconds) and speed-up comparisons among ACSMF-Simple [5] and our algorithm considering all the four filters for a text of size 299MB

m	k=2			k=3			k=4			k=5		
	Elapsed Time(s) of ACSMF-simple	Elapsed Time(s) of Our Algorithm	Speed up	Elapsed Time(s) of ACSMF-simple	Elapsed Time(s) of Our Algorithm	Speed up	Elapsed Time(s) of ACSMF-simple	Elapsed Time(s) of Our Algorithm	Speed up	Elapsed Time(s) of ACSMF-simple	Elapsed Time(s) of Our Algorithm	Speed up
500	9.18	2.681	3	8.701	2.77	3	8.701	2.77	3	10.997	3.541	3
700	9.698	2.658	4	10.032	2.97	3	10.032	2.97	3	10.19	3.481	3
900	8.506	2.86	3	9.092	3.27	3	9.092	3.27	3	10.758	5.374	2
1000	9.108	2.878	3	11.506	3.061	4	11.506	3.061	4	10.912	4.414	2
1600	9.162	2.589	4	8.479	2.707	3	8.479	2.707	3	12.516	3.362	4
1800	12.404	2.876	4	11.829	2.897	4	11.829	2.897	4	14.867	3.767	4
2000	11.209	2.769	4	11.535	2.842	4	11.535	2.842	4	13.063	3.485	4
2200	11.189	2.495	4	12.425	2.481	5	12.425	2.481	5	13.856	2.643	5
2400	12.555	2.794	4	12.503	2.777	5	12.503	2.777	5	12.812	3.365	4
2600	9.912	2.501	4	8.51	2.5	3	8.51	2.5	3	13.546	2.609	5
2750	13.307	2.51	5	11.572	2.507	5	11.572	2.507	5	15.158	2.518	6
2800	9.368	2.487	4	12.086	2.472	5	12.086	2.472	5	11.234	2.548	4
2900	11.137	2.667	4	12.318	2.531	5	12.318	2.531	5	13.588	2.475	5
3000	10.008	2.584	4	11.71	2.53	5	11.71	2.53	5	13.809	2.72	5

4.4 The Combined Algorithm

We have already described the two main components of our algorithm, namely, Procedure $ACPS\_FT$ and Procedure RSS_FT, which in fact calls the former. Now Procedure RSS_FT provides a reduced text $\mathcal{T}'$ (say) after filtering. At this point we can use any algorithm that can solve ACPM and apply it over $\mathcal{T}'$ and output the occurrences. Now, suppose we use Algorithm $\mathcal{A}$ at this stage which runs in $\mathcal{O}(f(|\mathcal{T}'|))$ time. Then, clearly, the overall running time of our approach is $\mathcal{O}(n) + \mathcal{O}(f(|\mathcal{T}'|))$. In our implementation we have used the recent algorithm of [5]. In particular, in [5], the authors have presented an approximate circular string matching algorithm with k-mismatches (ACSMF-Simple) via filtering. They have built a library for ACSMF-Simple algorithm. The library is freely available and can be found here: [1]. We only apply ACSMF-Simple on the reduced string.

5   Experimental Results

We have implemented our algorithm and conducted experiments in C++ using a GNU compiler with General Public License (GPL). As has been mentioned already above, our implementation uses the ACSMF-Simple [5]. ACSMF-Simple [5] has been implemented as library functions in the C programming language under *GNU/Linux* operating system.

We have used real genome data in our experiments as the text string, $\mathcal{T}$. This data has been collected from [2]. Here, we have taken *299MB* of data for our experiments.

We have conducted our experiments on a PowerEdge R820 rack serve PC with 6-core of Intel Xeon processor *E5-4600* product family and 64GB of RAM under GNU/Linux. With the help of the library used in [5], we have compared the running time of ACSMF-Simple of [5] and of our algorithm. Table 2 reports the elapsed time and speed-up comparisons for various pattern sizes ($500 \leq m \leq 3000$) and for various mismatch sizes ($2 \leq k \leq 5$). As can be seen from Table 2, our algorithm runs faster than ACSMF-Simple in all cases.

In order to analyze and understand the effect of our filters we have conducted further experiments. For space constraints, here, in Table 1, we only present the results of our algorithm where Filter 4 is omitted, i.e., Filters 1 through 3 are employed. As can be seen from Table 1, ACSMF-Simple is able to beat this version of our algorithm in a number of cases. This indicates that as more and more effective filters are imposed, our algorithm performs better. We believe after the application of two more filters from [4], we will get even better results.

6   Conclusions

In this paper, we have employed some effective lightweight filtering technique to reduce the search space of the Approximate Circular Pattern Matching (ACPM) problem. We have conducted experimental studies to show the effectiveness of our approach. Much of the speed of our algorithm comes from the fact that our filters are effective but extremely simple and lightweight.

Acknowledgement. Part of this research has been supported by an INSPIRE Strategic Partnership Award, administered by the British Council, Bangladesh for the project titled Advances in Algorithms for Next Generation Biological Sequences.

References

1. http://www.inf.kcl.ac.uk/research/projects/asmf/
2. http://hgdownload-test.cse.ucsc.edu/goldenPath/hg19/bigZips/
3. Allers, T., Mevarech, M.: Archaeal genetics – the third way. Nat. Rev. Genet. 6, 58–73 (2005)
4. Azim, M.A.R., Iliopoulos, C.S., Rahman, M.S., Samiruzzaman, M.: A fast and lightweight filter-based algorithm for circular pattern matching. In: Proceedings of the 5th ACM Conference on Bioinformatics, Computational Biology, and Health Informatics, BCB 2014, pp. 621–622. ACM, New York (2014)
5. Barton, C., Iliopoulos, C., Pissis, S.: Fast algorithms for approximate circular string matching. Algorithms for Molecular Biology 9(1), 9 (2014)
6. Chen, K., Huang, G., Lee, R.: Bit-parallel algorithms for exact circular string matching. Comput. J. (2013), doi:10.1093/comjnl/bxt023
7. Dulbecco, R., Vogt, M.: Evidence for a ring structure of polyoma virus DNA. Proc. Natl. Acad. Sci. 50(2), 236–243 (1963)
8. Fernandes, F., Pereira, L., Freitas, A.: CSA: An efficient algorithm to improve circular DNA multiple alignment. BMC Bioinformatics 10, 1–13 (2009)
9. Fredriksson, K., Grabowski, S.: Average-optimal string matching. J. Discrete Algorithms 7(4), 579–594 (2009)
10. Gusfield, D.: Algorithms on Strings, Trees and Sequences. Cambridge University Press, New York (1997)
11. Iliopoulos, C., Rahman, M.: Indexing circular patterns. In: Proceedings of the 2nd International Conference on Algorithms and Computation, pp. 46–57 (2008)
12. Lee, T., Na, J., Park, H., Park, K., Sim, J.: Finding optimal alignment and consensus of circular strings. In: Proceedings of the 21st Annual Conference on Combinatorial Pattern Matching, pp. 310–322 (2010)
13. Lin, J., Adjeroh, D.: All-against-all circular pattern matching. Comput. J. 55(7), 897–906 (2012)
14. Lipps, G.: Plasmids: Current Research and Future Trends. Caister Academic Press, Norfolk (2008)
15. Lothaire, M.: Applied Combinatorics on Words. Cambridge University Press, New York (2005)
16. Mosig, A., Hofacker, I., Stadler, P., Zell, A.: Comparative analysis of cyclic sequences: viroids and other small circular RNAs. In: German Conference on Bioinformatics. LNI, vol. 83, pp. 93–102 (2006)
17. Thanbichler, M., Wang, S., Shapiro, L.: The bacterial nucleoid: A highly organized and dynamic structure. J. Cell Biochem. 96(3), 506–521 (2005), http://dx.doi.org/10.1002/jcb.20519
18. Weil, R., Vinograd, J.: The cyclic helix and cyclic coil forms of polyoma viral DNA. Proc. Natl. Acad. Sci. 50(4), 730–738 (1963)

Fast Algorithms for Inferring Gene-Species Associations

Arkadiusz Betkier[1]([✉]), Paweł Szczęsny[2], and Paweł Górecki[1]

[1] Faculty of Mathematics, Informatics and Mechanics,
University of Warsaw, Warsaw, Poland
{arebet,gorecki}@mimuw.edu.pl

[2] Institute of Biochemistry and Biophysics, Polish Academy of Sciences,
Warsaw, Poland
szczesny@ibb.waw.pl

Abstract. Assessment of microbial biodiversity is typically made by sequencing either PCR-amplified marker genes or all genomic DNA from environmental samples. Both approaches rely on the similarity of the sequenced material to known entries in sequence databases. However, amplicons of non-marker genes are often used, when the research question aims at assessing both functional capabilities of a microbial community and its biodiversity. In such cases, a phylogenetic tree is constructed with known and metagenomic sequences, and expert assessment defines the taxonomic groups the amplicons belong to. Here, instead of relying on sequences, often missing, of non-marker genes, we use tree reconciliation to obtain a distribution of mappings between genes and species. We describe efficient algorithms for the reconstruction of gene-species mappings and a Monte-Carlo method for the inference of distributions for the cases when the number of optimal reconstructions is large. We provide a comparative study of different cost functions showing that the duplication-loss cost induces mappings of the highest quality. Further, we demonstrate the correctness of our approach using several datasets.

1 Introduction

Phylotype-centric studies of biodiversity of microbial communities have their obvious limitations - they do not allow reliable prediction of functional capabilities of the community [6]. Therefore, complex microbial communities are approached with functional analysis in mind. Sequence similarity searches against reference databases such as KEGG, COG, Pfam or SEED provide in-depth knowledge about their metabolism, biochemistry or ecology. However, given our limited knowledge about microbial biochemistry and the large number of different species in such communities, important metabolic phenotypes might not be visible in shotgun metagenomic studies. Therefore, researchers quite often use a different approach to address such issues - amplicons of important genes are used instead of undirected shotgun sequencing. However, these sequences can rarely be matched with species. We have only a few dozen well annotated universally

R. Harrison et al. (Eds.): ISBRA 2015, LNBI 9096, pp. 36–47, 2015.
DOI: 10.1007/978-3-319-19048-8_4

conserved protein coding marker genes, for example RecA [28]. For the rest of the genes, LCA-based methods such as MEGAN [14], or phylogeny-based methods such as MLTreeMap [26], or pplacer [19], have a fundamental issue: if species is not present in the reference database (i.e. genome is not known), sequence from this species cannot be accurately placed at the leaf level. In other words, such methods rely on annotated genes and do not use much richer microbial species trees, because for majority of these species genomic sequence is unknown.

While phylogenetic trees of prokaryotic genes have typically different topologies from species tree, there is a prominent coherent trend of tree-like evolution, that is substantially different from net-like trend produced by horizontal gene transfer (HGT) [23]. Therefore, it should be possible to reconcile gene trees with the species tree. While HGT is frequent in microbial world, it was shown that HGT rates between close species are on average higher than between distantly related ones [25]. This could mean that ambiguity stemming from HGT events should not significantly affect reconciliation up from genus level of species tree.

The concept of reconciliation was introduced by Goodman [9] and formalized by Page [21] in the context of reconciling a gene family tree with its species tree. In this model, any incongruence between gene and species trees is explained in terms of evolutionary events such as gene duplication, gene loss and speciation. Reconciliation is interpreted as the embedding of a gene tree into a species tree where these evolutionary events located in the species tree induce a biologically consistent scenario [11]. By counting gene duplication and loss events when reconciling trees, we can define a cost function, called a *duplication-loss cost* (DL) [30]. Formally, it is the minimal number of gene duplication and loss events required to reconcile a given gene tree with its species tree. Similarly, we define gene duplication (D) or deep coalescence (DC) [18,30].

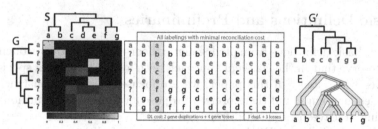

Fig. 1. An example of inference of gene-species mappings showing the heat map that represents the inferred distributions of species over the leaves of gene tree. *Top-left:* S is a species tree with 7 species: $a, b, \ldots, g$. G is a gene tree with 8 genes. There are 5 sequences in G that have undefined species assignment (marked with "?"). *Middle:* All 11 labelings that induce the minimal duplication-loss cost, such that every undefined label in G is replaced with a species from S. The heat map shows the distribution of species from the set of optimal labelings. *Right:* G' is obtained from G by using the second optimal labeling (marked with gray) E is the embedding of G' into S [5,11].

Tree reconciliation has been intensively studied in recent decades in many theoretical and practical contexts including supertree inference, error correction and HGT detection [2,7,8,12,15,17,25,27,30]. In this article we extend the concept of reconciliation by introducing partial mappings where the undefined value represents an unknown gene-species assignment. We propose to reconstruct the missing gene-species assignments by seeking for the labeling that extends the input labeling, and such that it minimizes a given cost. An example is depicted in Fig. 1. Similar concepts have been studied in several articles. In [20] authors proposed heuristic algorithms for the reconstruction of gene-species mappings under DC and a special case of binary gene trees with bijective leaf labellings. In [31] $O(|G||S|^2)$ time algorithm was presented for DC and DL and the analogous reconstruction problem under general leaf labellings, where G and S are the gene and species trees, respectively. In different biological context $O(|G||S|)$ time algorithm was developed for the duplication cost [3].

Our contribution. We propose a reconciliation-based formulation of the gene-species assignment problem for cost functions such as D, DL, DC and L (gene loss). Our algorithms run in $O(|G||S|)$ time if trees are binary. Thus, we improve known algorithm for DC and DL by a factor of $|S|$ and we propose new efficient algorithm for L. Our solution can be applied to non-binary trees in case of D or DC, then the time complexity is $O(|G||S|\Delta S\Delta G)$, where ΔG is the maximal out-degree of nodes from G. We also propose a Monte-Carlo approach to approximate the distribution of gene-species mappings by sampling the space of optimal reconstructions. Having this, we provide a comparative study of reconstructions for our cost functions. These algorithms have been implemented in a software package written in C++ that is publicly available. Our software can also solve instances composed of unrooted trees (which is not discussed in this article). In the last section we provide an experimental study showing the performance, and the quality of mappings' reconstruction for synthetic and empirical datasets.

2   Basic Definitions and Preliminaries

We provide basic definitions from phylogenetic theory and from the reconciliation model. Observe that in the classical approach both a gene tree and a species tree have leaves with labels. In this paper, we are focused on the reconstruction of gene-species mappings, and therefore, we propose an equivalent definition where the labeling of leaves is separated from gene trees.

Let G be a rooted tree. By L_G we denote the set of all leaves in a tree G. In this paper, a rooted tree is a model of a *gene tree*. We assume that every internal node of a rooted tree has at least two children. For a gene tree G, by ΔG we denote the maximal out-degree of its nodes. For instance, for a non-trivial (i.e., not single-noded) binary gene tree G, $\Delta G = 2$. By $|G|$ we denote the number of nodes present in G and by $\widehat{g}$ we denote the set of all children of a node g.

A *species tree* $S = \langle V_S, E_S \rangle$ is a rooted tree whose leaves are called *species*. For vertices $a, b \in V_S$, let $a \oplus b$ be the least common ancestor of a and b in S. We use the binary order relation $a \preceq b$ if b is a vertex on the path between a inclusively, and the root of S. For a gene tree G and a species tree S, a *(leaf) labeling from G to S*, is a function from the leaves of G into the species (leaves) present in S. Any leaf labeling $\lambda: L_G \to L_S$ can be extended into *the least common ancestor mapping*, or *lca-mapping*, $M_\lambda \colon V_G \to V_S$ defined as $M_\lambda(g) = \lambda(g)$ if g is a leaf, and $M_\lambda(g) = M_\lambda(c_1) \oplus M_\lambda(c_2) \oplus \cdots \oplus M_\lambda(c_n)$ where $\widehat{g} = \{c_1, c_2, \ldots, c_n\}$, otherwise.

Now, we define several known cost functions in a unified approach [10]. First, we introduce the notion of a scoring function ξ_K for $K \in \{D, DL, DC, L\}$ and a species tree S. For an internal node g with n children from a gene tree G and a sequence of species nodes $\{x_1, x_2, \ldots, x_n\}$ from S, $\xi_K(x_1, x_2, \ldots, x_n)$ is a contribution to a total cost K when all n children g are mapped into $x_1, x_2, \ldots, x_n$, respectively. We define the scoring functions for gene duplication (D), deep coalescence (DC), gene loss and gene duplication+loss (DL) costs:

- $\xi_D(x_1, x_2, \ldots, x_n) = \mathbb{1}(\exists i \in \{1, 2, \ldots, n\} \colon x_i = x_1 \oplus x_2 \oplus \cdots \oplus x_n)$,
- $\xi_{DC}(x_1, x_2, \ldots, x_n) = \sum_{i \in \{1,2,\ldots,n\}} ||x_1 \oplus x_2 \oplus \cdots \oplus x_n, x_i||$,
- $\xi_L(x, y) = ||x, y|| - 2(1 - \xi_D(x, y))$ (only for binary trees),
- $\xi_{DL}(x, y) = \xi_D(x, y) + \xi_L(x, y)$,

where $\mathbb{1}$ is the indicator function, that is, $\mathbb{1}(p)$ is 1 if p is satisfied and 0 otherwise, and $||x, y||$ denotes the length of the shortest path connecting nodes x and y. Now, we define the cost of reconciliation of a given gene tree G with labeling λ and a species tree S under cost $K \in \{D, L, DL, DC\}$:

$$K(G, S, \lambda) := \sum_{g \in V_G \setminus L_G, \; \widehat{g} = \{c_1, c_2, \ldots, c_n\}} \xi_K(M_\lambda(c_1), M_\lambda(c_2), \ldots, M_\lambda(c_n)).$$

For example, if $K = D$ then $D(G, S, \lambda)$ is the classical duplication cost [22].

Observe that the duplication-loss cost is usually defined for binary trees [11,21] due to complications with the biological interpretation of these events when a multifurcation is present in trees [29,4]. Our definitions, however, are general and they can be applied to any type of trees under assumption of that non-binary vertices refer to hard polytomies. Please refer to the literature for a more detailed study on the reconciliation model [18,21,30,32].

2.1   Problems: Resolving Unknown Gene-Species Mappings

We present several problems related to the reconstruction of gene-species mappings. To model undefined labels in gene trees we use the classical mathematical notion of a partial function. In other words, we express the problem of reconstruction of gene-species associations in terms of converting a partial function into a total one. For example, if $(a, (\perp, c))$ is a gene tree with one undefined label denoted by $\perp$ and $(a, (b, c))$ is a species tree, then the problem is to replace $\perp$ by a, b or c such that the total reconciliation cost is minimized (in this case b is the optimal choice for every cost function). See also Fig. 1 for a complex example.

Let G be a gene tree G and S a species tree. Any partial function $\phi \colon L_G \to L_S$ will be called a *partial (leaf) labeling* from G into S. Later on, we write $\phi(x) = \perp$ if ϕ is undefined for x. Now, we present problems related to the reconstruction of leaf labelings, i.e., total functions, from partial labelings. Below, we assume that $K \in \{D, L, DL, DC\}$.

Problem 1. Given a gene tree G, a species tree S and a partial labeling ϕ between G and S. Find a labeling λ such that (1) λ is a total function that extends ϕ, and (2) $K(G, S, \lambda)$ is minimal in the set of all labelings between G and S.

Such a labeling is called *optimal under K*. For given trees G, S and a partial labeling ϕ we denote the minimal cost introduced in Problem 1 by $K_{opt}(G, S, \phi)$.

Problem 2. Given trees G, S and a partial labeling ϕ between G and S. Find the number of optimal labelings under K that extends ϕ.

3   Methods

In this section we propose a polynomial time algorithm for the computation of optimal costs. Next, we discuss its properties in the context of DL and DC.

Given a gene tree G, a species tree S and $\phi\colon L_G \to L_S$ a partial labeling we show how to compute $K_{opt}(G, S, \phi)$. The dynamic programming formula has two components $\delta_K(g, s)$ and $\delta_K^\uparrow(g, s)$ where $g \in V_G$ and $s \in V_S$, such that

- $\delta_K(g, s)$ is the minimal cost $K(G|g, S|s, \psi)$ in the set of all labelings ψ extending ϕ and satisfying $M_\psi(g) = s$, where $G|g$ denotes the subtree of G rooted at g,
- $\delta_K^\uparrow(g, s)$ is the minimal cost $K(G|g, S|s, \psi) + d_\psi$ in the set of all labelings ψ extending ϕ and satisfying $M_\psi(g) \preceq s$; d_ψ is the number of additional deep coalescence events between $M_\psi(g)$ and s (i.e., the number of edges between $M_\psi(g)$ and s); these additional events are counted only when $K \neq$ D.

For δ_K we have the following formula:

$$\delta_K(g, s) = \begin{cases} 0 & \text{if } g \text{ and } s \text{ are leaves and } \phi(g) \in \{s, \perp\}, \\ \min\{\alpha, \beta\} & \text{if } g \text{ is not a leaf}, \\ +\infty & \text{otherwise}, \end{cases}$$

where, for $\omega_{DC} = \mathbb{1}(K \in \{L, DL, DC\})$ we have:

$$\alpha = \min_{\substack{p\colon \widehat{g} \to \widehat{s} \\ p \text{ is not a const. function}}} \sum_{c \in \widehat{g}} (\delta_K^\uparrow(c, p(c)) + \mathbb{1}(K = DC)),$$

$$\beta = \mathbb{1}(K \in \{D, DL\}) + \min_{\substack{p\colon \widehat{g} \to \widehat{s} \cup \{s\} \\ p(x)=s \text{ for some } x}} \sum_{c \in \widehat{g}} \begin{cases} \delta_K(c, s) & \text{if } p(c) = s, \\ \delta_K^\uparrow(c, p(c)) + \omega_{DC} & \text{if } p(c) \in \widehat{s}. \end{cases}$$

Functions p in above definitions denote all valid mapping assignments for the children of g. In particular, α represents the case when g is a speciation node [11], i.e., all children of g are mapped below s, and β represents the case when g is a duplication node, i.e., at least on child of g is mapped to s.

The formula for $\delta_K^\uparrow$ can be expressed as:

$$\delta_K^\uparrow(g,s) = \begin{cases} \delta_K(g,s) & \text{if } s \text{ is a leaf,} \\ \min\{\delta_K(g,s), \omega_{DC} + \min_{x \in \widehat{s}} \delta_K^\uparrow(g,x)\} & \text{otherwise.} \end{cases}$$

Theorem 1 (Correctness). *For a gene tree G, a species tree S, a partial labeling ϕ and every standard cost K, $K_{opt}(G,S,\phi) = \min_{s \in V_S} \delta_K(\text{root}(G), s)$.*

The proof of the next lemma provides details for efficient computation of δ_K.

Lemma 1. *For a fixed g and s, the values of α and β can be computed in $O(\Delta G \Delta S)$ time.*

Proof: A naive approach to computing these values is to enumerate all possible functions p present in these formulas. In the worst case scenario we have $\Delta S^{\Delta G}$ and $(\Delta S + 1)^{\Delta G}$ functions needed for α and β computation, respectively. Thus, the time complexity is $O((\Delta S + 1)^{\Delta G})$. However, when computing the optimal cost only, we can use a much more efficient algorithm.

For the computation of α, we need all values of $\delta_K^\uparrow$ for every pair from $\widehat{g} \times \widehat{s}$. Then, for every $c \in \widehat{g}$ we set $p(c)$ to be an element of $\arg \min_{x \in \widehat{s}} \delta_K^\uparrow(c,x)$. If p is a not constant function, then

$$\alpha = \sum_{c \in \widehat{g}} \delta_K^\uparrow(c, p(c)). \tag{1}$$

Otherwise, if p is a constant function, then for every c we have $p(c) = x$ for some $x \in \widehat{s}$. Let c' and x' be a pair of nodes that minimizes $\delta_K^\uparrow(c',x') - \delta_K^\uparrow(c',x)$ in the set of pairs $\widehat{g} \times \widehat{s} \setminus \{x\}$. It should be clear that the function p' obtained from p by setting the value of c' to be x' satisfies the equation (1). We conclude that α can be computed in $O(\Delta G \Delta S)$ time.

Similarly, we compute β in $O(\Delta G(\Delta S + 1))$ time. Details are omitted. $\square$

Theorem 2. *The time complexity of computing $K_{opt}(G,S,\phi)$ by the formula from Theorem 1 is $O(|G||S|\Delta S \Delta G)$.*

Proof. Computing δ_K and $\delta_K^\uparrow$ can be performed by bottom-up tree traversals that require in total $|G||S|$ steps. By Lemma 1 each step requires $O(\Delta S \Delta G)$ time.

Observe that in the case of binary trees we have an algorithm that runs in $O(|G||S|)$ time. The dynamic programming formula can be extended by backtracking to infer one optimal labeling (Problem 1). The time complexity of such an extended algorithm is $O(|G||S|\Delta S \Delta G)$. However, when inferring the number of optimal labelings (Problem 2) every function p present in formulas for α and β has to be enumerated. Therefore, the time complexity of such an algorithm (Problem 2) is $O(|G||S|(\Delta S + 1)^{\Delta G})$.

In the case of DC, we can propose a better solution. Observe that if a partial labeling has an empty image, i.e., every gene leaf has undefined label, then optimal labelings under DC, that extend such a labeling, are constant functions that yield the optimal cost 0. To generalize this property, let us call a subtree G' of G *unlabeled* under partial labeling ϕ if $\phi(g) = \bot$ for every leaf g of G'. Then:

Lemma 2. *Let* $\phi: L_G \to L_S$ *be a partial labeling. Let a subtree G' of G be unlabeled under ϕ. Then, for every optimal labeling under DC that extends ϕ, leaves of G' have the same label.*

We conclude that the computation of optimal DC cost can be solved as follows: (1) compress G: let G_ϕ be a gene tree G obtained from G by replacing every maximal unlabeled subtree of G with a single-noded unlabeled tree, and (2) compute optimal DC cost for G_ϕ and S with the labeling adjusted to capture the compression of G. The procedure of compression requires one traversal of G. Thus, the time complexity is $O(|G| + |S||G_\phi|\Delta S\Delta G_\phi)$.

3.1   Inferring Gene-Species Distributions

From the practical point of view, the crucial problem is the inference of gene-species mappings (Problem 1). It can be solved efficiently by our algorithms when inferring just one labeling. However, in practice the goal would be rather to find all possible optimal labelings. As already mentioned, it may be difficult due to potentially large number of optimal solutions. Therefore, instead of computing all possible optimal labelings, we propose to apply the Monte Carlo method: (1) repeat random sampling of the space of optimal labelings, and (2) aggregate inferred labelings into the set of distributions of species associated with the leaves of a gene tree. To guarantee proper approximation of species' distribution, the first step has to be performed under the assumption that every optimal labeling has the same sampling probability. We solved this problem by computing the number of possible optimal variants assigned with every pair g and s, when computing δ_K and $\delta_K^\uparrow$. For example, such variants are represented by functions p in formulas for α and β. Then, the sampling is performed by top-down traversal of the gene tree and by choosing randomly one variant at each level from a uniform probability distribution determined by the counts of variants.

The result of aggregation is formally defined as follows. Let us assume that $X = \{\phi_1, \phi_2, \dots, \phi_n\}$ is the collection of optimal labelings inferred by our sampling method, i.e., for every i, we have $\phi_i: L_G \to L_S$. Then, for every leaf $g \in L_G$ and a species s we define the species frequency distribution at g by $p_g(s, X) = |\{i: \phi_i(g) = s\}|/n$. It should be clear that for a fixed g, $p_g(\cdot, X)$ is a probability distribution. Such distributions called *gene-species distributions* can be presented in the form of a heat map (see Fig. 1).

4   Experimental Evaluation

All the experiments were performed on a server with 256GB RAM and 8 AMD Opteron processors. The computer program `mgremap` is written in C++ and is freely available at `http://bioputer.mimuw.edu.pl/gorecki/mgremap`.

4.1   Reconstruction Quality and Runtime Analysis

Assume that we have a species tree and a gene tree with a labeling without unknown labels. There is a natural way to validate the reconstruction quality: first mask some labels of the gene tree, then reconstruct them by using our algorithms and check how many labels were correctly reconstructed.

We generated 650 species trees of size 10^3. Each tree was generated by the following procedure: start from a list of 10^3 single-noded trees representing species, replace two randomly chosen trees T and T' with a single tree (T, T') and repeat until the list consists of one tree. It can be shown that such a process is equivalent to the classical Yule-Harding model for rooted tree shapes [13]. Then, for each species tree S, we generated one gene tree G with labeling λ. We ensure that the distribution of dissimilarity measure, defined as $DC(G, S, \lambda) - 999$, in our dataset is uniformly distributed over the interval $[0, 3800]$. Note that the measure 0 is equivalent to tree isomorphism.

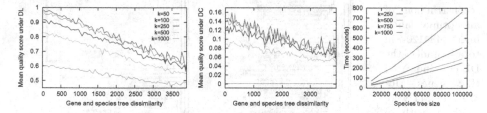

Fig. 2. *Left and middle:* Mean quality score for the reconstructions of gene species mappings under DL (left) and DC (middle) cost functions. The quality score on the Y axis represents the correctness of gene-species mapping reconstruction, e.g., if the quality score is 1.0 then every unknown label was correctly reconstructed. The parameter k denotes how many labels were set to be unknown in the input labeling of a gene tree. Note that $k = 1000$ denotes the situation when all leaves are undefined. The Y axis has a different range in both diagrams. *Right:* Runtime performance for species trees of size $1000, 2000, \ldots 10^5$, a fixed gene tree of with 1000 leaves, and masking parameter k. For every instance our program inferred 4000 optimal labelings for DL.

For each $\langle G, S, \lambda \rangle$ from our dataset, we generated 20 partial labelings by the masking procedure: set k values of λ to be undefined for $k \in \{50, 100, 250, 500, 1000\}$, in particular $k = 1000$ means that every leaf has an undefined label. Then, for each partial labeling ϕ, we computed gene-species distributions from 10^4 optimal mappings obtained by the MC method for the DC and DL costs. Finally, for a fixed $\langle G, S, \lambda \rangle$, a partial labeling ϕ with k undefined labels and their gene-species distributions, the quality score is the expected value of the probability that the undefined labels will be correctly reconstructed. Formally, if p are the gene-species distributions induced by a set X of optimal labelings that extend ϕ then the quality score is: $\zeta(G, S, \lambda, \phi, p) = \frac{1}{k} \sum_{\phi(g)=\perp} p_g(\lambda(g), X)$ (see Fig. 2).

In the left and middle diagrams of Fig. 2 we summarize the experimental evaluation. For readability reasons, all 650 pairs of the input trees were split into bins such that each bin represents pairs whose DC score is between d and $d + 50$, for d starting from 0. For each k we show mean values of quality score

Fig. 3. *Left:* a gene tree of 100 proteins similar to mcrA from Methanobrevibacter ruminantium with 9 unknown gene-species labels M1-M9. *Right:* A part of the SILVA species tree with the species present in the reconstructed gene-species distributions. The reconstruction of all optimal labelings (484120 in total) under the DL cost indicated that genes M6-M9 and M4 were resolved with a unique species assignment, while the remaining four genes have species assignments uniformly distributed over the leaves of several clusters from the species tree.

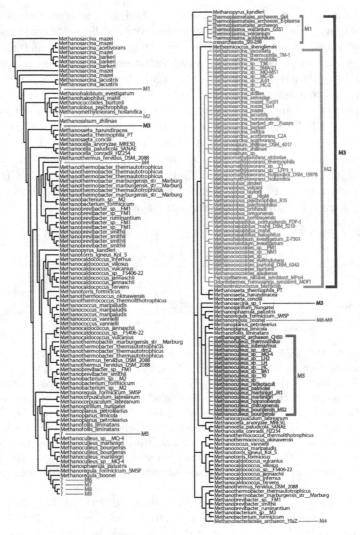

in these bins. In both diagrams we see trends that can be summarized: the quality of reconstructions is better when the input trees are similar and when the number of unknown labels is low. However, even for partial labelings with high number of undefined values, our algorithms can still properly reconstruct the majority of labels under DL. In general, our experiments confirmed that DL can be successfully applied to reconstruct gene-species associations even in the case when all leaves have unknown labels. On the other hand, the DC cost induces generally low quality mappings. It can be explained by Lemma 2: leaves of every unlabeled subtree of a gene tree will be mapped into the same species under DC. For example in the middle diagram for DC, when the whole gene tree is unlabeled the constant labeling is optimal yielding the mean score (almost) 0.0. Thus, the bijective labelings of the input trees are usually incorrectly reconstructed under DC and, in consequence, the quality score is low.

The runtime analysis is depicted in the rightmost panel of Fig. 2.

4.2 Empirical Dataset Evaluation

To test the quality we have attempted to reproduce a typical scenario of amplicon analysis. We have selected mcrA gene, that has been proposed as a marker gene in the phylogenetic analysis of archeal methanogen populations [16]. Luton and coworkers had seen a similarity between topologies of mcrA tree and 16S rRNA tree from corresponding species, which in principle should make the case easier.

First 100 proteins similar to mcrA from *Methanobrevibacter ruminantium* were selected using BLAST [1]. The list contained genes from uncultured archeons. The gene tree was built using program proml from phylip package and tested with a tree containing over 1400 known *Euryarchaeota* species from SILVA database [24]. In total, we have attempted to resolve mappings of 9 unknown sequences out of 100. A typical manual interpretation of the gene tree alone, depicted in Fig. 3, would lead to mostly unambiguous assignment of all unknown sequences at least at the genus level. However our method assigned only half of the sequences unambiguously. The rest have mostly a uniform distribution spanning sometimes several genuses. Cross-checking of the results with databases of known genomes resulted in identification of a single error. One of the sequences (M1) was assigned to a group consisting mainly of *Thermoplasmaceae* species. This is clearly wrong as this taxonomic group does not seem to posses mcrA gene nor methanogenic function. The other sequences have no such issues.

Despite large differences in tree sizes the final assignment did not differ much in size from gene tree. While synthetic tests presented above show high level of accuracy of our method, the evaluation on empirical data indicate that biological interpretation of results should be important part of the process. Depending on a situation, interpretation of ambiguity could be either HGT, or selective pressure or simply understudied branch of species tree.

5   Conclusion and Future Outlook

In this article we proposed the first reconciliation-based approach to infer gene-species mappings from partial gene tree labelings. We studied properties of this approach and proposed efficient algorithms for the optimal cost computation and gene-species assignment inference under several cost functions such as DL or DC. We showed that the proposed algorithms implemented in the computer program mgremap are able to infer optimal gene-species mappings even for large input trees within minutes. The quality validation shows that DL cost performs significantly better that DC cost. We also provided a theoretical explanation of this phenomenon. Finally, the results from empirical tests indicate that our approach is able to strengthen the taxonomic assignment of metagenomic sequences where it can be done unambiguously and identify sequences with ambiguous taxonomic context, which should not be attempted to classify at species or genus level.

This article is focused on the efficient analysis of a single gene family under classical duplication related models. Our next step is to extend the model by considering cost functions with HGTs. Such models, however, are usually computationally hard. Further extensions include methods for the analysis of whole metagenomic samples that may contain sequences from many gene families.

Acknowledgments. Support to all authors was provided by the grant of National Science Centre (NCN, 2011/01/B/ST6/02777).

References

1. Altschul, S.F., Madden, T.L., Schäffer, A.A., Zhang, J., Zhang, Z., Miller, W., Lipman, D.J.: Gapped BLAST and PSI-BLAST: a new generation of protein database search programs. Nucleic Acids Research 25(17), 3389–3402 (1997)
2. Arvestad, L., Lagergren, J., Sennblad, B.: The gene evolution model and computing its associated probabilities. Journal of ACM 56(2) (2009)
3. Bafna, V., Hannenhalli, S., Rice, K., Vawter, L.: Ligand-Receptor pairing via tree comparison. Journal of Computational Biology 7, 59–70 (2000)
4. Berglund-Sonnhammer, A.-C., Steffansson, P., Betts, M.J., Liberles, D.A.: Optimal gene trees from sequences and species trees using a soft interpretation of parsimony. Journal of Molecular Evolution 63(2), 240–250 (2006)
5. Bonizzoni, P., Vedova, G.D., Dondi, R.: Reconciling a gene tree to a species tree under the duplication cost model. Theoretical Computer Science 347(1-2), 36–53 (2005), doi:10.1016/j.tcs.2005.05.016
6. Dinsdale, E.A., et al.: Functional metagenomic profiling of nine biomes. Nature 452(7187), 629–632 (2008)
7. Doyon, J.-P., Chauve, C., Hamel, S.: Space of gene/species tree reconciliations and parsimonious models. Journal of Computational Biology 16 (2009)
8. Durand, D., Halldórsson, B.V., Vernot, B.: A hybrid micro-macroevolutionary approach to gene tree reconstruction. Journal of Computational Biology 13(2), 320–335 (2006)
9. Goodman, M., Czelusniak, J., Moore, G.W., Romero-Herrera, A.E., Matsuda, G.: Fitting the gene lineage into its species lineage, a parsimony strategy illustrated by cladograms constructed from globin sequences. Systematic Zoology 28(2), 132–163 (1979)
10. Górecki, P., Eulenstein, O., Tiuryn, J.: Unrooted tree reconciliation: A unified approach. IEEE/ACM Transactions on Computational Biology and Bioinformatics 10(2), 522–536 (2013)
11. Górecki, P., Tiuryn, J.: DLS-trees: A model of evolutionary scenarios. Theoretical Computer Science 359(1-3), 378–399 (2006)
12. Hallett, M.T., Lagergren, J.: Efficient algorithms for lateral gene transfer problems. In: RECOMB, pp. 149–156 (2001)
13. Harding, E.F.: The probabilities of rooted tree-shapes generated by random bifurcation. Advances in Applied Probability 3(1), 44–77 (1971)
14. Huson, D.H., Auch, A.F., Qi, J., Schuster, S.C.: MEGAN analysis of metagenomic data. Genome Research 17(3), 377–386 (2007)
15. Lafond, M., Swenson, K.M., El-Mabrouk, N.: An optimal reconciliation algorithm for gene trees with polytomies. In: Raphael, B., Tang, J. (eds.) WABI 2012. LNCS, vol. 7534, pp. 106–122. Springer, Heidelberg (2012)

16. Luton, P.E., Wayne, J.M., Sharp, R.J., Riley, P.W.: The mcrA gene as an alternative to 16S rRNA in the phylogenetic analysis of methanogen populations in landfill. Microbiology 148(11), 3521–3530 (2002)
17. Ma, B., Li, M., Zhang, L.: From gene trees to species trees. SIAM Journal on Computing 30(3), 729–752 (2000)
18. Maddison, W.P.: Gene trees in species trees. Systematic Biology 46, 523–536 (1997)
19. Matsen, F.A., Kodner, R.B., Armbrust, E.V.: pplacer: linear time maximum-likelihood and Bayesian phylogenetic placement of sequences onto a fixed reference tree. BMC Bioinformatics 11(1), 538 (2010)
20. O'Meara, B.C.: New heuristic methods for joint species delimitation and species tree inference. Systematic Biology 59, 59–73 (2010)
21. Page, R.D.M.: Maps between trees and cladistic analysis of historical associations among genes, organisms, and areas. Syst. Biol. 43(1), 58–77 (1994)
22. Page, R.D.M., Charleston, M.A.: From gene to organismal phylogeny: reconciled trees and the gene tree/species tree problem. Molecular Phylogenetics and Evolution 7, 231–240 (1997)
23. Puigbo, P., Wolf, Y.I., Koonin, E.V.: The tree and net components of prokaryote evolution. Genome Biology and Evolution 2, 745–756 (2010)
24. Quast, C., Pruesse, E., Yilmaz, P., Gerken, J., Schweer, T., Yarza, P., Peplies, J., Glöckner, F.O.: The SILVA ribosomal RNA gene database project: improved data processing and web-based tools. Nucleic Acids Research 41(D1), D590–D596 (2013)
25. Sjöstrand, J., Tofigh, A., Daubin, V., Arvestad, L., Sennblad, B., Lagergren, J.: A Bayesian method for analyzing lateral gene transfer. Systematic Biology (2014)
26. Stark, M., Berger, S.A., Stamatakis, A., von Mering, C.: MLTreeMap - accurate maximum likelihood placement of environmental DNA sequences into taxonomic and functional reference phylogenies. BMC Genomics 11(1), 461 (2010)
27. Stolzer, M., Lai, H., Xu, M., Sathaye, D., Vernot, B., Durand, D.: Inferring duplications, losses, transfers and incomplete lineage sorting with nonbinary species trees. Bioinformatics 28(18), i409–i415 (2012)
28. Thompson, C.C., Thompson, F.L., Vandemeulebroecke, K., Hoste, B., Dawyndt, P., Swings, J.: Use of recA as an alternative phylogenetic marker in the family vibrionaceae. International Journal of Systematic and Evolutionary Microbiology 54(3), 919–924 (2004)
29. Vernot, B., Stolzer, M., Goldman, A., Durand, D.: Reconciliation with non-binary species trees. Journal of Computational Biology 15(8), 981–1006 (2008)
30. Zhang, L.: From gene trees to species trees II: Species tree inference by minimizing deep coalescence events. IEEE/ACM Transactions on Computational Biology and Bioinformatics 8, 1685–1691 (2011)
31. Zhang, L., Cui, Y.: An efficient method for DNA-based species assignment via gene tree and species tree reconciliation. In: Moulton, V., Singh, M. (eds.) WABI 2010. LNCS, vol. 6293, pp. 300–311. Springer, Heidelberg (2010)
32. Zheng, Y., Zhang, L.: Reconciliation with non-binary gene trees revisited. In: Sharan, R. (ed.) RECOMB 2014. LNCS, vol. 8394, pp. 418–432. Springer, Heidelberg (2014)

Couplet Supertree Based Species Tree Estimation

Sourya Bhattacharyya[✉] and Jayanta Mukhopadhyay

Department of Computer Science and Engineering,
Indian Institute of Technology, Kharagpur, WB 721302, India
sourya.bhatta@gmail.com, jay@cse.iitkgp.ernet.in

Abstract. Inference of a species tree from multi-locus gene trees having topological incongruence due to incomplete lineage sorting (ILS), is currently performed by either consensus (supertree), parsimony analysis (minimizing deep coalescence), or statistical methods. However, statistical approaches involve huge computational complexity. Accuracy of approximation heuristics used in either consensus or parsimony analysis, also varies considerably. We propose *COSPEDSpec*, a novel two stage species tree estimation method, combining both consensus and parsimony approaches. First stage uses our earlier proposed couplet supertree technique *COSPEDTree* [2] [3], whereas the second stage proposes a greedy heuristic to refine a non-binary (unresolved) supertree into a binary species tree. During each iteration, it reduces the number of extra lineages between the current species tree and the input gene trees, thus modeling ILS as the cause of gene tree / species tree incongruence. COSPEDSpec incurs time and space complexity lower or equal to the reference methods. For large scale datasets having hundreds of taxa and thousands of gene trees, COSPEDSpec produces species trees with lower branch dissimilarities and much less computation.

1 Introduction

Gene trees are constructed by sampling individual genes among a group of taxa, and subsequently employing phylogenetic reconstruction methods [7]. Rapid increase of molecular phylogenetic data provides a set $\mathcal{G}$ of M (> 1) gene trees covering a set of N taxa. However, these gene trees often associate conflicting topologies and branch lengths, due to independent site specific evolution of individual gene sequences. Genealogical discordance among these M trees can be so high that no single gene tree topology predominates [7]. So, topology of the final species tree S may considerably deviate even from the most frequent gene tree. Such discordance of gene trees with the species tree occurs due to one of the following three evolutionary processes: 1) Horizontal Gene Transfer (HGT), 2) Gene duplication / loss, and 3) Incomplete Lineage Sorting (ILS) or deep coalescence (DC) [22]. Here we focus on constructing a species tree S from $\mathcal{G}$, when the gene tree discordance occurs due to ILS, which is the failure of two or more lineages in a population to coalesce. So, at least one of the lineages first

© Springer International Publishing Switzerland 2015
R. Harrison et al. (Eds.): ISBRA 2015, LNBI 9096, pp. 48–59, 2015.
DOI: 10.1007/978-3-319-19048-8_5

coalesces with a lineage from a less closely related population [7]. For example, suppose the true species tree S is (X, (Y, (Z, W))), where lineages Z and W coalesce at time t_1, and subsequently the lineage Y coalesces at time t_2. However, in a gene tree $G \in \mathcal{G}$, suppose the lineages Y and W first coalesce at time t_2, and the lineage Z coalesces with them at time t ($t > t_2 > t_1$, assuming time increases into the past). In this case, ILS causes discordance between G and S.

Currently, ILS based species tree estimation is done either by concatenation [24] [6], or by separate analysis [30] techniques. The former infers species tree using sequence alignment, but does not consider gene tree variabilities with respect to ILS. Separate analysis methods use input tree topologies for species tree estimation. They can employ summary or consensus clade analysis [14] [31] [32]; subtree decomposition and consensus determination such as ASTRAL [20]; average rank and coalescence time analysis as in GLASS [21], STEAC [18], maximum tree [17], iGLASS [12], iSTEAC [11], shallowest divergence (SD) [19], STAR [18]; parsimony analysis by minimizing the number of deep coalescence between S and $\mathcal{G}$, as mentioned in Phylonet [30] [32] [1], iGTP [4], Notung [8]; minimizing the sum of Robinson-Foulds (RF) distance [23] between S and $\mathcal{G}$ as in mulRF [5], etc. Statistical modeling based species tree estimation techniques like STEM [13] and MP-EST [16] (using maximum likelihood), BEST [9], *BEAST [10], BBCA [33], BUCKy [15] (using Bayesian statistic), are statistically consistent, but involve huge computational complexity, thus mostly applicable on small datasets involving $\approx$ 20 - 30 taxa and < 100 gene trees. Most of the summary based methods except [32] [1] do not support incomplete (less than N taxa) non-binary input gene trees. They employ additional heuristics to refine input trees into rooted binary format, which may reduce performance.

We propose a novel two stage method COSPEDSpec, which uses both consensus and parsimony techniques for species tree construction. First stage produces a supertree S' from $\mathcal{G}$, using our previously implemented couplet (taxa pair) supertree algorithm COSPEDTree [2] [3]. However, resulting tree S' may not be completely resolved [2], thus may not form a binary species tree. So, the second stage refines S' to form a fully resolved binary species tree S. It uses a novel greedy heuristic *minimum normalized sum of deep coalescence (MNDC)*, by iteratively reducing the sum of extra lineages between the inferred species tree S'' and the input set of gene trees $\mathcal{G}$, at successive iterations. COSPEDSpec works on rooted input gene trees, to produce a rooted species tree as its output. It supports incomplete or non-binary gene trees. We show that for large datasets involving high number of gene trees and taxa, COSPEDSpec produces species trees mostly with lower branch dissimilarities than reference methods. COSPEDSpec involves time complexity of O $\left(N^3 + MN^2 \lg N\right)$, and storage complexity of O $\left(N^2\right)$, both of which are equal or lower than existing species tree construction methods.

Generation of the first stage supertree S' from input $\mathcal{G}$ can be found in our earlier COSPEDTree algorithm [2] [3]. Section 2 discusses the refinement of S' using MNDC criterion. Section 3 describes its performance analysis.

2   Refinement of S' into Binary Tree S

Directly applying basic COSPEDTree [2] algorithm on the set of gene trees G may produce non-binary species tree S', as shown in Fig. 1(a) and Fig. 1(b). Although multispecies coalescent model [7] does not assume strict binary species tree, the datasets we have experimented associate binary gene trees. So, we propose a refinement of S' into a binary species tree S. Such a refined species tree is shown in Fig. 1(c).

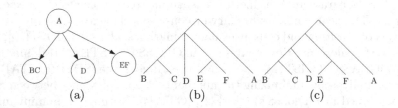

(a)   (b)   (c)

Fig. 1. Example of non binary tree represented (a) as cluster, (b) as tree, (c) an example of possible bifurcation based refinement

Producing a strict binary tree S requires generating bifurcation among the taxa clusters underlying any of the multifurcation instance of S'. Example tree shown in Fig. 1(c) indicates that introducing a bifurcation (speciation) between the clusters (D) and (B,C), eventually resolves the original tri-furcation (shown in Fig. 1(b)) by producing a bifurcation among the clusters (E,F) and (D, (B,C)). We propose an agglomerative clustering technique to generate such bifurcation. Its principle is to find a pair of taxa clusters (such as (D) and (B,C)) which are closest compared to all other cluster pairs. The closeness between a pair of taxa cluster is determined by a novel distance function, termed as the *normalized deep coalescence (NDC)*, between them. The cluster pair having *minimum value of normalized deep coalescence (MNDC)* is termed as closest. They are inserted as children of a newly introduced bifurcation (speciation) node. In the current example, a bifurcation (D,(B,C)) is introduced to form a rooted binary subtree, and the clusters (D) and (B,C) are inserted as its children. The process is repeated until the original multi-furcation is resolved into a bifurcation (in the current example, bifurcation among (E,F) and (D, (B,C))). Below we first elaborate the NDC based distance function between individual taxa cluster pairs, and subsequently describe the agglomerative clustering using MNDC criterion.

NDC Between Individual Taxa Clusters: Let us denote for a phylogenetic tree T, $V(T)$, $E(T)$, and $L(T)$ as its set of vertices, edges, and the leaves, respectively. $Clade_T(v)$ denotes the subtree of a tree T rooted at a node $v \in V(T)$. Set of leaves in $Clade_T(v)$ is denoted as $Cluster_T(v)$. For a set of taxa $A \subseteq L(T)$, let $LCA_T(A)$ denotes the least (most recent) common ancestor of all the taxa in A, with respect to the tree T. With such definitions, suppose $g \in V(G)$ and

$s \in V(S)$ are two internal nodes of the respective gene and species trees G ($\in$ $\mathcal{G}$) and S, such that $g = LCA_G(Cluster_S(s))$. Then, number of leaves under the subtree rooted at g is $Cluster_G(g)$. Number of extra lineages with respect to $s \in V(S)$ and this gene tree G is $XL(s, G) = |Cluster_G(g)| - |Cluster_S(s)|$, where $|.|$ denotes the cardinality of a set. Total number of extra lineages (also called the deep coalescence cost) between G and S is $XL(S, G) = \sum_{s \in V(S)} XL(s, G)$. For a set of gene trees $\mathcal{G}$, the number of extra lineages between S and $\mathcal{G}$ is $XL(S, \mathcal{G}) = \sum_{G \in \mathcal{G}} XL(S, G)$.

Following above notations, $LCA_G(C_x \cup C_y)$ denotes the least common ancestor node of the union of taxa clusters C_x and C_y in the gene tree $G \in \mathcal{G}$. Suppose we denote this node as G_{xy}. If $|C_x|$ and $|C_y|$ respectively denote the cardinality of taxa clusters C_x and C_y, the node G_{xy} associates following number of extra lineages, apart from the taxa clusters C_x and C_y:

$$XL(G_{xy}, G) = Cluster_G(G_{xy}) - |C_x| - |C_y| \tag{1}$$

We denote by $D(C_x, C_y)$, the sum of extra lineages $XL(G_{xy}, G)$ computed for all the gene trees, as the following:

$$D(C_x, C_y) = \sum_{G \in \mathcal{G}} XL(G_{xy}, G) \tag{2}$$

Now suppose the non-binary species tree S' has a multi-furcating internal speciation node, having k (> 2) taxa clusters $C_1, \ldots, C_k$ under it. For a particular cluster C_i ($1 \leq i \leq k$), we define $D(C_i, :)$, the *sum of distances from C_i to all other clusters*, as following:

$$D(C_i, :) = \sum_{1 \leq j \leq k, i \neq j} D(C_i, C_j) \tag{3}$$

Using above sum for individual clusters C_i ($1 \leq i \leq k$), *normalized deep coalescence* (NDC) $D^N(C_x, C_y)$ between individual cluster pairs C_x and C_y is computed as:

$$D^N(C_x, C_y) = \frac{D(C_x, C_y)}{D(C_x, :) + D(C_y, :)} \tag{4}$$

MNDC Criterion Based Agglomerative Clustering: From the NDC functions for individual cluster pairs, the pair (C_x, C_y) is said to satisfy *minimum NDC* (MNDC) criterion, provided:

$$D^N(C_x, C_y) = \min_{\forall i, j, 1 \leq i, j \leq k, i \neq j} D^N(C_i, C_j) \tag{5}$$

Suggested binary refinement is carried out by iteratively selecting a cluster pair (C_x, C_y) satisfying the MNDC criterion among the k (> 2) taxa clusters $C_1, \ldots, C_k$ under the multi-furcating internal speciation node. One new internal node is added in the species tree, and selected clusters C_x and C_y are placed as its children. This produces a bifurcation.

Lemma 1. *Above mentioned MNDC criterion based binary refinement maximally reduces the number of extra lineages with respect to the current species tree S' and the input gene tree set $\mathcal{G}$, during every iteration. That is, for a particular iteration, reduction in $XL(S',\mathcal{G})$ is maximum.*

Proof. Suppose, the species tree S' (generated from COSPEDTree) contains a multi-furcating internal speciation node having k taxa clusters $C_1, C_2, \ldots, C_k$ as its children. First we prove the lemma for $k = 3$, and later prove for any k. Considering three clusters C_1, C_2 and C_3, suppose the cluster pair (C_1, C_2) satisfies MNDC criterion. So,

$$D^N(C_1, C_2) < D^N(C_1, C_3)$$

$$\implies \frac{D(C_1, C_2)}{D(C_1,:) + D(C_2,:)} < \frac{D(C_1, C_3)}{D(C_1,:) + D(C_3,:)} \text{ (Using Eq. 4)}$$

$$\implies \frac{D(C_1, C_2)}{2D(C_1, C_2) + D(C_1, C_3) + D(C_2, C_3)}$$

$$< \frac{D(C_1, C_3)}{2D(C_1, C_3) + D(C_2, C_3) + D(C_1, C_2)} \implies D(C_1, C_2) < D(C_1, C_3)$$

(Since $\dfrac{X}{X + Z} < \dfrac{Y}{Y + Z}$ means $X < Y$, for positive X, Y, Z

where $X = D(C_1, C_2), Y = D(C_1, C_3), Z = D(C_1, C_2) + D(C_1, C_3) + D(C_2, C_3)$)

$$\implies \sum_{G \in \mathcal{G}} XL(G_{12}, G) < \sum_{G \in \mathcal{G}} XL(G_{13}, G)$$

Where G_{xy} indicates $LCA_G(C_x \cup C_y)$ for the gene tree $G \in \mathcal{G}$. Similarly, we can show

$$\sum_{G \in \mathcal{G}} XL(G_{12}, G) < \sum_{G \in \mathcal{G}} XL(G_{23}, G)$$

So, creating a speciation node and inserting the clusters (C_1, C_2) as its children, produces lower sum of extra lineages (with respect to $\mathcal{G}$) compared to merging other cluster pairs in the current step.

For $k > 3$, we can similarly show that if (C_1, C_2) satisfies MNDC, $XL(G_{12}, \mathcal{G})$ is minimum of all $XL(G_{ij}, \mathcal{G})$ (where i, j denote cluster indices; $1 \le i, j \le k$, $i \ne j$). Thus, MNDC criterion reduces the extra lineages maximally.   □

To continue successive iterations for binary refinement, previously agglomerated cluster pair (C_x, C_y) is treated as a single cluster (say it is denoted as C_{xy}). We first adjust the distances from all other taxa clusters C_i $(1 \le i \le k, i \ne x, y)$ to this new cluster C_{xy} as

$D(C_i, C_{xy}) = \max(D(C_i, C_x), D(C_i, C_y), D(C_x, C_y))$

where max() indicates maximum operation. Employing such heuristic is motivated by the following lemma:

Lemma 2. *For any phylogenetic tree t, and any three distinct taxa clusters X, Y, $Z \subseteq L(t)$, $LCA_t(Z, X \cup Y)$ is any two of the following: 1) $LCA_t(Z, X)$, 2) $LCA_t(Z, Y)$, and 3) $LCA_t(X, Y)$.*

Proof. We can easily verify it from any of the trees in Fig. 1. $LCA_t(Z, X \cup Y)$ denotes the root of the subtree containing the triplet (X, Y, Z). So, any two of the given expressions will map to the root. $\qquad\square$

Above lemma indicates that the node $LCA_G(C_i \cup C_x \cup C_y)$ is identical to any of the following three nodes: 1) $LCA_G(C_i \cup C_x)$, 2) $LCA_G(C_i \cup C_y)$, and 3) $LCA_G(C_x \cup C_y)$. So, the number of extra lineages $D(C_i, C_{xy})$ will be any one of the three quantities: 1) $D(C_i, C_x)$, 2) $D(C_i, C_y)$, or 3) $D(C_x, C_y)$. Here we have used the maximum of them as the new approximation of $D(C_i, C_{xy})$. Such approximation saves $O(M)$ computation at each iteration, where $M = |\mathcal{G}|$. Following such adjustment, the normalized distance values (D^N) are recomputed according to the Eq. 4. New pairs of clusters are put under a new speciation node, until the original multi-furcation is completely resolved.

The non-binary tree S' may have more than one internal multi-furcating nodes. To resolve S' into a binary tree S, proposed refinement is applied on the multi-furcating nodes occurring during the postorder traversal of S'. As individual refinement steps do not create any new multi-furcation, complete postorder traversal refines S' into a strict binary species tree S.

Table 1. Comparison of time and space complexity between COSPEDSpec and reference methods. $M = |\mathcal{G}|$. N is the number of taxa. Q is the size of largest population.

Method	Time Complexity	Space Complexity
SMRT [6]	$O(N^5)$	$O(N^3)$
ASTRAL [20]	$O(N^4 M^3)$	$O(N^4)$
iGLASS [12]	$O(NMQ^2 + MN^2)$	$O(N^2 M)$
iGTP [4] and SD [19]	$O(N^4 M)$	$O(N^2 M)$
Phylonet [30] [32]	$O(N^2 M^2)$	$O(N^2)$
mulRF [5]	$O(N^3 M)$	$O(N^2 M)$
MP-EST [16], BBCA [33]	$O(N^3 M)$	$O(N^3)$
COSPEDSpec	$O(N^3 + MN^2 \lg N)$	$O(N^2)$

Computational Complexity: COSPEDTree [2] [3] involves $O(N^3)$ and $O(N^2)$ time and space complexities for N taxa. COSPEDSpec does not introduce additional storage complexity. Postorder traversal associated with the proposed binary refinement involves processing at most $O(N)$ internal multi-furcating nodes. For each such node $n \in V(S)$, computation of $D^N(C_x, C_y)$ for individual pair of taxa clusters (C_x, C_y) within $Clade_S(n)$, requires $O(M \lg N)$ time. Here, $M = |\mathcal{G}|$. The factor $\lg N$ is for finding the $LCA(C_x, C_y)$ in a gene tree G, where G contains at most N taxa. As $|V(S)| \approx O(N)$, refinements for all multi-furcating nodes require $O(N^2)$ time to compute D^N values. Thus, complexity for the refinement of S' is $O(MN^2 \lg N)$. So, overall time complexity of COSPEDSpec is $O(N^3 + MN^2 \lg N)$. Comparison of time and space complexities between COSPEDSpec and the reference approaches are summarized in Table 1.

3   Experimental Results

COSPEDSpec is implemented in Python (version 2.7). Phylogenetic library Dendropy [29] is used for reading and processing tree datasets. Evolutionary relations between individual couplets in the COSPEDTree [2] algorithm were computed using the default rooting configuration provided in individual datasets.

3.1   Datasets Used

Simulated Mammalian Dataset: We have used the simulated 37-taxon Mammalian dataset with 447 gene trees, and the model species tree, as reported in [20] [26]. We have used the simulated gene trees provided by [20]. There, different degrees of ILS were modeled by scaling up (2X and 5X) or down (0.2X and 0.5X) the branch lengths of the model species tree. Another species tree without any such branch length scaling (denoted as 1X) was used to denote the default ILS condition. For Mammalian 0.5X dataset, number of gene trees is 16000. Rest of the datasets contain 4000 gene trees each. Such difference in the count of gene trees is due to varying number of bootstrap replicates in gene tree simulation.

Simulated 100 Taxa Dataset: We have also used the simulated dataset of 100 taxa [31], containing ten different gene tree sets. Individual set consists of 25 different gene trees. As reported in [31], gene trees associate high topological dissimilarities among themselves due to high degree of ILS. We have executed COSPEDSpec and the reference approaches in all the copies, and reported results for each of them separately.

Biological Dataset: We have analyzed five biological datasets for performance comparison: 1) *Apicomplexan dataset* [30] [14] containing eight species and 268 gene trees; 2) *Yeast dataset 1* [30] [24] containing seven species and 106 gene trees; 3) *Mammalian dataset* [26] containing 37 species and 447 gene trees; 4) *Placental Mammal dataset* [18] [27] containing 54 species and 6000 genes; and 5) *Yeast dataset 2* [25] of 23 species and 1070 gene trees. The *Mammalian* [26] dataset contains 440 distinct gene tree topologies out of 447 gene trees. As reported in [26], [20], such difference in topologies is due to high degree of ILS, without any recombination or other evolutionary processes. The *Yeast dataset 2* [25] is also unique in the sense that no input gene tree topology exactly matches with the model species tree topology.

3.2   Performance Measures

For evaluation of species trees obtained from different approaches, we have used following performance measures:

A) *Robinson-Foulds (RF)* distance [23] between the inferred and the model species tree, counting the number of bipartitions present in one of the trees, but not in both. We have normalized the RF values by dividing the bipartition count with (2N-6), where N is the number of taxa. Species tree having lower RF is topologically closer to the model species tree.

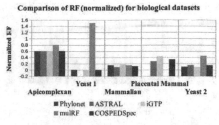
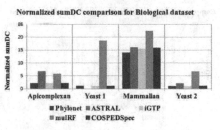

(a) RF measure comparison (b) DC measure comparison

Fig. 2. Comparison of RF and DC measures between COSPEDSpec and reference approaches, when executed with biological datasets. Both Phylonet and MulRF could not parse Placental Mammal dataset. We could not compute DC values for Placental Mammal dataset, since the dataset could not be parsed using Phylonet (used for DC value computation). ASTRAL could not process Yeast 1 dataset since input gene trees contain multi-furcation. Negative scale in y axis is used to mark the instances of zero RF values.

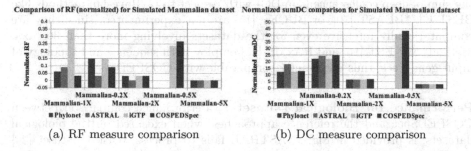

(a) RF measure comparison (b) DC measure comparison

Fig. 3. Comparison of RF and DC measures between COSPEDSpec and reference approaches, when executed with simulated mammalian datasets. For the Mammalian-0.5X dataset, both Phylonet and ASTRAL could not converge to a solution. Negative scale in y axis is used to mark the instances of zero RF values.

B) *Sum of deep coalescence count (sumDC)* [19] [30], or the sum of extra lineages $XL(S, \mathcal{G})$, computed using the routine available in Phylonet [30] package. The value was normalized by dividing it with the number of input gene trees M. Lower sumDC indicates better species tree. However, a species tree depicting lowest sumDC may not be always the true species tree.

3.3 Performance Comparison

Performance of COSPEDSpec with respect to above mentioned measures is benchmarked with the species tree estimation methods ASTRAL [20], Phylonet [30], mulRF [5], and iGTP [4]. Only the heuristic (default) versions of Phylonet [30] and ASTRAL [20] were tested, since their exact versions are applicable to

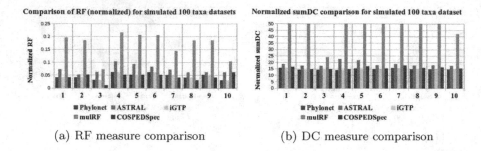

(a) RF measure comparison   (b) DC measure comparison

Fig. 4. Comparison of RF and DC measures between COSPEDSpec and reference approaches, when executed with simulated 100 taxa datasets. X axis denotes ten gene tree datasets numbered 1 to 10.

at most 20 taxa. ASTRAL [20] was executed using its default settings, thus not using any extra bipartitions generated from MP-EST and concatenation analysis. Default rooting was employed for the output trees of ASTRAL, to compute the sum of deep coalescence (sumDC) with respect to G. Bayesian methods like BEST [9], *BEAST [10], or BUCKy [15] are not experimented, since they are computationally intensive even for the datasets involving more than 30 taxa. Methods MP-EST [16] and RAxML [28] employ both sequence and topology of input gene trees, thus not exactly comparable with COSPEDSpec.

Performance on Biological Datasets: Performance comparison between COSPEDSpec and the reference approaches, when executed on the biological datasets, is provided in Fig. 2. ASTRAL fails to process Yeast 1 dataset [24] since it does not support non-binary gene trees. COSPEDSpec, on the other hand, supports non-binary or incomplete gene trees. Both Phylonet and mulRF packages exhibit parsing error for the Placental Mammal dataset. So we could not compute the sumDC measure for this dataset. Overall we find that, COSPED-Spec exhibits lowest or second lowest RF values for all datasets. In terms of sumDC values, COSPEDSpec is only behind Phylonet and iGTP, both of which employ parsimony technique to achieve low sumDC.

Performance on Simulated Mammalian Dataset: In this case, the technique mulRF was excluded from performance comparison, since it required more than a day for processing such a large number of trees. Results in Fig. 3 show that COSPEDSpec exhibits best performance across all the measures for the Mammalian 1X dataset, which best resembles to the true biological dataset. For other datasets, only ASTRAL consistently produces trees with lower RF than COSPEDSpec. With respect to sumDC, COSPEDSpec performs better or equal to the parsimonious approaches iGTP and Phylonet for almost all datasets.

Further, for the Mammalian 0.5X dataset containing 16000 gene trees, both ASTRAL and Phylonet having time complexities proportionate to O (M^3) and O (M^2), respectively, could not generate a solution in a day. COSPEDSpec, with a time complexity proportionate to O (M), quickly produces the species tree (in about 20 minutes), thus proving its utility in processing large datasets.

Performance on Simulated 100 Taxa Dataset: Performance comparison between COSPEDSpec and the reference methods, with respect to the simulated 100 taxa dataset is shown in Fig. 4. Results for all ten gene tree sets have been reported. COSPEDSpec produces lowest RF values for the majority of these sets. Only the methods iGTP and Phylonet are comparable with COSPEDSpec, in terms of the RF values. Considering the measure sumDC, COSPEDSpec performs better than ASTRAL, and equal to the parsimonious methods Phylonet and iGTP.

Discussion: Overall, the technique mulRF [5] based on minimizing the RF value between S and G, is computationally intensive for large datasets; such minimization does not work well for high number of taxa, as shown in Fig. 4. Parsimony approaches iGTP [4] and Phylonet [30] aiming minimum sumDC, may converge to local minima as the number of taxa increases. Further, lower sumDC values with respect to G, do not always indicate lower RF with respect to the model species tree. ASTRAL [20] produces species tree maximizing the similarities with input gene tree quartets. For increasing taxa, maximizing such quartet similarities may often lead to suboptimal solution, as found in the results for 100 taxa datasets. Both ASTRAL and Phylonet involves high time complexity with respect to large number of gene trees. Further, ASTRAL does not support non-binary gene trees. On the other hand, greedy heuristics employed in COSPEDSpec, produce better performance as the number of gene trees or taxa increase. One disadvantage of COSPEDSpec is that, the underlying supertree technique COSPEDTree [2] resolves couplets preferably with strict consensus and most frequent relations (with respect to G). If such relations are not supported in the final species tree S, topological performance of S may be low. However, inclusion of such non-consensus relations cannot be predicted from input tree topologies, and is thus equally probable for other reference approaches as well. Overall, COSPEDSpec involves low computation and high topological similarities with respect to the model species tree. So it is applicable for large biological datasets.

Executable: Executable of COSPEDSpec is provided in
http://facweb.iitkgp.ernet.in/~jay/phtree/COSPEDSpec/cospedspec.html

Acknowledgments. The first author acknowledges Tata Consultancy Services (TCS) for providing the research scholarship.

References

1. Bayzid, M.S., Warnow, T.: Estimating optimal species trees from incomplete gene trees under deep coalescence. Journal of Computational Biology 19(6), 591–605 (2012)
2. Bhattacharyya, S., Mukherjee, J.: Cospedtree: Couplet supertree by equivalence partitioning of taxa set and dag formation. IEEE/ACM Trans. Comp. Biol. Bioinfo. 1, 1 (2014), doi:10.1109/TCBB.2014.2366778 (preprints)
3. Bhattacharyya, S., Mukhopadhyay, J.: Couplet supertree by equivalence partitioning of taxa set and dag formation. pp. 259–268. 5th ACM Conference on Bioinformatics, Computational Biology and Health Informatics (ACM-BCB) (2014)
4. Chaudhary, R., Bansal, M.S., Wehe, A., Fernández-Baca, D., Eulenstein, O.: igtp: a software package for large-scale gene tree parsimony analysis. BMC Bioinformatics. 23(574), 1–7 (2010)
5. Chaudhary, R., Burleigh, J.G., Fernández-Baca, D.: Inferring species trees from incongruent multi-copy gene trees using the robinson-foulds distance. Algorithms for Molecular Biology. 8(1(28)), 1–12 (2013)
6. DeGiorgio, M., Degnan, J.H.: Fast and consistent estimation of species trees using supermatrix rooted triples. Mol. Biol Evol. 27(3), 552–569 (2010)
7. Degnan, J.H., Rosenberg, N.A.: Gene tree discordance, phylogenetic inference and the multispecies coalescent. Trends in Ecology and Evolution 24(6), 332–340 (2009)
8. Durand, D., Halldorsson, B.V., Vernot, B.: A hybrid micro-macroevolutionary approach to gene tree reconstruction. Journal of Computational Biology 13(2), 320–335 (2005)
9. Edwards, S.V., Liu, L., Pearl, D.K.: High-resolution species trees without concatenation. PNAS 104(14), 5936–5941 (2007)
10. Heled, J., Drummond, A.J.: Bayesian inference of species trees from multilocus data. Mol Biol Evol. 27(3), 570–580 (2010)
11. Helmkamp, L.J., Jewett, E.M., Rosenberg, N.A.: Improvements to a class of distance matrix methods for inferring species trees from gene trees. Journal of Computational Biology 19(6), 632–649 (2012)
12. Jewett, E.M., Rosenberg, N.A.: iglass: An improvement to the glass method for estimating species trees from gene trees. Journal of Computational Biology 19(3), 293–315 (2012)
13. Kubatko, L.S., Carstens, B.C., Knowles, L.: Stem: species tree estimation using maximum likelihood for gene trees under coalescence. Bioinformatics 25(7), 971–973 (2009)
14. Kuo, C.H., Wares, J.P., Kissinger, J.C.: The apicomplexan whole-genome phylogeny: An analysis of incongurence among gene trees. Mol Biol Evol 25(12), 2689–2698 (2008)
15. Larget, B.R., Kotha, S.K., Dewey, C.N., Ané, C.: Bucky: Gene tree / species tree reconciliation with bayesian concordance analysis. Bioinformatics 26(22), 2910–2911 (2010)
16. Liu, L., Yu, L., Edwards, S.V.: A maximum pseudo-likelihood approach for estimating species trees under the coalescent model. BMC Evolutionary Biology. 10(302), 1–18 (2010)
17. Liu, L., Yu, L., Pearl, D.K.: Maximum tree: a consistent estimator of the species tree. J. Math. Biol. 60(1), 95–106 (2010)
18. Liu, L., Yu, L., Pearl, D.K., Edwards, S.V.: Estimating species phylogenies using coalescence times among sequences. Syst. Biol. 58(5), 468–477 (2009)

19. Maddison, W.P., Knowles, L.L.: Inferring phylogeny despite incomplete lineage sorting. Syst Biol 55(1), 21–30 (2006)
20. Mirarab, S., Reaz, R., Bayzid, M.S., Zimmermann, T., Swenson, M.S., Warnow, T.: Astral: genome-scale coalescent-based species tree estimation. Bioinformatics 30(17), i541–i548 (2014)
21. Mossel, E., Roch, S.: Incomplete lineage sorting: Consistent phylogeny estimation from multiple loci. IEEE/ACM Trans Comp Biol Bioinfo 7(1), 166–171 (2010)
22. Nakhleh, L.: Computational approaches to species phylogeny inference and gene tree reconciliation. Trends in Ecology and Evolution 28(12), 719–728 (2013)
23. Robinson, D.R., Foulds, L.R.: Comparison of phylogenetic trees. Mathematical Biosciences 53(1-2), 131–147 (1981)
24. Rokas, A., Williams, B., King, N., Carroll, S.: Genome-scale approaches to resolving incongruence in molecular phylogenies. Nature 425, 798–804 (2003)
25. Salichos, L., Rokas, A.: Inferring ancient divergences requires genes with strong phylogenetic signals. Nature 497(7449), 327–333 (2013)
26. Song, S., Liu, L., Edwards, S.V., Wu, S.: Resolving conflict in eutherian mammal phylogeny using phylogenomics and the multispecies coalescent model. Proc Natl Acad Sci USA. 109(37), 14942–14947 (2012)
27. Springer, M.S., Burk-Herrick, A., Meredith, R., Eizirik, E., Teeling, E., O'Brien, S.J., Murphy, W.J.: The adequacy of morphology for reconstructing the early history of placental mammals. Syst. Biol. 56(4), 673–684 (2007)
28. Stamatakis, A.: Raxml-vi-hpc: maximum likelihood-based phylogenetic analyses with thousands of taxa and mixed models. Bioinformatics 22(21), 2688–2690 (2006)
29. Sukumaran, J., Holder, M.T.: Dendropy: a python library for phylogenetic computing. Bioinformatics 26(12), 1569–1571 (2000)
30. Than, C., Nakhleh, L.: Species tree inference by minimizing deep coalescences. PLOS Computational Biology 5(9), 1–12 (2009)
31. Yang, J., Warnow, T.: Fast and accurate methods for phylogenomic analyses. BMC Bioinformatics 12(9), 1–12 (2011)
32. Yu, Y., Warnow, T., Nakhleh, L.: Algorithms for mdc-based multi-locus phylogeny inference: Beyond rooted binary gene trees on single alleles. Journal of Computational Biology 18(11), 1543–1559 (2011)
33. Zimmermann, T., Mirarab, S., Warnow, T.: Bbca: Improving the scalability of *beast using random binning. BMC Genomics. 15 (Suppl 6)(S11), 1–9 (2014)

A Novel Computational Method for Deriving Protein Secondary Structure Topologies Using Cryo-EM Density Maps and Multiple Secondary Structure Predictions

Abhishek Biswas, Desh Ranjan, Mohammad Zubair, and Jing He[✉]

Department of Computer Science, Old Dominion University, Norfolk, VA 23525, USA
jhe@cs.odu.edu

Abstract. A key idea in de novo secondary structure topology determination methods is to calculate an optimal mapping between the observed secondary structure traces in a Cryo-EM density image and the predicted secondary structures on the protein sequence. The problem becomes much more complicated in presence of multiple secondary structure predictions for the protein sequence (for example those predicted by different prediction methods). We present a novel computational method that elegantly and efficiently solves the problem of dealing with multiple secondary structure predictions and calculating the optimal mapping. The proposed method uses a two-step approach – it first uses the consensus positions of the secondary structures to produce top K topologies, and then it uses a dynamic programming method to find the optimal placement for the secondary structure traces of the density image. The method was tested using twelve proteins of three types. We observed that the rank of the true topologies is consistently improved with the use of multiple secondary structure predictions over single prediction. The results show that the algorithm is robust and works well even in presence of errors/misses in predicted secondary structures from the image or the sequence. The results also show that the algorithm is efficient and is able to handle proteins with as many as thirty-three helices.

Keywords: Cryo-EM · Dynamic Programming · Graph · Image · Protein · Secondary Structure

1 Introduction

The field of electron cryomicroscopy (Cryo-EM) has gone through dramatic growth in the last few decades, and has become a major technique in structure determination of large molecular complexes [8] . Unlike X-ray crystallography and Nucleic Magnetic Resonance (NMR), Cryo-EM is particularly suitable for large molecular complexes such as viruses, ribosomes and membrane-bound ion channels [9-11]. Density maps with high resolution (2-5 Å), the atomic structure can be derived directly, since the backbone and large side chains are mostly resolved. However, it is computationally challenging to derive atomic structures from the medium resolution maps (5-10 Å). Two major methods have been previously used to derive atomic structures from the

© Springer International Publishing Switzerland 2015
R. Harrison et al. (Eds.): ISBRA 2015, LNBI 9096, pp. 60–71, 2015.
DOI: 10.1007/978-3-319-19048-8_6

medium resolution maps. The first is to fit a known atomic structure in the Cryo-EM map as a component using rigid-body or dynamic flexible fitting [12, 13]. The other is to use a template atomic structure from a homologous protein to build and evaluate potential models [14, 15]. The limitation here is the need for atomic structures that are either components of or homologous to the protein of atypical size. When there is no template structure with sufficient similarity, one must devise and use de novo methods. These methods derive the structure from the intrinsic relationship among the secondary structures visible in the density map, such as α-helices and β-sheets.

Although it is not possible to distinguish the amino acids at medium resolutions, secondary structures such as α-helices (red lines in Fig. 1A) and β-sheets (blue density voxels in Fig. 1A) can be identified [18-23]. We have recently developed *StrandTwister*, a method to predict the location of β-strands through the analysis of β-sheet twist [24]. A helix detected from a Cryo-EM image can be represented as a line, referred here as an α-trace, that corresponds to the central axis of a helix (red lines in Fig. 1B). Similarly, a β-strand can be represented as a β-trace that corresponds to the central line of the β-strand (Fig. 1B). Secondary structure traces (SSTs) refers to the set of α-traces and β-traces detected from the 3-demensional (3D) image. In order to know how a protein chain threads through SSTs, it is necessary to know which secondary structure trace is near the N-terminal of the protein chain and which trace follows next.

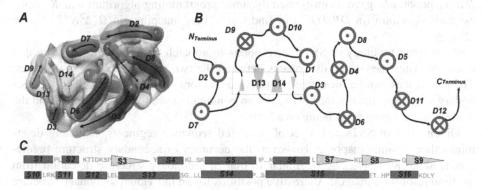

Fig. 1. Secondary structures and topology. (A) The density map (gray) was simulated to 10 Å resolution using protein 3PBA from the Protein Data Bank (PDB) and EMAN software [1]. The secondary structure traces (red: helix sticks, blue: sheet, purple: β-strands) were detected using *SSETracer* [2] and *StrandTwister* [6], and viewed using Chimera [17]. For clear viewing, only those at the front of the structure are labeled. Arrows: the direction of the protein sequence; (B) The true topology of SSTs (arrow, cross and dot for directions); (C) red rectangles: helix segments; blue triangles (S_3, S_7, S_8, and S_9): β-strands; "...": loops longer than two amino acids.

In order to help us determine the threading of the protein chain through the SSTs, we first use a computational method, such as Jpred [5], to predict the subsequences (sequence segments) of the protein sequence that are likely secondary structures and then map the SSTs to these subsequences. A *topology* of the SSTs refers to their order with respect to the protein sequence and the direction of each helix or strand. For example in Fig. 1, $D_1, D_2 ..., D_{14}$ represent the SSTs and $S_1, S_2, ... S_{16}$ represent the subsequences on the protein chain that correspond to secondary structures. In this case, *SSETracer* was able to detect twelve helices (red sticks in Fig. 1A) and one sheet (blue in Fig. 1A) from the density map. *StrandTwister* was then applied to the β-sheet density and it detected two (purple stick Fig. 1A and D_{13}, D_{14} in Fig. 1B) of the four β-strands. For example, Fig. 1 presents a possible topology for the SSTs (it happens to be the true topology of this known protein structure). Each stick/trace $D_j, j = 1, ..., 14$ corresponds to a sequence segment S_i, $i = 1, ..., 16$. α-traces correspond to α-helices on the sequence, and β-traces correspond to β-strands on the sequence. Four sequence segments S_3, S_7, S_8, and S_9 are β-strands of a β-sheet. The true topology maps SSTs $(D_2, D_7, D_9, D_{10}, D_1, D_{13}, D_{14}, D_3, D_6, D_4, D_8, D_5, D_{11}, D_{12})$ to $(S_1, S_2, S_4, S_5, S_6, S_7, S_8, S_{10}, S_{11}, S_{12}, S_{13}, S_{14}, S_{15}, S_{16})$. Observe that the two β-strands S_3 *and* S_9 were note detected in the image. Also, note that there are two possible directions when mapping sequence segment S_i to D_j (arrows of Fig. 1A and dot/cross in Fig. 1B). We have shown previously that finding the optimal mapping between SSTs and sequence segments is an NP-hard problem [25]. A naïve approach to find the optimal solution requires $\Omega(N! \, 2^N)$ time, where N is the number of SSTs. A dynamic programming algorithm has been previously devised to find the optimal match in $O(N^2 2^N)$ computation time, reduced from $O(N! \, 2^N)$. In a general case in which M sequence segments are mapped to N SSTs (assuming $M \geq N, \Delta = M - N$), we previously gave a constrained dynamic programming algorithm and K shortest path algorithm (in *DP-TOSS*) to find top K best mappings in $O(\Delta^2 N^2 2^N)$ time [26].

An optimal topology of SSTs corresponds to a match with the optimal score that often evaluates the overall differences between the two sets of secondary structures. The differences can be measured with various factors such as the length of the secondary structures, the distance between two consecutive secondary structures, and the likelihood for amino acids being on a loop.

Given a set of SSTs and a set of predicted sequence segments, matching determines their optimal mapping. However, the accuracy of secondary structure prediction is about 80%, [7, 27-29], similar to that of secondary structure detection from medium resolution images. Alternative positions for an individual secondary structure are often needed due to the inaccuracy in the detection or prediction (Fig 2.). However, this approach faces huge computational cost. Let N be the number of secondary structures in a protein. Suppose there are three alternative positions for each helix segment on the sequence and four alternative positions for each of the SSTs, then there are $3^N 4^N$ possible pairs of secondary structure sets to be matched. The total number of possible matches will be $3^N 4^N N! \, 2^N$, since there are $N! \, 2^N$ different ways (or different topologies) to match a given pair.

Designing de novo methods for determining secondary structure topology from Cryo-EM data is a challenging problem. Although a few de novo methods have been developed, the efficiency and accuracy of these methods leave room for substantial improvement. One such method, *Gorgon*, formulates the topology problem into a graph matching problem and searches for possible topologies using A* search [30]. Another method, *DP-TOSS*, formulates the topology problem as a shortest path problem in a graph and uses the K shortest paths algorithm in combination of dynamic programming to produce a set of K top-ranked topologies [3]. *DP-TOSS* shows improved time and accuracy over Gorgon, particularly when working with large proteins [3].

Fig. 2. Secondary structure predictions from multiple servers. The amino acid sequence of protein 2XVV (PDB ID) is labeled at the outermost circle. The positions of helices are shown as red rectangles from outer to inner circles as the true position of the secondary structures obtained from PDB, the prediction using SYMPRED[4] , JPred[5], PSIPRED[7] and PREDATOR[16] respectively. The α-traces detected from density map using SSETracer are shown in the center.

However, both methods address the mapping problem rather than the placement problem. One either has to submit the best estimated secondary structure positions to *DP-TOSS* or Gorgon, or to run either programs multiple times on alternative positions that are produced from multiple secondary structure prediction servers. We previously attempted a dynamic graph approach in which the alternative positions are handled in the graph update process [31]. This approach yielded, on average, running time that was about 34% lower than a naïve way. In this paper, we present a new more effective two-step approach. The consensus positions of the secondary structures will be used in the first step to obtain top K topologies using *DP-TOSS*. For each such topology, an efficient dynamic programming algorithm is devised to find the optimal placement of SSTs. To the best of our knowledge, this is the first algorithm that handles multiple predicted secondary structures. Moreover, the results show the algorithm is efficient in terms of running time and improves the ranking of the true topology.

2   Methods

2.1   The Secondary Structure Mapping Problem

Suppose that there are N_α helices and N_β β-strands detected from a 3D image and $N = N_\alpha + N_\beta$. Suppose that there are M_α helices and M_β β-strands predicted from the amino acid sequence of the protein, and $M = M_\alpha + M_\beta$. For simplicity in description, we assume $M_\alpha = N_\alpha$ and $M_\beta = N_\beta$ and consequently $M = N$. Our actual algorithm and implementation handle the case where $M \neq N$. Let the sequence segments of the secondary structures be $\{S_1, S_2, \ldots, S_N\}$, where S_i denotes the ith sequence segment from the N-terminus. Let the SSTs detected from the density map be $\{D_1, D_2, \ldots, D_N\}$, and $N = N_\alpha + N_\beta$. For convenience, we let $D_1, D_2, \ldots, D_{N_\alpha}$ be the α-traces, and $D_{N_\alpha+1}, D_{N_\alpha+2}, \ldots, D_{N_\alpha+N_\beta}$ be the β-traces. The secondary structure mapping problem is to find a mapping σ from $\{S_1, S_2, \ldots, S_N\}$ to $\{D_1, D_2, \ldots, D_N\}$ such that two criteria are satisfied. (1) Both S_i and $D_{\sigma(i)}$ correspond to a helix or both correspond to a β-strand; (2) The mapping score is optimal. A variety of factors have been considered to score a mapping. The length of a helix segment is represented as the number of amino acids involved in the secondary structure. It can be converted to the axial length of the helix in 3D using 1.5Å rise between two amino acids. Therefore the length of the helix can be used in comparison during mapping. The length of a loop between two consecutive helices can also be considered in scoring a mapping. A rough estimate of the loop length between two sticks is the Euclidean distance between the two end points of two sticks. However, a better estimation is to measure the length along skeleton image between the two end points [32-34]. Loop score that measures the likelihood of a loop and other empirical constraints have also been used [35]. The scoring function used in this paper consists of skeleton length between two secondary structure traces, the length of a secondary structure and the loop length.

Given a specific set of secondary structure traces and a specific set of predicted secondary structure sequence segments, the best match is determined through the mapping process. In order to cut down the computation, we took a two-step approach in this paper. In the first step, we used the consensus sequence segments predicted using SYMPRED [4] as $(S_1, S_2, \ldots, S_M)$, and the detected SSTs as $\{D_1, D_2, \ldots, D_N\}$. The idea is to use the best estimation of the secondary structure positions in the first step to obtain a small number of possible topologies. We applied *DP-TOSS*, a dynamic programming approach built for a topology graph, to obtain the top ranked topologies [3]. For each possible topology, the best placement of the secondary structures will be searched in the second step.

2.2   Dynamic Programming for Finding the Optimal Placement

When there is a small pool of highly ranked topologies, it is possible to identify the optimal placement of secondary structures for each such topology. The idea is to enrich the ranking of the true topology using the optimally placed secondary structure positions. In this paper we show that there is an efficient dynamic programming method to find the optimal placement as long as the topology is given.

Let us represent the alternative sequence segments for secondary structure as the following. Let (S_i, α_i^l) be the lth alternative segment on the amino acid sequence for secondary structure i, where $l = 1, 2, ..., q, i = 1, 2, ..., M$. For a given topology, the order of SSTs and the direction of each trace are known. Let $(S_1, S_2, ..., S_N)$ be mapped to $(D_{\sigma(1)}, D_{\sigma(2)}, ..., D_{\sigma(N)})$. The optimal placement of secondary structures on the sequence is to find a placement of the sequence segments $(S_1, \alpha_1^{l_1}), (S_2, \alpha_2^{l_2}), ..., (S_N, \alpha_N^{l_N})$, $1 \le l_1, l_2, ..., l_N \le q$, such that the score of mapping $(S_1, \alpha_1^{l_1}), (S_2, \alpha_2^{l_2}), ..., (S_N, \alpha_N^{l_N})$ to $(D_{\sigma(1)}, D_{\sigma(2)}, ..., D_{\sigma(N)})$ minimized.

A naïve way to find the best placement of a topology is to exhaustively score q^N different ways to map a set of alternative sequence segments to the set of SSTs. A better way is to use a dynamic programming where we store and reuse information.

Let $g(i, k)$ denote the best cost that can be obtained when $(S_1, S_2, ..., S_i)$ is mapped to $(D_{\sigma(1)}, D_{\sigma(2)}, ..., D_{\sigma(i)})$ with the kth placement α_i^k used for S_i. Then for any position $\alpha_{\sigma(i+1)}^{k'}$ of S_{i+1}, $g(i+1, k')$ is only affected by the values $g(i, k)$, where $k = 1, 2, ..., q$, and the score obtained from the relative positioning of the ith mapped segment and the $(i+1)^{th}$ mapped segment. More precisely, for $k' \in \{1, 2, ..., q\}$,

$$
\begin{aligned}
g(i+1, k') &= \min_{k \in \{1,2,...,q\}} \Big(g(i, k) + \Big| l\left(S_{(i+1)}, \alpha_{(i+1)}^{k'}\right) - l(D_{\sigma(i+1)}) \Big| \\
&+ \Big| d\left((S_i, \alpha_i^k) - (S_{(i+1)}, \alpha_{(i+1)}^{k'})\right) - \delta(D_{\sigma(i)}, D_{\sigma(i+1)}) \Big| \Big)
\end{aligned}
$$

Note that $l(D_{\sigma(i+1)})$ measures the length of the secondary structure trace $D_{\sigma(i+1)}$ and $\delta(D_{\sigma(i)}, D_{\sigma(i+1)})$ measures the skeleton length between $D_{\sigma(i)}$ and $D_{\sigma(i+1)}$ in the 3D image. Ideally, $\delta(D_{\sigma(i)}, D_{\sigma(i+1)})$ corresponds to the length of the loop connecting the two secondary structures and $d(a - b)$ measures the loop length between two consecutive secondary structures a and b on the sequence.

2.3   Secondary Structure Predictions from Multiple Servers

We submitted protein sequences to five online servers (SYMPRED[4] , JPred[5], PSIPRED[7], PREDATOR[16], and Sable [36]) to perform secondary structure predictions. Since there are always differences in the above predictions, we derive an initial set of positions of secondary structures. The initial positions include the predicted positions using SYMPRED together with major predicted difference using other four methods. Such initial positions were used to obtain initial topologies using *DP-TOSS*. Alternative positions of each secondary structure were generated based on the results from five secondary structure predictions. An optimal placement of SSTs was searched among alternatives for each of the top ranked topologies.

3   Results

The accuracy and efficiency of the two-step approach were tested using three kinds of data: helix-only proteins (five), α-β proteins (five) and Cryo-EM proteins (two). The three datasets represent increasing level of difficulty in the data. The helix-only proteins are the largest in the dataset, ranging from 207 amino acids (2XB5) to 585 amino acids (2XVV) in length. They are a good test case for the efficiency of our method. Proteins with β-sheets are generally more challenging than helix-only proteins in terms of accuracy. Firstly, detection of β-sheets is generally more challenging than the detection of helices. Secondly, the close spacing of β-strands makes it more challenging to identify the correct topology. True atomic structures of the proteins were downloaded from the PDB. For the helix-only proteins and the α-β proteins, each atomic structure was used to simulate a density map at 10Å resolution using EMAN software [1]. Helices were detected from such density maps using *SSETracer* [2], and are represented as α-traces (sticks in Fig. 3).

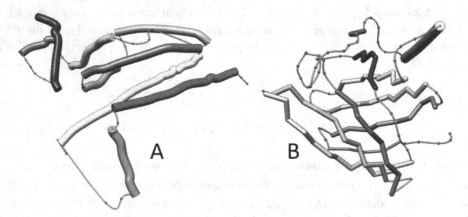

Fig. 3. The true topology derived from the two-step approach. The true topology is ranked the 40th for protein 2XB5 (PDB ID) in (A) and the 2nd for protein 1BJ7 (PDB ID). It is shown in rainbow color with blue corresponding to the N-terminal of the chain. The SSTs (sticks) were detected using *SSETracer* [2]and the connection was selected from skeleton points using *DP-TOSS*[3].

In order to test if our two-step approach works for the proteins with β-sheets, we first used the β-traces derived from the true structure for those proteins in the α-β dataset. The computationally detected β-traces were then used in the test involving Cryo-EM proteins. For each of the proteins in the test, multiple secondary structure predictions were performed and the predicted positions of secondary structures were used as the input information from protein sequence. To find the top 1000 ranked topologies, the initial secondary structure positions were used to match with the SSTs.

Each of the initial 1000 topologies was reevaluated using the proposed dynamic programming placement method in searching for the optimal placement. The 1000 optimally placed topologies were sorted based on the mapping score. The rank of the true topology was compared among three methods: the one that uses the true positions of the secondary structures on the sequence, the one that uses consensus (SYMPRED) prediction, and the one that uses multiple secondary structure predictions with the dynamic programming algorithm searching for an optimal placement.

As an example for 2XB5, there are thirteen helices, out of which SSETracer detected nine α-traces from the density image (Fig. 3 & Table 1). Four short helices were not detected. The detected helices may be shorter/longer or shifted. The consensus server (SYMPRED) predicted ten of the thirteen helices from the amino acid sequence. Using the predicted positions provided by SYMPRED, the true topology was ranked the 977th. Yet, when more alternative predictions are available from five prediction servers, our dynamic programming algorithm is fast enough to explore more alternative positions. It takes 6.94 seconds to generate the top 1000 topologies and to find each optimal placement for all 1000 topologies. The true topology was ranked the 40th (Fig. 3), much improved from 977th when only SYMPRED was used. When the true sequence position of secondary structures is used, the true topology was derived by our method *DP-TOSS* [3] as the 11th (Table 1). Although there is inaccuracy in both secondary structure predictions and the SST detection, Rank_d (40th) is close to Rank_t (11th), both near the top of the huge solution space for possible topologies. This suggests that the true topology can be ranked near the top even when two sets of non-perfect data are matched.

Ultimately the rank of the true topology is determined by the overall similarity of the two sets rather than a few individuals, although they may affect to some extent. Similar message was suggested from the test using an α-β protein 1BJ7 (Fig. 3 and Table 1). In this case, the true topology is ranked as the 2nd when multiple secondary structure predictions and optimal placement were used, much improved from the rank (>1000) when one prediction is used. It is even better than the rank (4th) when the true secondary structure positions were used.

We applied the two-step approach to two experimentally derived cryo-EM density map, (EMDB_5030 and EMDB_1780) that were downloaded from Electron Microscopy Data Bank (EMDB). Each density map corresponds to an atomic structure, and therefore can be used to test the accuracy of our approach. We extracted the density component corresponding to chain R of the protein for EMDB_5030 and chain K for EMDB_1780 respectively (Table 1, section 3). In the case of EMDB_5030, all three helices and three β-strands were detected using *SSETracer* and *StrandTwister*. The true topology was ranked 55th when multiple secondary structure predictions and dynamic programming placement were used. Surprisingly, the rank (47th) is even better than the rank derived using true secondary structure positions on the protein sequence. This is probably due to the existence of inaccuracy in the SST detected from the density image. Although the two Cryo-EM proteins are smaller than most other proteins in the test, they are the first two cases demonstrated the success in topology determination directly using computationally obtained β-traces and multiple secondary structure predictions. Overall, our two-step approach shows improved ranking of the true topology for nine of eleven tested proteins, when it is compared to the method that uses a consensus secondary structure prediction. For the rest two cases (2KZX, 1OZ9), the rank of the true topology is 1st and the 10th respectively, already near the top of the list.

Table 1. The rank of the true topology and run-time of the two-step approach

PID[a]	#α-Helices[b]	#β-Strands[c]	#α-stk/ β-stk[d]	# SS(α-hlx/ β-strand)[e]	Rank_t[f]	Rank_c[g]	Rank_d[h]	Time (sec)[i]
Helix-only Proteins								
2XB5	13	0	9/0	10/0	11	977	40	6.94
3ACW	17	0	14/0	14/0	408	>1000	485	15.73
3ODS	21	0	16/0	17/0	12	198	22	977.18
1Z1L	23	0	15/0	17/0	157	>1000	568	14.19
2XVV	33	0	19/0	27/0	21	>1000	87	1013.0
α-Helix & β-Sheet Proteins								
1BJ7	1	9	1/9	1/9	4	>1000	2	7.48
2L6M	2	3	2/3	3/3	6	>1000	42	37.34
2KZX	3	3	3/3	3/5	10	10	10	14.33
1J1L	4	5	4/5	4/5	16	16	14	16.89
1OZ9	5	5	5/4	5/3	2	1	1	2.69
Cryo-EM Proteins								
3FIN_R (5030)[*]	3	3	3/3	3/3	55	97	47	2.11
3IZ6_K (1780)[*]	3	5	2/5	2/5	2	6	2	12.56

[a]The PDB ID with chain; [*] EMDB ID of the experimentally derived Cryo-EM map;
[b]The number of helices in the true structure;
[c]The number of β-strands in the true structure;
[d]The number of α-traces/β-traces detected from the 3D image;
[e]The number of helices/β-strands predicted by SYMPRED;
[f]The rank of the true topology using the true sequence position of secondary structures;
[g]The rank of the true topology using consensus secondary structure position SYMPRED;
[h]The rank of the true topology using multiple secondary structure predictions with dynamic programming algorithm for optimal placement;
[i]The run-time (in seconds) of the two-step approach. It includes the time to generate top 1000 topologies and the   total time to find each optimal placement for 1000 topologies;

The execution time of the algorithm is dominated by the first step as the dynamic programming step requires a relatively short execution time. The total execution time to determine the best topology for protein 2XVV took a total of 1013 seconds out of which *DP-TOSS* took 1002 seconds (Table 1). The dynamic programming algorithm scales linearly to the product of the number of secondary structures and the number of possible alternatives for each secondary structure. The experiments in this paper were executed on a 2x Intel Xenon E5-2660 v2, 2.2GHz server machine.

4   Conclusions

Due to inaccuracy in the estimation of secondary structures, determination of the topology for SSTs requires the exploration of alternatives. Effective methods are needed to explore the large solution space resulted from the alternatives. We propose a dynamic programming algorithm to find the optimal placement when a topology is given. This algorithm is combined with our previous mapping algorithm and the shortest K paths algorithm to form a two-step approach. A test using twelve proteins shows that the two-step approach improves the ranking of the true topology as it is compared to using single consensus prediction. We demonstrate for the first time that computationally detected helices and β-strands from an experimentally derived Cryo-EM density image can be combined with multiple secondary structure predictions to rank the true topology near the top of the list. Our previous methods were mostly tested using the true positions of secondary structures. Now we have made a big step ahead by establishing an efficient algorithm to address the increased computation due to the alternatives.

Acknowledgements. The work in this paper is partially supported by NSF DBI-1356621, M&S fellowship and FP3 fund of the Old Dominion University. Authors' contribution: AB, DR, MZ, and JH developed the algorithm, AB implemented and conducted test. We thank Dong Si and Maryam Arab for the help in *SSETracer* and the preparation of the Cryo-EM data.

References

[1] Ludtke, S.J., Baldwin, P.R., Chiu, W.: EMAN: Semi-automated software for high resolution single particle reconstructions. J. Struct. Biol. 128(1), 82–97 (1999)

[2] Si, D., He, J.: Beta-sheet Detection and Representation from Medium Resolution Cryo-EM Density Maps. In: BCB 2013: Proceedings of ACM Conference on Bioinformatics, Computational Biology and Biomedical Informatics, Washington, D.C, September 22-25 (2013)

[3] Al Nasr, K., Ranjan, D., Zubair, M., et al.: Solving the Secondary Structure Matching Problem in Cryo-EM De Novo Modeling Using a Constrained k-Shortest Path Graph Algorithm. IEEE/ACM Transactions on Computational Biology and Bioinformatics 11(2), 419–430 (2014)

[4] Simossis, V.A., Heringa, J.: The influence of gapped positions in multiple sequence alignments on secondary structure prediction methods. Computational Biology and Chemistry 28(5-6), 351–366 (2004)

[5] Cuff, J.A., Clamp, M.E., Siddiqui, A.S., et al.: JPred: a consensus secondary structure prediction server. Bioinformatics 14(10), 892–893 (1998)

[6] Si, D., He, J.: Tracing beta-strands using strandtwister from cryo-EM density maps at medium resolutions. Structure 22(11), 1665–1676 (2014)

[7] McGuffin, L.J., Bryson, K., Jones, D.T.: The PSIPRED protein structure prediction server. Bioinformatics 16(4), 404–405 (2000)

[8] Chiu, W., Baker, M.L., Jiang, W., et al.: Electron cryomicroscopy of biological machines at subnanometer resolution. Structure 13(3), 363–372 (2005)

[9] Anger, A.M., Armache, J.P., Berninghausen, O., et al.: Structures of the human and Drosophila 80S ribosome. Nature 497(7447), 80–85 (2013)

[10] Jiang, W., Baker, M.L., Jakana, J., et al.: Backbone structure of the infectious epsilon15 virus capsid revealed by electron cryomicroscopy. Nature 451(7182), 1130–1134 (2008)

[11] Zhang, X.K., Ge, P., Yu, X.K., et al.: Cryo-EM structure of the mature dengue virus at 3.5-angstrom resolution. Nature Structural & Molecular Biology 20(1), 105–110 (2013)

[12] Chan, K.-Y., Trabuco, L.G., Schreiner, E., et al.: Cryo-Electron Microscopy Modeling by the Molecular Dynamics Flexible Fitting Method. Biopolymers 97(9), 678–686 (2012)

[13] Schröder, G.F., Brunger, A.T., Levitt, M.: Combining Efficient Conformational Sampling with a Deformable Elastic Network Model Facilitates Structure Refinement at Low Resolution. Structure 15(12), 1630–1641 (2007)

[14] Lasker, K., Forster, F., Bohn, S., et al.: Molecular architecture of the 26S proteasome holocomplex determined by an integrative approach. Proc. Natl. Acad. Sci. USA 109(5), 1380–1387 (2012)

[15] Zhang, J., Baker, M.L., Schroder, G.F., et al.: Mechanism of folding chamber closure in a group II chaperonin. Nature 463(7279), 379–383 (2010)

[16] Frishman, D., Argos, P.: Seventy-five percent accuracy in protein secondary structure prediction. Proteins: Structure, Function, and Bioinformatics 27(3), 329–335 (1997)

[17] Pettersen, E.F., Goddard, T.D., Huang, C.C., et al.: UCSF Chimera–a visualization system for exploratory research and analysis. J. Comput. Chem. 25(13), 1605–1612 (2004)

[18] Baker, M.L., Ju, T., Chiu, W.: Identification of secondary structure elements in intermediate-resolution density maps. Structure 15(1), 7–19 (2007)

[19] Del Palu, A., He, J., Pontelli, E., et al.: Identification of Alpha-Helices from Low Resolution Protein Density Maps. In: Proceeding of Computational Systems Bioinformatics Conference(CSB), pp. 89–98 (2006)

[20] Jiang, W., Baker, M.L., Ludtke, S.J., et al.: Bridging the information gap: computational tools for intermediate resolution structure interpretation. J. Mol. Biol. 308(5), 1033–1044 (2001)

[21] Kong, Y., Ma, J.: A structural-informatics approach for mining beta-sheets: locating sheets in intermediate-resolution density maps. J. Mol. Biol. 332(2), 399–413 (2003)

[22] Rusu, M., Wriggers, W.: Evolutionary bidirectional expansion for the tracing of alpha helices in cryo-electron microscopy reconstructions. J. Struct. Biol. 177(2), 410–419 (2012)

[23] Si, D., Ji, S., Nasr, K.A., et al.: A machine learning approach for the identification of protein secondary structure elements from electron cryo-microscopy density maps. Biopolymers 97(9), 698–708 (2012)

[24] Si, D., He, J.: Tracing Beta Strands Using StrandTwister from Cryo-EM Density Maps at Medium Resolutions. Structure (October 8, 2014)

[25] Al Nasr, K., Ranjan, D., Zubair, M., et al.: Ranking Valid Topologies of the Secondary Structure Elements Using a Constraint Graph. Journal of Bioinformatics and Computational Biology 09(03), 415–430 (2011)

[26] Al Nasr, K., Ranjan, D., Zubair, M., et al.: Solving the secondary structure matching problem in de novo modeling using a constrained K-shortest path graph algorithm. IEEE Transaction of Computational Biology and Bioinformatics 11(2), 419–430 (2014)

[27] Pollastri, G., McLysaght, A.: Porter: a new, accurate server for protein secondary structure prediction. Bioinformatics 21(8), 1719–1720 (2005)

[28] Przybylski, D., Rost, B.: Alignments grow, secondary structure prediction improves. Proteins 46(2), 197–205 (2002)

[29] Ward, J.J., McGuffin, L.J., Buxton, B.F., et al.: Secondary structure prediction with support vector machines. Bioinformatics 19(13), 1650–1655 (2003)

[30] Baker, M.L., Baker, M.R., Hryc, C.F., et al.: Gorgon and pathwalking: macromolecular modeling tools for subnanometer resolution density maps. Biopolymers 97(9), 655–668 (2012)

[31] Biswas, A., Si, D., Al Nasr, K., et al.: Improved efficiency in cryo-EM secondary structure topology determination from inaccurate data. J. Bioinform. Comput. Biol. 10(3), 1242006 (2012)

[32] McKnight, A., Si, D., Al Nasr, K., et al.: Estimating loop length from CryoEM images at medium resolutions. BMC Structural Biology 13(suppl. 1), S5 (2013)

[33] Al Nasr, K., Liu, C., Rwebangira, M., et al.: Intensity-Based Skeletonization of CryoEM Grayscale Images Using a True Segmentation-Free Algorithm. IEEE/ACM Trans. Comput. Biol. Bioinform. 10(5), 1289–1298 (2013)

[34] Abeysinghe, S., Ju, T., Baker, M.L., et al.: Shape modeling and matching in identifying 3D protein structures. Computer Aided-design 40, 708–720 (2008)

[35] Lindert, S., Staritzbichler, R., Wötzel, N., et al.: EM-fold: De novo folding of alpha-helical proteins guided by intermediate-resolution electron microscopy density maps. Structure 17(7), 990–1003 (2009)

[36] Adamczak, R., Porollo, A., Meller, J.: Combining prediction of secondary structure and solvent accessibility in proteins. Proteins: Structure, Function, and Bioinformatics 59(3), 467–475 (2005)

Managing Reproducible Computational Experiments with Curated Proteins in KINARI-2

John C. Bowers[1], Rose Tharail John[1], and Ileana Streinu[1,2(✉)]

[1] School of Computer Science, University of Massachusetts, Amherst, MA, USA
{jbowers,rosejohn}@cs.umass.edu
[2] Department of Computer Science, Smith College, Northampton, MA, USA
istreinu@smith.edu, streinu@cs.umass.edu

Abstract. KINARI-2 is the second release of the web server KINARI-Web for rigidity and flexibility of biomolecules. Besides incorporating new web technologies and making substantially improved tools available to the user, KINARI-2 is designed to automatically ensure the reproducibility of its computational experiments. It is also designed to facilitate incorporating third-party software into computational pipelines and to simplify the process of large scale validation of its underlying model through comprehensive comparisons with other competing coarse-grained models. In this paper we describe the underlying architecture of the new system, as it pertains to experiment management and reproducibility.

1 Introduction

Modeling Protein Flexibility. Structure, flexibility and motion are the key elements that relate a protein to its function. Yet current experimental methods and simulation techniques to determine flexibility parameters and to induce large scale motions are expensive, difficult and insufficiently developed. The recent successes in simulating fast motions [7] do not scale when applied to these much more challenging types of conformational changes. A different approach to understanding large-scale domain motions is to use coarse-grained models.

One of the best studied ones is the Gaussian Network Model (GNM), for which servers [16] and third-party software [9] are available. A different coarse-grained approach, pioneered by Thorpe and collaborators [14,13], relies on mathematical results from combinatorial rigidity theory.

These coarse-grained graph-based models for rigidity analysis of protein structures have been around for over 15 years and several implementations of this method are available. Several implementations exist, such as the stand alone executable ProFlex-FIRST [8], or the web server FlexWeb http://flexweb.asu.edu. Yet, the model is still far from providing convincing evidence that the biologically

This project is supported by NSF CCF-1319366, NSF UBM-1129194 and NIH/NIGMS 1R01GM109456.

© Springer International Publishing Switzerland 2015
R. Harrison et al. (Eds.): ISBRA 2015, LNBI 9096, pp. 72–83, 2015.
DOI: 10.1007/978-3-319-19048-8_7

relevant (structural and functional) information that has been demonstrated on a handful of protein structures can be automatically extracted on a larger scale. This is due, primarily, to the lack of large scale validation efforts. For instance, several studies have pointed out the sensitivity of the method to cut-off values and choice of parameters. Inconsistencies in the reported results obtained by different implementations have also been identified recently, and they are difficult to explain without the ability to easily reproduce the computational experiments that generated them.

KINARI-Web Server. Similar to FIRST in that it uses the rigidity based approach, KINARI-Web (http://kinari.cs.umass.edu) is a server for rigidity and flexibility analysis of biomolecules developed in the group of the senior author of this paper. There are however several differences in the underlying modeling and in the algorithms used in KINARI, compared to the other available implementations. FIRST used a strict, pre-determined scheme for building a graph from a molecule, in which the main parameter that the user can vary is a cut-off value for the inclusion of hydrogen bonds. This value is based on a bond energy calculated with the Mayo formula [10]. FIRST does not curate the protein and the user has to preprocess a PDB file prior to submitting it for the rigidity analysis. By contrast, KINARI included from the very beginning several tools for curating the protein and the option of altering the mechanical model for experimenting with its parameters. KINARI relies on several third-party, freely available protein curation tools such as Reduce (for adding Hydrogen atoms to protein structures determined by X-ray crystallography) or HBPLUS (for computing the Hydrogen bonds), as well as in-house implementations of other relevant atomic interactions (such as hydrophobics) and energies. In addition, KINARI-Web offers an integrated JMol visualizer and it returns to the user the files produced during curation. This facilitates the process of building the right model and computationally experimenting with it, and it allows for a better degree of *reproducibility* of the results. Indeed, FIRST's results are often impossible to reproduce without direct access to the specific, typically not publicly available, curated protein file that was submitted for analysis.

Reproducibility of KINARI-Web Experiments. The purpose of offering these options in KINARI is the desire to make rigidity analysis a *computationally reproducible process*. Our goal is to have a system in which the knowledge of the PDB id of the protein or access to it in a PDB-formatted file, plus knowledge of the parameters used during the modeling process would make it possible for any reader of a paper describing a rigidity analysis experiment to reproduce it. Towards this goal, KINARI also offered access to the recorded options selected during the curation process in a configuration file that was made available at the end of the process.

However, with the increased use of the system and several new applications being integrated into KINARI, we reached the limitations of the initial design. This paper describes the new, redesigned user and experiment infrastructure underlying KINARI-2, whose goal is to permit full reproducibility of KINARI-integrated computational experiments. In addition, we are improving and extending the curation process, and have designed a series of templates (based on the Model-View-Controller software design paradigm) that will facilitate adding new modules to KINARI which incorporate third-party software. The need for this option is described next.

Validation. A major goal of this line of research, independent of the software tool that one uses, is a thorough validation of the rigidity-based coarse-grained modeling approach. The two implemented versions of FIRST already used different algorithms. These groups, and other researchers [14,13,5], reported on a number of studies on specific proteins, where the rigid cluster decomposition results obtained computationally matched protein flexibility properties observed in lab experiments. However, more recent, slightly larger scale studies such as [17,6], observed that the method is sensitive (among others) to the placement of hydrogen bonds, and that there is no universal cut-off value for the hydrogen bond energy which would give biologically meaningful results for all the proteins in a specific dataset. These studies point to the need for systematic and comprehensive validation of rigidity analysis results, and of building benchmarking datasets to assist with this goal.

One approach is to compare the results obtained through rigidity analysis with other coarse-grained approaches, for which validation studies have been conducted. These validation studies can themselves be by comparison with other approaches, but in the end the results of a cluster (or domain) decomposition should have been compared with biologically relevant properties of specific protein datasets. Moreover, we do not want just to make one series of runs on a system versus another and compare them. We want to provide a tool which is easy to extend for any kind of cluster-decomposition method and for any available dataset. Thus, from the very beginning we set as our goal to develop a system in which all computational experiments would be fully reproducible.

Reproducible Protein Dilution Experiments. We illustrate the need for such an infrastructure through a case study performed in our group and reported in a companion paper [4]. We have developed there a method for visually comparing the rigidity analysis results on sets of proteins related by some common type of computational experiment. Ultimately, of course, the goal is to compare the cluster decompositions for any experiment on any family of proteins, but this is, algorithmically, a very challenging task. To get started, we have chosen two benchmark applications: Dilution and Mutations on proteins.

The first one is a simplified model for *protein unfolding*. It is one of the first applications to demonstrate the usefulness of rigidity analysis and has been described in [13,12]. Several subsequent protein dilution studies were conducted by other groups. With the existing tools provided by FIRST, the results are visualized and reported using a 1D comparison plot called a *dilution plot*. Our visualization method proposed in [4] is much more intuitive, being based on

3D structures. Since our dilution process relies on a different algorithmic kernel (KINARI versus FIRST), we wanted to make sure that its results are compatible with those reported in the literature. For this purpose, we have re-implemented the 1D dilution plots and ran the dilution experiments (in KINARI) on data for which dilution experiments done with FIRST were reported in the literature [17,6]. In this way, we came very quickly across a situation where strikingly different results were obtained. Tracing back the source of the discrepancy, we found a large difference between the number of hydrogen bonds calculated by us using HBPLUS [11] (which is third party software integrated with KINARI to perform these calculations) and the number reported in [6] for the same protein. Running a different software tool [15] for placing hydrogen bonds still did not justify the discrepancy. We conjectured several possible scenarios that could explain the results, but the lack of access to the precise data file on which the outlier experiment of [6] was conducted made it impossible to settle to a definitive explanation and remedy. More details about this case study appear in [3], with an abstract in [4].

This experience prompted us to consider providing, with KINARI, an infrastructure where *reproducible computational experiments* could be carried by all interested users. The purpose of this paper is to present the design of such an infrastructure in the new, extended and improved web server KINARI-2, planned to be publicly released by the end of the year.

Reproducibility of Protein Data Curation. It is well known that the data deposited in the Protein Data Bank is not of uniform quality: resolution and B-factors are parameters recorded with each file which may help with judging the quality of the experimental data deposited in the PDB. A number of entries in the PDB have been declared obsolete and replaced by others. Tools for checking the quality of crystallographic data are also available, such as MolProbity [2]. However, the curation process for PDB data required as preprocessing prior to rigidity analysis is not only about the acuracy of the molecular model. Kinari Curation includes several steps that are performed with third party software, and different software performing the same task may produce different results. These steps include: adding the hydrogen atoms if the data comes from an X-ray crystallography experiment; pruning the hydrogen bonds according to a user-selected cut-off value; selecting the model from among several available in a file containing data from an NMR experiment; computing the hydrogen bonds and hydrophobic interactions; building a biological assembly or, possibly, a small crystal, etc. Wihout precisely recording the entire sequence of steps performed during a curation experiment, the reproducibility of a subsequent rigidity analysis experiment may be compromised. Therefore, we are placing maximum emphasis on the management and reproducibility of curation experiments in KINARI.

2   Methods and Design

The new KINARI is based on a system to manage users and their experiments in such a way that: (a) user privacy is guaranteed (we do not require registration nor verification of the user's identity); (b) it is easy to get started and resume experiments (a new user account can be set up at any time, and, within a certain time span, the user can return and still find around, in that account, the previously computed data); (c) the user can download all the files resulting from the computation done in KINARI, including a readable configuration file that keeps track of all the actions performed on the input PDB file, and (d) the user can return, upload the previously saved files from some unfinished experiment, and resume the experiment. Since we do not retain the user experiment data indefinitely (the temporary storage is cleared automatically during routine maintenance procedures), this ensures that the users can conduct longer experiments in several sessions, and protects them against other unwanted interruptions such as those due to network connectivity disruptions.

Besides this basic user and experiment management system, our new design has built-in capabilities for extending the system with new applications, which correspond to experiment types. Each experiment consists in running one of these apps. A series of experiments can then be either manually or automatically streamed into an automatically executed sequence, thus permitting the design of larger scale experiments on single molecules or on large datasets.

The server side application is implemented in PHP and Python and invokes JMol and external binaries. It is hosted on an Apache web server. The user interface is written in HTML5, CSS, JavaScript, JQuery and JsMol scripting.

We describe now the overall structure of the system, and focus afterwards on the infrastructure for new application, experiment and step design.

2.1   System Design

The infrastructure of KINARI-2 is organized as a collection of applications (apps), each of which is responsible for performing a particular experiment, or set of computational tasks. The web server also has a Main component which serves to log users into the system and provide tools for managing a user's ongoing or completed experiments. When the user first logs in she is presented with a list of her experiments and from there can resume an ongoing experiment, delete experiments, or start new experiments. Once the user chooses to either resume an ongoing experiment or create a new experiment, control is handed off to the appropriate app.

For organizational purposes we group apps into *domains* based on what type of data the application operates on. For instance, the Gaussian Network Model (GNM) app included now in KINARI-2 operates on biomolecules, and is grouped within the "Biomolecules" domain. An overview of this structure is shown in Fig. 1.

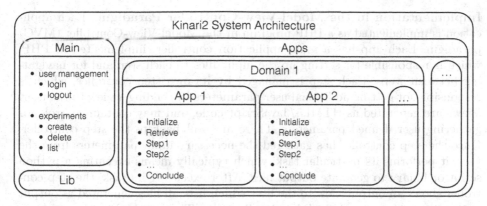

Fig. 1. The architecture of the system

2.2 Application Design

Each application is broken into a series of logical steps. For instance, the GNM app example discussed in Sec. 3 performs five steps: Input Retrieve, which retrieves a PDB file from the user; Process PDB for GNM, which processes the PDB in order to produce the appropriate input files for the gnm-domdecomp.e binary [9]; Run GNM, which runs the gnm-domdecomp.e binary to output a domain decomposition file; Prepare GNM for JMol, which takes the domain decomposition file produced by gnm-domdecomp.e and produces a JMol script for visualizing the results; and Visualize GNM, which displays a JMol visualization of the domain decomposition to the user.

The Control Flow Graph. In the GNM app example, the control flow of the experiment is linear–each step follows directly from the previous. In other words, this simple app is obtained by pipelining several steps. However, for certain applications this is not the case. In all applications the control flow between steps forms a connected directed acyclic graph (DAG) with a single initial step (of in-degree 0) and a single final step (of out-degree 0). The initial step is typically an input retrieval step for obtaining a PDB or other input file. This can be obtained directly from the user using a file upload or copy-and-paste, or the user can select to retrieve a file from a publicly available database, such as the RCSB [1], or select from prebuilt datasets that we curate in-house. All applications end with a Conclusion step, which gives the user the option to download the entire history of the concluded experiment. The result can then be transferred and serve as input to another experiment.

If the out-degree of a step is greater than one, then the step is a *branching step*. Here the "next step" may be determined by a user parameter or a computation. For instance, in the Biomolecule Curation app, after the Summary step which gives the user a summary of the contents of the file they selected in the Input Retrieval step, the next step depends on whether the PDB file was obtained using X-Ray Crystallography or NMR.

Implementation in the Model-View-Controller Paradigm. Each application is implemented as a PHP program in the Model-View-Controller (MVC) paradigm. Each app has a main application controller, implemented in PHP, which is responsible for setting up the input files to each step and for navigating between steps. Each step is managed by its own step controller, which is responsible for getting additional user parameters by sending a view to the user. Views are generated as HTML5/JavaScript code, and may contain controls for gathering user defined parameters. These are sent back to the step controller. Once the step controller has gathered the necessary input parameters from the user, it performs its particular task, which typically means executing a python script or binary to generate output files. After executing its task, the step controller shows an output view to the user which lists all files generated as output of the step and may include JMol visualizations. The user is then given the option of continuing to the next step. The state of an experiment is maintained by an Experiment Configuration File, which is a record of all tasks performed by the application for the current Experiment. The Experiment Configuration File is designed to store all information needed to reproduce an experiment from scratch. See Sec. 2.3.

Navigation. The steps performed during a particular experiment form a path in the control flow graph of the application. This *history path* is shown to the user in the gray navigation bar across the top of the screen. See Fig. 2. When the user clicks on a previous step the main controller backtracks the experiment to the state as it existed before the selected step was originally performed. In other words, all data output by that step and any subsequent steps are erased as if the subsequent steps had not been performed. This allows the user to tinker with parameters at any step, see the results, and backtrack as necessary to fine tune an experiment.

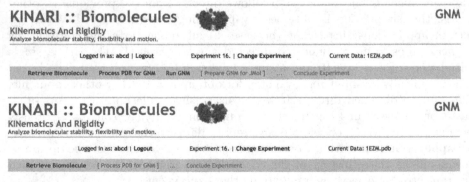

Fig. 2. Example of navigation in the GNM application case study. Top: the user is on the "Prepare GNM for JMol" step. The current step is displayed in red and the previous steps are highlighted in blue. The user backtracks to a previous step by clicking on it. Backtracking removes the steps that occurred after the selected backtrack step. Bottom: the steps in the history after the user backtracks to the "Process PDB for GNM step".

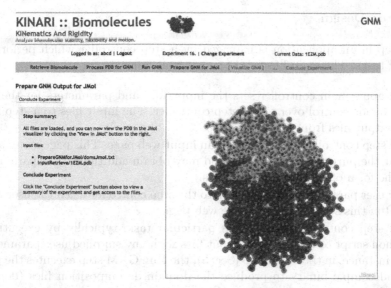

Fig. 3. Screenshots of the visualization step of the GNM application. The gray navigation bar across the top allows the user to backtrack the experiment to a previous step.

2.3 Managing an Experiment

An application manages an experiment from start to finish. The goal of our system design is to enable complete reproducibility for each experiment. To do this we require two main components, an experiment folder for storing all inputs and outputs for all steps of the experiment, and an experiment configuration file detailing the parameters used to perform each step. This allows an experiment to be rerun from start to finish using exactly the same parameters and allows each step to be independently verified. The final Conclusion step in each application gives the user the option of downloading the Experiment Folder which contains the Experiment Configuration File.

Experiment Folder. The system creates a new Experiment Folder each time the user starts a new experiment. The folder contains the Experiment Configuration File which stores metadata about the experiment, as well as a separate output folder for each completed step in the current experiment. Any output files produced by a step are placed in this folder. These output files may then be used as input files to future steps.

Experiment Configuration File and Reproducibility. Each step performed by the user is recorded in an Experiment Configuration file (stored as an XML document). The file stores the current state of the experiment which includes the entire history path of the experiment. For each step we record what input files and parameters were used by the step and what output files were created.

2.4   Step Design

Each step in an application is managed by a step controller which performs a series of tasks. The main lifecycle of a step is as follows.

1. The app's main controller sets the input files and parameters for the step and hands control over to the step controller. The input files to a step must be output files from a previous step.
2. The step controller shows the user an input web page. This page summarizes what the purpose of the step is, and may obtain additional user parameters in the form of HTML input controls.
3. Any user parameters are sent back to the step controller when the user clicks a "Run this step." button on the web page.
4. The step controller performs its particular task, typically by executing a python script or binary on its input files with any supplied user parameters. For instance, in the GNM app (Sec. 3), the Run GNM step executes the gnm-domdecomp.e binary to produce the domain decomposition files (dom.txt and dom1.txt).
5. The output of (4) is written to the server's file system in the step's output folder.
6. The input files, user parameters from (3), and output files generated by (4) are recorded in the Experiment Configuration File.
7. The user is shown an output page that states that the step was completed and allows the user to download the output files from the step. This page also shows a "Go to next step" button.
8. When the user clicks the "Go to next step" button, the app controller is invoked to load the next step controller and start the process over.

Figure 4 shows a flowchart of the steps above.

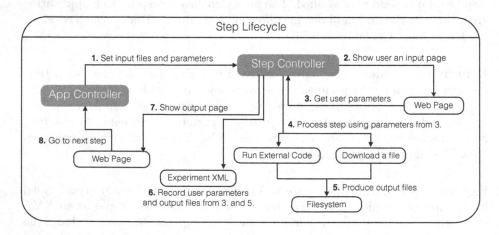

Fig. 4. The lifecycle of a step in an application

3   Results

We illustrate the described methodology with a simple application implemented in our new system, and with a sketch of some representative steps from the larger Curation application, which was redesigned both for guaranteeing reproducibility and for adding much improved visualization tools.

GNM: An Example of a Pipelined Application Running Third-party Software. Figure 5 illustrates an entire application, chosen to be as simple as possible for illustration purposes. This application starts by retrieving a protein, then runs the GNM domain decomposition program [9] on it, and ends with a visualization of the results in a JMol applet. This application is obtained by sequentially executing the following five steps. Each step except the first one retrieves its necessary files from previous steps, and returns the results in a step-specific folder from which future steps can retrieve them.

1. Input Retrieve: gets an unprocessed PDB file using user input (either from the RCSB database, or from a file upload, etc.)
2. Process PDB for GNM: runs a python script on the PDB file to create the correct inputs for the external, third-party GNM application.
3. Run GNM: runs the gnm program on the files produced in the previous step.
4. Prepare GNM for JMol: runs a python script on the domain file produced by the previous step to obtain a JMol script for visualizing the results of GNM.
5. Visualize GNM: loads the JMol script produced by the previous step and shows the outcome of the script in a JMol visualizer.

The red arrows in Fig. 5 show that the input of each step is given by the output of previous steps. For instance the output file produced by the Run GNM step is given as input to the Prepare GNM for JMol step. It should be noted that the input files to a given step may be output files from any previous step, not just the step immediately prior. The mapping of output files from prior steps to input files for each step is saved as part of the Experiment Configuration File.

Curation: An Example of a Branching Application. The Curation application starts by retrieving the file, after which a Summary of the biomolecule is computed. This extracts the experimental method and branches into a step that extracts for curation a single model from the PDB file (if the method was NMR) or into a step that performs an operation specific to proteins obtained through X-ray crystallography, such as the placing of the missing Hydrogen atoms. JsMol vsualization with step and molecule specific options are also provided on many Curation steps. For instance, the user will have the option of choosing between several methods for computing the Hydrogen bonds. For lack of space, we do not pursue in detail the description of the entire Curation application.

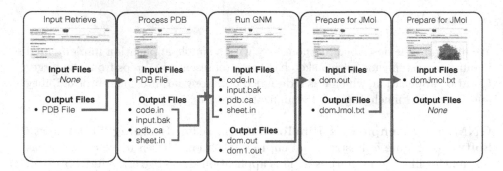

Fig. 5. An example application that runs and visualizes the output of the GNM program on a protein. Each box is a step in the application. The input files and output files for each step are listed. In each case the input files for a step are given by output files from previous steps (denoted by the red arrows).

4 Conclusion

We described the structure of KINARI-2, a web server for conducting reproducible computational experiments on biomolecular data from the PDB or other sources. The ultimate goal of KINARI-2 is to facilitate rigidity analysis and flexibility calculation experiments, and to contribute to the validation of the rigidity analysis method by providing tools for comparing its results with those obtained by other methods. The system architecture presented here has been implemented and tested. We are currently converting the previous KINARI-Web apps to the new system, ensuring full reproducibility of those experiments and extending the system with new and substantially improved tools, which will be described elsewhere as they are completed.

Authors' Contributions. IS conceived the project and the overall design of the system, and oversaw the project. RTJ implemented the first prototype, and JB redesigned and restructured it into the current version. JB and IS wrote the paper.

References

1. Berman, H., Westbrook, J., Feng, Z., Gilliland, G., Bhat, T., Weissig, H., Shindyalov, I., Bourne, P.: The protein data bank. Nucleic Acids Research 28, 235–242 (2000)
2. Chen, V.B., Arendall, W.B., Headd, J.J., Keedy, D.A., Immormino, R.M., Kapral, G.J., Murray, L.W., Richardson, D.C.: Molprobity: all-atom structure validation for macromolecular crystallography. Acta Crystallographica Section D-biological Crystallography 66, 12–21 (2010)
3. Flynn, E.: Comparing and visualizing in silico protein rigidity: Methods and applications. Honors thesis, Smith College (2014)

4. Flynn, E., Streinu, I.: Comparing rigid domain decompositions for dilution and mutation analysis of proteins. Submitted to this Conference (2015)
5. Fulle, S., Gohlke, H.: Constraint counting on RNA structures: linking flexibility and function. Methods 42, 181–188 (2009)
6. Heal, J., Jimenez-Roldan, J., Wells, S., Freedman, R., Römer, R.: Inhibition of HIV-1 protease: the rigidity perspective. Bioinformatics 28, 350–357 (2012)
7. Klepeis, J.L., Lindorff-Larsen, K., Dror, R.O., Shaw, D.E.: Long-timescale molecular dynamics simulations of protein structure and function. Current Opinion in Structural Biology 19(2), 120–127 (2009)
8. Kuhn, L.A.: First installation and user's guide (floppy inclusion and rigid substructure topography, version 4.0) (2004),
http://www.kuhnlab.bmb.msu.edu/software/proflex/index.html
9. Kundu, S., Sorensen, D., Phillips Jr., G.N.: Automatic decomposition of proteins by a gaussian network model. Proteins: Structure, Function, and Bioinformatics 57, 725–733 (2004)
10. Mayo, S., Dahiyat, B., Gordon, D.: Automated design of the surface positions of protein helices. Protein Science 6, 1333–1337 (1997)
11. McDonald, I., Thornton, J.: Satisfying hydrogen bonding potential in proteins. Journal of Molecular Biology 238, 777–793 (1994)
12. Rader, A., Anderson, G., Isin, B., Khorana, H., Bahar, I., Klein-Seetharaman, J.: Identification of core amino acids stabilizing rhodopsin. PNAS 101, 7246–7251 (2004)
13. Rader, A.J., Hespenheide, B.M., Kuhn, L.A., Thorpe, M.F.: Protein unfolding: Rigidity lost. Proceedings of the National Academy of Sciences 99(6), 3540–3545 (2002)
14. Rader, D.J.A., Kuhn, L., Thorpe, M.: Protein flexibility predictions using graph theory. Proteins 44, 150–165 (2001)
15. Richardson, J., Richardson, D.: Bndlst,
http://kinemage.biochem.duke.edu/software/utilities.php
16. Wang, L., Rader, A., Liu, X., Jursa, C., Chen, S., Karimi, H., Bahar, I.: oGNM: online computation of structural dynamics using the Gaussian Network Model. Nucleic Acids Research 34, W24–W31 (2006)
17. Wells, S., Jimenez-Roldan, J., Römer, R.: Comparative analysis of rigidity across protein families. Physical Biology 6 (2009)

Protein Crystallization Screening Using Associative Experimental Design

İmren Dinç[1(✉)], Marc L. Pusey[2], and Ramazan S. Aygün[1]

[1] Department of Computer Science, University of Alabama in Huntsville,
Huntsville, AL 35899, USA
{id0002,aygunr}@uah.edu, marc.pusey@ixpressgenes.com
[2] iXpressGenes, Inc., 601 Genome Way, Huntsville,
AL 35806, USA

Abstract. Protein crystallization remains a highly empirical process. The purpose of protein crystallization screening is the determination of the main factors of importance leading to protein crystallization. One of the major problems about determining these factors is that screening is often expanded to many hundreds or thousands of conditions to maximize combinatorial chemical space coverage for a successful (crystalline) outcome. In this paper, we propose a new experimental design method called "Associative Experimental Design (AED)" that provides a list of screening factors that are likely to lead to higher scoring outcomes or crystals by analyzing preliminary experimental results. We have tested AED on Nucleoside diphosphate kinase, HAD superfamily hydrolase, and nucleoside kinase proteins derived from the hyperthermophile Thermococcus thioreducens [1]. After obtaining the candidate novel conditions, we have confirmed that AED method yielded high scoring crystals after experimenting in a wet lab.

Keywords: Associative Experimental Design · Protein Crystallization · Screening · Experimental Design

1 Introduction

Protein crystallization is the process of formation of 3-dimensional structure of a protein. One of the significant difficulties in macromolecular crystallization is setting up the parameters that yield a single large crystal for X-ray data collection [2], [3]. The major difficulty in this process is the trial of abundance of parameters with significant number of possible values. Physical, chemical and biochemical factors such as type of precipitants, type of salts, ionic strengths, pH value of the buffer, temperature of the environment, and genetic modifications of the protein affect the crystallization process significantly [4]. Because each protein has a unique primary structure, it is quite challenging to predict the parameters of the experiment that can yield crystal for a specific protein [2]. Theoretically, it is possible to crystallize a protein in a specific solution under certain conditions; however, it may not be possible to crystallize in practice [5].

© Springer International Publishing Switzerland 2015
R. Harrison et al. (Eds.): ISBRA 2015, LNBI 9096, pp. 84–95, 2015.
DOI: 10.1007/978-3-319-19048-8_8

This means that we can generate hundreds or thousands of conditions to maximize combinatorial chemical space coverage hoping for a crystalline outcome. However, setting up huge number of experiments is not feasible in terms of cost and time.

Basically, there are two main approaches to set parameters for protein crystallization experiments [6], [7]: 1) incomplete factorial design (IFD) [8], [9] or sparse matrix sampling (SMS) [3], [10], and 2) grid screening (GS) of crystallization conditions [11]. The first approach has been widely used by commercial companies such as Hampton Research, Emerald Biostructures, etc. [6]. Carter et al. proposed to set parameters of protein crystallization experiments using incomplete factorial design in their study [8]. The main goal of incomplete factorial design experiments is to identify important factors of the experiments and to produce much less number of experiments compared to full factorial design experiments. The IFD is a very effective method as experts may not afford to set up many experiments or they may not have enough resources to carry out those many experiments [12]. The basic idea of IFD is that after identifying important factors of the experiments; balanced experiments are generated in terms of factors. In the sparse matrix sampling [3], parameters of the experiments are set using fewer major reagents (i.e., pH values, type of precipitants, type of salts, etc.) as in IFD. It can be considered as an optimized version of IFD. In SMS, values of type of salts, pH, and type of precipitants are selected based on past experiences, and these variables are mostly favorable for protein crystallization experiments. The reagents occur based on their frequency in the sparse matrix [10]. This idea was commercialized by Hampton Research [13]. Grid screening of crystallization conditions [11] is an early method that tries possible different solutions exhaustively until the experiments succeed. This takes more time and effort compared to IFD and SMS. However, it could be the only solution for some of the proteins that have never or rarely been crystallized.

In the literature, there are also some optimization methods based on IFD and GS [12], [14], [13]. We do not intend to give detail about these methods in this paper. These studies in macromolecular crystallization generally try to optimize available conditions changing one or few parameters in the chemical space such as pH, concentration of precipitant, etc. For example, Snell et. al optimized the conditions to change the pH value of the buffer and the weight of the precipitant rather than just changing one value at a time by visualizing the result [13]. The traditional optimization techniques do not consider new combinations of reagents [12]. In this paper, we propose a new experimental design method called Associative Experimental Design (AED) that generates candidate conditions by analyzing preliminary experimental data. This existing data is analyzed to determine which screening factors are most likely to lead to higher scoring outcomes, crystals. Unlike IFD, AED generates unbalanced experiments for protein crystallization that may include novel conditions. This means AED is not a typical optimization method for crystallization conditions. In the literature, optimization steps usually include changing the pH value, concentration, weight of precipitants and salts. The AED method finds small but effective number of conditions that may lead to crystallization. The main idea of the AED method is to generate novel conditions for crystallization by keeping at least two reagents from promising conditions. Basically, the AED analyzes other possible interactions between reagents to determine new crystallization conditions. In this study, we have generated candidate conditions for Nucleoside diphosphate kinase, HAD superfamily hydrolase, and nucleoside kinase proteins using preliminary experimental results. After obtaining the candidate novel conditions, we have confirmed that AED method yielded high scoring crystals after experimenting in a wet lab.

The rest of the paper is structured as follows. Background information is provided in Section 2. The proposed method, "Associative Experimental Design (AED)," is explained in Section 3. Experimental results are provided in Section 4. Finally, our paper is concluded with the last section.

2   Background

In this section, we provide some information about the phase diagram, which is a useful diagram for setting up protein crystallization experiments. We develop AED based on the phase diagram, and we believe that a brief explanation of it would help reader to understand the problem domain and our method. In addition, we are going to provide some brief explanation of Hampton scores in this section, since we are going to refer those scores throughout the paper.

2.1   Phase Diagram

In chemistry, a phase diagram is a graphical representation of different phases (solid, liquid and gas) of a substance with respect to temperature and pressure. In structural biochemistry, a phase diagram mostly represents solubility curve of a protein with respect to some parameters such as precipitant, pH, etc. Since the proteins can grow only in supersaturated solutions, it is important to locate solubility curve based on these parameters [5], [15]. Thus, the phase diagram is useful to set parameters for the experiments properly for X-ray diffraction studies [16]. Figure 1 shows a visual representation of a phase diagram.

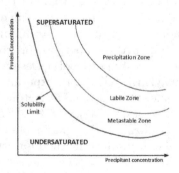

Fig. 1. Phase Diagram

The phase diagram mainly has two main zones: undersaturated region and supersaturated region. The supersaturated region consists of three subdivisions as can be seen in Figure 1. The first region is labile zone, where nuclei of protein crystals can form and continue growing its structure if the certain conditions are provided. Once the nucleation starts, protein crystals start using the nutrients of the solution, which will reduce the concentration of the solution. While the concentration of the solution reduces, the solution will be in metastable region. In this region, protein crystal may continue to grow up to its concentration equal to the solubility limit, if there are nuclei that have formed before. In other words, new nuclei cannot form in that region [4], [2]. If the supersaturation is too high, amorphous precipitates can also appear in precipitation zone instead of crystals, which is not a desirable outcome for crystallization process [5]. Furthermore, the amorphous precipitates do not yield crystals, when they complete their formations. Since nucleation can only occur in labile zone, AED focuses the conditions that fall into that region. Detailed information will be provided in Section 3.

2.2 Hampton Scoring

Hampton scoring is used to evaluate the growth of the protein during the crystallization experiments. In Hampton scoring, there are 9 scores from 1 to 9. In most of the experiments, a score that is greater than 7 is desired by the crystallographers, although scores between 5-7 are also classified as crystals. It should be noted that mostly crystals that have either score 8 or 9 are able to provide sufficient information about their 3D structures. Table 1 shows the list of Hampton scores. The brief explanations of some scores[1] are provided below.

Table 1. List of Hampton scores

Score	Outcome	Score	Outcome
1	Clear solution	6	1D needles
2	Phase change (oiling out)	7	2D plates
3	Regular granular precipitate:	8	3D crystals small, $< 200\mu m$
4	Birefringent precipitate or bright spots	9	3D crystals large, $> 200\mu m$
5	Spheroids, dendrites, urchins		

In this study, we focused on scores from 4 to 7 to generate novel conditions using AED method. The details about AED are provided in Section 3.

---

[1] http://hamptonresearch.com/tips.aspx

3 Proposed Method: Associative Experimental Design(AED)

3.1 Motivation

In this research, we have generated some crystal screens for a few specific proteins using preliminary crystal screen data with their Hampton scores. We use 3 different proteins to test our approach. There are 86 different crystal screens in our dataset for the protein $Tt189$ without considering the conditions having multiple types of salts or precipitants. This data set contains 9 different salt concentration values, 23 different type of salts, 7 different pH values, 45 different precipitant concentration values, 85 different precipitants, 3 different protein concentration values, where the concentrations and pH values are continuous data and the other features are categorical data. (Note that type of buffer is not considered, since it is generally correlated with pH value.) If we consider full factorial design, it means that we need to set up approximately 16,627,275 different experiments for a single protein based on this dataset without considering the continuity of some of the variables and this is not feasible. In this research, our goal is to generate less number of conditions rather than 16.6M that is more likely to form a crystal. To achieve this goal, we proposed a method called "Associative Experimental Design (AED)."

3.2 Method

Associative experimental design generates a new set of experiments by analyzing the scores of experiments already evaluated in the lab. We use almost the same scores from 1 to 9 provided in Table 1. Since we are using trace fluorescent labeling (TFL) [17], a score of 4 is assigned to outcomes giving "bright spot" lead conditions as an exceptional case.

We start with the notation for screening conditions and scores. Let

$$D = \{(C_i, H_i) \mid (C_1, H_1), (C_2, H_2), ..., (C_n, H_n)\} \tag{1}$$

be our dataset consisting of the pairs that include features of the conditions C_i and their scores H_i for the ith solution in the dataset. For simplicity we discarded conditions that have more than one type of salt or precipitant. We only focused on three main components of the remaining conditions: type of precipitant, type of salt and pH value of the solution, while separating their concentrations. Let

$$C_i = \{S_i[sc_i], pH_i, P_i[pc_i]\} \tag{2}$$

be the set of all the features of ith crystal screen where i is $1 \leq i \leq n$, n is the number of samples in our dataset, $S_i[sc_i]$ represents type of salt with the concentration of sc_i, pH_i value represents the pH of ith solution, and $P_i[pc_i]$ represents type of precipitant with the concentration of pc_i. Let R be a subset of D that contains the crystal screen pairs having a score greater than or equal to low_H and less than or equal to $high_H$:

$$R = \{(C_i, H_i) \mid (C_i, H_i) \in D, low_H \leq H_i \leq high_H, 1 \leq i \leq n\} \tag{3}$$

In our preliminary experiments, we set $low_H = 4$ and $high_H = 7$. Thus, the samples that have a score of 8 or 9 are excluded to generate unbiased conditions for the proteins. Similarly, for simplicity the samples that have score from 1 to 3 have not been included in the result set.

The AED analysis process consists of two major stages. In the first stage, we process the data to reduce its size as we stated before. Let

$$R_c = \{C_i \mid (C_i, H_i) \in R\} \tag{4}$$

be the set of conditions of R, where $SC_i = \{sc_1, sc_2, ..., sc_k\}$ represents the all unique concentration values of the ith salt, and $PC_i = \{pc_1, pc_2, ..., pc_k\}$ represents the all unique concentration values of ith precipitant. Then, we compare each C_i and C_j condition pairs where $i \neq j$ in R_C. If there is a common component between C_i and C_j, then we generate the candidate conditions set Z based on these two sets. For example, let $C_i = \{S_i[SC_i], pH_i, P_i[PC_i]\}$ and $C_j = \{S_j[SC_j], pH_j, P_j[PC_j]\}$ where $S_i = S_j$ (i.e., the type of salt is common in C_i and C_j). We generate two new conditions Z by switching the other components among each other. Thus,

$$Z = \{\{S_i[SC_i], \mathbf{pH_j}, \mathbf{P_i}[\mathbf{PC_i}]\}, \{S_i[SC_i], \mathbf{pH_i}, \mathbf{P_j}[\mathbf{PC_j}]\}\} \tag{5}$$

is the set of candidate crystal screens for the pair C_i and C_j. Similarly, candidate screens can be generated where pH value or precipitant is common between the pairs as well. After we generate candidate combinations using these components, we remove conditions that are replicated or are already in the training data. In the second stage of our method, we assign unique values of concentrations, generate SC_i and PC_i, and unique type of buffers that were used in the preliminary data to generate finalized crystal screens. At the end, we merge generated results from two stages of the method. The identified significant factors are output and used to generate condition screens with factor concentrations varied over the indicated ranges from the analysis. These screens are then used to prepare a new plate. Since we are comparing each condition with the remaining conditions to find the common agent, the complexity of our algorithm is $O(n^2)$ where $n = |R|$. Considering today plate sizes(up to 1536-well plate), we do not expect n is a very large number. Therefore, this implies $O(n^2)$ is a reasonable time for this problem. Figure 2 shows the flow diagram of AED.

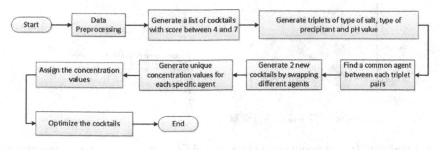

Fig. 2. Flow diagram of AED

Sample Scenario. Figure 3 shows the scores from four experiments from a commercial screen. The figure shows a partial graph of scores for pH value of 6.5. These conditions led to four scores: 1, 1, 4, and 4. As it can be seen, none of the conditions lead to a good crystallization condition. Our *AED* method finds the common reagent between solutions that could lead crystallization conditions. In this scenario, there are only two promising conditions (with score 4): $[Zn(O_2CCH_3)_2, PEG\,8K, pH = 6.5]$ and $[(NH_4)_2SO_4, PEG\,MME\,5K, pH = 6.5]$. The *AED* draws a rectangle where these conditions (with score 4) are the corners of this rectangle (Figure 4). The other corners represent the candidate conditions. There are two possible conditions for this scenario. One of them $([(NH_4)_2SO_4, PEG\,8K, pH = 6.5])$ already appeared in the commercial screen with a low score. When we generate the experiment for the other condition $([Zn(O_2CCH_3)_2, PEG\,MME\,5K, pH = 6.5])$, we were able to get a score of 7 after optimizations. The experiments have not been conducted for others since they were not on the corners of conditions with promising scores.

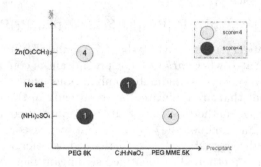

Fig. 3. Preliminary screen results

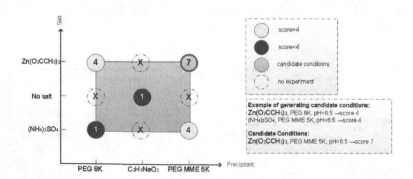

Fig. 4. Candidate (green node) conditions that *AED* generated based on preliminary data

We wanted to check that AED is able to generate novel conditions (leading to crystallization) that do not appear in any commercial screen. A question was where to draw the distinction between identical, similar, or different screen conditions in comparison to those present in the original or all commercially available screens. Using the C6 webtool [18], an exact match to an existing commercial screen condition gives a score of 0. Variations on that condition (change in one or both precipitant concentrations, or the buffer and/or pH), give scores > 0, ranging to 1 for completely different conditions. The C6 web tool gives the top 10 matches to the input conditions. Our usual first pass optimization approach to a lead condition, having precipitants A and B, is to use four solutions; one at 100% A and B, one at 50% A and 100% B, one at 100% A and 50% B, and one at 50% each A and B. The buffer is unchanged for all four conditions. Using the C6 webtool the greatest difference between the starting and optimization conditions is for the 50% A and B, with a score of 0.269, using a reference condition of 0.5M ammonium sulfate, 30% PEG 4K, 0.1M Tris-HCl pH 8.5. This is rounded to 0.3 for our threshold score for a different screen condition. Scores > 0 but ≤ 0.3 are taken to be similar to an existing screen condition, with a score of 0 indicating identity.

4 Experiments

4.1 Dataset

The Associated Experimental Design (AED) approach was evaluated using proteins derived from the hyperthermophile Thermococcus thioreducens [1]. Six crystallization screening plates, three using TFL'd and three unlabeled protein, all with the Hampton Research High Throughput screen (HR2-130) had been set up for each of these proteins as part of a separate experimental program. For this preliminary test the scores of the results from the second (of the three) plates for the TFL'd protein were used, as this also includes scores of potential cryptic leads indicated by TFL. One was a difficult crystallizer (Tt106, annotated as a nucleotide kinase) with no conditions giving needles, plates or 3D crystals; one a moderate crystallizer (Tt82, annotated as a HAD superfamily hydrolase), with one condition giving 2D plates but none giving needles or 3D crystals; one an easy crystallizer (Tt189, annotated as a nucleotide diphosphate kinase) having five conditions that gave 3D crystals).

4.2 Results and Discussion

The crystallization screen components that were determined to have the greatest positive effect were determined by the AED software, and a 96 condition optimization screen generated using those components for each protein. Optimization was in 96 well sitting drop plates, with the protein being TFL'd to facilitate results analysis. The successful conditions were identified and scored. Those conditions giving 2D and 3D crystals were then used to search the C6 database [18] for similar conditions across all commercially available screens as a determination of their uniqueness. As the optimization screens had different concentration ratios for the same precipitant pairs, each ratio where a hit was obtained was searched and the lowest C6 score was used.

The moderate and difficult proteins, $Tt82$ and $Tt106$ respectively, were subjected to a second round of optimization based on the results from the first. In the case of $Tt82$ the second round was a grid screen around a condition that gave an aggregated mass of plates. Many of the second round optimization wells also showed clusters of plates. However, in one case a single plate was observed. Although not pursued, the plate clusters could be excellent starting material for seeded crystallizations, both with the original and first stage optimization screening conditions. The second optimization round for protein $Tt106$ used ionic liquids as an additive [19], with the lead conditions selected from those outcomes giving "bright spots" in the first round. Within one week one family of conditions had 3D crystals, Figure 5. Novelty of the second round conditions was determined from the grid screen condition for $Tt82$, while it was based on the parent condition for $Tt106$. Additional lead conditions were apparent in the optimization screens for $Tt82$ and $Tt106$.

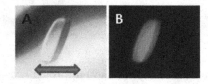

Fig. 5. White light (A) and fluorescent images (B) of second round optimization crystals of $Tt106$. Crystallization conditions: 0.2M Na/K Tartrate, 0.75M Ammonium Sulfate, 0.1 M NaCitrate, pH 5.6, 0.1M 1-hexyl, 3-methyl imidazolium chloride. Scale bar is $300\mu m$. All images are to the same scale.

Table 2. Summary of Experiments

Protein Annotated Function	$HSHT$ Screen [b] Score = 7	Optimize Screen Score = 8, 9	Novel Cond. vs HSHT Screen*	Novel Cond. vs All Screens*
$Tt189$ (Nucleoside diphosphate kinase)	0 / 2	5 / 3	5	4
$Tt82$ (HAD superfamily hydrolase)	1 / 1	0 / 1	2	2
$Tt106$ (Nucleoside kinase)	0 / 0	0 / 1	1	1
* Using C6 tool for scores of 7, 8, & 9 threshold value of 0.3				

[b] $HSHT$:Hampton Screen High-Throughput.

The results are summarized in Table 2. The numerical values in the first two columns after the protein name refer to the number of conditions with that score in the original screening experiment (numerator) vs. those with that score in the optimization screen (denominator). For example, (0/2) indicates two novel crystallization conditions with the score of 7 (for Tt189), which did not have score 7 in the original screening experiment. The third column lists the number of optimization conditions that are novel compared to the original screen, while the last column lists those that are novel compared to all available screens. All found conditions were judged to be novel compared to the original screen on the basis of our cutoff score criteria. For $Tt189$, one optimization condition was identical to an existing commercial screen condition.

5 Conclusion and Future Work

According to Table 2, AED generated 7 novel conditions compared to commercially available conditions for 3 different proteins derived from the hyperthermophile Thermococcus thioreducens [1]. The results obtained indicate that scored results from commercially available screens can be analyzed, and that components that may contribute to the crystallization of the macromolecule can be derived. Not surprisingly, a number of novel conditions were found for the facile crystallizer ($Tt189$). However, conditions were also found for both the moderate and difficult crystallizers, one of which had not shown any results of needles or better in the original screens ($Tt106$). For all three proteins crystallization conditions were obtained that were novel combinations of the identified factors.

These results show that AED is an efficient tool to generate novel conditions based on existing experimental results, which helps to save time and resources, as well as facilitating more rapid progress. In the future, we plan to include the conditions that have scores from 1 to 3 into AED analysis. Thus, we may generate novel conditions that may yield a successful outcome. We are also going to work on the correlation between original pair of conditions and candidate conditions by analyzing their scores. By using the scores of the original pairs, we plan to rank the candidate conditions to determine the conditions for a 96-well plate.

Acknowledgments. This research was supported by National Institutes of Health (GM090453) grant.

References

1. Pikuta, E.V., Marsic, D., Itoh, T., Bej, A.K., Tang, J., Whitman, W.B., Ng, J.D., Garriott, O.K., Hoover, R.B.: Thermococcus thioreducens sp. nov., a novel hyperthermophilic, obligately sulfur-reducing archaeon from a deep-sea hydrothermal vent. International Journal of Systematic and Evolutionary Microbiology 57(7), 1612–1618 (2007)

2. McPherson, A., Gavira, J.A.: Introduction to protein crystallization. Acta Crystallographica Section F: Structural Biology Communications 70(1), 2–20 (2014)
3. Jancarik, J., Kim, S.-H.: Sparse matrix sampling: a screening method for crystallization of proteins. Journal of Applied Crystallography 24(4), 409–411 (1991)
4. McPherson, A.: Crystallization of Biological Macromolecules. Cold Spring Harbor Laboratory Press (1999), http://books.google.com/books?id=EDNRAAAAMAAJ
5. Asherie, N.: Protein crystallization and phase diagrams. Methods 34(3), 266–272 (2004)
6. Stevens, R.C.: High-throughput protein crystallization. Current Opinion in Structural Biology 10(5), 558–563 (2000)
7. Brodersen, D.E., Andersen, G.R., Andersen, C.B.F.: Mimer: an automated spreadsheet-based crystallization screening system. Acta Crystallographica Section F 69(7), 815–820 (2013), http://dx.doi.org/10.1107/S1744309113014425, doi:10.1107/S1744309113014425
8. Carter Jr., C.W., Carter, C.W.: Protein crystallization using incomplete factorial experiments. J. Biol. Chem. 254(23), 12219–12223 (1979)
9. Abergel, C., Moulard, M., Moreau, H., Loret, E., Cambillau, C., Fontecilla-Camps, J.C.: Systematic use of the incomplete factorial approach in the design of protein crystallization experiments. Journal of Biological Chemistry 266(30), 20131–20138 (1991)
10. Doudna, J.A., Grosshans, C., Gooding, A., Kundrot, C.E.: Crystallization of ribozymes and small rna motifs by a sparse matrix approach. Proceedings of the National Academy of Sciences 90(16), 7829–7833 (1993)
11. Cox, M.J., Weber, P.C.: An investigation of protein crystallization parameters using successive automated grid searches (sags). Journal of Crystal Growth 90(1), 318–324 (1988)
12. Luft, J.R., Newman, J., Snell, E.H.: Crystallization screening: the influence of history on current practice. Structural Biology and Crystallization Communications 70(7), 835–853 (2014)
13. Snell, E.H., Nagel, R.M., Wojtaszcyk, A., O'Neill, H., Wolfley, J.L., Luft, J.R.: The application and use of chemical space mapping to interpret crystallization screening results. Acta Crystallographica Section D: Biological Crystallography 64(12), 1240–1249 (2008)
14. McPherson, A., Cudney, B.: Optimization of crystallization conditions for biological macromolecules. Acta Crystallographica Section F 70(11), 1445–1467 (2014), http://dx.doi.org/10.1107/S2053230X14019670, doi:10.1107/S2053230X14019670
15. Yang, H., Rasmuson, Å.C.: Phase equilibrium and mechanisms of crystallization in liquid–liquid phase separating system. Fluid Phase Equilibria 385, 120–128 (2015)
16. Baumgartner, K., Galm, L., Nötzold, J., Sigloch, H., Morgenstern, J., Schleining, K., Suhm, S., Oelmeier, S.A., Hubbuch, J.: Determination of protein phase diagrams by microbatch experiments: Exploring the influence of precipitants and ph. International Journal of Pharmaceutics 479(1), 28–40 (2015)
17. Forsythe, E., Achari, A., Pusey, M.L.: Trace fluorescent labeling for high-throughput crystallography. Acta Crystallographica Section D: Biological Crystallography 62(3), 339–346 (2006)

18. Newman, J., Fazio, V.J., Lawson, B., Peat, T.S.: The c6 web tool: a resource for the rational selection of crystallization conditions. Crystal Growth & Design 10(6), 2785–2792 (2010)

19. Pusey, M.L., Paley, M.S., Turner, M.B., Rogers, R.D.: Protein crystallization using room temperature ionic liquids. Crystal Growth & Design 7(4), 787–793 (2007)

DAM: A Bayesian Method for Detecting Genome-wide Associations on Multiple Diseases

Xuan Guo[1], Jing Zhang[2](✉), Zhipeng Cai[1], Ding-Zhu Du[3], and Yi Pan[1,4](✉)

[1] Department of Computer Science, Georgia State University, Atlanta, GA, USA
[2] Department of Mathematics and Statistics, Georgia State University, Atlanta, GA, USA
jing.maria.zhang@gmail.com, yipan@gsu.edu
Department of Computer Science, The Univ. of Texas at Dallas, Richardson, TX, USA
[3] Department of Biology, Georgia State University, Atlanta, GA, USA

Abstract. Taking the advantage of high-throughput single nucleotide polymorphism (SNP) genotyping technology, large genome-wide association studies (GWASs) have been considered to hold promise for unraveling complex relationships between genotypes and phenotypes. Current multi-locus-based methods are insufficient to detect interactions with diverse genetic effects on multifarious diseases. In addition, statistic tests for high order epistasis (≥ 2 SNPs) raise huge computational and analytical challenges because the computation increases exponentially as the growth of the cardinality of SNPs combinations. In this paper, we provide a simple, fast and powerful method, DAM, using Bayesian inference to detect genome-wide multi-locus epistatic interactions on multiple diseases. Experimental results on simulated data demonstrate that our method is powerful and efficient. We also apply DAM on two GWAS datasets from WTCCC, *i.e.* Rheumatoid Arthritis and Type 1 Diabetes, and identify some novel findings. Therefore, we believe that our method is suitable and effective for the full-scale analysis of multi-disease-related interactions in GWASs.

Keywords: Bayesian inference · Genome-wide association studies · Genetic factors · Epistasis

1 Introduction

Genome-wide association study (GWAS) has been proved to be a powerful genomic and statistical inference tool. The goal is to identify genetic susceptibility through statistical tests on associations between a trait of interests and the genetic information of unrelated individuals [1]. In genetics, genotype-phenotype association studies have established that single nucleotide polymorphisms (SNPs) [2], one type of genetic variants, are associated with a variety of diseases [3]. The current primary analysis paradigm for GWAS is dominated by the analysis on susceptibility of individual SNPs to one disease a time, which might only explain a small part of genetic causal effects and relations for multiple complex diseases [4]. The word, epistasis, has been defined generally as

© Springer International Publishing Switzerland 2015
R. Harrison et al. (Eds.): ISBRA 2015, LNBI 9096, pp. 96–107, 2015.
DOI: 10.1007/978-3-319-19048-8_9

the interaction among different genes [5]. Many studies [6] have demonstrated that the epistasis is an important contributor to genetic variation in complex diseases, such as asthma [7][8], breast cancer [9], diabetes[10], coronary heart disease [11], and obesity [12]. In this article, we consider epistatic interactions as the statistically significant associations of d-SNP modules ($d \geq 2$) with multiple phenotypes [13].

Recently, the problem of detecting high-order genome-wide epistatic interaction for case-control data has attracted extensive research interests. Generally, there are two challenges in mapping genome-wide associations for multiple diseases on large GWAS dataset [14]: the first is arose from the heavy computational burden, $i.e.$ the number of association patterns increases exponentially as the order of interaction goes up. For example, around 6.25×10^{11} statistical tests are required to detect pairwise interactions for a dataset with 500,000 SNPs. The second challenge is that existing approaches lack statistical powers for searching high-order multi-locus models of disease. Because of the huge number of hypotheses and the limited sample size, a large proportion of significant associations are expected to be false positives. Many computational algorithms have been proposed to overcome the above difficulties. More details about these tools can be found in a recent survey [15]. To the best of our knowledge, current epistasis detecting tools are only capable of identifying interactions on GWAS data with two groups, $i.e.$ case-control studies. Thus, they are incompetent to discover genetic factors with diverse effects on multiple diseases. Moreover, they lose the benefit of alleviating deficiency of statistical powers by pooling different disease samples together.

In this paper, we design and implement a Bayesian inference method for Detecting genome-wide Association on Multiple diseases, named DAM, to address above challenges. DAM employs Markov Chain Monte Carlo (MCMC) sampling based on the Bayesian variable partition model, and makes use of stepwise condition evaluation to identify significant disease(s)-specific interactions. It first generates a candidate set of SNPs based on our Bayesian variable partition model by applying Metropolis-Hastings (MH) algorithm. A stepwise evaluation of association is engaged to further detect the genetic effect types for each interaction. Systematic experiments on both simulated and real GWAS datasets demonstrate that our method is feasible for identify multi-locus interaction on GWAS datasets and enriches some novel, significant high-order epistatic interactions with specialties on various diseases.

2   Method

2.1   Notations

Suppose a GWAS dataset D has M diallelic SNPs and N samples. In general, bi-allelic genetic markers use uppercase letters (e.g. A, B,...) to denote major alleles and lowercase letters (e.g. a, b) to denote minor alleles. For encoding three genotypes, one popular way is to use $\{1, 2, 3\}$ to represent $\{aa, Aa, AA\}$, respectively. For a GWAS dataset with L groups, it includes one shared control group and $L-1$ case groups. We use $N^{(L)}$ denotes the number of controls (*i.e.* normal individuals) and $N^{(i)}$ denotes the number of cases (*i.e.* disease individuals) in i-th groups ($i = 1 \ldots L - 1$). X is utilized to indicate the ordered set of SNPs, and x_i represents i-th SNP in X.

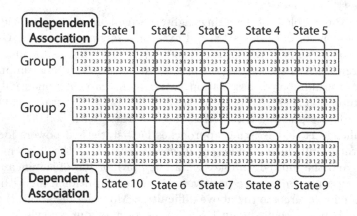

Fig. 1. Illustration for 10 states on 3 groups

For a set of L groups, there are B_L partitions, and here we also refer partition to state. Let S denote the set of states, and s_k is the k-th state with $|s_k|$ non-empty sets of groups. In general, the M markers are assigned into $2B_L$ states, and all states belong to two categories: $s_{k_1} \in \{s_1, \ldots, s_{B_L}\}$ indicates SNP markers contributing independently to the phenotypes, and $s_{k_2} \in \{s_{B_L+1}, \ldots, s_{2B_L}\}$ indicates SNP markers that jointly influence the phenotypes. An example for a three groups dataset with 10 possible states is showed in Figure 1, where states 1 to 5 indicate that SNPs are independently associated with certain phenotypes, and states 6 to 10 indicate that SNPs are dependently associated with the phenotypes. In our experiments, group 1 and 2 are cases and group 3 is control. Since we want to identify SNPs associated with phenotypes, SNPs in states 2 to 5 and states 6 to 10 are the desired ones with disease associations. Let $I = (I_1, \ldots, I_M)$ record the memberships of SNP with $I_m \in \{1, \ldots, 2B_L\}$, $\mathbb{M}_k$ denote the number of SNP markers in k-th state ($\sum_{k=1}^{2B_L} \mathbb{M}_k = M$), and $D^{(k)}$ denote genotypes of SNPs in k-th state.

2.2   Bayesian Variable Partition Model

Consider a categorical variable X, which can be sampled at t different states $\{\Delta_1, \Delta_2, \ldots, \Delta_t\}$ with t different distribution $\{\Theta_1, \Theta_2, \ldots, \Theta_t\}$, where Θ_k is the distribution of X at k-th state. The model describing the sums of independently and identically distributed mixture categorical variables at different states is referred as a 'multinomial model', meaning that it can be partitioned into t inseparable multinomial models. Consider a model for a vector of M categorical variables $X = \{x_1, x_2, \ldots, x_M\}$. If all variables are independent, the model can be simply treated as the union of M univariate multinomial models. If interactions exist among multiple variables, a new model with a single variable by collapsing the interacting variables can replace the model for these multiple variables. The sample space of the collapsed variable is the product of the sample spaces of the variables before collapsing. Bayesian variable partition model (BVP) is a multinomial model based on Bayesian theorem. The likelihood for the multinomial model by given i-th state is

$$P(D_m|\Delta_k) = \int P(D_m|\Delta_k, \Theta_k)d\Theta_k$$

$$= \int_{\theta_1, \theta_2, \ldots, \theta_g} P(D_m|\theta_1, \theta_2, \ldots, \theta_g)P(\theta_1, \theta_2, \ldots, \theta_g)dp \qquad (1)$$

where D_m is the observation for the categorical variable x_m, and g is the number of category value for the variable x_m. We set $P(\Theta = (\theta_1, \theta_2, \ldots, \theta_g))$ to be Dirichlet distribution $Dir(\alpha_1, \alpha_2, \ldots, \alpha_g)$; then we can have a closed form for Equation 1:

$$P(D_m|\Delta_k) = \int_{\theta_1, \theta_2, \ldots, \theta_g} P(D_m|\theta_1, \theta_2, \ldots, \theta_g)P(\theta_1, \theta_2, \ldots, \theta_g)dp$$

$$= \int_{\theta_1, \theta_2, \ldots, \theta_g} \frac{1}{B(\alpha_1, \alpha_2, \ldots, \alpha_g)} \prod_{i=1}^{g} p_i^{n_i + a_i - 1}dp$$

$$= \left(\prod_{i=1}^{g} \frac{\Gamma(n_i + \alpha_i)}{\Gamma(\alpha_i)} \right) \frac{\Gamma(|\alpha|)}{\Gamma(\mathbb{N} + |\alpha|)} \qquad (2)$$

where $\mathbb{N}$ is the total number of observations, and $|\alpha|$ is the sum of $(\alpha_1, \alpha_2, \ldots, \alpha_g)$. Suppose the vector I is a vector of membership of state for categorical variable vector X, we obtain the posterior distribution of I as

$$P(I|D) \propto \left(\prod_{m=1}^{M} P(D_m|I) \right) P(I) \qquad (3)$$

Based on Bayesian theorem, we describe the specific Bayesian variable partition model for genome-wide association mapping as follows. For the SNPs independently associated with phenotypes, we use $\Theta_{k_1} = ((\theta_{m1}^{(\omega)}, \theta_{m2}^{(\omega)}, \theta_{m3}^{(\omega)}) : \omega \in \{1, 2, \ldots, |s_k|\}, I_{x_m} \in \{1, \ldots, B_L\})$ to denote the genotype frequencies of SNP x_m in k_1 states. Note that SNP with membership value in $\{1, \ldots, B_L\}$ does not have interaction with other SNPs. The likelihood of D^{k_1} from BVP model is that

$$P(D^{(k_1)}|\Theta_{k_1}) = \prod_{I_{x_m} = k_1}^{|s_{k_1}|} \prod_{\omega=1}^{3} \prod_{i=1}^{3} (\theta_{mi}^{(\omega)})^{n_{mi}^{(\omega)}}, \qquad (4)$$

where $\left\{n_{m1}^{(\omega)}, n_{m2}^{(\omega)}, n_{m3}^{(\omega)}\right\}$ are genotype counts of SNP x_m in ω-th subset in k_1-th state. Similar to the above assumption, we set Θ_{k_1} to be a Dirichlet distribution $Dir(\alpha)$ with parameter $\alpha = (\alpha_1, \alpha_2, \alpha_3)$, we integrate out Θ_{k_1} and obtain the marginal probability:

$$P(D^{(k_1)}|I) = \prod_{I_{x_m}=k_1}^{|s_{k_1}|} \prod_{\omega=1} \left(\left(\prod_{i=1}^{3} \frac{\Gamma(n_{mi}^{(\omega)} + \alpha_i^{(\omega)})}{\Gamma(\alpha_i^{(\omega)})} \right) \frac{\Gamma(|\alpha^{(\omega)}|)}{\Gamma(\mathbb{N}_{k_1,\omega} + |\alpha^{(\omega)}|)} \right) \quad (5)$$

where $\mathbb{N}_{k_1,\omega}$ is the count of individuals in groups belonging to ω-th subset of k_1-th state, and $|\alpha|$ represents the sum of all elements in α.

SNP markers in state $\{s_{B_L}, s_{B_L+1}, \ldots, s_{2B_L}\}$ influence the disease statues through interactions. Thus, we concatenate $\mathbb{M}_{k_2}$ SNPs into a single categorical variable to resolve the interactions ($B_L+1 \le k_2 \le 2B_L$). Note that there are $3^{\mathbb{M}_{k_2}}$ possible concatenated genotype combinations. Let $\Theta_{k_2} = ((\phi_1^{(\omega)}, \phi_2^{(\omega)}, \ldots, \phi_{3^{\mathbb{M}_{k_2}}}^{(\omega)}) :$ $\omega = \{1, 2, \ldots, |s_{k_2}|\})$ be the concatenated genotype frequencies over $\mathbb{M}_{k_2}$ SNPs in $s_{k_2} \in \{s_{B_L+1}, \ldots, s_{2B_L}\}$. Similarly, we use a Dirichlet prior $Dir(\beta)$ for Θ_{k_2}, $\beta = (\beta_1, \beta_2, \ldots, \beta_{3^{\mathbb{M}_{k_2}}})$. According to Equation 2, we obtain the marginal probability:

$$P(D^{(k_2)}|I) = \prod_{\omega=1}^{|s_{k_2}|} \left(\left(\prod_{i=1}^{3^{\mathbb{M}_{k_2}}} \frac{\Gamma(n_i^{(\omega)}) + \beta_i^{(\omega)}}{\Gamma(\beta_i^{(\omega)})} \right) \frac{\Gamma(|\beta^{(\omega)}|)}{\Gamma(\mathbb{N}_{k_2,\omega} + |\beta^{(\omega)}|)} \right) \quad (6)$$

where $\mathbb{N}_{k_2,\omega}$ is the count of individuals belonging to ω-th subset k_2-th state and $n_i^{(\omega)}$ is the count of i-th concatenated genotype combinations in ω-th subset in k_2-th state.

Combining Equation 3, 5 and 6 , we obtain the posterior distribution of I as

$$P(I|D) \propto \left(\prod_{k_1=1}^{B_L} P(D^{(k_1)}|I) \right) \left(\prod_{k_2=(B_L+1)}^{2B_L} P(D^{(k_2)}|I) \right) P(I) \quad (7)$$

In BVP, we set $P(I) \propto \prod_{k=1}^{2B_L} p_k^{\mathbb{M}_k}$ to embed the prior knowledge of the proportions of SNP associating with certain phenotypes. In our experiments with three groups, we set $p_k = 0.001, k \in \{2, \ldots, 10\}$, and $\alpha_i = \beta_j = 0.5, \forall i, j$.

2.3   MCMC Sampling

We apply MCMC method to sample the indicator I from the distribution in Equation 7. According to the prior $P(I)$, DAM first initializes I, then use the Metropolis-Hastings (MH) algorithm [16] to construct a MCMC to update I. Three types of updating strategies are used: (i) randomly change a SNP's state, (ii) randomly exchange two SNPs' states between $(s_1, \ldots, s_{2B_L})$, or (iii) randomly shuffle the state labels between $\{s_{B_L+1}, \ldots, s_{2B_L}\}$. At each iteration, the acceptance of new indicator based on the MH ratio, a Gamma functions. DAM records the entire accepted indicator after the burn-in process, and represent it as the posterior distribution of single disease-related SNPs and interactions associated with multiple diseases. The number of iteration in burn-in process is fixed to $10M$ and the number of sampling iteration is set to M^2 in our experiments. We also apply a distance constraint that the physical distance between two SNPs in multi-locus module is at least 1Mb. This constraint is used to avoid associations that might be attributed to the LD effects [5].

2.4   Evaluation of Interaction

With the candidate SNPs generated by MCMC sampling, we apply the χ^2 statistic and its conditional test to measure the significance for a dependent SNP association. Let $\mathbb{A} = (x_1, x_2, \ldots, x_d : k)$ denote an SNP module $\mathbb{A}$ with d SNPs in k-th state. We denote its χ^2 statistic as $\chi^2(x_1, x_2, \ldots, x_d : k)$ and the conditional χ^2 statistic as $\chi^2(x_1, x_2, \ldots, x_d | x_{c_1}, x_{c_2}, \ldots, x_{c_{d'}} : k)$ by given a module $\mathbb{A}$ and a subset of it, $(x_{c_1}, x_{c_2}, \ldots, x_{c_{d'}})$ with d' SNPs. The χ^2 statistic can be calculated as

$$\chi^2(x_1, x_2, \ldots, x_d : k) = \sum_{i=1}^{|s_k|} \sum_{j=1}^{3^d} \frac{(n_{g_i,j} - e_{g_i,j})^2}{e_{g_i,j}} \tag{8}$$

where g_i is the i-th genotype combination for d SNPs, $n_{g_i,j}$ is the number of individuals having i-th genotype combination in j-th subset in k-th state, and $e_{g_i,j}$ is the corresponding expected value. The degrees of freedom for Equation 8 is $(|s_k| - 1) \cdot (3^d - 1)$. The conditional independent test via χ^2 statistic is defined as follows

$$\chi^2(x_1, x_2, \ldots, x_d | x_{c_1}, x_{c_2}, \ldots, x_{c_{d'}} : k) = \sum_{\iota=1}^{3^{d'}} \sum_{i=1}^{|s_k|} \sum_{j=1}^{3^{d-d'}} \frac{(n_{g_i,j}^{(\iota)} - e_{g_i,j}^{(\iota)})^2}{e_{g_i,j}^{(\iota)}} \tag{9}$$

where we calculate χ^2 statistic separately for each genotype combination from $\mathbb{A} - \mathbb{A}'$. The degrees of freedom for Equation 9 is $3^{d'} \cdot (|s_k| - 1) \cdot (3^{d-d'} - 1)$. In order to avoid redundant SNPs in a SNP module indicating that conditional independence model fits better, we define an epistatic interaction $(d \leq 2)$ as a compact significant SNP module with definition 1.

Definition 1 *A SNPs module $\mathbb{A} = (x_1, x_2, \ldots, x_d : k)$ is considered as a compact significant interaction by given the significant level α_d, if it meets the following three conditions:*
(1) the p-value of $\chi^2(x_1, x_2, \ldots, x_d : k) \leq \alpha_d$;
(2) the p-value of $\chi^2(x_1, x_2, \ldots, x_d : k) < \forall$ p-value of $\chi^2(x_1, x_2, \ldots, x_d : k')$, $k \neq k'$ and $k' \in \{1, 2, \ldots, |S|\}$;
(3) the p-value of $\chi^2(x_1, x_2, \ldots, x_d | x_{c_1}, x_{c_2}, \ldots, x_{c_{d'}} : k) \leq \alpha_d$ for $\forall \mathbb{A}' = (x_{c_1}, x_{c_2}, \ldots, x_{c_{d'}} : k)$ whose p-value $\leq \alpha_{d'}$.

Based on definition 1, we develop a stepwise algorithm to search for top-f significant d-locus compact significant interactions, where the searching space only includes the SNP markers generated by MCMC sampling. We assume that one SNP can only participate in one significant interaction in one state. So for the SNP markers with state in $\{s_1, \ldots, s_{B_L}\}$, we first searches all the modules with just one SNP based on definition 1, then the algorithm recursively tests all the possible combinations by setting the module size with one more SNP. For the SNPs reported as jointly contributing to the disease risk, we calculate the p-value under different states and use the conditional test if part of SNPs already reported as significant. All SNPs with significant marginal associations after a Bonferroni correction are reported in a list $\mathbb{L}$. The algorithm recursively searches the interaction space with larger module size until d reaches user preset value. We add all novel d-way interactions (*i.e.*, no SNPs has been reported earlier) that are significant after the Bonferroni correction for $2B_L \cdot \binom{M}{d}$ tests. For the interactions whose subsets have been reported as compact significant, we use the conditional independent test, and put the interaction in $\mathbb{L}$ if it is still significant after Bonferroni correction of $2B_L \cdot \binom{M}{d} \cdot \binom{d}{d'}$ tests.

3   Results

To the best of our knowledge, DAM is the first method to detect associations on multiple diseases, so we first give definitions of 8 simulated multi-disease models and the power metric measurement, and then evaluate the effectiveness of our method. The false positive rate of DAM is showed in Supplementary Material. We also apply DAM on two real GWAS datasets, Rheumatoid Arthritis (RA) and Type 1 Diabetes (T1D), and we find not only the results reported by other literatures but also some novel interesting interactions. DAM (in Java) is conducted on a 64-bit Windows 8 platform with 1.8 GHz Intel CPU and 4 GB RAM.

3.1   Experimental Design

Data simulation To evaluate the effectiveness of DAM, we perform extensive simulation experiments using eight disease models with one- and two-locus associations on three groups. The genotypes of unassociated SNP are generated by the same procedure used in previous studies [14] with Minor Allele Frequencies (MAFs) sampled from $[0.05, 0.5]$. The odds tables for eight models are showed in Table 1 in Supplementary Material. Model 5, 6, 7, and 8 are the extensions of Model 1, 2, 3, and 4, respectively. The settings for four datasets are showed in Table 2 in Supplementary Material. In a setting, all models are using the same MAF $\in \{0.1, 0.2, 0.4\}$, we generate 100 replicas per setting. Therefore, by given a MAF, a dataset contains at most 8 associations labeled as Ep 1 to 8. Note that in model 5 there are 7 associations, because the combination of three 2-locus models does not exist when MAF $= 0.1$. Each simulated replica contained $M = 1000$ SNPs. The sizes of three groups are set to $(1000, 1000, 2000)$ or $(2000, 2000, 4000)$, where the first two groups are considered as case groups and the third one is control group. More details about model simulation can be found in Supplementary Material.

Statistical power In the evaluation of performances on simulated data, 100 datasets are generated for each setting. The measure of discrimination power is defined as the fraction of 100 datasets on which the ground-truth associations are identified as compact and significant by DAM.

3.2 Single-Locus Disease Models

Test results are illustrated in Figure 2 in Supplementary Material for SNPs contributing independently to the disease risks. We can find that DAM is able to report nearly 100% of embedded single SNP associations under most settings. Carefully examining the results, we found that some SNPs are incorrectly assigned to a state by MCMC sampling, although they do have significant association with the phenotypes. After the stepwise evaluation, most mistakenly labeled SNPs are corrected.

3.3 Two-Locus Disease Models

Test results for SNPs contributing jointly to the disease risks are illustrated in Figure 2. We can find that DAM is able to report nearly 100% of embedded interactions for dataset 1 and 2. It also obtained nearly full power when MAF is 0.1 for dataset 1, 2, and 4. Similar to the results on single-locus disease models, after stepwise procedure, more interactions were assigned to correct states.

3.4 Experiments on WTCCC Data

We have applied DAM to analyze data from the WTCCC (3999 cases in total and 3004 shared controls) on two common human diseases: Rheumatoid Arthritis (RA), Type 1 Diabetes (T1D), where RA is treated as group 1, T1D is treated as group 3, and control group is group 3. The procedure of quality control is the same as presented in the [14]. After SNP filtration the dataset contains 333,739 high quality SNPs. DAM ran about 36 hours, for a total of 1×10^{11} iterations. Because the importance of the MHC region in chromosome 6 with respect to infection, inflammation, autoimmunity, and transplant medicine has been heavily reported [17] [18] [19], we concentrate on the results by DAM on Chromosome 6. The posterior probabilities for SNP on Chromosome 6 are showed in Figure 3 in Supplementary Material and Figure 3.

Recent studies [18] [20] has shown that both T1D and RA strongly associated with the MHC region via single-locus association mapping, which is also verified by our results that a large portion of SNPs' posterior probabilities greater than 0.5 spreading in the region $28, 477, 797 - 33, 448, 354$. Comparing results from state 6 to state 7, we can find that many SNPs contributing to RA are not located inside the MHC region, while the SNPs associated with T1D gather in MHC region. We select top 50 SNPs according to their posterior probabilities and

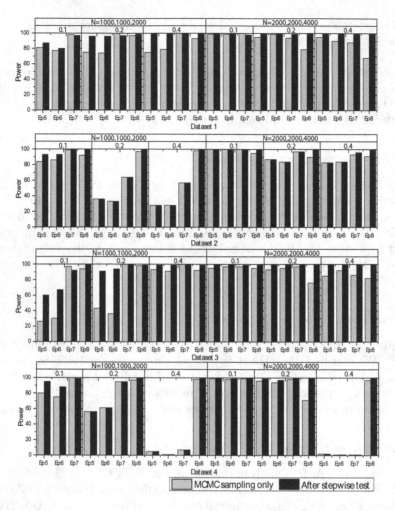

Fig. 2. Performance comparison between DAM with MCMC sampling only and with stepwise test on simulated disease datasets 1-4 embedded with Joint effect SNPs. Note that the combination of model 5 with other three 2-locus models does not exist when MAF = 0.1

analyze them with the stepwise evaluation procedure introduced in Section 2.4. Table 1 summarizes some novel findings of the significant interactions with p-values adjusted by 1.61×10^{18} for three loci and 1.93×10^{23} for four loci interactions, respectively. Take the four-locus interaction (rs1977, rs707974, rs10755544, rs2322635) for example. rs1977 is located inside gene BTN3A2, which encodes a member of the immunoglobulin superfamily that may be involved in the adaptive immune response. rs707974 is in gene GPANK1, encoding a protein which plays a role in immunity. rs10755544 is at the upstream of gene KHDRBS2, which is thought to involve SH2 domain binding and protein heterodimerization activity. rs2322635 is located in gene BCKDHB for encoding branched-chain keto acid

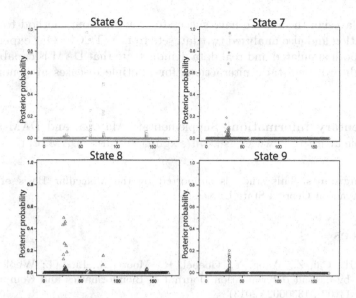

Fig. 3. Posterior probabilities of SNPs on chromosome 6. States 6 to 9 indicate joint association probabilities per SNP. X-axis indicates the chromosomal position (Mb), y-axis shows the posterior probability.

dehydrogenase, which is a multienzyme complex associated with the inner membrane of mitochondria. BTN3A2 has been shown to associate with T1D in [21]. And mutations in the BCKD gene, BCKDHA, is also known to result in maple syrup urine disease, which is related to T1D [22].

Table 1. Significant interactions obtained from theWTCCC data. Following each SNP is its location.

State Index	DAM p-value	SNP 1	SNP 2	SNP 3	SNP 4
6	1.35E-26	rs4634439	rs707974	rs4236164	rs2322635
6	1.61E-26	rs6931858	rs707974	rs10755544	rs3805878
7	3.31E-26	rs1977	rs707974	rs10755544	rs2322635
7	1.86E-35	rs3117425	rs1150753	rs239494	
7	5.79E-24	rs200481	rs1150753	rs12194665	

4 Conclusions

The large number of SNPs genotyped in genome-wide case-control studies poses a great computational challenge in the identification of gene-gene interactions. During the last few years, many computational and statistical tools are developed to finding gene-gene interactions for data with only two groups, *i.e.* case and control groups. In this paper, we present a method, named "DAM", to address the computation and statistical power issues for multiple diseases GWASs based on Bayesian theory. We have successfully applied our method to systematic simulation and also analyzed two datasets from WTCCC. Our experimental results on both simulated and real data demonstrate that DAM is capable of detecting high order epistatic interactions for multiple diseases at genome-wide scale.

Supplementary Information. Supplementary Material and DAM software are available at http://www.cs.gsu.edu/~xguo9/research/DAM.html

Acknowledgments. This study is supported by the Molecular Basis of Disease (MBD) program at Georgia State University.

References

1. Sabaa, H., Cai, Z., Wang, Y., Goebel, R., Moore, S., Lin, G.: Whole genome identity-by-descent determination. Journal of Bioinformatics and Computational Biology 11(02), 1350002 (2013)
2. He, Y., Zhang, Z., Peng, X., Wu, F., Wang, J.: De novo assembly methods for next generation sequencing data. Tsinghua Science and Technology 18(5), 500–514 (2013)
3. Peter, K., Hunter, D.J.: Genetic risk prediction: Are we there yet? The New England Journal of Medicine 360(17), 1701–1703 (2009)
4. He, Q., Lin, D.Y.: A variable selection method for genome-wide association studies. Bioinformatics 27(1), 1–8 (2011)
5. Cordell, H.J.: Epistasis: what it means, what it doesn't mean, and statistical methods to detect it in humans. Human Molecular Genetics 11(20), 2463–2468 (2002)
6. Liu, W., Chen, L.: Community detection in disease-gene network based on principal component analysis. Tsinghua Science and Technology 18(5), 454–461 (2013)
7. Cai, Z., Sabaa, H., Wang, Y., Goebel, R., Wang, Z., Xu, J., Stothard, P., Lin, G.: Most parsimonious haplotype allele sharing determination. BMC Bioinformatics 10(1), 115 (2009)
8. Cheng, Y., Sabaa, H., Cai, Z., Goebel, R., Lin, G.: Efficient haplotype inference algorithms in one whole genome scan for pedigree data with non-genotyped founders. Acta Mathematicae Applicatae Sinica, English Series 25(3), 477–488 (2009)
9. Ritchie, M.D., Hahn, L.W., Roodi, N., Bailey, L.R., Dupont, W.D., Parl, F.F., Moore, J.H.: Multifactor-dimensionality reduction reveals high-order interactions among estrogen-metabolism genes in sporadic breast cancer. The American Journal of Human Genetics 69, 138–147 (2001)
10. Wang, Y., Cai, Z., Stothard, P., Moore, S., Goebel, R., Wang, L., Lin, G.: Fast accurate missing snp genotype local imputation. BMC Research Notes 5(1), 404 (2012)
11. Nelson, M., Kardia, S., Ferrell, R., Sing, C.: A combinatorial partitioning method to identify multilocus genotypic partitions that predict quantitative trait variation. Genome Research 11(3), 458–470 (2001)
12. Cordell, H.J.: Detecting gene-gene interactions that underlie human diseases. Nat. Rev. Genet. 10, 392–404 (2009)
13. Wang, Y., Liu, G., Feng, M., Wong, L.: An empirical comparison of several recent epistatic interaction detection methods. Bioinformatics 27(21), 2936–2943 (2011)

14. Guo, X., Meng, Y., Yu, N., Pan, Y.: Cloud computing for detecting high-order genome-wide epistatic interaction via dynamic clustering. BMC Bioinformatics 15(1), 102 (2014)

15. Guo, X., Yu, N., Gu, F., Ding, X., Wang, J., Pan, Y.: Genome-wide interaction-based association of human diseases-a survey. Tsinghua Science and Technology 19(6), 596–616 (2014)

16. Liu, J.S.: Monte Carlo strategies in scientific computing. Springer (2008)

17. Lechler, R., Warrens, A.N.: HLA in Health and Disease. Academic Press (2000)

18. Wan, X., Yang, C., Yang, Q., Xue, H., Fan, X., Tang, N.L., Yu, W.: Boost: A fast approach to detecting gene-gene interactions in genome-wide case-control studies. The American Journal of Human Genetics 87(3), 325–340 (2010)

19. Zhang, J., Wu, Z., Gao, C., Zhang, M.Q.: High-order interactions in rheumatoid arthritis detected by bayesian method using genome-wide association studies data. American Medical Journal 3(1), 56–66 (2012)

20. Zeggini, E., Weedon, M.N., Lindgren, C.M., Frayling, T.M., Elliott, K.S., Lango, H., Timpson, N.J., Perry, J., Rayner, N.W., Freathy, R.M., et al.: Wellcome trust case control consortium (wtccc), mccarthy mi, hattersley at: Replication of genome-wide association signals in uk samples reveals risk loci for type 2 diabetes. Science 316(5829), 1336–1341 (2007)

21. Viken, M.K., Blomhoff, A., Olsson, M., Akselsen, H., Pociot, F., Nerup, J., Kockum, I., Cambon-Thomsen, A., Thorsby, E., Undlien, D., et al.: Reproducible association with type 1 diabetes in the extended class i region of the major histocompatibility complex. Genes and Immunity 10(4), 323–333 (2009)

22. Henneke, M., Flaschker, N., Helbling, C., Müller, M., Schadewaldt, P., Gärtner, J., Wendel, U.: Identification of twelve novel mutations in patients with classic and variant forms of maple syrup urine disease. Information Theory 1993 22(5), 417–417 (2003)

MINED: An Efficient Mutual Information Based Epistasis Detection Method to Improve Quantitative Genetic Trait Prediction

Dan He[1](✉), Zhanyong Wang[2], and Laxmi Parida[1]

[1] IBM T.J. Watson Research, Yorktown Heights, NY, USA
{dhe,parida}@us.ibm.com
[2] Dept. of Comp. Sci., Univ. of California Los Angeles, Los Angeles, CA 90095, USA
zywang@cs.ucla.edu

Abstract. Whole genome prediction of complex phenotypic traits using high-density genotyping arrays has attracted a great deal of attention, as it is very relevant to plant and animal breeding. More effective breeding strategies can be developed based on a more accurate prediction. Most of the existing work considers an additive model on single markers, or genotypes only. In this work, we studied the problem of epistasis detection for genetic trait prediction, where different alleles, or genes, can interact with each other. We have developed a novel method MINED to detect significant pairwise epistasis effects that contribute most to prediction performance. A dynamic thresholding and a sampling strategy allow very efficient detection, and it is generally 20 to 30 times faster than an exhaustive search. In our experiments on real plant data sets, MINED is able to capture the pairwise epistasis effects that improve the prediction. We show it achieves better prediction accuracy than the state-of-the-art methods. To our knowledge, MINED is the first algorithm to detect epistasis in the genetic trait prediction problem. We further proposed a constrained version of MINED that converts the epistasis detection problem into a Weighted Maximum Independent Set problem. We show that Constrained-MINED is able to improve the prediction accuracy even more.

Keywords: Genetic trait prediction · Mutual information · Epistasis · Weighted maximum independent set

1 Introduction

Whole genome prediction of complex phenotypic traits using high-density geno-typing arrays is an important computational problem, as it is relevant in the fields of plant and animal breeding as well as genetic epidemiology [12,16,4]. Given a set of biallelic molecular markers, such as SNPs (Single-nucleotide polymorphisms) for variant sites in the genome of a collection of plant, animal or human samples, with genotype values encoded as $\{0, 1, 2\}$ for each variant site, the goal is to predict the quantitative trait values by simultaneously modeling all marker effects. The traits

© Springer International Publishing Switzerland 2015
R. Harrison et al. (Eds.): ISBRA 2015, LNBI 9096, pp. 108–124, 2015.
DOI: 10.1007/978-3-319-19048-8_10

are typically physical properties of the samples, such as height, weight, size, etc. We show an example of the problem in Figure 1. More accurate genetic trait prediction can help to develop more effective breeding strategies for both plants and animals and therefore can save cost and effort for the breeding companies.

Training Data ●——→ New Data

Plant Lines	Genotype	Trait	Plant Lines	Genotype	Trait
line 1	0020202021002···	13.1	line $n+1$	0002002012200···	12.1
line 2	0020002002001···	9.4	line $n+2$	2000202010200···	10.3
line 3	2002000102001···	10.3	line $n+3$	2012020020220···	9.9
line 4	0020200002202···	11.5	line $n+4$	2002002010000···	10.7
⋮	⋮	⋮	⋮	⋮	⋮
line n	0200202001202···	12.6	line $n+m$	0000201010202···	11.8

Fig. 1. An example of the quantitative genetic trait prediction. On the left side is the training data, where each row is a sample, each column of the genotype matrix is a feature, the trait is the target variable. We build a predictive model on the training data, and then use the model to predict the trait values of the new data, or test data, which is of the same format.

A widely used algorithm for the genetic trait prediction problem is *rrBLUP* (Ridge-Regression BLUP) [12,21]. The algorithm assumes all the markers contribute to the trait value more or less, and it builds an additive linear regression model by fitting the genotypes for all the markers on the trait being studied. It fits the coefficient computed for each marker, which can be considered as a measure of the importance of the marker. The rrBLUP method has the benefits of the underlying hypothesis of normal distribution of the trait value and the marker effects (well suited for highly polygenic traits). It is quick to compute, robust, and is one of the most used models in whole genome prediction. Its performance is as good as or better than other popular predictive models such as Elastic-Net, Lasso, Ridge Regression [18,3], Bayes A, Bayes B [12], Bayes Cπ [9], and Bayesian Lasso [10,13], as well as other machine learning methods.

Epistasis is the phenomenon where different alleles, or genes, can interact with each other. The problem of epistasis detection has been widely studied in GWAS (Genome Wide Association Studies). Exhaustive search of all possible epistasis interactions is infeasible even for a small number of markers. Greedy strategies [14,11,5,22,24,6] have been applied to detect epistasis effects where a subset of high-marginal effect markers, which are markers that contribute to the trait themselves, are first selected. Then the test is conducted either between all the markers in this subset or between the markers in this subset and the remaining markers. These strategies, however, miss all the possible epistasis between the low-marginal effect markers, which are shown to exist [8]. Xiang et al. [23] proposed an optimal algorithm to efficiently detect epistasis without conducting an extensive search. A data structure is created to effectively prune interactions that are potentially insignificant. In [1], a lasso for hierarchical interactions is proposed, which again, considers interactions where one or both involved variables are marginally important. Therefore, it does not meet our requirement of epistasis where both involved variables might be marginally not important.

These existing methods all target epistasis detection in GWAS. In this work, we study the problem of detecting significant pairwise epistasis effects for genetic trait prediction. As the genetic trait prediction problem is usually modeled as a linear regression problem, a traditional approach for interactions is a multiplicative model. The pairwise interaction between two genotypes can be modeled as the product of them, which is considered as a new feature in the linear regression model. As the number of possible epistasis effects is far more than the number of samples, naturally a feature selection method is used to select the significant effects.

Recently, He et al. [7] proposed a feature selection method for genetic trait prediction, based on the motivation that not necessarily all the marker effects contribute to the trait values. A feature selection followed by prediction algorithms with cross validation determines the set of marker effects that contribute most to the prediction. Indeed, unlike the traditional feature selection problem where generally a very small set of features are selected, here a relatively large set of features need to be considered, often much larger than the number of samples. Note that, for complex traits, it is known that many loci contribute to the traits. Cross validation is applied to determine the set of loci that contribute the most.

In this work, we proposed an efficient method MINED (**M**utual **IN**formation based **E**pistasis **D**etection) to select significant pairwise epistasis effects. To avoid an exhaustive search of all pairs of interactions, MINED applies a sampling strategy combined with a dynamic thresholding strategy to efficiently detect significant epistasis effects. Our experiments show that MINED is much more efficient than an exhaustive search without loss of accuracy and it is able to effectively capture epistasis effects that can improve prediction performance. We show it achieves better prediction accuracy than the state-of-the-art methods. To our knowledge, MINED is the first algorithm to detect epistasis for genetic trait prediction problem. We further proposed a constrained version of MINED, which converts the epistasis detection problem into a Weighted Maximum Independent Set problem. We show that constrained-MINED is able to improve the prediction accuracy even more.

2   Preliminaries

The genetic trait prediction problem is defined as follows. Given n training samples, each with $m \gg n$ genotype values (we use "feature", "marker", "genotype", "SNP" interchangeably) and a trait value, and a set of n' test samples each with the same set of genotype values but without trait value, the task is to train a predictive model from the training samples to predict the trait value or phenotype of each test sample based on their genotype values. Let Y be the trait value of the training samples. The problem is usually represented as the following linear regression model:

$$Y = \beta_0 + \sum_{i=1}^{m} \beta_i X_i + e_l \tag{1}$$

where X_i is the i-th genotype value, m is the total number of genotypes, β_i is the regression coefficient for the i-th genotype and e_l is the error term.

The pairwise interaction between two genotypes X_i and X_j is modeled as the product of the two genotype values. Therefore, with the traditional representation, the linear regression model with pairwise epistasis interactions is modified as follows:

$$Y = \beta_0 + \sum_{i=1}^{m} \beta_i X_i + \sum_{i,j}^{m} \alpha_{i,j} X_i X_j + e_l \tag{2}$$

where $X_i X_j$ is the product of the genotype values of the i-th and j-th genotype and it denotes the interaction of the two genotypes while $\alpha_{i,j}$ represents the coefficient for the interaction.

Unlike the model in Eq. 1, which has $O(m)$ features (single marker effects only), the epistasis model in Eq. 2 has $O(m^2)$ features (both single marker effects and pairwise epistasis effects). It is unrealistic to consider all $O(m^2)$ features from both complexity and performance perspectives. MINT [7], a mutual information based transductive feature selection method, has been shown to have good performance for genetic trait prediction. MINT selects features based on a MRMR criterion (Maximum Relevance and Minimum Redundancy), namely the selected features maximize their relevances to the target variable (trait in our problem setting) while minimizing the redundancy among the features themselves. Both relevances and redundancies are computed as mutual information. When the relevance is computed, MINT uses only the training data as target variable. However, when the redundancy is computed, MINT uses both training and test data as only features (genotypes) are involved and MINT assumes the features of test data is known *apriori*. MINT selects features by an incremental greedy search. Given t the number of target features, MINT works in the following two stages: (1) First it ranks all the features by relevance and then selects the top-k most relevant features, where $m >> k >> t$. This is based on the assumption that most of the features have low relevance scores and including them does not help improve the prediction performance. (2) Secondly, it selects the features, one at a time, by maximizing an objective function defined in Equation 3, where S_{r-1} is the set of selected features at step $r - 1$, X is the set of top-k most relevant features, x_j is the j-th feature, c is the trait, $I(A, B)$ is the mutual information between A and B. It is shown that the transductive strategy can usually lead to better selection performance.

$$\max_{x_j \in X - S_{r-1}} \left(I(x_j; c) - \frac{1}{r - 1} \sum_{x_i \in S_{r-1}} I(x_j; x_i) \right) \tag{3}$$

Thus a naive pipeline is to first generate all $O(m^2)$ features, then apply a feature selection method to select a subset of important features. However, generating all $O(m^2)$ features is infeasible for large data sets. Therefore, a challenging task is to avoid exhaustive search of the $O(m^2)$ interactions. This is described in the next section.

For two given vectors X, Y, their mutual information is computed as follows:

$$I(X,Y) = \sum_{y \in Y} \sum_{x \in X} p(x,y) log\left(\frac{p(x,y)}{p(x)p(y)}\right), \qquad (4)$$

where $p(x)$ is the marginal probability $p(X = x)$ and $p(x,y)$ is the joint probability $p(X = x, Y = y)$. For vectors with discreet values, we can easily compute $p(x), p(y), p(x,y)$ by considering the frequency of the corresponding values. For continuous values, the summation in the above formula should be replaced with integral, as follows:

$$I(X,Y) = \int_Y \int_X p(x,y) log\left(\frac{p(x,y)}{p(x)p(y)}\right) dxdy.$$

As the trait values are continuous, we perform discretization on the trait values. We first compute the z-score of phenotype value for each sample as $\frac{x-\mu}{\delta^2}$. Then, we assign discretized values to samples according to their z-score using the following formula:

$$discretized\ value = \begin{cases} -1 & \text{if z-score} < \text{-1} \\ 1 & \text{if z-score} > 1 \\ 0 & \text{otherwise} \end{cases}$$

3   Methods

3.1   MINED: Mutual Information Based Epistasis Detection

To efficiently detect epistasis effects for genetic trait prediction, we need to address the following two issues:

Problem 1. How to select a mutual information threshold to determine if an epistasis interaction is significant?

Problem 1 is challenging since using all significant interactions may not necessarily lead to the best prediction performance. Thus this is significantly different from the GWAS scenario, where all significant epistasis effects are reported. Therefore, if the threshold is too high, we may miss some important effects. If the threshold is too low, too many interactions may be included leading to both poor prediction power and poor computational complexity. To address this issue, we update the threshold dynamically to only keep a top set of most significant interactions. The threshold is initially set low to allow considering relatively less significant interactions. As we collect more and more significant interactions, the threshold is increased such that less significant interactions are pruned more efficiently. See Section 3.2 for further details.

Problem 2. How to utilize the mutual information threshold to prune the interactions that are potentially not significant?

To avoid exhaustive search in GWAS, both greedy and optimal strategies have been proposed. However, as we use mutual information, we can not adapt the optimal methods from Xiang et al. [23], which rely on an F-test score. Problem 2 is challenging since the objective is to prune the interactions without even computing their relevance scores. We observed that when a SNP is involved in a significant interaction, it is very likely that the SNP is involved in multiple significant interactions. Therefore, for each SNP, if we sample a small set of interactions where the SNP is involved, it is likely that we can capture some significant interactions the SNP is involved in. We also observed the relevance scores of all the interactions that a SNP is involved in follow a truncated normal distribution, as shown in Figure 2, similar to the hypothesis in rrBLUP that the single marker effects follow a normal distribution. Then based on the sampled interactions, we can estimate the probability that the SNP is involved in at least one significant interaction to further determine if the SNP should be thoroughly investigated for all possible interactions that it is involved in. See the following sections for further details.

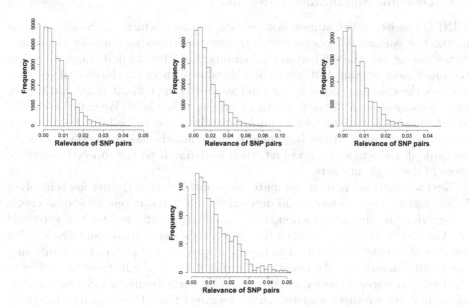

Fig. 2. The histogram of the relevance scores for all the pairwise interactions of four randomly sampled SNPs for the Maize Flint data set [16]. They all follow a truncated normal distribution.

In summary, we propose a novel framework **MINED** (**M**utual **IN**formation based **E**pistasis **D**etection) for pairwise epistasis effects detection, where we conduct a sampling for every single marker and compute the probability of the marker involved in at least one significant epistasis effect. If the probability is higher than a certain threshold, we check all the interactions between the current marker and all the other markers. In the meanwhile, we maintain a set of top features and insert any new significant interactions into the set by removing the feature with the lowest relevance score. The significance threshold is dynamically updated as the lowest relevance score from the set of the top features. Finally after all markers are checked, we select the target features from the set of most relevant features using the objective function defined in Equation 3. As the number of single markers is much smaller than the number of pairwise interactions and most of the markers have small probabilities, our method is in general very time-efficient. For example, for data sets with 20,000 markers and 200 samples, our method finishes in less than half an hour while an exhaustive search takes over 15 hours.

Notice that the goal of MINED is to select a set of significant epistasis effects efficiently by maximizing the prediction accuracy. Whether the prediction performance based on epistasis effects is superior to the performance based on just the single marker effects, and how much improvement epistasis effects can lead to, depend on the quality of the epistasis effects themselves.

3.2   Dynamic Significance Threshold

MINED consists of two stages: first we rank features, which can be either single markers or epistasis effects by relevance score and then we consider the redundancy among only the top features. Assuming that after the first stage, we collect the top-k most relevant features to do redundancy check for the second stage, as we rank the epistasis effects by their relevances to the trait, it is natural to take the relevance score R of the top-k-th feature as a threshold. We call this threshold R the *significance threshold* and k generally needs to be large to guarantee a good performance. At the beginning we have not checked any interactions. So we rank all the single markers first, then initialize R as the top-k-th relevance score of the single markers.

Next we scan each marker, compute the probability that the marker is involved in any significant interaction and determine what interactions we should check. As we show in the next section, the probability is affected by the threshold R. The larger R is, the smaller the probability tends to be and the smaller number of interactions we need to check. Therefore, we update this significance threshold dynamically. We maintain the sorted list of the features according to their relevance scores (notice we consider both epistasis effects and single marker effects). When we check an interaction, we insert the interaction into the top-k feature set if its relevance score is better than R and we remove the last feature from the list. If the interaction does not have a higher relevance score than R, we do not change the list and we say the interaction is *pruned*. We then set the threshold R as the relevance score of the current k-th feature. We keep on updating the threshold as we insert more interactions, while keeping the order of the list according to the relevance scores. Obviously, the threshold becomes higher and higher, and it becomes easier and easier to prune the remaining interactions as shown in the next section.

We next show a running example. Assuming the current top-6 feature set has scores [0.8, 0.5, 0.3, 0.3, 0.2, 0.1], the input features have scores 0.05, 0.4, 0.15,

0.19. The threshold value is 0.1, the smallest score in the top-6 feature set. The first input feature with score 0.05 will be pruned as $0.05 < 0.1$. The second input feature will be kept as $0.4 > 0.1$. The top-6 feature set will be updated as [0.8, 0.5, 0.4, 0.3, 0.3, 0.2] and the threshold will be updated as 0.2. Therefore, the last two input features with score 0.15 and 0.19 will be all pruned. If we do not update the threshold, both of them will be selected.

3.3 Compute Marker Probability

Given the current significance threshold R, a naive strategy is to search all possible interactions for each marker and select the interactions that have score higher than R. However, this is equivalent to an exhaustive search and is infeasible for large data sets. In MINED, for each marker, we compute the probability that it is involved in some significant epistasis interactions (we call this probability the *significance probability*). If the probability is high enough, we continue exploring all its possible interactions. Otherwise we ignore this marker, or *prune* this marker such that all the interactions where the marker is involved in are also pruned. To compute the probability efficiently, we conduct a sampling such that exhaustive search of all possible interactions can be avoided.

The motivation of applying sampling to estimate the probability is based on two observations: first of all, when a SNP is involved in a significant interaction, it is very likely that the SNP is involved in many significant interactions. Wei et al. [20] reported that SLC2A9 gene interacted with multiple loci across the genome, indicating the observation is common. To further validate this observation, we conduct an exhaustive search for all pairs of interactions on three plant data sets Maize [16], Rice [25] and Pine [15]. We rank the interactions by their relevance scores. Then we consider the top 20,000 interactions with the highest relevance scores as significant interactions, and we plot the histogram for the number of significant interactions that each SNP is involved in. Due to space limits, we only state the results: out of the top-20,000 interactions around 85% SNPs are involved in more than 5 significant interactions. The histogram for Maize data set is shown in Figure 3.

Secondly, we observe that the relevance score follows a truncated normal distribution. Then if we sample f relevance scores, where $f \ll m$, the number of genotypes, we can fit the normal distribution to estimate the mean and the standard deviation. Using this distribution, and given the total number of features as m, we compute the probability of seeing at least one significant relevance score out of the $m - 1$ possible interactions, where a score is significant if it is higher than the current significance threshold R. If the probability is high, we check all the $m - 1$ interactions for this marker. If not, we do not need to further check this marker and we can safely *prune* this marker. Obviously, the larger R, the lower the probability and this is the motivation for the dynamic thresholding strategy to keep on updating R such that the markers can be pruned more efficiently. Therefore, it is reasonable to conduct a sampling to estimate the probability that a SNP is involved in some significant epistasis interaction.

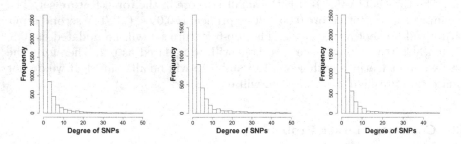

Fig. 3. The histogram for the number of significant interactions out of the top-20,000 interactions that each SNP is involved in (degree of SNPs) for the three traits of the Maize Dent data set [16]. We can see over 85% SNPs are involved in more than 5 significant interactions, namely the degree of SNPs is greater than 5.

Based on the above two observations, we check the markers one by one, and, when we compute the significance probability of the current maker, we randomly sample f markers from the set of remaining markers and compute the relevance score of the interactions between the marker and the f sampled markers, where $f \ll m$ and m is the number of original features. If the probability is higher than a threshold P, we will check the interactions between the marker and all the remaining markers. Otherwise we move to the next marker. f needs to be small to guarantee efficiency but can not be too small in order to capture significant interactions. We set f empirically and we observed that $f = 0.01 \times m$ generally achieves a good balance of efficiency and effectiveness. In order to capture as many significant epistasis interactions as possible, we generally use a small value for the significance probability s, say 0.001 (notice this threshold is the probability threshold and is different from the threshold R, which is the relevance score threshold, or the so-called significance threshold). And our experiments show that even with such a small threshold, the markers and interactions can be pruned efficiently. The pesuedoucode of the algorithm MINED is shown in Algorithm 1.

3.4   Epistasis Detection with Constraints

As we show later in the experiments, epistasis effects can be highly redundant, since one SNP can be involved in multiple interactions and these interactions are likely to have higher redundancy. Therefore, even though we keep a large k for the set of top-k most relevant features, the epistasis effects in this set can be still highly redundant. Thus considering redundancy later in a set of highly redundant features will not help improve the prediction. To address this problem, we need to increase the value of k to be large enough such that enough relatively unique epistasis effects can be captured. However, given the extremely large number of possible epistasis effects, for example, around 900 million epistasis effects for the Maize Dent data set, lots of redundant epistasis effects have high relevance score. Thus k needs to be very large, which makes selecting the final set of features very

---

Algorithm 1. The pseudocode for the algorithm MINED to select a set of target features considering both single marker effects and pairwise epistasis effects.

Input: Original marker set X, size of target feature set t, size of the most relevant features k, a probability threshold s

Output: Set of target markers

1: Rank the markers in X in decreasing order of their relevance scores as $x_1, x_2, \ldots$
2: Threshold $R \leftarrow score(x_k)$
3: **for** $x_i \in X$ **do**
4:     Randomly sample a set of f distinct x_j's XM from X
5:     $prob \leftarrow estimateProb(x_j, XM, t)$
6:     **if** $prob > s$ or $\exists score(x_i, x_j) > t$ for $x_j \in XM$ **then**
7:         Compute $score(x_i, x_h)$ for $1 \leq h \leq d$ and $h \neq x$
8:         **if** $score(x_i, x_h) > t$ **then**
9:             Update the top-k set by inserting the new feature (x_i, x_h)
10:             $R \leftarrow score(x_k)$
11:         **end if**
12:     **end if**
13: **end for**
14: Select a final set of t features using Equation 3

---

inefficient, as the complexity to select the final set of features is $O(kt)$, where t is the number of targeted features.

To address this problem, we further propose an algorithm *Constrained-MINED*, where we set up a *constraint threshold* n such that one marker can be involved in at most n different interactions. We call two epistasis effects sharing the same marker as an *overlap*. The rationale is that when two interactions share the same marker, it is more likely that they have higher redundancy. Therefore, if we add a constraint on the number of epistasis effects that can share the same marker, we can probably reduce the redundancy of the selected epistasis effects. This strategy leads to a great computational advantage: we do not need to select a very large k for the top-k most relevant features. The epistasis effects selected using this strategy are naturally less redundant as we force these effects to not overlap. We next give more details of the algorithm.

We set up an *overlap threshold* n such that one SNP can be involved in at most n interactions and we can rewrite the linear regression model with constraints as the following:

$$Y = \beta_0 + \sum_{i=1}^{d} \beta_i X_i + \sum_{i=1}^{d} \sum_{j=1}^{d} \alpha_{i,j} I_{i,j} X_i X_j + e$$

$$\sum_{i=1}^{d} I_{i,j} \leq n \ for \ 1 \leq j \leq d$$

$$\sum_{j=1}^{d} I_{i,j} \leq n \ for \ 1 \leq i \leq d$$

$$I_{i,j} = \{0, 1\}$$

where $I_{i,j}$ is an indicator function of value either 0 or 1, indicating if we select the interaction $X_i X_j$ or not. We call the set of selected interactions as *constraint-based interactions*.

We construct an *interaction graph*, where the nodes are interactions, the edges indicate the two associated nodes overlap by some SNPs. For example, for three SNPs A, B and C, the edge between an interaction (AB) and an interaction (AC) indicates the two interactions overlap as they share a common SNP A. There is a weight associated with each node, which corresponds to the significance score of the interaction. We show an example in Figure 4. If we set the overlap threshold as 1, namely one SNP can be involved in only one interactions, we can select two nodes AC and BD which are disjoint and the sum of their weights 0.7 is the maximum sum of weights we can obtain.

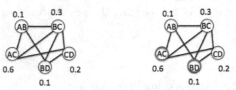

Fig. 4. An example of interaction graph, where the constraint, or the overlap threshold is set to 1, namely one SNP can not appear in more than one interaction.

With the above graph representation, the set of interactions with maximum sum of significance score is converted to a Weighted Maximum Independent Set (WMIS) problem. The WMIS problem seeks to select a set of nodes from a graph to form an independent set, where all the nodes are not adjacent, such that the sum of the weights on the nodes is maximized. As all the nodes are not adjacent in the independent set, all selected interactions are guaranteed non-overlapping. This is equivalent to using an overlap threshold of 1. When the overlap threshold n is greater than 1, we allow the degree of the connectivity of the selected nodes to be no greater than n, and we call the new problem Weighted Maximum n-Independent Set (WMNIS) problem.

The WMIS problem is well-known to be NP-complete and therefore so is WM-NIS problem. Many greedy methods have been proposed for the WMIS problem [2] [17] [19]. We next developed a greedy algorithm for the WMNIS problem. The basic idea is that we select the interactions according to their relevance scores, where we select interactions with higher scores first. We maintain a count for each single marker. Once an interaction is selected, we increase the counts of both markers by 1. When we select an interaction, if either of its single markers has count greater than the threshold n, we skip this interaction and move to the next one. The pesuedoucode of the algorithm MINED is shown in Algorithm 2.

4   Experimental Results

4.1   Maize Data

We tested the performance of MINED on three real data sets. The first data set is the Maize data set [16], which consists of two maize diversity panels with

Algorithm 2. The pseudocode for the algorithm Constrained-MINED to select a set of target features considering both single marker effects and pairwise
epistasis effects.

Input: Original marker set X, size of target feature set t, size of the most relevant
    features k, a probability threshold s, constraint threshold n

Output: Set of target markers

1: Rank the markers in X in decreasing order of their relevance scores as $x_1, x_2, \ldots$
2: Threshold $R \leftarrow score(x_k)$
3: **for** $x_i \in X$ **do**
4:    **if** $countDegree(x_i) \leq n$ **then**
5:       Randomly sample a set of f distinct x_j's XM from X
6:       $prob \leftarrow estimateProb(x_j, XM, t)$
7:       **if** $prob > s$ or $\exists score(x_i, x_j) > t$ for $x_j \in XM$ **then**
8:          Compute $score(x_i, x_h)$ for $1 \leq h \leq d$ and $h \neq x$
9:          **if** $countDegree(x_h) \leq n$ and $score(x_i, x_h) > t$ **then**
10:            Update the top-k set by inserting the new feature (x_i, x_h)
11:            $R \leftarrow score(x_k)$
12:            $degree(x_i) \leftarrow degree(x_i) + 1$
13:            $degree(x_h) \leftarrow degree(x_h) + 1$
14:          **end if**
15:       **end if**
16:    **end if**
17: **end for**
18: Select a final set of t features using Equation 3

---

300 Flint and 300 Dent lines developed for the European CornFed program. The
two panels, Flint and Dent, were genotyped using a 50k SNP array, which after
removing SNPs with high rate of missing markers and high average heterozy-
gosity, yielded 29,094 and 30,027 SNPs respectively. Both of them contain 261
samples and three traits. In all experiments, we perform 10-fold cross-validations
and measure the average coefficient of determination r^2 (computed as the square
of Pearson's correlation coefficient) between the true and the predicted outputs,
where higher r^2 indicates better performance. We use only r^2 here as it is the
most common evaluation metric for genetic trait prediction problem.

As rrBLUP [12] is one of the most widely used methods for genetic trait prediction and generally achieves better or equal results compared with other regression methods such as Bayesian methods, Lasso, Elastic Net, Random Forests
and Boosting, we select it as our baseline method and we apply it on the selected
features for prediction. For MINED and Contraint-MINED, we do grid search
to determine the number of features to be used. Notice we have two types of
features, single markers and pairwise epistasis effects. Epistasis effects are extra
information for the model, so we consider using both types of features and we
seek for the best combination of them. In our algorithms, we set the probability
threshold s as 0.001, the size for the set of most relevant features k as 5,000, the
constraint threshold n as 5. These parameters are selected empirically to achieve
a good balance between the efficiency and the accuracy. We also conducted a

Table 1. Performance (average r^2) of rrBLUP, Bayesian, SVR (Support Vector Regression), Lasso, exhaustive epistasis model, MINED followed by rrBLUP (MINED + rrBLUP), and C-MINED (Constrained-MINED) followed by rrBLUP (Constrained-MINED + rrBLUP) on the three phenotypes of Dent and Flint data sets.

					Dent		
Phenotype	rrBLUP	Bayesian	SVR	Lasso	Exhaustive	MINED	C-MINED (Improvements over rrBLUP)
TASS	0.590	0.591	0.565	0.591	0.551	0.590	**0.596** (1%)
DMC	0.552	0.563	**0.567**	0.563	0.548	0.552	0.563 (2%)
DM_Yield	0.321	0.321	0.327	0.321	0.321	0.356	**0.356** (11%)
					Flint		
Phenotype	rrBLUP	Bayesian	SVR	Lasso	Exhaustive	MINED	C-MINED
TASS	0.470	0.471	0.467	0.471	0.48	0.476	**0.482** (3%)
DMC	0.301	0.305	0.302	0.3	0.308	0.316	**0.316** (5%)
DM_Yield	0.057	0.062	0.073	0.058	0.054	**0.096**	0.078 (37%)

grid search and vary t, the number of selected epistasis effects to be used for prediction as 0, 500, 1000 and 2000. Notice we allow not using any epistasis effects by setting $t = 0$ if including them deteriorate the performance. Therefore the performance of MINED+rrBLUP is guaranteed to be no worse than rrBLUP along. The same strategy can be applied to any other regression methods such that MINED followed by any method is guaranteed to be no worse than the method itself, due to the extra information from the epistasis effects and the grid search to determine if such extra information is useful or not. We compare the performance of our methods with that of other different popular methods and we show the improvements of Constrained-MINED over the baseline method rrBLUP. Notice that for genetic trait prediction, a 5% improvement on r^2 is considered as significant. As we can see in our experiments, Constrained-MINED is able to make significant prediction improvements in most of the cases.

We show the results in Table 1, and we can see that both MINED and Constrained-MINED achieve better performance than other methods in most cases. Notice both methods achieve better performance than the exhaustive epistasis model where all epistasis effects are considered in Equation 2. This clearly indicates that considering all epistasis effects is not only very expensive (we had used our super computer Blue Gene which has tens of thousands of nodes as the rrBLUP model needs to handle billions of features at the same time), but also often leads to poorer performance. Both MINED and Constrained-MINED finished in around 1,500 seconds.

One needs to emphasize that the goal of MINED is to capture significant epistasis effects by maximizing the prediction accuracy using cross validation. We can see that MINED is indeed effective on this, but it can not guarantee that these effects will improve the prediction, which depends on the effects themselves. When the effects themselves are poor, for example, when they are highly redundant, it's unlikely that MINED can improve the prediction. However, when the effects do improve the prediction, MINED is able to capture them efficiently and our experiments show that for most of the data sets epistasis effects help to improve the prediction.

Table 2. Performance (average r^2) of rrBLUP, Bayesian, SVR (Support Vector Regression), Lasso, exhaustive epistasis model, MINED followed by rrBLUP (MINED + rrBLUP), and C-MINED (Constrained-MINED) followed by rrBLUP (Constrained-MINED + rrBLUP) on the two phenotypes of Rice data set.

Phenotype	rrBLUP	Bayesian	SVR	Lasso	Exhaustive	MINED	C-MINED (Improvements over rrBLUP)
Pericarp color	0.409	0.378	0.428	0.393	0.486	0.443	**0.528** (29%)
Protein content	0.192	0.174	0.212	0.173	0.111	0.229	**0.229** (19%)

Table 3. Performance (average r^2) of rrBLUP, MINED followed by rrBLUP (MINED + rrBLUP), Constrained-MINED followed by rrBLUP (Constrained-MINED + rrBLUP) on five randomly selected phenotypes of Rice data set.

Phenotype	rrBLUP	MINED	Constrained-MINED (Improvements over rrBLUP)
Flowering time at Faridpur	0.282	0.282	**0.291** (3%)
Flowering time at Aberdeen	0.343	0.344	**0.351** (2%)
FT ratio of Faridpur/Aberdeen	0.204	0.251	**0.251** (23%)
Culm habit	0.488	0.488	**0.488** (0%)
Flag leaf length	0.281	0.281	**0.301** (7%)

Table 4. Performance (average r^2) of rrBLUP, MINED, Constrained-MINED, on four randomly selected phenotypes of Pine data set.

Phenotype	rrBLUP	MINED	Constrained-MINED (Improvements over rrBLUP)
BD	0.07	**0.098**	0.093 (33%)
BLC	0.24	0.245	**0.252** (5%)
CWAC	0.23	0.233	**0.233** (1.3%)
CWAL	0.15	0.154	**0.166** (11%)

As we can see in Table 1, Constrained-MINED achieves better performance than MINED for all cases except for Flint trait DM_Yield, indicating that limiting the overlaps of the epistasis effects does help to reduce the redundancy of the selected effects resulting in improved prediction power.

4.2   Rice Data

Next we consider the second data set, the Asian rice, *Oryza sativa*, data set [25]. This data set was based on 44,100 SNP variants from 413 accessions of O. sativa, taken from 82 countries containing 34 phenotypes. The data sets have 36,901 markers and 413 samples. As the rice data set is much bigger than the Maize data, we compare our methods against all other methods for only two randomly selected phenotypes: "Pericarp.color" and "Protein.content". We also randomly selected another five phenotypes. As we have shown that rrBLUP is as good as other methods, we compare our methods with rrBLUP only for these five phenotypes. We use the same parameter setting for our methods and grid search is conducted by our methods to select the number of epistasis effects to be used. Average results of 10 fold cross-validation are shown in Table 2 and 3. We see that for most of the phenotypes, both MINED and Constrained-MINED achieves better performance than the other methods and rrBLUP, indicating both methods are able to identify significant epistasis effects and including these epistasis effects helps to improve the prediction. This again indicates that our methods are efficient in capturing significant epistasis effects. The results also show again that whether the prediction can be improved using the epistasis effects depends on the data set itself. For phenotype "Culm habit", including epistasis effects does not improve the prediction, and out of seven phenotypes, Constrained-MINED achieves better performance for four phenotypes compared with MINED. The improvement is especially significant for phenotype "Pericarp.color", indicating that the constraint strategy is in general effective. Both MINED and Constrained-MINED finished in around 3,000 seconds.

4.3 Pine Data

Finally we test the performance of our method on the third data set, the Loblolly Pine data set [15], which contains 17 de-regressed phenotypes for the 926 samples, each with 4,854 genotypes. Average results of 10 fold cross-validation for four randomly selected phenotypes are shown in Table 4, and again we compare our methods with rrBLUP only. We use the same parameter setting for our methods, and grid search is conducted by our methods to select the number of epistasis effects to be used. We see that both MINED and Constrained-MINED achieves better performance than rrBLUP does for all phenotypes. Constrained-MINED achieves the best results for three phenotypes. For phenotype "BD", MINED achieves better results. These results are consistent with the results on the previous two data sets, again illustrating the effectiveness of our methods. Both MINED and Constrained-MINED finished in around 4,800 seconds.

Although we only tested the performance of our methods for genetic trait prediction problem, they can be applied directly to prediction or classification problems of other domains when the interactions of the features need to be modeled.

5 Conclusions

In this work, we proposed an efficient mutual information based pairwise epistasis detection method MINED for the genetic trait prediction problem. To our knowledge, this is the first algorithm to detect epistasis for genetic trait prediction problem. The method applies a sampling strategy as well as a dynamic thresholding strategy to avoid exhaustive search. We show that MINED is much faster than an exhaustive search. We also show that the method is effective in capturing the truly significant epistasis effects. However, whether it is able to improve the prediction performance depends on the quality of the epistasis effects. In our future work, we would like to improve the method to better handle low quality epistasis effects, for example, epistasis effects that are highly redundant. We would also like to extend the framework to handle multi-way interactions, where more than two SNPs are involved.

References

1. Bien, J., Taylor, J., Tibshirani, R., et al.: A lasso for hierarchical interactions. The Annals of Statistics 41(3), 1111–1141 (2013)
2. Brendel, W., Amer, M., Todorovic, S.: Multiobject tracking as maximum weight independent set. In: 2011 IEEE Conference on Computer Vision and Pattern Recognition (CVPR), pp. 1273–1280. IEEE (2011)
3. Chen, S.S., Donoho, D.L., Saunders, M.A.: Atomic decomposition by basis pursuit. Atomic decomposition by basis pursuit. SIAM Journal on Scientific Computing 20, 33–61 (1998)
4. Cleveland, M.A., Hickey, J.M., Forni, S.: A common dataset for genomic analysis of livestock populations. G3: Genes— Genomes— Genetics 2(4), 429–435 (2012)
5. Cook, N.R., Zee, R.Y.L., Ridker, P.M.: Tree and spline based association analysis of gene–gene interaction models for ischemic stroke. Statistics in Medicine 23(9), 1439–1453 (2004)
6. Fang, G., Haznadar, M., Wang, W., Yu, H., Steinbach, M., Church, T.R., Oetting, W.S., Van Ness, B., Kumar, V.: High-order snp combinations associated with complex diseases: efficient discovery, statistical power and functional interactions. PloS One 7(4), e33531 (2012)
7. He, D., Rish, I., Haws, D., Teyssedre, S., Karaman, Z., Parida, L.: Mint: Mutual information based transductive feature selection for genetic trait prediction. arXiv preprint arXiv:1310.1659 (2013)
8. Kilpatrick, J.R.: Methods for detecting multi-locus genotype-phenotype association. PhD thesis, Rice University (2009)
9. Kizilkaya, K., Fernando, R.L., Garrick, D.J.: Genomic prediction of simulated multibreed and purebred performance using observed fifty thousand single nucleotide polymorphism genotypes. Journal of Animal Science 88(2), 544–551 (2010)
10. Legarra, A., Robert-Granié, C., Croiseau, P., Guillaume, F., Fritz, S., et al.: Improved lasso for genomic selection. Genetics Research 93(1), 77 (2011)
11. Marchini, J., Donnelly, P., Cardon, L.R.: Genome-wide strategies for detecting multiple loci that influence complex diseases. Nature Genetics 37(4), 413–417 (2005)
12. Meuwissen, T.H.E., Hayes, B.J., Goddard, M.E.: Prediction of total genetic value using genome-wide dense marker maps. Genetics 157, 1819–1829 (2001)
13. Park, T., Casella, G.: The bayesian lasso. Journal of the American Statistical Association 103, 681–686 (2008)
14. Pattin, K.A., White, B.C., Barney, N., Gui, J., Nelson, H.H., Kelsey, K.T., Andrew, A.S., Karagas, M.R., Moore, J.H.: A computationally efficient hypothesis testing method for epistasis analysis using multifactor dimensionality reduction. Genetic Epidemiology 33(1), 87–94 (2009)
15. Resende, M.F.R., Muñoz, P., Resende, M.D.V., Garrick, D.J., Fernando, R.L., Davis, J.M., Jokela, E.J., Martin, T.A., Peter, G.F., Kirst, M.: Accuracy of genomic selection methods in a standard data set of loblolly pine (pinus taeda l.). Genetics 190(4), 1503–1510 (2012)
16. Rincent, R., Laloë, D., Nicolas, S., Altmann, T., Brunel, D., Revilla, P., Rodriguez, V.M.: Maximizing the reliability of genomic selection by optimizing the calibration set of reference individuals: Comparison of methods in two diverse groups of maize inbreds (zea mays l.). Genetics 192(2), 715–728 (2012)
17. Sakai, S., Togasaki, M., Yamazaki, K.: A note on greedy algorithms for the maximum weighted independent set problem. Discrete Applied Mathematics 126(2), 313–322 (2003)

18. Tibshirani, R.: Regression shrinkage and selection via the lasso. Journal of the Royal Statistical Society, Series B 58, 267–288 (1994)
19. Valiente, G.: A new simple algorithm for the maximum-weight independent set problem on circle graphs. In: Ibaraki, T., Katoh, N., Ono, H. (eds.) ISAAC 2003. LNCS, vol. 2906, pp. 129–137. Springer, Heidelberg (2003)
20. Wei, W., Hemani, G., Hicks, A.A., Vitart, V., Cabrera-Cardenas, C., Navarro, P., Huffman, J., Hayward, C., Knott, S.A., Rudan, I., et al.: Characterisation of genome-wide association epistasis signals for serum uric acid in human population isolates. PloS One 6(8), e23836 (2011)
21. Whittaker, J.C., Thompson, R., Denham, M.C.: Marker-assisted selection using ridge regression. Genet. Res. 75, 249–252 (2000)
22. Yang, C., He, Z., Wan, X., Yang, Q., Xue, H., Yu, W.: Snpharvester: a filtering-based approach for detecting epistatic interactions in genome-wide association studies. Bioinformatics 25(4), 504–511 (2009)
23. Zhang, X., Huang, S., Zou, F., Wang, W.: Team: efficient two-locus epistasis tests in human genome-wide association study. Bioinformatics 26(12), i217–i227 (2010)
24. Zhang, Y., Liu, J.S.: Bayesian inference of epistatic interactions in case-control studies. Nature Genetics 39(9), 1167–1173 (2007)
25. Zhao, K., Tung, C.-W., Eizenga, G.C., Wright, M.H., Ali, L., Price, A.H., Norton, G.J., Islam, M.R., Reynolds, A., Mezey, J., et al.: Genome-wide association mapping reveals a rich genetic architecture of complex traits in oryza sativa. Nature Communications 2, 467 (2011)

Domain Adaptation with Logistic Regression for the Task of Splice Site Prediction

Nic Herndon[✉] and Doina Caragea

Computing and Information Sciences, Kansas State University
234 Nichols Hall, Manhattan, KS 66506, USA
{nherndon,dcaragea}@ksu.edu

Abstract. Supervised classifiers are highly dependent on abundant labeled training data. Alternatives for addressing the lack of labeled data include: labeling data (but this is costly and time consuming); training classifiers with abundant data from another domain (however, the classification accuracy usually decreases as the distance between domains increases); or complementing the limited labeled data with abundant unlabeled data from the same domain and learning semi-supervised classifiers (but the unlabeled data can mislead the classifier). A better alternative is to use both the abundant labeled data from a source domain and the limited labeled data from the target domain to train classifiers in a domain adaptation setting. We propose such a classifier, based on logistic regression, and evaluate it for the task of splice site prediction – a difficult and essential step in gene prediction. Our classifier achieved high accuracy, with highest areas under the precision-recall curve between 50.83% and 82.61%.

Keywords: Domain adaptation · Logistic regression · Splice site prediction · Imbalanced data

1 Introduction

The adoption of next generation sequencing (NGS) technologies a few years ago has led to both opportunities and challenges. The NGS technologies have made it affordable to sequence new organisms but have also produced a large volume of data that need to be organized, analyzed, and interpreted to create or improve, for example, genome assemblies or genome annotations. For genome annotation a major task is to accurately identify the splice sites – the regions of DNA that separate the exons from introns (donor splice sites), and the introns from exons (acceptor splice sites). The majority of the donor and acceptor splice sites, also known as canonical sites, are the GT and AG dimers, respectively, but not all GT, AG dimers are splice sites. Only about 1% or less of them are [23], making the splice site prediction a difficult task.

NGS technologies have also enabled better gene predictions through programs that assemble short RNA-Seq reads into transcripts and then align them against the genome. For example, TWINSCAN [16] and CONTRAST [11] model the

© Springer International Publishing Switzerland 2015
R. Harrison et al. (Eds.): ISBRA 2015, LNBI 9096, pp. 125–137, 2015.
DOI: 10.1007/978-3-319-19048-8_11

entire transcript structure as well as conservation in related species. However, transcript assemblies from RNA-Seq reads are not error proof, and should be subjected to independent validation [27].

Machine learning algorithms, which have been successfully applied on many biological problems, including gene prediction, could be seen as alternative tools that can help with such validation. For example, support vector machines (SVM) have been used for problems such as identification of translation initiation sites [19,30], labeling gene expression profiles as malign or benign [20], *ab initio* gene prediction [4], and protein function prediction [5], while hidden Markov models have been used for *ab initio* gene predictions [15,26], among others.

However, supervised machine learning algorithms require large amounts of labeled data to learn accurate classifiers. Yet, for many biological problems, including splice site prediction, labeled data may not be available for an organism of interest. An option would be to label enough data from the target domain for a supervised target classifier, but this is time consuming and costly. Another option is to complement the limited labeled data with abundant unlabeled data from the same target domain and learn semi-supervised classifiers. However, it can happen that a classifier is degraded by the unlabeled data [7]. Assuming that labeled data can can be plentifully available for a different, but closely related model organism (for example, a newly sequenced organism is generally scarce in labeled data, while a related, well-studied model organism is rich in labeled data), another option is to learn a classifier from the related organism. Nevertheless, using a classifier trained on labeled data from the related problem to classify unlabeled data for the problem of interest does not always produce accurate predictions.

A better alternative is to learn a classifier in a domain adaptation framework. In this setting, the large corpus of labeled data from the related, well studied organism is used in conjunction with available labeled data from the new organism to produce an accurate classifier for the latter.

Towards this goal, we propose a domain adaptation approach, presented in Sect. 3.3, based on the supervised logistic regression classifier described in Sect. 3.1. This approach is simple, yet highly accurate. When trained on a source organism, *C.elegans*, and one of four target organisms, *C.remanei*, *P.pacificus*, *D.melanogaster*, and *A.thaliana*, with data described in Sect. 3.4, this algorithm achieved high accuracy, with highest areas under the precision-recall curve between 50.83% for distant domains and 82.61% for closely related domains, as shown in Sect. 4.

2   Related Work

Most of the approaches addressing splice site prediction involve supervised learning. For example, Li *et al.* [18] proposed a method that used the discriminating power of each position in the DNA sequence around the splice site, estimated using the chi-square test. They used a support vector machine algorithm with a radial basis function kernel that combines the scaled component features, the nucleotide frequencies at conserved sites, and the correlative information of two sites, to train a classifier for the human genome. Baten et al. [3], Sonnenburg et al. [23], and Zhang et al. [29], also proposed supervised support vector machine classifiers, while Baten et al. [2] proposed a method using a hidden Markov model, Cai et al. [6] proposed a Bayesian network algorithm, and Arita, Tsuda, and Asai [1] proposed a method using Bahadur expansion truncated at the second order. However, one major drawback of these supervised algorithms is that they typically require large amounts of labeled data to train a classifier.

An alternative, when the amount of labeled data is not enough for learning a supervised classifier, is to use the limited amount of labeled data in conjunction with abundant unlabeled data to learn a semi-supervised classifier. However, semi-supervised classifiers could be misled by the unlabeled data, especially when there is hardly any labeled data [7]. For example, if during the first iteration one or more instances are misclassified, the semi-supervised algorithm will be skewed towards the mislabeled instances in subsequent iterations. Another deficiency of semi-supervised classifiers is that their accuracy decreases as the imbalance between classes increases. Stanescu and Caragea [25] studied the effects of imbalanced data on semi-supervised algorithms and found that although self-training that adds only positive instances in the semi-supervised iterations achieved the best results out of the methods evaluated, oversampling and ensemble learning are better options when the positive-to-negative ratio is about 1:99. In their subsequent study [24], they evaluated several ensemble-based semi-supervised learning approaches, out of which, again, a self-training ensemble with only positive instances produced the best results. However, the highest area under precision-recall curve for the best classifier was 54.78%.

Another option that addresses the lack of abundant labeled data needed with supervised algorithms is to use domain adaptation. This approach has been successfully applied to other problems even when the base learning algorithms used in domain adaptation make simplifying assumptions, such as features' independence. For instance, in text classification, Dai et al. [9] proposed an iterative algorithm derived from naïve Bayes that uses expectation-maximization for classifying text documents into top categories. This algorithm performed better than supervised SVM and naïve Bayes classifiers when tested on datasets from Newsgroups, SRAA and Reuters. A similar domain adaptation algorithm proposed by Tan et al. [28], identified and used only the generalizable features from the source domain, in conjunction with unlabeled data from the target domain. It produced promising results for several target domains when evaluated on the task of sentiment analysis.

Even though domain adaptation has been used with good results in other domains, there are only a few domain adaptation methods proposed for biological problems. For example, Herndon and Caragea [14] modified the algorithm proposed by Tan et al. [28], by using a small amount of labeled data from the target domain and incorporating self-training. Although this modified algorithm produced promising results on the task of protein localization, it performed poorly on the splice site prediction data. The updated version of that algorithm, [13], implemented further changes, such as normalizing the counts for the prior and likelihood, using mutual information in selecting generalizable features, and representing the DNA sequences with location aware features. With these changes, the method produced promising results on the task of splice site prediction, with values for the highest area under the precision-recall curve between 43.20% for distant domains and 78.01% for related domains. A recent approach for splice site prediction, Giannoulis *et al.* [12], proposed a modified version of the *k*-means clustering algorithm that took into account the commonalities between the source and target domains for splice site prediction. While this method seems promising, in its current version, it was less accurate than the method in [13] – with the best values for the area under receiver operating characteristic curve below 70%. The best results for the task of splice site prediction, *up until now*, were obtained with a support vector machine classifier proposed by Schweikert *et al.* [22] (which used a weighted degree kernel proposed by Rätsch *et al.* [21]), especially when the source and target domain were not closely related.

3   Methods and Materials

In this section, we present the three logistic regression classifiers that we use in our experiments. We describe them in the context of a binary classification task since splice site prediction is a binary classification problem. The first classifier, proposed by Le Cessie and Van Houwelingen [17] , is a supervised logistic regression classifier. We will use this as a baseline for our domain adaptation classifiers. The second classifier uses a method proposed by Chelba and Acero [8] for maximum entropy models. This is a logistic regression classifier for the domain adaptation setting. The third classifier is our proposed classifier for the domain adaptation setting.

3.1   Logistic Regression with Regularized Parameters

Given a set of training instances generated independently $X \in \mathbb{R}^{m \times n}$ and their corresponding labels $y \in \mathcal{Y}^m$, $\mathcal{Y} = \{0,1\}$, where m is the number of training instances and n is the number of features, logistic regression models the posterior probability as

$$p(y \mid x; \theta) = \begin{cases} g(\theta^T x) & \text{, if } y = 1 \\ 1 - g(\theta^T x) & \text{, if } y = 0 \end{cases} = \left[g(\theta^T x) \right]^y \cdot \left[1 - g(\theta^T x) \right]^{1-y}$$

where $g(\cdot)$ is the logistic function $g(\theta^T x) = \frac{1}{1+e^{-\theta^T x}}$.

With this model, the log likelihood can be written as a function of the parameters θ as follows:

$$l(\theta) = \log \prod_{i=1}^{m} p(y_i \mid x_i; \theta) = \log \prod_{i=1}^{m} \left[g(\theta^T x_i) \right]^{y_i} \cdot \left[1 - g(\theta^T x_i) \right]^{1-y_i}$$

$$= \sum_{i=1}^{m} \left[y_i \log g(\theta^T x_i) + (1 - y_i) \log \left(1 - g(\theta^T x_i) \right) \right]$$

The parameters are estimated by maximizing the log likelihood, usually using maximum entropy models, after a regularization term, with parameter λ, is introduced to penalize large values of θ:

$$\theta = \arg\max_{\theta}\left\{\sum_{i=1}^{m}\left[y_i \log g(\theta^T x_i) + (1 - y_i) \log\left(1 - g(\theta^T x_i)\right)\right] - \lambda\|\theta\|^2\right\}$$

Note that x_i is the ith row in X, in our case, the ith DNA sequence in the training data set, y_i is the ith element of y, i.e., the corresponding label of x_i, and $x_{i0} = 1, \forall i \in \{1, 2, \ldots, m\}$ such that $\theta^T x_i = \theta_0 + \sum_{j=1}^{n} \theta_j x_{ij}$.

3.2 Logistic Regression for Domain Adaptation Setting with Modified Regularization Term

The method proposed by [8] for maximum entropy models involves modifying the optimization function. First, this method learns a model for the source domain, θ_S, by using the training instances from the source domain, (X_S, y_S), where $X_S \in \mathbb{R}^{m_S \times n}$ and $y_S \in \mathcal{Y}^{m_S}$ (note that the subscripts indicate the domain, with S for the source, and T – in the subsequent equations – for the target).

$$\theta_S = \arg\max_{\theta_S}\left\{\sum_{i=1}^{m_S}\left[y_i \log g(\theta_S^T x_i) + (1 - y_i) \log\left(1 - g(\theta_S^T x_i)\right)\right] - \lambda_S\|\theta_S\|^2\right\}$$

Then, using the source model to constrain the target model, learn a model of the target domain, θ_T, by using the training instances from the target domain, (X_T, y_T), where $X_T \in \mathbb{R}^{m_T \times n}$ and $y_T \in \mathcal{Y}^{m_T}$, but with the following modified optimization function:

$$\theta_T = \arg\max_{\theta_T}\left\{\sum_{i=1}^{m_T}\left[y_i \log g(\theta_T^T x_i) + (1 - y_i) \log\left(1 - g(\theta_T^T x_i)\right)\right] - \lambda_T\|\theta_T - \theta_S\|^2\right\}$$

3.3 Logistic Regression for Domain Adaptation Setting with Convex Combination of Posterior Probabilities

The method we are proposing uses a convex combination of two logistic regression classifiers – one trained on the source data, and the other trained on the target data. First, we learn a model for the source domain and a model for the target domain, using the training instances from the source domain, (X_S, y_S) and from the target domain, (X_T, y_T), respectively:

$$\theta_S = \arg\max_{\theta_S}\left\{\sum_{i=1}^{m_S}\left[y_i \log g(\theta_S^T x_i) + (1 - y_i) \log\left(1 - g(\theta_S^T x_i)\right)\right] - \lambda_S\|\theta_S\|^2\right\}$$

$$\theta_T = \arg\max_{\theta_T}\left\{\sum_{i=1}^{m_T}\left[y_i \log g(\theta_T^T x_i) + (1 - y_i) \log\left(1 - g(\theta_T^T x_i)\right)\right] - \lambda_T\|\theta_T\|^2\right\}$$

Then, using these models, we approximate the posterior probability for every instance x from the test set of the target domain as a normalized convex combination of the posterior probabilities for the source and target domains:

$$p(y \mid x; \delta) \propto (1 - \delta) \cdot p_S(y \mid x; \theta_S) + \delta \cdot p_T(y \mid x; \theta_T) \tag{1}$$

where $\delta \in [0, 1]$ is a parameter that shifts the weight from the source domain to the target domain depending on the distance between these domains, and the amount of target data available. This parameter is optimized using the training instances of the target domain, (X_T, y_T), as validation set.

3.4   Data Set

We evaluated our proposed algorithm on the splice site dataset[1] first introduced in [22]. This contains DNA sequences from five organisms, *C.elegans* used as the source domain and four other organisms at increasing evolutionary distance from it, *C.remanei*, *P.pacificus*, *D.melanogaster*, and *A.thaliana*, as target domains. Each instance is a 141 nucleotides long DNA sequence, with the AG dimer at the sixty-first position, along with a label that indicates whether this AG dimer is an acceptor splice site or not. In each file 1% of the instances are positive, i.e., the AG dimer at 61st position is an acceptor splice site, with small variations (variance is 0.01), while the remaining instances are negative. The data from the target organisms is split into three folds (by the authors who published the data in [22]) to obtain unbiased estimates for the classifier performance. Similar to [22], for our experiments, we used the training set of 100,000 instances from *C.elegans*, and the three folds of 2,500, 6,500, 16,000, and 40,000 labeled instances from the other organisms, and for testing, three folds of 20,000 instances each, from the target organisms. This allows us to compare our results with the previous state-of-the-art results on this dataset in [22]. Note that although the dataset we used only has acceptor splice sites, the problem of predicting donor splice sites can be addressed with the same approach.

3.5   Data Preparation and Experimental Setup

We use two similar representations for the data. In one of them, we convert each DNA sequence into a set of features that represent the nucleotides present in the sequence at each position, and the trimer at each position. For example, given a DNA sequence starting with `AAGATTCGC...` and label -1 we represent it as `A,A,G,A,T,T,C,G,C,...,AAG,AGA,GAT,...,-1`.

With these features we create a compact representation of a balanced combination of simple features in each DNA sequence, i.e., the 1-mers, and more complex features – features that capture the correlation between the nucleotides, i.e., the 3-mers. However, when the training data has a small number of instances, the trimers lead to a set of sparse features which can result in decreased

---

classification accuracy. Therefore, in the other representation we keep only the nucleotide features. For an example DNA sequence starting with AAGATTCGC... and label -1 we represent it as A,A,G,A,T,T,C,G,C,...,-1.

We use these representations for two reasons. First, with these representations we achieved good results in [13] with a naïve Bayes classifier in a domain adaptation setting. And second, this allows us to compare the results of our proposed method with our previous results.

To find the optimal parameters' values we first did a grid search for λ, using the baseline, supervised logistic classifier, with $\lambda = 10^x, x \in \{-8, -6, \ldots, 4\}$, trained with data from source and target domains. For these datasets we got the best results when $\lambda = 1,000$. Therefore, for our proposed algorithm we set λ_S and λ_T to 1,000, and did a grid search for δ with values from $\{0.1, 0.2, \ldots, 0.9\}$, while for our implementation of the method proposed in [8] we set λ_S to 1,000 and did a grid search for λ_T with $\lambda_T = 10^x, x \in \{-8 - 7, \ldots, 4\}$. We tuned λ_T for the method in [8], as λ_T controls the trade-off between source and target parameters, and thus it is similar to the δ parameter for our proposed method.

For the domain adaptation setting we trained on source and target data, while for the baseline classifiers, the supervised logistic regression, in one setting we trained on source, and in another setting we trained on each of the labeled target data set sizes: 2,500, 6,500, 16,000, and 40,000. To evaluate the classifiers we tested them on the test target data from the corresponding fold. We expect the results of the baseline, logistic regression classifier trained on each of the target labeled data sets to be the lower bound for our proposed method trained on the source data and that corresponding target labeled data, since we believe that adding data from a related organism should produce a better classifier.

All results are reported as averages over three random train-test splits to ensure that our results are unbiased. To evaluate the classifiers we used the area under the precision-recall curve (auPRC) for the positive class, since the data are so highly imbalanced [10].

With this experimental setup we wanted to evaluate:

1. The influence of the following factors on the performance of the classifier:

    (a) The features used: nucleotides, or nucleotides and trimers.
    (b) The amount of target labeled data: from 2,500, 6,500, 16,000 to 40,000 instances.
    (c) The evolutionary distance between the source and target organisms.
    (d) The weight assigned to the target data through the δ parameter in Equation 1.

2. The performance of the two domain adaptation classifiers derived from the supervised logistic regression classifier (the method proposed by [8], and our proposed method), compared to other domain adaptation classifiers for the task of splice site prediction, namely, the SVM classifier proposed by [22] and the naïve Bayes classifier proposed by [13].

Table 1. auPRC values for the minority (i.e., positive) class for four target organisms based on the number of labeled target instances used for training: 2,500, 6,500, 16,000, and 40,000. The top five rows in each subtable show the auPRC values for the classifiers trained with nucleotide features (N), while the bottom five rows show the values for the classifiers trained with nucleotide and trimer features (N&T). The LR_SL classifier is the baseline logistic regression classifier trained on 100,000 instances from the source domain, *C.elegans* (first and seventh rows), and target labeled data (second and eighth rows). The LR_cc and LR_reg domain adaptation classifiers are trained on a combination of source labeled and target labeled data, while the NB domain adaptation classifier is trained on a combination of source labeled, target labeled, and target unlabeled data. We show for comparison with our classifier (the one with blue cell text) the values for the best overall classifier in [22], $SVM_{S,T}$, (listed in these subtables as SVM), the values for our implementation of the LR_reg classifier proposed in [8], and the values for the best overall classifier in [13], A1, (listed in these subtables as NB). Note that the SVM classifier used different features. The best average values for each target dataset size is shown in bold. We would like to highlight that our classifier always performed better than the baseline classifier, and performed better in 9 out of 16 cases than the SVM classifier – the best classifier out of the three domain adaptation classifiers used for comparison with our classifier. We couldn't check if the differences between our classifier and the SVM classifier are statistically significant, as we did not have the performance results per-fold for the SVM classifier(only average performance values were available in [22]).

Feat.	Classifier	2,500	6,500	16,000	40,000
	LR_SL_S		77.63±1.37		
	LR_SL_T	31.07±8.72	54.20±3.97	65.73±2.76	72.93±1.70
N	LR_cc	77.64±1.39	77.75±1.25	77.88±1.42	78.10±1.15
	LR_reg	16.30±7.70	40.87±3.26	49.07±0.93	58.37±2.63
	NB	59.18±1.17	63.10±1.23	63.95±2.08	63.80±1.41
	SVM	77.06±2.13	77.80±2.89	77.89±0.29	79.02±0.09
	LR_SL_S		81.37±2.27		
	LR_SL_T	26.93±9.91	55.26±2.21	68.30±1.91	77.33±2.78
N&T	LR_cc	**81.39±2.30**	**81.47±2.19**	**81.78±2.08**	**82.61±2.00**
	LR_reg	2.30±1.05	14.50±4.68	40.10±3.72	63.53±7.10
	NB	45.29±2.62	72.00±4.16	74.83±4.32	77.07±4.45

(a) *C.remanei*

Feat.	Classifier	2,500	6,500	16,000	40,000
	LR_SL_S		64.20±1.91		
	LR_SL_T	29.87±3.58	49.03±4.90	59.93±2.74	69.10±2.25
N	LR_cc	64.70±1.85	65.31±2.10	66.76±0.89	70.18±2.12
	LR_reg	18.00±3.83	32.73±2.69	40.73±4.30	55.73±1.62
	NB	45.32±2.68	49.82±2.58	52.09±2.04	54.62±1.51
	SVM	**64.72±3.75**	**66.39±0.66**	68.44±0.67	71.00±0.38
	LR_SL_S		62.37±0.84		
	LR_SL_T	28.40±4.49	49.67±2.83	62.97±3.32	74.60±2.85
N&T	LR_cc	64.18±1.10	65.49±1.84	**69.76±2.08**	**75.82±2.00**
	LR_reg	4.37±1.76	14.50±4.86	38.23±6.54	63.70±5.28
	NB	20.21±1.17	53.29±3.08	62.33±3.60	69.88±4.04

(b) *P.pacificus*

Table 1. *(Continued)*

Feat.	Classifier	2,500	6,500	16,000	40,000
N	LR_SL$_S$	35.87±2.32			
	LR_SL$_T$	19.97±3.48	31.80±3.86	42.37±2.15	50.53±1.80
	LR_cc	39.70±2.82	**42.19±3.41**	49.72±2.01	53.43±0.89
	LR_reg	11.33±1.36	22.80±2.60	27.30±3.92	42.67±0.76
	NB	33.31±3.71	36.43±2.18	40.32±2.04	42.37±1.51
	SVM	**40.80±2.18**	37.87±3.77	**52.33±0.91**	**58.17±1.50**
N&T	LR_SL$_S$	32.23±2.76			
	LR_SL$_T$	15.07±4.11	28.30±5.45	44.67±3.23	38.43±32.36
	LR_cc	37.24±2.20	40.93±3.79	50.54±3.91	45.89±22.25
	LR_reg	3.40±1.82	8.37±2.48	21.20±2.85	26.50±22.44
	NB	25.83±2.35	32.58±5.83	39.10±1.82	47.49±3.44

(c) *D.melanogaster*

Feat.	Classifier	2,500	6,500	16,000	40,000
N	LR_SL$_S$	16.93±0.21			
	LR_SL$_T$	13.87±2.63	26.03±3.29	38.43±6.18	49.33±4.07
	LR_cc	20.67±0.58	27.19±1.30	40.56±3.26	49.75±2.82
	LR_reg	8.50±2.08	17.93±4.72	23.30±2.35	39.10±4.97
	NB	18.46±1.13	25.04±0.72	31.47±3.56	36.95±3.39
	SVM	**24.21±3.41**	**27.30±1.46**	38.49±1.59	49.75±1.46
N&T	LR_SL$_S$	14.07±0.31			
	LR_SL$_T$	8.87±1.84	21.10±4.45	38.53±8.08	49.77±2.77
	LR_cc	16.42±1.20	26.44±2.49	**41.35±6.49**	**50.83±2.28**
	LR_reg	2.50±0.10	8.27±1.60	20.03±3.36	30.27±2.57
	NB	3.99±0.43	13.96±2.42	33.62±6.31	43.20±3.78

(d) *A.thaliana*

4 Results and Discussion

Table 1 shows the auPRC values of the minority class when using our proposed domain adaptation with logistic regression classifier and, for comparison, when using the supervised logistic regression classifiers (trained on source or target), the logistic regression for domain adaptation classifier proposed by [8], the naïve Bayes classifier for domain adaptation from our previous work [13], and the best overall SVM classifier for domain adaptation proposed by [22], SVM$_{S,T}$. Based on these results, we make the following observations:

1. In terms of the factors that influence the performance of the classifier:

   (a) **Features:** our proposed classifier performed better with nucleotide and trimer features, when the source and target domains are closely related and the classifier has more target labeled data available. However, as the distance between the source and target domains increases, our algorithm performs better with nucleotide features when there is little target labeled data. This conforms with our previous results [13], and with our intuition (see Section 3.5): since trimers generate a sparse set of features, they lead to decreased classification accuracy when there are a small number of target training instances.

(b) **Amount of target labeled training data**: the more target training data used by the classifier the better the classifier performs. This makes sense, as more sample data describes more closely the distribution.

(c) **Distance between domains**: as the distance between the source and target domains increases the contribution of the source data decreases. It is interesting to note though that based on these results the splice site prediction problem seems to be more difficult for more complex organisms. For all dataset sizes there is a common trend of decreasing auPRC values as the complexity of the organisms increases, from *C.remanei*, *P. pacificus*, *D.melanogaster*, to *A.thaliana*, as shown in Table 1. We believe this is a major reason that helps explain the decreased auPRC values for all classifiers, for these organisms, respectively, i.e., in general auPRC for *C.remanei* > *P. pacificus* > *D.melanogaster* > *A.thaliana*.

(d) **Weight assigned to target data**: Intuitively, we expect δ to be closer to one when the source and target domain are more distantly related, and closer to zero otherwise. The results conform with our intuition, with δ between 0.1 and 0.6 for *C.remanei*, between 0.7 and 0.8 for *P.pacificus*, between 0.8 and 0.9 for *D.melanogaster*, and 0.9 for *A.thaliana*.

2. In terms of performance, the method proposed by [8] produced worse results than the supervised logistic regression classifier trained on the target data. We believe that these poor results are due to this method's modified optimization function, which constrains the values of the parameters for the target domain to be close to the values of the parameters for the source domain. In addition, this method performed worse than the domain adaptation naïve Bayes classifier proposed in our previous work [13], except for two cases (when using nucleotides as features, the target domains are *D.melanogaster*, and *A.thaliana*, and the algorithms are trained on 40,000 target instances).

Our proposed method produced better average results than the supervised logistic classifier trained on either the source or the target domain in *every* case of the 16 we evaluated. This confirms our hypothesis that augmenting a small labeled dataset from the target domain with a large labeled dataset from a closely related source domain improves the accuracy of the classifier. In addition, this method outperformed the domain adaptation naïve Bayes classifier proposed in our previous work [13], as well as the method proposed by [8] in every case, and outperformed the best overall domain adaptation SVM classifier proposed by [22] in 9 out of the 16 cases. Based on these results we would recommend using our proposed method over the domain adaptation SVM classifier when the source and target domains are closely related, or when there is quite a bit of labeled data for the target domain. However, when there are only very little labeled data for the target domain and the domains are more distantly related, we would recommend using the SVM algorithm proposed by [22].

5   Conclusions and Future Work

In this paper, we compared two domain adaptation algorithms derived from the supervised logistic regression classifier for the task of splice site prediction. One of these algorithms is our implementation of the method proposed in [8], in which the optimization function is modified. With this approach, a model for the source domain is learned first, and then a model for the target domain is learned with the target parameters' values constrained to be close to the source parameters' values through the optimization function. The other algorithm is our proposed method that uses a convex combination of a supervised logistic regression classifier trained on the source data and a supervised logistic regression classifier trained on the target data to approximate the posterior probability for every instance from the test set of the target domain.

We evaluated these classifiers on four target domains of increasing distance from the source domain. While the method proposed by [8] performed worse in most cases than the domain adaptation naïve Bayes classifier proposed in our previous work [13], our newly proposed method outperformed the best overall domain adaptation SVM classifier [22] in 9 out of the 16 cases. Our empirical evaluation of these classifiers also provided evidence that the task of splice site prediction becomes more difficult as the complexity of the organism increases.

In future work, we would like to explore ways to improve the accuracy of the classifier, even with these highly imbalanced data. For example, we would like to randomly split the negative instances to create smaller balanced data sets. Then, we would train an ensemble of classifiers with the method we proposed in this paper. Furthermore, we would like to evaluate the effectiveness of our proposed method on other problems that can be addressed in a domain adaptation framework, e.g. text classification problems, sentiment analysis.

Acknowledgments.. This work was supported by an Institutional Development Award (IDeA) from the National Institute of General Medical Sciences of the National Institutes of Health under grant number P20GM103418. The content is solely the responsibility of the authors and does not necessarily represent the official views of the National Institute of General Medical Sciences or the National Institutes of Health. The computing for this project was performed on the Beocat Research Cluster at Kansas State University, which is funded in part by grants MRI-1126709, CC-NIE-1341026, MRI-1429316, CC-IIE-1440548.

References

1. Arita, M., Tsuda, K., Asai, K.: Modeling splicing sites with pairwise correlations. Bioinformatics 18(suppl 2), S27–S34 (2002)
2. Baten, A.K.M.A., Halgamuge, S.K., Chang, B., Wickramarachchi, N.: Biological Sequence Data Preprocessing for Classification: A Case Study in Splice Site Identification. In: Liu, D., Fei, S., Hou, Z., Zhang, H., Sun, C. (eds.) ISNN 2007, Part II. LNCS, vol. 4492, pp. 1221–1230. Springer, Heidelberg (2007)

3. Baten, A.K.M.A., Chang, B.C.H., Halgamuge, S.K., Li, J.: Splice site identification using probabilistic parameters and svm classification. BMC Bioinformatics 7(suppl 5), S15 (2006)
4. Bernal, A., Crammer, K., Hatzigeorgiou, A., Pereira, F.: Global Discriminative Learning for Higher-Accuracy Computational Gene Prediction. PLoS Comput. Biol. 3(3), e54 (2007)
5. Brown, M.P.S., Grundy, W.N., Lin, D., Cristianini, N., Sugnet, C., Furey Jr., T.S., Ares, M., Haussler, D.: Knowledge-based Analysis of Microarray Gene Expression Data Using Support Vector Machines. PNAS 97(1), 262–267 (2000)
6. Cai, D., Delcher, A., Kao, B., Kasif, S.: Modeling splice sites with Bayes networks. Bioinformatics 16(2), 152–158 (2000)
7. Catal, C., Diri, B.: Unlabelled extra data do not always mean extra performance for semi-supervised fault prediction. Expert Systems 26(5), 458–471 (2009); Wiley Online Library
8. Chelba, C., Acero, A.: Adaptation of maximum entropy capitalizer: Little data can help a lot. Computer Speech & Language 20(4), 382–399 (2006)
9. Dai, W., Xue, G.R., Yang, Q., Yu, Y.: Transferring Naïve Bayes Classifiers for Text Classification. In: Proceedings of the 22nd AAAI Conference on Artificial Intelligence (2007)
10. Davis, J., Goadrich, M.: The relationship between Precision-Recall and ROC curves. In: Proceedings of the Twenty Third International Conference on Machine Learning, pp. 233–240. ACM (2006)
11. Gross, S.S., Do, C.B., Sirota, M., Batzoglou, S.: Contrast: a discriminative, phylogeny-free approach to multiple informant de novo gene prediction. Genome Biology 8(12), R269 (2007)
12. Giannoulis, G., Krithara, A., Karatsalos, C., Paliouras, G.: Splice site recognition using transfer learning. In: Likas, A., Blekas, K., Kalles, D. (eds.) SETN 2014. LNCS (LNAI), vol. 8445, pp. 341–353. Springer, Heidelberg (2014)
13. Herndon, N., Caragea, D.: Empirical Study of Domain Adaptation with Naïve Bayes on the Task of Splice Site Prediction. In: Proceedings of the 5th International Conference on Bioinformatics Models, Methods and Algorithms, pp. 57–67 (2014)
14. Herndon, N., Caragea, D.: Predicting Protein Localization Using a Domain Adaptation Approach. In: FernÁndez Chimeno, M., Fernandes, P.L., Alvarez, S., Stacey, D., Solé-Casals, J., Fred, A., Gamboa, H. (eds.) BIOSTEC 2013. CCIS, vol. 452, pp. 191–206. Springer, Heidelberg (2014)
15. Hubbard, T.J., Park, J.: Fold recognition and ab initio structure predictions using hidden markov models and β-strand pair potentials. Proteins: Structure, Function, and Bioinformatics 23(3), 398–402 (1995)
16. Korf, I., Flicek, P., Duan, D., Brent, M.R.: Integrating genomic homology into gene structure prediction. Bioinformatics 17(suppl. 1), S140–S148 (2001)
17. Le Cessie, S., Van Houwelingen, J.C.: Ridge estimators in logistic regression. Applied Statistics, 191–201 (1992)
18. Li, J.L., Wang, L.F., Wang, H.Y., Bai, L.Y., Yuan, Z.M.: High-accuracy splice site prediction based on sequence component and position features. Genet. Mol. Res. 11(3), 3432–3451 (2012)
19. Müller, K.-R., Mika, S., Rätsch, G., Tsuda, S., Schölkopf, B.: An Introduction to Kernel-Based learning Algorithms. IEEE Transactions on Neural Networks 12(2), 181–202 (2001)
20. Noble, W.S.: What is a support vector machine? Nat. Biotech. 24(12), 1565–1567 (2006)

21. Rätsch, G., Sonnenburg, S., Srinivasan, J., Witte, H., Müller, K.-R., Sommer, R., Schölkopf, B.: Improving the C. elegans genome annotation using machine learning. PLoS Computational Biology 3, e20 (2007)
22. Schweikert, G., Widmer, C., Schölkopf, B., Rätsch, G.: An Empirical Analysis of Domain Adaptation Algorithms for Genomic Sequence Analysis. In: NIPS 2008, pp. 1433–1440 (2008)
23. Sonnenburg, S., Schweikert, G., Philips, P., Behr, J., Rätsch, G.: Accurate Splice site Prediction Using Support Vector Machines. BMC Bioinformatics 8(suppl.10), 1–16 (2007)
24. Stanescu, A., Caragea, D.: Ensemble-based semi-supervised learning approaches for imbalanced splice site datasets. In: Proceedings of the 6th IEEE International Conference on Bioinformatics and Biomedicine, BIBM 2014, pp. 432–437 (2014)
25. Stanescu, A., Caragea, D.: Semi-supervised self-training approaches for imbalanced splice site datasets. In: Proceedings of the 6th International Conference on Bioinformatics and Computational Biology, BICoB 2014, pp. 131–136 (2014)
26. Stanke, M., Waack, S.: Gene prediction with a hidden markov model and a new intron submodel. Bioinformatics 19(suppl 2), ii215–ii225 (2003)
27. Steijger, T., Abril, J.F., Engström, P.G., Kokocinski, F., Hubbard, T.J., Guigó, R., Harrow, J., Bertone, P., RGASP Consortium, et al.: Assessment of transcript reconstruction methods for rna-seq. Nature Methods 10(12), 1177–1184 (2013)
28. Tan, S., Cheng, X., Wang, Y., Xu, H.: Adapting Naïve Bayes to Domain Adaptation for Sentiment Analysis. In: Boughanem, M., Berrut, C., Mothe, J., Soule-Dupuy, C. (eds.) ECIR 2009. LNCS, vol. 5478, pp. 337–349. Springer, Heidelberg (2009)
29. Zhang, Y., Chu, C.H., Chen, Y., Zha, H., Ji, X.: Splice site prediction using support vector machines with a Bayes kernel. Expert Syst. Appl. 30(1), 73–81 (2006)
30. Zien, A., Rätsch, G., Mika, S., Schölkopf, B., Lengauer, T., Müller, K.-R.: Engineering support vector machine kernels that recognize translation initiation sites. Bioinformatics 16(9), 799–807 (2000)

A Stacking-Based Approach to Identify Translated Upstream Open Reading Frames in *Arabidopsis Thaliana*

Qiwen Hu[1], Catharina Merchante[2], Anna N. Stepanova[2],
Jose M. Alonso[2], and Steffen Heber[1(⊠)]

[1]Bioinformatics Research Center, North Carolina State University,
Raleigh, NC 27606, USA
[2]Department of Plant and Microbial Biology, North Carolina State University,
Raleigh, NC 27606, USA
{qhu,atstepan,jmalonso,sheber}@ncsu.edu, merchante@uma.es

Abstract. Upstream open reading frames (uORFs) are open reading frames located within the 5' UTR of an mRNA. It is believed that translated uORFs reduce the translational efficiency of the main coding region, and play an important role in gene regulation. However, only few uORFs are experimentally characterized. In this paper, we use ribosome footprinting together with a stacking-based classification approach to identify translated uORFs in *Arabidopsis thaliana*. Our approach resulted in a set of 5360 potentially translated uORFs in 2051 genes. GO terms enriched in uORF-containing genes include gene regulation, signal transduction and metabolic pathway. The identified uORFs occur with a higher frequency in multi-isoform genes, and many uORFs are affected by alternative transcript start sites or alternative splicing events.

Keywords: uORF · Translation · Ribosome footprinting · Stacking · Classification · *Arabidopsis thaliana*

1 Introduction

Upstream open reading frames (uORFs) are open reading frames that appear in the 5' untranslated region (UTR) of an mRNA. Studies have shown that uORFs are often involved in the regulation of the downstream main open reading frame [1-3]. It is estimated that the *Arabidopsis* genome encodes more than 20,000 uORFs [4, 5]. However, only few uORFs are experimentally characterized, and in most cases it is unknown what biological function they have, and if they are translated [6-8]. Lab-based identification of functional uORFs is time-consuming (~ 4 man-months per gene), and so far, only few uORFs have been directly characterized through forward genetic analysis at the whole plant level [9, 10]. A comprehensive identification of translated uORFs via mass spectrometry has been challenging due to the short length of the encoded proteins [5]. Several studies have predicted functional uORFs based on evolutionary conservation [11, 12]. For example, in [12], the authors have developed a BLAST-based algorithm to identify conserved uORFs across eudicots species and

© Springer International Publishing Switzerland 2015 2015
R. Harrison et al. (Eds.): ISBRA 2015, LNBI 9096, pp. 138–149, 2015.
DOI: 10.1007/978-3-319-19048-8_12

reported 18 novel uORFs in *Arabidopsis thaliana*. Unfortunately, conserved uORFs only account for a small part of the uORFs in the *Arabidopsis* genome - currently, the TAIR database lists only about 70 conserved uORFs [5]. The biological function and translation status of most uORFs is still unknown.

Recently, ribosome footprinting (RF) has been developed to investigate translation via deep sequencing of ribosome protected mRNA fragments (ribosome footprints) [13]. The RF technique is able to provide experimental evidence for translation initiation sites (TISs) and uORFs. For example, Fritsch and colleagues recorded the coverage of ribosome footprints upstream of annotated TISs and trained a neural network to detect novel TISs. Their experiment identified 2994 novel uORFs in the human genome [14]. A similar study has also been performed in mouse [15].

In this paper, we use a stacking-based classification approach that combines RF data with additional genome information to identify translated uORFs in *Arabidopsis thaliana*. Using this approach, we found 5360 translated uORFs that occur in 2051 genes. In a preliminary analysis of the predicted uORFs we found that the enriched GO terms of the uORF-containing genes include gene regulation, signal transduction and metabolic pathway, and that uORFs are prevalent in multi-isoform genes.

2   Material and Methods

Our approach consists of five steps: (1) First, we aligned ribosome footprints and the corresponding mRNA reads to the genome sequence of *Arabidopsis thaliana,* and assigned the aligned sequences to uORFs and the annotated main coding sequences. (2) For each uORF, we extracted 12 features for subsequent analyses from our dataset. (3) Then, we used *k*-means clustering to construct a training dataset. (4) We trained five different base-level classifiers. (5) Finally, we used a stacking approach to combine the results of the base classifiers in order to achieve more accurate results. A detailed description of the individual steps is given below.

2.1   Data Preparation

Ribosome footprints and RNA-seq data were generated from *Arabidopsis thaliana* wildtype using the Illumina HiSeq2000 platform. We analyzed over 90 million reads from two biological replicates. First, we performed quality control and removed adaptor sequences and low quality reads using the FASTX-Toolkit (http://hannonlab. cshl.edu/fastx_toolkit/). The resulting reads were aligned to the genome sequence of *Arabidopsis thaliana* using Tophat [16]. Reads that mapped to multiple genomic positions, as well as reads with length smaller than 25bp, or larger than 40bp, were discarded. The remaining reads were assigned to transcript regions using custom perl scripts. Genome and transcript sequences, as well as gene annotation were downloaded from The Arabidopsis Information Resource (TAIR, version 10, http://www.arabidopsis.org/). We generated an exhaustive list of uORFs, where each uORF corresponds to a sequence of start and stop codons (start: ATG, stop: TAG, TAA, TGA) interrupted by one or more additional codons. All generated uORFs start in the 5'UTR, but they might extend beyond this transcript region. We excluded uORFs that consist of only a start and a stop codon (but no additional codons) from our analysis, because such uORFs are not considered to be functional [17]. This resulted in 29629 uORFs in 7831 genes.

2.2   Feature Extraction

Several features of functional uORFs are known to have an impact on the translation of the downstream coding sequence (CDS) [18]. Those features include the length of uORF and the distance between the uORF and its CDS. It has been shown that ribosome footprints tend to accumulate at the translation initiation sites, as a consequence, more ribosome footprints align to the start site of the translated open reading frames [15, 19]. According to these observations, we extracted features that characterize the distribution of ribosome footprints in the neighborhood of a uORF, and measure their relative position with respect to the CDS. The following part describes the individual features in detail.

Denote $u(1),u(2),\ldots u(l)$ the sequence positions of an uORF with length l in a transcript t. We assume that the CDS starts at position s in t. For $i=1,\ldots,l$ we denote the number of ribosome footprints mapping to $u(i)$ by $c(u(i))$. We computed 12 features for each possible uORF:

1) Distance from uORF start to the start of CDS: $Ds = s - u(1)$.
2) Distance from uORF end to CDS: $De = s - u(l)$.
3) Length of uORF: l.
4) Distance from uORF start to the nearest peak of the ribosome density curve: Dp.

Assume the number of ribosome footprints aligned to the positions of a gene with length n were counted as $(c(1), c(2),\ldots, c(n))$. We use a kernel smoother to estimate the ribosome density curve:

$$f_h(x) = \frac{1}{nh} \sum_{i=1}^{n} \frac{K_h(x - c(i))}{h},$$

where K_h is the kernel function and h is the smoothing parameter (bandwidth). We used the R function density with kernel function Gaussian and bandwidth 5. Peaks $p(1),p(2),\ldots,p(k)$ of the density curve indicate positions where ribosome footprints accumulate. We have

$$Dp = \min(\,|\,p(i) - u(1)\,|\,), i = 1,2,\ldots,k$$

5) Ribosome density of a uORF: Den.

$$Den = \frac{\sum_{i=u(1)}^{u(l)} c(i)}{l}$$

6) Maximum local ribosome density of uORF (window size 3): Den_max.

The local ribosome density is calculated using a sliding window of size three along the uORF region. Den_max is the maximum value of the resulting local ribosome densities.

7) Minimum local ribosome density of uORF (window size 3): *Den_min*.

Den_min is the minimum value of all local ribosome densities.

8) Ribosome density for the region left of uORF: *Den_left*.

The ribosome density upstream of uORF indicates ribosome loading before start codon. We chose 15bp as it is about the half length of ribosome in *Arabidopsis*. We have

$$Den\_left = \frac{\sum_{i=u(i)-15}^{u(i)} c(i)}{15},$$

9) Ribosome density for the region right of uORF: *Den_right*.

$$Den\_right = \frac{\sum_{i=w(i)}^{u(i)+15} c(i)}{15},$$

10) Variance of ribosome footprints distribution along uORF region: *Var*.

$$Var = \frac{1}{i} \sum_{i=u(i)}^{u(i)} (c(i) - \mu)^2,$$

where μ is the mean value of c(*i*) in the uORF region.

11) Ribosome density of UTR region: *Den_utr*.

Assume the utr region extends from position *a* to position *b* on a transcript, we have

$$Den\_utr = \sum_{i=a}^{b} \frac{c(i)}{b - a + 1},$$

12) Ribosome density of CDS: *Den_cds*.

Assume the CDS region extends from position *n* to position *k* on a transcript, we have

$$Den\_cds = \sum_{i=n}^{k} \frac{c(i)}{k - n + 1},$$

2.3   Training Set Construction

We performed *k*-means clustering to identify groups of similar uORFs. The resulting clusters were characterized with respect to their translation behavior. A detailed description is given in Section 3.1. The training set is constructed based on our clustering result. The positive class (translated uORFs) is chosen from cluster 2.1 (see Figure 2). The uORFs in this cluster show the characteristic distribution of ribosome footprints in a canonically translated uORF. There are 76 uORFs in this cluster. The negative class is randomly selected from cluster 1 and 2.2. The uORFs in these clusters exhibit a small ribosome footprints density, or ribosome footprints accumulate far from the translation start site. We selected 76 records from both classes, resulting in a training set with 152 records.

2.4   Base-Level Classifier

Our classification model is developed based on the positive and negative classes described above. We use five base-level classification algorithms: a k-nearest neighbor classifier, a support vector machine (SVM), a decision tree, a Naïve Bayes classifier, and a neural network. We have chosen these five algorithms because they used the different classification strategies. Each classifier was tuned and the model with lowest error rate was used. The performance of the different classifiers was evaluated by a leave-one-out cross validation.

2.5   Stacking

Stacking is a method that combines the predictions of several base-level learning algorithms by a meta-level learning algorithm in order to improve predictive accuracy. It has been shown that stacking can combine the expertise of different base-level classifiers while reducing their bias [21]. We refer the reader to [20, 21] for a detailed description of the stacking framework. Here, we use stacking to combine the results from different classifiers and the features from the uORF regions in the training set. First, we use our training set to train the base-level classifier. We record the results of the base-level classifiers and use them together with the extracted features of the training set to generate a meta-level k-nearest neighbor classifier. Finally, we use leave-one-out cross validation to evaluate the performance of our approach. Suppose our training dataset D consists of N records D(1), ...,D(N). For each record D(i), $i=1,...,N$, we train the base-level and meta-level classifiers using D - D(i), and we evaluate their performance using D(i). The algorithm is given below.

Algorithm: Stacking Classification (LOOCV)

  1: **for all** data points j **do**
  2:   **for all** base-level classifiers BC **do**
  3:     Train model BC based on training set D - D(j)
  4:     Use BC to predict class labels of D - D(j)
  5:   **done**
  6:   Combine prediction results of base-level classifiers with features of D - D(j)
  7:   Train meta-level classifier MC based on combined dataset.
  8:   Use MC to predict D(j)
  9: **done**
10: Calculate accuracy, precision, recall and f-score.

---

3   Results

3.1   Cluster Analysis

To identify groups of similar uORFs, we performed a cluster analysis using k-means clustering and Euclidean distance. We restricted our analysis to well-expressed genes that contain one single uORF in their 5' UTR. To determine a suitable number of clusters k, we used the average silhouette value [22]. The silhouette value measures the fit of a data point within its cluster in comparison with neighboring clusters. Silhouette values are in the range of -1 to 1. A silhouette value close to 1 indicates that a data point is in an appropriate cluster, while a silhouette value close to -1 indicates that it might be erroneously assigned. Figure 1 shows the average silhouette value for different numbers of clusters k. The average silhouette values for $k=2, \ldots, 6$ clusters are similar and clearly larger than average silhouette values for $k>6$.

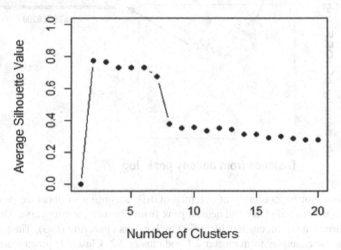

Fig. 1. Average silhouette value for different numbers of clusters k

To identify translated uORFs, we have focused on the joint distribution of *Den* and *Dp* feature values - these two features are the most important features that determine the translation status of uORFs [15, 19]. Figure 2 shows a contour plot of our dataset. The Figure shows two clusters that coincide with the cluster reported by 2-means clustering: cluster 1 (green diamond at the bottom of Figure 2) consists of uORFs for which only few ribosomal footprints have been detected. In contrast, the uORFs in cluster 2 show ample ribosomal footprints. 3-means clustering splits cluster 2 into two clusters: cluster 2.2 with large *Dp* and smaller *Dens* values (points with black triangle) and cluster 2.1 with small *Dp* and large *Dens* values (points with red circle); cluster 1 remains unchanged. For k-means clustering with $k=4, \ldots, 6$, the cluster 2.1 and 2.2 remained unchanged, while cluster 1 is subdivided. Therefore, we decided to choose $k=3$ clusters for our subsequent analyses.

To learn the characteristics of the uORFs in the different clusters, we analyzed their features. There are no ribosome footprints in the uORFs from cluster 1, and we hypothesize that the uORFs in this group are not translated. The group accounts for about 65% of the total uORFs in our dataset. The uORFs from cluster 2.1 and 2.2 have a positive footprint density *Den*, but we observed a significant differences in the variable *Dp*. *Dp* is significantly larger in cluster 2.2 indicating that fewer ribosomal footprints accumulate at the start codon of the corresponding uORFs. This is inconsistent with translated open reading frames [10,15]. In addition, we analyzed:

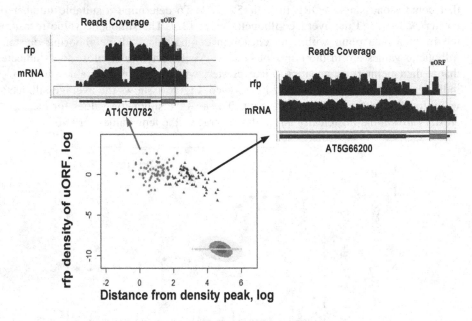

Fig. 2. Contour plot for clustering of potential uORFs according to ribosome density and distance between the start of uORF and nearest peak from ribosome density curve. Different shape of points indicates different clusters identified by *k*-means clustering (*k*=3). The reads coverage plots above show examples from cluster 2.1 and cluster 2.2. Cluster 1: green diamond. Cluster 2.1: red points. Cluster 2.2: black triangle. The background color indicates the density of points, a darker color indicates a higher point density.

1) Experimentally verified uORFs: there are two experimentally verified uORFs whose genes are well expressed and translated in our dataset. Both uORFs belong to cluster 2.1.

2) GO terms of uORF-containing genes: we used AgriGO [23] to identify overrepresented GO terms. Genes in cluster 2.1 show terms such as biological regulation (GO:0065007), metabolic process (GO:0008152) and cellular process (GO:0009987). This is consistent with the GO term annotation of currently known uORF-containing genes [5-7, 24]. We did not find overrepresented GO terms for cluster 1 and 2.2.

3) *Den_min*: *Den_min* indicates the local coverage of ribosome footprints in a uORF region. A *Den_min* value larger than 0 indicates the continuous translation of ribosomes in the region. Ideally, a well translated uORF should show *Den_min*=0 for only a small fraction of its length. We found a significant difference of this value between cluster 2.1 and cluster 2.2 (Figure 3). For cluster 2.1, about 20% uORFs have *Den_min*=0, whereas in cluster 2.2 we have 74%.

4) Ribosome density in the first 6bp region immediately after the start codon: we checked the footprint density immediately after the start codon (Figure 3). Our analysis indicates that ~70% of uORFs in cluster 2.2 do not show any footprints in this region. In contrast, all uORFs in cluster 2.1 show a non-zero footprint density in this region.

Based on these observations we chose the uORFs in cluster 2.1 as templates for translated uORFs and used them as positive class to train our classifiers. After inspecting representative examples from cluster 2.2 (Figure 2), we hypothesize that some of the uORFs in this cluster might use a different, non-canonical translation start site.

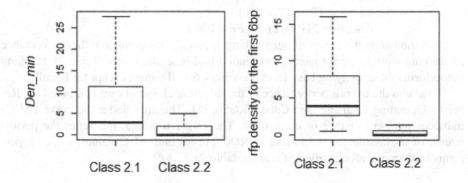

Fig. 3. Distribution of minimal local density and ribosome footprints density in the first 6bp region immediately after the start codon for uORFs in cluster 2.1 and 2.2

3.2   Performance Evaluation

To evaluate the performance of each base-level classifier, we compared the data points of the training set that are identified correctly by a specific algorithm. We found a large overlap between algorithms, however, each classifier also detects a certain proportion of the data which is not detected by the other algorithms (Table 1). The results suggest that each classifier has its own expertise for classifying uORFs correctly, and stacking the classifiers may improve their performance.

Table 1. Intersection of correctly classified data points for the different classification algorithms. The numbers on the main diagonal in the table indicate the total number of correctly classified data points for the individual classifier. The off-diagonal numbers is the number of data that correctly identified and overlapped between two algorithms. The number in the brackets is the overlap percentage for the classifiers in rows. SVM: support vector machine, DT: decision tree, NB: Naïve Bayes, NN: neural network, KNN: k-nearest neighbor.

	SVM	DT	NB	NN	KNN
SVM	115	111(97%)	87(78%)	108(97%)	105(95%)
DT	111(82%)	136	96(71%)	122(90%)	111(82%)
NB	87(83%)	96(91%)	105	94 (90%)	86(82%)
NN	108(83%)	122(94%)	94(72%)	130	107(82%)
KNN	105(91%)	111(97%)	86(75%)	107(93%)	115

To evaluate the performance of these classifiers, we performed a leave-one-out cross validation and calculated accuracy, precision, recall and f-score. Denote TP, FN, TN and FP is the number of true positives, false negatives, true negatives and false positives. We have

Accuracy = (TP + TN) / (TP + FN + TN + FP),
Precision = TP / (TP + FP),
Recall = TP / (TP + FN),
F-score = 2TP / (2TP + FP + FN).

To demonstrate the power of our stacking approach, we compared the performance of stacking with the performance of the individual base classifiers (Table 2). Stacking outperforms the underlying base-level classifiers for all values except for Recall.

To assess the overall performance of the different classifiers we computed the Receiver Operating Characteristic Curve (ROC)[25]. The area under the curve (AUC) indicates the performance of a classifier. The larger the area, the better the performance of a classifier is. According to ROC curves and AUC values, stacking performs best among all classifiers (Figure 3, table 2).

Table 2. Overall performance for each classifier. AUC: area under the curve

Algorithms	Accuracy	Precision	Recall	F-score	AUC
KNN	0.76	0.63	0.84	0.72	0.85
SVM	0.76	0.74	0.77	0.75	0.85
Decision Tree	0.89	0.84	**0.94**	0.89	0.86
NaiveBayes	0.69	0.54	0.77	0.64	0.86
Neural network	0.86	0.87	0.85	0.86	0.92
Stacking with KNN	**0.90**	**0.95**	0.87	**0.91**	**0.94**

3.3   Translated uORFs in *Arabidopsis thaliana*

Our stacking approach identified 5360 translated uORFs. The identified uORFs occur in 2051 genes, which account for about 6% of the all annotated genes in *Arabidopsis thaliana*. Likely, this number is an under-estimation since about 30% of uORF-containing genes were not transcribed in our experiment. Remarkably, the majority of translated uORFs occurs in multi-isoform genes. When comparing single- and multi-isoform genes with respect to the occurrence of translated uORFs, we found a significant difference (p-value < 2.2e-16, Fisher's exact test); only 3.4% of the single-isoform genes contain translated uORFs, whereas about 19% of the multi-isoform genes contain translated uORFs. About 15% of these uORFs do not occur in all transcripts generated by the multi-isoform gene, see Table 3 for a detailed breakdown. We hypothesize, that in some cases alternative transcription start sites, or alternative splicing events (AS events), might regulate presence and absence of translated uORFs.

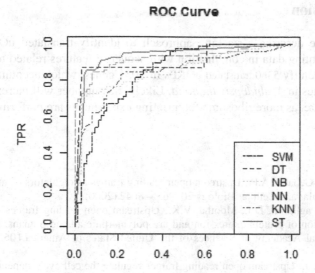

Fig. 4. ROC curves for different classifiers. SVM support vector machine, DT decision tree, NB: Naïve Bayes, NN neural network, KNN k-nearest neighbor, ST stacking with KNN

The majority (~58%) of the genes that contain translated uORFs contain only one single uORF. However, there are few genes that include up to 30 uORFs. To further characterize uORF-containing genes, we performed a GO-term analysis. Enriched Go terms include catalytic activity, binding, transferase activity, phosphotransferase activity, kinase activity and transcription regulator activity. Our results provide ample candidates for experimental characterization and functional analysis of uORFs.

Table 3. Potential uORFs identified in the genome of Arabidopsis. uORF level indicates the number of uORFs identified. Gene level indicates the number of genes that contains potential uORFs. There are totally 27717 single-isoform genes and 5885 multi-isoform genes.

Translated uORFs in *Arabidopsis* genome	uORF Level	Gene Level	Percentage of total genes
Total uORFs identified	5360	2051	6.10%
uORFs in multi-isoform genes	3783	1121	3.34%
uORFs affected by AS events	580	293	0.87%
uORFs not affected by AS events	3203	828	2.46%
uORFs in single-isoform genes	1577	930	2.77%

4   Conclusion

In this paper, we describe a stacking approach to identify translated uORFs using ribosome footprinting data in combination with sequence features related to functional uORFs. We identify 5360 translated uORFs in 2051 genes, which account for 6% of all annotated genes in *Arabidopsis thaliana*. Likely, this number will increase significantly in the future, as more ribosome footprinting experiments are performed.

References

1. Morris, D.R., Geballe, A.P.: Upstream open reading frames as regulators of mRNA translation. Molecular and Cellular Biology 20, 8635–8642 (2000)
2. Calvo, S.E., Pagliarini, D.J., Mootha, V.K.: Upstream open reading frames cause widespread reduction of protein expression and are polymorphic among humans. Proceedings of the National Academy of Sciences of the United States of America 106, 7507–7512 (2009)
3. Jeon, S., Kim, J.: Upstream open reading frames regulate the cell cycle-dependent expression of the RNA helicase Rok1 in Saccharomyces cerevisiae. FEBS Letters 584, 4593–4598 (2010)
4. Kim, B.H., Cai, X., Vaughn, J.N., von Arnim, A.G.: On the functions of the h subunit of eukaryotic initiation factor 3 in late stages of translation initiation. Genome Biology 8, R60 (2007)
5. von Arnim, A.G., Jia, Q., Vaughn, J.N.: Regulation of plant translation by upstream open reading frames. Plant Science: an International Journal of Experimental Plant Biology 214, 1–12 (2014)
6. Imai, A., Hanzawa, Y., Komura, M., Yamamoto, K.T., Komeda, Y., Takahashi, T.: The dwarf phenotype of the Arabidopsis acl5 mutant is suppressed by a mutation in an upstream ORF of a bHLH gene. Development 133, 3575–3585 (2006)
7. Alatorre-Cobos, F., Cruz-Ramirez, A., Hayden, C.A., Perez-Torres, C.A., Chauvin, A.L., Ibarra-Laclette, E., Alva-Cortes, E., Jorgensen, R.A., Herrera-Estrella, L.: Translational regulation of Arabidopsis XIPOTL1 is modulated by phosphocholine levels via the phylogenetically conserved upstream open reading frame 30. Journal of Experimental Botany 63, 5203–5221 (2012)
8. Ebina, I., Takemoto-Tsutsumi, M., Watanabe, S., Koyama, H., Endo, Y., Kimata, K., Igarashi, T., Murakami, K., Kudo, R., Ohsumi, A., Noh, A.L., Takahashi, H., Naito, S., Onouchi, H.: Identification of novel Arabidopsis thaliana upstream open reading frames that control expression of the main coding sequences in a peptide sequence-dependent manner. Nucleic Acids Research 43, 1562–1576 (2015)
9. Hanfrey, C., Franceschetti, M., Mayer, M.J., Illingworth, C., Michael, A.J.: Abrogation of upstream open reading frame-mediated translational control of a plant S-adenosylmethionine decarboxylase results in polyamine disruption and growth perturbations. The Journal of Biological Chemistry 277, 44131–44139 (2002)
10. Selpi, B.C.H., Kemp, G.J., Sarv, J., Kristiansson, E., Sunnerhagen, P.: Predicting functional upstream open reading frames in Saccharomyces cerevisiae. BMC Bioinformatics 10, 451 (2009)
11. Cvijovic, M., Dalevi, D., Bilsland, E., Kemp, G.J., Sunnerhagen, P.: Identification of putative regulatory upstream ORFs in the yeast genome using heuristics and evolutionary conservation. BMC Bioinformatics 8, 295 (2007)

12. Takahashi, H., Takahashi, A., Naito, S., Onouchi, H.: BAIUCAS: a novel BLAST-based algorithm for the identification of upstream open reading frames with conserved amino acid sequences and its application to the Arabidopsis thaliana genome. Bioinformatics 28, 2231–2241 (2012)
13. Ingolia, N.T., Ghaemmaghami, S., Newman, J.R., Weissman, J.S.: Genome-wide analysis in vivo of translation with nucleotide resolution using ribosome profiling. Science 324, 218–223 (2009)
14. Fritsch, C., Herrmann, A., Nothnagel, M., Szafranski, K., Huse, K., Schumann, F., Schreiber, S., Platzer, M., Krawczak, M., Hampe, J., Brosch, M.: Genome-wide search for novel human uORFs and N-terminal protein extensions using ribosomal footprinting. Genome Research 22, 2208–2218 (2012)
15. Ingolia, N.T., Lareau, L.F., Weissman, J.S.: Ribosome profiling of mouse embryonic stem cells reveals the complexity and dynamics of mammalian proteomes. Cell 147, 789–802 (2011)
16. Trapnell, C., Pachter, L., Salzberg, S.L.: TopHat: discovering splice junctions with RNA-Seq. Bioinformatics 25, 1105–1111 (2009)
17. Andrews, S.J., Rothnagel, J.A.: Emerging evidence for functional peptides encoded by short open reading frames. Nat. Rev. Genet. 15, 193–204 (2014)
18. Vilela, C., McCarthy, J.E.: Regulation of fungal gene expression via short open reading frames in the mRNA 5'untranslated region. Molecular Microbiology 49, 859–867 (2003)
19. Juntawong, P., Girke, T., Bazin, J., Bailey-Serres, J.: Translational dynamics revealed by genome-wide profiling of ribosome footprints in Arabidopsis. Proceedings of the National Academy of Sciences of the United States of America 111, E203–E212 (2014)
20. Saso Dzeroski, B.Z.: Is Combining Classifiers with Stacking Better than Selecting the Best One. Machine Learning 54, 255–273 (2004)
21. Wolpert, D.H.: Stacked generalization. Neural Networks 5, 241–259 (1992)
22. Rousseeuw, P.J.: Silhouettes: a Graphical Aid to the Interpretation and Validation of Cluster Analysis. Computational and Applied Mathematics, 53–65 (1987)
23. Du, Z., Zhou, X., Ling, Y., Zhang, Z., Su, Z.: agriGO: a GO analysis toolkit for the agricultural community. Nucleic Acids Research 38, W64–W70 (2010)
24. Tabuchi, T., Okada, T., Azuma, T., Nanmori, T., Yasuda, T.: Posttranscriptional regulation by the upstream open reading frame of the phosphoethanolamine N-methyltransferase gene. Bioscience, Biotechnology, and Biochemistry 70, 2330–2334 (2006)
25. Fawcett, T.: An introduction to ROC analysis. Pattern Recogn. Lett. 27, 861–874 (2006)

PRESS-PLOT: An Online Server for Protein Structural Analysis and Evaluation with Residue-level Virtual Angle Correlation Plots

Yuanyuan Huang[1,3], Kejue Jia[1], Robert Jernigan[1,2], and Zhijun Wu[1,3(✉)]

[1] Program of Bioinformatics and Computational Biology, Iowa State University,
Ames, IA, USA
zhijun@iastate.edu
[2] Department of Biochemistry, Biophysics and Molecular Biology, Iowa State
University, Ames, IA, USA
[3] Department of Mathematics, Iowa State University, Ames, IA, USA

Abstract. Based on the data and tools in PRESS, especially the residue-level virtual angle correlation plots, a web-server called PRESS-PLOT is further developed for easy access and display of the plots for structural analysis and evaluation. A structure to be analyzed and evaluated can be submitted to the server by either giving its structural ID in PDB or uploading its structural file in the PDB format. The residue-level virtual bond angles and torsion angles of the structure are then computed. The neighboring virtual bond angle and torsion angle pairs are displayed as scattered points in a 2D graph and compared against the 2D contour map of the density distribution of such angle pairs in known protein structures, as given in the background of the 2D graph. The virtual angle pairs that can be analyzed and evaluated include α-τ and τ-β angle pairs as they appear in either general structures or specific secondary structures such as α-helices, β-sheets, or their turns. As a justification of PRESS-PLOT, more than 1000 obsoleted structures (with lower resolutions) in PDB are evaluated using PRESS-PLOT and compared with their current superseded versions (with higher resolutions). The results show that PRESS-PLOT distinguishes high-quality structures (the current ones) from low-quality structures (the obsoleted ones) clearly in its angle correlation plots. The PRESS-PLOT server can be accessed online at [http://pidd.math.iastate.edu/press/].

Keywords: Protein structural bioinformatics · Protein residue distances and angles · Statistical structural analysis · Online servers for structural evaluation

1 Introduction

With the enormous number of protein structures already determined and deposited in PDB, statistical learning becomes not just a necessary but also feasi-

Z. Wu—This work is partially supported by the NIH/NIGMS grant R01GM081680 and by the NSF/DMS grant DMS0914354.

ble and revitalizing tool for structural bioinformatics: Many structural properties
can now be surveyed statistically in the database of known protein structures.
The distributions or correlations of these properties in the structures can be
computed for structural inferences. They provide a wealth of information for re-
covering general structural properties beyond individual experimental outcomes.
They can be based to develop computational tools for structural analysis as well
as structural determination including structural assessment, refinement, and prediction [1].

The atomic-level structural properties of proteins, such as the backbone torsion angles ϕ, ψ, and ω, which are among the main determinants of a protein
fold, have been well studied and understood based on either chemistry knowledge or statistical analysis. For example, it is well known that the allowed range
of ω angle is very restrictive, while ϕ and ψ angles are closely correlated to each
other. The latter is a key indicator for the correct fold of a structure, and is
often demonstrated via a so-called Ramachandran Plot, a 2D contour map of
the density distribution of the ϕ-ψ angle pairs in known protein structures. The
Ramachanduan Plot has been widely adopted for structural analysis and evaluation, with its 2D contour map used as a reference for the correct formation of
the ϕ-ψ angle pairs in the structure [2,3].

Structural properties similar to those at atomic level can also be found at
residue level such as the distances between two neighbouring residues (called
virtual bonds); the angles formed by three residues in sequence (called virtual
bond angles); and the torsion angles of four residues in sequence (called virtual
torsion angles) (see Fig. 1). They can be as important as those at atomic level
for structural analysis and evaluation, especially when reduced models for proteins are considered with residues used as basic units [4]. Due to the difficulty
of measuring the residue distances and angles, either experimentally or theoretically, a statistical approach to the study of these properties becomes crucial and
necessary. Much work has been done along this line in the past [5,6,7,8,9]. In
particular, Huang et al [10] have conducted a detailed survey on residue-level
protein structural properties using a large set of known protein structures in
PDB. An R package called PRESS (Protein REsidual-level Structural Statistics) is released for the access to the structural properties calculated and to
the structural analysis tools developed [11]. Among the analysis tools developed
is a set of so-called residue-level virtual angle correlation plots, with a similar
nature of Ramachadran Plot for atomic-level angle correlations. These residue-
level angle correlation plots contain 2D contour maps of density distributions
of certain virtual bond angle and torsion angle pairs in the surveyed structures.
They can be used to analyze and evaluate any given protein structures, either
experimentally determined or theoretically predicted, with the 2D contour maps
used as references for the correct formation of the virtual angle pairs in the
structures. These angle correlation plots provide a unique and valuable set of
tools for residue-level structural analysis and assessment, and are expected to
have a useful impact in current protein modeling practices.

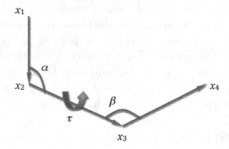

Fig. 1. Residue-level virtual angles. Assume that four residues in sequence are located at x_1, x_2, x_3, x_4. Then, the distances between the neighboring residues are called virtual bonds; the angles formed by three connected residues such as α and β are called virtual bond angles; and the dihedral angles formed by four connected residues such as τ are called virtual torsion angles.

Following Huang et al [10,11], this work is to develop a web-server called PRESS-PLOT for easy access and display of the virtual angle correlation plots in PRESS, especially for easy online access for WWW (World Wide Web) users. A structure to be analyzed and evaluated can be submitted to the server by either giving its structural ID in PDB or uploading its structural file in the PDB format. The residue-level virtual bond angles and torsion angles of the structure are then computed. The neighboring virtual bond angle and torsion angle pairs are displayed as scattered points in a 2D graph and compared against the 2D contour map of the density distribution of such angle pairs in known protein structures, as given in the background of the 2D graph. The virtual angle pairs that can be analyzed and evaluated include α-τ and τ-β angle pairs as they appear in either general structures or specific secondary structures such as α-helices, β-sheets, or their turns. As a justification of PRESS-PLOT, more than 1000 obsoleted structures (with lower resolutions) in PDB are evaluated using PRESS-PLOT and compared with their current superseded versions (with higher resolutions). The results show that PRESS-PLOT distinguishes high-quality structures (the current ones) from low-quality structures (the obsoleted ones) clearly in its angle correlation plots. The PRESS-PLOT server can be accessed online at [http://pidd.math.iastate.edu/press/].

2   Implementation

PRESS-PLOT is derived from PRESS structural data and functions for structural analysis and evaluation using residue-level virtual angle correlation plots. Different from PRESS, PRESS-PLOT is focused on structural assessment. It has a web interface for online access. It also evaluates the virtual angle correlations for specific as well as general secondary structures. The development of PRESS-PLOT is motivated by the successful application of residue-level virtual angle correlation plots to structural assessment and justified by extensive testings on a large set of current vs. obsoleded structures in PDB.

Structural Data

As in PRESS, a total of 1052 X-ray crystallography structures are downloaded from PDB, with resolution $\leq 1.5\text{Å}$, sequence similarity $\leq 30\%$, and only single chains. The angle sequences α-τ-β for all four residue sequences in the structures are calculated and stored in a database named **ATA-database**. Each record in the database contains the following information:

$$\boxed{\text{ID} \mid R_1, S_1 \mid R_2, S_2 \mid R_3, S_3 \mid R_4, S_4 \mid \alpha \mid \tau \mid \beta \mid SS}$$

where ID is the structural ID in PDB, R_j is the type of the jth residue in the sequence, S_j is the secondary structure type of R_j, α, τ, β are the corresponding virtual bond and torsion angles, and SS is the type of the secondary structure of the whole residue sequence. The last item is determined by the following rules: A four residue sequence R_1-R_2-R_3-R_4 is considered to be in

1. α-helix: if R_1, R_2, R_3, R_4 are in α-helix
2. head of α-helix: if R_2, R_3, R_4 are in α-helix
3. tail of α-helix: if R_1, R_2, R_3 are in α-helix
4. β-sheet: if R_1, R_2, R_3, R_4 are in β-sheet
5. head of β-sheet: if R_2, R_3, R_4 are in β-sheet
6. tail of β-sheet: if R_1, R_2, R_3 are in β-sheet

where the secondary structure type of each residue is identified by using the program DSSP [12]. With the identification of the secondary structure type, PRESS-PLOT is capable of evaluating the virtual angle correlations when they are in specific types of secondary structures, while PRESS evaluates the correlations without specifying the secondary structure types of the angle pairs.

Calculation of Virtual Angles

The virtual bond angles and torsion angles are calculated using standard trigonometric relations, given the positions of the residues in the residue sequences. In all the calculations, the position of the backbone C_α of each residue is used to represent the position of the residue.

Let R_1, R_2, R_3 be a sequence of three residues located at positions x_1, x_2, x_3 (see Fig. 1). Let $u = x_2 - x_1$, $v = x_3 - x_2$. Then,

$$\|u + v\|^2 = \|u\|^2 + \|v\|^2 - 2\|u\|\|v\|\cos\alpha$$

where $\|\cdot\|$ is the Euclidean norm, and α is the virtual bond angle of this sequence.

Let R_1, R_2, R_3, R_4 be a sequence of four residues located at positions x_1, x_2, x_3, x_4 (see Fig. 1). Let $u = x_2 - x_1$, $v = x_3 - x_2$, $w = x_4 - x_3$. Then,

$$\|u + v + w\|^2 = \|u\|^2 + \|v\|^2 + \|w\|^2$$

$$-2\|u\|\|v\|\cos\alpha - 2\|v\|\|w\|\cos\beta - 2\|u\|\|w\|\cos\theta$$

where $\cos\theta = \sin\alpha\sin\beta\cos\tau - \cos\alpha\cos\beta$, and α, β, τ are the virtual bond and torsion angles of this sequence.

All α-τ and τ-β angle pairs in the downloaded structures are collected from the **ATA-database**. Let $[0°, 180°]$ be divided into 90 small bins for α. Let $[0°, 360°]$ be divided into 180 small bins for τ. Multiply the two intervals to form a 2D subspace $[0°, 180°] \times [0°, 360°]$. The 2D subspace then consists of 90×180 squares. The density of the α-τ angle pair in any of these squares is defined as the number of the α-τ angle pairs in that square divided by the total number of α-τ angle pairs collected. The density distribution of the τ-β angle pairs is calculated in a similar way. The calculations are also similar for the density distributions of these angle pairs when they are in certain types of secondary structures. All the above calculations are carried out in R with BIO3D as a library [13].

Plot of Density Maps

The 2D countour maps of the density distributions of α-τ and τ-β angle pairs are ploted in 2D α-τ and τ-β planes, respectively. The maps are displayed in a special graphical form similar to that for the Ramachadran Plot: Each map has three different density regions, with high 50%, 75%, and 90% of density, called **most favored**, **favored**, and **allowed** regions, and plotted in dark, less dark, and light colors, respectively. The region with lower 10% density is called **disallowed** region and colored in white (see Fig. 2). The maps for the distributions in certain secondary structure conditions are plotted similarly, with the density percentages adjusted slightly for those different density regions.

Web Interface

PRESS-PLOT is a web-based integrated online service dedicated to protein structural assessment. It helps the user to visualize the quality of a given structure in terms of its residue-level virtual angle correlations. PRESS-PLOT integrates web pages and server-side programs in a one-step query workbench, making it easy to submit queries and acquire results. It allows the user to assess a structure and display all the plots from any devices with internet connection without the need of downloading and installing any large software and complicated library dependencies. The service can be accessed anonymously without registering or providing any personal information. Each user will be assigned a query session so that multiple requests can be handled in parallel and independently. The query results can also be downloaded in different formats for future use. ([http://pidd.math.iastate.edu/press/]).

PRESS-PLOT can be broken down into two major components: front-end dynamic web pages and back-end computing components (Fig. 3). The front-end web pages are designed in MVC (Model, View, and Control) pattern, which provides a high refactoring ability and is also simple for maintenance. The result generated by PRESS-PLOT is graphic-based data. It is important that any result is presented to the user immediately. For a faster query response, AJAX (Asynchronous JavaScript and XML) technique is adopted on the web pages. It allows the web pages to update the result without refreshing all the elements on the pages. The back-end computing components are composed of two sub-units, query handling unit and computing unit. The query handling unit is responsible for pre-processing and transferring user queries to computing unit. After the results are generated, it also renders the results and outputs plots onto the web pages. The query handling unit is implemented in PHP, one of the most popular and widely supported scripting languages. The computing unit implements the core computing functions. It accepts the query information from the query handling unit, computes the virtual angle data for the input structure, and generates the final graphical results. It is implemented in R, an open source environment for statistical computing.

3   Results

PRESS-PLOT is developed to provide an online server for structural assessment using the PRESS virtual angle correlation plots. In addition, it further extends the PRESS angle correlation plots to angle pairs in specific secondary structures, which can be more accurate for specific structural types and practical for more detailed structural analysis. PRESS-PLOT is tested on a large set of structures in PDB, showing that higher-resolution structures in general have better evaluations in PRESS-PLOT angle correlation plots.

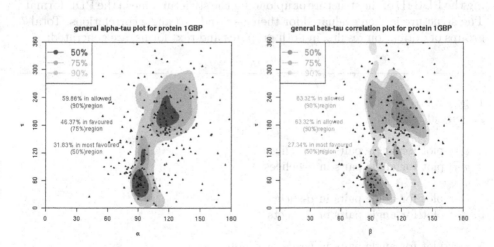

Fig. 2. Virtual angle correlation plots. The α-τ and τ-β angle correlation plots for structure 1GBP, where there are three different density regions: **most favored**, **favored**, and **allowed**, corresponding to high 50%, 75%, and 90% of α-τ and τ-β density, respectively. The rest of the area is called **disallowed** region, with lower 10% of density.

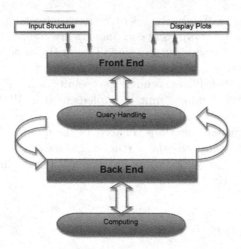

Fig. 3. PRESS-PLOT organization. PRESS-PLOT can be broken down into two parts: The front end and the back end. The front end takes the user's input structure and passes it to the query handling unit of the back end. The latter carries out preprocessing and directs the structure to the computing unit of the back end for required calculation and plot generation. The query handling unit takes the final results from the computing unit and renders them to the front end for display.

Display Functions

A structure to be evaluated can be submitted to PRESS-PLOT by either providing the PDB ID of the structure or uploading the structural file in the PDB format. The structure is then evaluated for their α-τ and τ-β angle correlations. Total 7 groups of evaluation results, in both graphics and text forms, are generated:

---

1. general α-τ plot
2. general τ-β plot

3. α-τ plot for angle pairs in α-helices
4. τ-β plot for angle pairs in α-helices

5. α-τ plot for angle pairs in β-sheets
6. τ-β plot for angle pairs in β-sheets

7. α-τ plot for angle pairs in heads of α-helices
8. τ-β plot for angle pairs in heads of α-helices

9. α-τ plot for angle pairs in heads of β-sheets
10. τ-β plot for angle pairs in heads of β-sheets

11. α-τ plot for angle pairs in tails of α-helices
12. τ-β plot for angle pairs in tails of α-helices

13. α-τ plot for angle pairs in tails of β-sheets
14. τ-β plot for angle pairs in tails of β-sheets

The first group of results is displayed in the window as default. The remaining groups are listed as small icons in the bottom of the window and can be selected to show in the window. Each plot shows the corresponding type of angle pairs in the given structure as scattered points in the corresponding density map. The percentages of the points in different density regions are summarized in the graph. Fig. 2 shows the general α-τ and τ-β correlation plots for a structure 1GBP. Examples for plots for specific secondary structures can be found at the server page.

In the first group of plots, all α-τ (τ-β) angle pairs of the given structure are calculated and ploted as scattered points in the α-τ (τ-β) plane. The background of the α-τ (τ-β) plane is the contour map of the density distribution of the α-τ (τ-β) angle pairs in general structures that include all types of secondary structures. If the percentages of the α-τ (τ-β) angle pairs of the given structure in **most favored**, **favored**, and **allowed** regions are around or above 50%, 75%, and 90%, respectively, the structure is considered to be well formed in terms of α-τ (τ-β) angle correlations.

In the second group of plots, all α-τ (τ-β) angle pairs in α-helices of the given structure are calculated and plotted as scattered points in the α-τ (τ-β) plane. The background of the α-τ (τ-β) plane is the contour map of the density distribution of the α-τ (τ-β) angle pairs in α-helices. Likewise, in the third group of plots, all α-τ (τ-β) angle pairs in β-sheets of the given structure are calculated and plotted as scattered points in the α-τ (τ-β) plane. The background of the α-τ (τ-β) plane is the contour map of the density distribution of the α-τ (τ-β) angle pairs in β-sheets. The remaining groups of plots are generated similarly for α-τ (τ-β) angle pairs in heads or tails of α-helices or β-sheets.

Testings

PRESS-PLOT is applied to evaluating a large set of obsoleted structures in PDB. The results are compared with those for the current superseded structures. Up to early 2012, there are total 1,654 obsoleted protein structures superseded by their succesors according to a report from PDB [14]. For each pair of obsoleted and replaced structures, the percentages of the virtual angle pairs in **most favored**, **favored**, and **allowed** regions of the virtual angle correlation plots are examined. The average percentages for the structural pairs with RMSD values in between 0 and 1Å, 1 and 3Å, 3 and 5Å, and beyond 5Å are calculated and summarized in Table 1 and 2. The structural pairs for which the RMSD values cannot be computed due to various reasons are considered as a separate group (RMSD: NA).

Table 1 and 2 show the average percentages of α-τ and τ-β angle pairs in general secondary structures, respectively. These results show that the current structures with higher resolutions than their previous ones all have, in average, higher percentages of virtual angle pairs in the high-density regions of virtual angle correlation plots, which implies that PRESS-PLOT can distinguish low quality structures from high quality ones very well. In particular, for the structure pairs with RMSD values in between 1 and 3Å, the differences in these percentages between the superseded and obsoleted ones are the most notable. A simple explanation is that if two structures are very similar (with RMSD < 1Å), their virtual angle correlations are certainly expected to be about the same, and therefore, their PRESS-PLOT evaluations would be similar. On the other hand, if two structures are very different (with RMSD > 3Å), they may differ in their tertiary structures but still have similar secondary structures and hence similar local structures. The latter would keep the virtual angle correlations of the two structures similar.

Table 1. Assessments of α-τ correlation plots on PDB structures. The structures are grouped according to the RMSD values of the obsolete vs. superseded structural pairs. For each group of structures, the average percentages of their α-τ angle pairs in different density regions in the α-τ correlation plots are summarized. Table legends: RMSD – RMSD range for obsoleted and superceded structural pairs; size – # of structural pairs with given RMSD range; obsX% – average percentage of obsoleted structures in high X% region; supX% – average percentage of superceded structures in high X% region.

RMSD	size	obs90%	sup90%	obs75%	sup75%	obs50%	sup50%
NA	922	86.41	88.42	71.42	73.75	47.19	49.38
(0Å, 1Å)	542	89.56	89.69	75.36	75.51	51.00	51.16
(1Å, 3Å)	37	83.04	86.39	68.66	72.11	44.06	48.52
(3Å, 5Å)	17	86.42	86.95	69.26	71.20	43.47	43.23
(5Å, ∞)	136	84.09	86.31	67.69	69.84	42.62	45.03

The above statistics are further demonstrated by using boxplots in Fig. 4, where the values at four different quartiles of the percentages are plotted for the structural pairs with RMSD values between 1 and 3Å. These plots show that our conclusions above are also valid even in terms of the quartile values including the medians of the percentages. In addition, we have also calculated, for all the structures, their average percentages of α-τ and τ-β angle pairs in different density regions of the angle correlation plots in different secondary structures, including α-helices, β-sheets, heads and tails of α-helices and β-sheets. In a similar fashion, we have also compared these average percentages for all obsoleted and superseded structural pairs. All the results (not shown) are consistent with the above results on general secondary structures.

Table 2. Assessments of τ-β correlation plots on PDB structures. The structures are grouped according to the RMSD values of the obsolete vs. superceded structural pairs. For each group of structures, the average percentages of their τ-β angle pairs in different density regions in the τ-β correlation plots are summarized. Table legends: RMSD – RMSD range for obsoleted and superceded structural pairs; size – # of structural pairs with given RMSD range; obsX% – average percentage of obsoleted structures in high X% region; supX% – average percentage of superceded structures in high X% region.

RMSD	size	obs90%	sup90%	obs75%	sup75%	obs50%	sup50%
NA	922	86.44	88.27	71.83	74.01	48.03	50.22
(0Å,1Å)	542	89.60	89.76	75.35	75.57	51.86	52.00
(1Å, 3Å)	37	82.95	86.14	68.65	72.68	44.42	49.53
(3Å, 5Å)	17	85.54	86.15	68.81	71.12	45.12	46.08
(5Å, ∞)	136	83.86	84.94	67.60	69.70	43.22	45.72

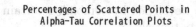

Percentages of Scattered Points in Alpha-Tau Correlation Plots   **Percentages of Scattered Points in Tau-Beta Correlation Plots**

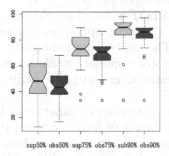 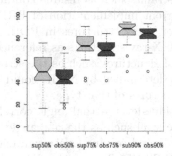

Fig. 4. Assessment of α-τ and τ-β correlation plots. For the structural pairs with RMSD values between 1 and 3Å and for each density region of the α-τ (or τ-β) correlation plot, the obsoleted and superseded structures are each divided into four quartiles, in terms of their percentages of α-τ (or τ-β) angle pairs in the region. The percentage values at these quartiles are then plotted in a boxplot form. The plots show that the values at all these quartiles including the medians of the percentages for the supersede structures (in orange color) are consistently higher than those for the obsoleted ones (in green color) in all different density regions.

4   Discussion

Atomic-level structural analysis tools such as the Ramachandran Plot have been used successfully for protein structural analysis and evaluation. Residue-level structural properties are as important as those at atomic level for protein modeling but are more difficult to measure. PRESS-PLOT has provided an extremely valuable set of tools to analyze and evaluate protein structures based on their residue-level virtual angle correlations. The effectiveness of the tools are clearly demonstrated in their ability to distinguish the low resolution obsoleted structures from their superseded high-resolution counter parts.

PRESS-PLOT is derived from the PRESS angle-based structural assessment function, but it has more detailed analysis on the angle correlations: It examines the virtual angle pairs in specific secondary part of the structure as well as the whole structure, extending the correlation plots from original two to fourteen. PRESS-PLOT utilizes various advanced web technologies and makes it possible for the users to get access to the PRESS-PLOT structural evaluation tools easily from anywhere on the internet, with zero software installation or command typing efforts. The users can submit their structures and obtain the PRESS-PLOT evaluations immedaitely in both graphics and text forms.

PRESS-PLOT is the most effective for secondary structure assessment, because the virtual angle correlations are short-range restrictions (within four connected residues) and relate directly to the correct fold of the secondary structures. If there are two structures with the same secondary structural components, but different tertiary orders, their PRESS-PLOT evaluations would be about the same, beacuse their local virtual angle correlations would remain the same. Tools for tertiary structural assessment may be developed by combining certain long range constraints such as residue contact potentials [5].

The current implementation of PRESS-PLOT is based on the survey on a large set of X-ray structures in PDB, and therefore, applies to general structures, with X-ray structures as references. The implementation based on a special type of structures, such as the structures of a special protein family or the structures determined by NMR, could be interesting and particularly effective for the structures of that type.

The residue-level virtual angle correlations are not as restrictive as those at the atomic level such as the ϕ-ψ angle correlations in Ramachandran Plot. For both atomic and residue-level accuracies, one may use Ramachandran Plot as well as PRESS-PLOT as a pair of complementary assessment tools. After all, the PRESS-PLOT assessment is statistically based. The results need to be examined with caution: There could be exceptions: some angle pairs in **most favored** regions may not be really favored in a particular structure; some in **disallowed** regions may be just due to a special arrangement in that structure.

Author's Contributions

YH collected the data, wrote the programs for statistical calculations and plot generations, performed the testings, and drafted the paper. KJ was responsible for the development of the web server and graphics design. RJ and ZW analyzed the data and interpreted the results. YH, KJ, and ZW finished the writing.

References

1. Gu, J., Bourne, P.: Structural Bioinformatics. Wiley-Blackwell, Hoboken (2009)
2. Ramachandran, G.N., Sasisekharan, V.: Conformation of polypeptides and proteins. Adv. Protein Chem. 23, 283–437 (1968)

3. Laskowski, R.A., MacArthur, M.W., Moss, D.S., Thornton, J.M.: PROCHECK: a program to check the stereochemical quality of protein structures. J. Appl. Cryst. 26, 283–291 (1993)
4. Kolinski, A., Skolnick, J.: Reduced models of proteins and their applications. Polymer 45, 511–524 (2004)
5. Miyazawa, S., Jernigan, R.L.: Estimation of effective inter-residue contact energies from protein crystal structures: quasi-chemical approximation. Macromolecules 18, 534–552 (1985)
6. Sippl, M.J.: Calculation of conformational ensembles from potentials of mean force. J. Mol. Biol. 213, 859–883 (1990)
7. Kuszewski, J., Gronenborn, A.M., Clore, G.M.: Improving the quality of NMR and crystallographic protein structures by means of a conformational database potential derived from structure databases. Protein Science 5, 1067–1080 (1996)
8. Wu, D., Jernigan, R., Wu, Z.: Refinement of NMR-determined protein structures with database derived mean force potentials. Proteins: Structure, Function, Bioinformatics 68, 232–242 (2007)
9. Wu, D., Cui, F., Jernigan, R., Wu, Z.: PIDD: A protein inter-atomic distance distribution database. Nucleic Acid Research 35, D202–D207 (2007)
10. Huang, Y., Bonett, S., Kloczkowski, A., Jernigan, R., Wu, Z.: Statistical measures on protein residue level structural properties. Journal of Structural and Functional Genomics 12, 119–136 (2011)
11. Huang, Y., Bonett, S., Kloczkowski, A., Jernigan, R., Wu, Z.: PRESS: A software package for exploring protein residue level structural statistics. Journal of Bioinformatics and Computational Biology 10, 1–30 (2012)
12. Kabsch, W., Sander, C.: Dictionary of protein secondary structure: Pattern recognition of hydrogen-bonded and geometrical features. Biopolymers 22, 2577–2637 (1983)
13. Grant, B., Rodrigues, A., ElSawy, K., McCammon, J.A., Caves, L.: Bio3D: An R package for the comparative analysis of protein structures. Bioinformatics 22, 2695–2696 (2006)
14. PDB 2012 report on obsoleted structures (2012), ftp://ftp.wwpdb.org/pub/pdb/data/status/obsolete.dat

Calcium Ion Fluctuations Alter Channel Gating in a Stochastic Luminal Calcium Release Site Model

Hao Ji[1], Yaohang Li[1], and Seth H. Weinberg[2(✉)]

[1] Department of Computer Science, Old Dominion University, Norfolk, VA, USA
{hji,yaohang}@cs.odu.edu
[2] Virginia Modeling, Analysis and Simulation Center (VMASC),
Old Dominion University, Norfolk, VA, USA
sweinber@odu.edu

Abstract. Stochasticity and small system size effects in complex biochemical reaction networks can greatly alter transient and steady-state system properties. A common approach to modeling reaction networks, which accounts for system size, is the chemical master equation that governs the dynamics of the joint probability distribution for molecular copy number. However, calculation of the stationary distribution is often prohibitive, due to the large state-space associated with most biochemical reaction networks. Here, we analyze a network representing a luminal calcium release site model and investigate to what extent small system size effects and calcium fluctuations, driven by ion channel gating, influx and diffusion, alter steady-state ion channel properties including open probability. For a physiological ion channel gating model and number of channels, the state-space may be between approximately $10^6 - 10^8$ elements, and a novel modified block power method is used to solve the associated dominant eigenvector problem required to calculate the stationary distribution. We demonstrate that both small local cytosolic domain volume and a small number of ion channels drive calcium fluctuations that result in deviation from the corresponding model that neglects small system size effects.

Keywords: Systems biology · Chemical master equation · Fluctuation · Calcium · Ion channel · Stationary distribution · Eigenvector · Block power method

1 Introduction

In a biochemical reaction network, the copy number of the molecules in the system randomly fluctuates due to the random timing of individual reactions [1]. When the system size is small, concentration or density fluctuations are large in amplitude, and these fluctuations may alter steady-state system properties. In particular, when reactions are higher than first-order, the expected value calculated from the stationary distribution of a discrete system representation (that

© Springer International Publishing Switzerland 2015
R. Harrison et al. (Eds.): ISBRA 2015, LNBI 9096, pp. 162–174, 2015.
DOI: 10.1007/978-3-319-19048-8_14

accounts for fluctuations and small system size) may disagree with the steady-state value calculated from the corresponding continuous system representation (that neglects fluctuations and ignores system size effects) [2].

In many cell types, calcium (Ca^{2+}) is a key signaling molecule that drives important physiological functions, such as neurotransmitter release and myocyte contraction [3]. Ca^{2+} signaling is often localized in "spatially restricted" domains of small volume, or Ca^{2+} microdomains. Ca^{2+} influx into microdomains often occurs via Ca^{2+}-regulated Ca^{2+} channels, and the number of Ca^{2+} channels is often small. For example, in cardiomyocytes, localized Ca^{2+} signaling occurs in dyadic subspaces, estimated to contain between $10 - 100$ Ca^{2+}-activated channels[4]. Accounting for stochasticity in Ca^{2+} channel gating, i.e., transitions between open, closed, and inactivated channel states, due to the small number of ion channels, is important and necessary to reproduce many aspects of subcellular Ca^{2+} signaling [5]. However, due to small domain volume ($0.001 - 0.1$ μm^3) and resting Ca^{2+} concentration ($[Ca^{2+}]$, 0.1 μM), the expected number of Ca^{2+} ions is also typically very small ($0.06 - 6$ Ca^{2+} ions). Yet the influence of stochasticity due to Ca^{2+} ion fluctuations is not as well understood.

A common approach to modeling biochemical reaction networks that accounts for system size is the chemical master equation that governs the joint probability distribution for molecular copy number [6]. In prior work, Weinberg and Smith utilized this approach to investigate the influence of $[Ca^{2+}]$ fluctuations in minimal Ca^{2+} microdomain model, comprised of two-state Ca^{2+} channels, activated by local domain Ca^{2+} [7]. Here, we expand on this prior work to include a more physiological number of channels, channel gating model, which accounts for both Ca^{2+}-dependent activation and inactivation, and both cytosolic and luminal Ca^{2+} domains. With this increasing level of physiological detail, the associated state-space for the luminal Ca^{2+} release site model contains between $10^6 - 10^8$ elements. A novel modified block power method is used to solve the associated dominant eigenvector problem required to calculate the stationary distribution. Our paper is organized as follows: In Section 2, we briefly present the chemical master equation and calculation of the stationary distribution in a chemical reaction network. In Section 3, we describe the luminal Ca^{2+} release site model. In Section 4, we illustrate how accounting for stochasticity influences steady-state channel gating. We conclude with a brief discussion of our results in Section 5.

2 Chemical Reaction Network

2.1 Chemical Master Equation

We follow the general notation for representing a biochemical reaction network as presented in the excellent review by Goutsias and Jenkinson [6]. We describe the biochemical interactions in a system between N molecular species $X_1, X_2, \ldots, X_N$ via M reactions,

$$\sum_{n \in \mathcal{N}} \nu_{nm} X_n \to \sum_{n \in \mathcal{N}} \nu'_{nm} X_n, \quad m \in \mathcal{M}, \tag{1}$$

where $\mathcal{N} = \{1, 2, \ldots, N\}$ and $\mathcal{M} = \{1, 2, \ldots, M\}$. ν_{nm} and ν'_{nm} are the *stochiometric coefficients* that describe the number of molecules of the n-th species consumed or produced in the m-th reaction. We collect the *net stoichiometric coefficients* in the $N \times M$ matrix $\mathbb{S} = (s_{nm})$, where $s_{nm} = \nu_{nm} - \nu'_{nm}$.

In a Markov reaction network, the probability of a reaction occurring only depends on the current system state, and further, to first-order

$$\Pr\{\text{reaction } m \text{ occurs within } [t, t + dt) | \boldsymbol{X}(t) = \boldsymbol{x}\} = \pi_m(\boldsymbol{x}) dt, \tag{2}$$

where vector $\boldsymbol{X}(t) = (X_1(t), X_2(t), \ldots, X_N(t))$, $\boldsymbol{x} = (x_1, x_2, \ldots, x_n)$ is a known system state, and $\pi_m(\boldsymbol{x})$ is a state-dependent function called the *propensity function*, associated with the m-th reaction. The joint probability distribution $p_{\boldsymbol{X}}(t)$ is governed by the partial differential equation, known as the *chemical master equation*,

$$\frac{\partial p_{\boldsymbol{X}}(\boldsymbol{x}; t)}{\partial t} = \sum_{m \in \mathcal{M}} \{\pi_m(\boldsymbol{x} - \boldsymbol{s}_m) p_{\boldsymbol{X}}(\boldsymbol{x} - \boldsymbol{s}_m; t) - \pi_m(\boldsymbol{x}) p_{\boldsymbol{X}}(\boldsymbol{x}; t)\}, \tag{3}$$

where $\boldsymbol{s}_m$ is the m-th column of matrix $\mathbb{S}$.

If we index the elements in state space $\mathcal{X}$, then the master equation can be expressed as a linear system of coupled first-order differential equations

$$\frac{d\boldsymbol{p}(t)}{dt} = \boldsymbol{p}(t) \mathbb{Q}, \tag{4}$$

where $\boldsymbol{p}(t)$ is a $1 \times K$ vector containing the probabilities $p_{\boldsymbol{X}}(\boldsymbol{x}; t)$, $\boldsymbol{x} \in \mathcal{X}$, $\mathbb{Q} = (q_{ij})$ is a large $K \times K$ sparse matrix, known as the *infinitesimal generator matrix*, whose structure can be determined directly from the master equation, and K is the cardinality of (number of elements in) state space $\mathcal{X}$.

If a stationary distribution exists, then at steady-state, $\boldsymbol{p}^{ss} \mathbb{Q} = 0$, which can also be found by solution of $\boldsymbol{p}^{ss} \mathbb{P} = \boldsymbol{p}^{ss}$, where $\mathbb{P} = \mathbb{Q} \Delta t + \mathbb{I}$, $\mathbb{I}$ is the identity matrix of appropriate size, and Δt is sufficiently small that the probability of two transitions taking place in time Δt is negligible, i.e., matrix $\mathbb{P}$ is stochastic [8]. We have essentially discretized the continuous-time Markov chain into a discrete-time Markov chain with transition matrix $\mathbb{P}$. To guarantee that $\mathbb{P}$ is stochastic, $0 < \Delta t < (\max_i |q_{ii}|)^{-1}$, and specifically, for numerical considerations discussed in Stewart [8], we define $\Delta t = 0.99 (\max_i |q_{ii}|)^{-1}$. The stationary joint distribution $p_{\boldsymbol{X}}^{ss}$ can be determined from $\boldsymbol{p}^{ss}$, which is calculated as described in the following section.

2.2   Numerical Calculation of Dominant Eigenvector

A Markov chain converges to a stationary distribution provided that it is aperiodic and irreducible. Let $\lambda_1, \lambda_2, \ldots, \lambda_K$ be the eigenvalues of the transition matrix $\mathbb{P}$ of the Markov chain, where $|\lambda_1| = 1 \geq |\lambda_2| \geq \ldots \geq |\lambda_K|$, and $v_1, v_2, \ldots, v_K$ are the corresponding eigenvectors. The stationary distribution of the Markov chain corresponds to the principle eigenvector v_1. Although many numerical methods

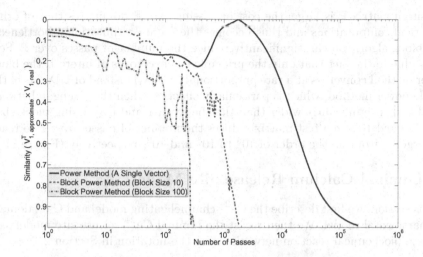

Fig. 1. Convergence of the simple power method and the block power method ($k = 10$ and 100) on a test matrix of size 1000×1000. The block power method significantly reduces the number of passes over the transition matrix $\mathbb{P}$ and accelerates convergence to the stationary distribution.

can be used to calculate v_1, when the transition matrix $\mathbb{P}$ is large, the power method, with a convergence rate proportional to $1/|\lambda_2|$, is typically the most feasible method with the least memory requirement.

To calculate the stationary distribution of the luminal Ca^{2+} release site model with over a million states, we develop a modified block power method. The block power method was originally designed to estimate multiple dominating eigenvalues/eigenvectors, where the convergence depends on the eigengap between the k-th and $(k + 1)$-th eigenvalues, for a given block size k ($1 < k \ll K$) [9]. Here, we are only interested in the principle eigenvector, and therefore the eigengap between $\lambda_1 = 1$ and λ_k governs the convergence speed.

Starting with a $K \times k$ orthogonal initial matrix $X^{(0)}$, the block power iteration generates the matrices sequence $\left\{ X^{(i)} \right\}_{i=0}^{\infty}$ by defining

$$X^{(i)} := \mathbb{P} X^{(i-1)}, \quad i = 1, 2, \ldots \tag{5}$$

For each iteration, the top-k eigenvectors of $\mathbb{P}$ are approximated by eigendecomposing a small $k \times k$ matrix $\mathbb{B}^{(i)^T} \mathbb{P} \mathbb{B}^{(i)}$, where $\mathbb{B}^{(i)}$ is a basis of the range space of $X^{(i)}$. More specifically, by performing block power iteration, the range space of $X^{(i)}$ becomes an approximate space capturing the dominant information of matrix $\mathbb{P}$. We construct the matrix $\mathbb{B}^{(i)^T} \mathbb{P} \mathbb{B}^{(i)}$ to project matrix $\mathbb{P}$ into the range space of $X^{(i)}$ and compute its eigenvectors $U^{(i)}$. Then, the top-k eigenvectors of matrix $\mathbb{P}$ can be approximated effectively through a simple matrix multiplication $\mathbb{B}^{(i)} U^{(i)}$ and the largest is selected as the approximate principle eigenvector.

Compared to the simple power method, the block power method has important advantages in handling very large transition matrices. Firstly, when the transition matrix $\mathbb{P}$ is large, the cost of passing over $\mathbb{P}$ dominates that of other numerical computations and thus becomes the main computational bottleneck. The block algorithm can significantly reduce the number of passes over $\mathbb{P}$. Secondly, due to the fact that only the principle eigenvector is of interest, the block power method converges at a rate proportional to $1/|\lambda_k|$ instead of $1/|\lambda_2|$ as in the simple power method, which is particularly effective when the eigengap between 1 and $|\lambda_k|$ is significantly wider than that between 1 and $|\lambda_2|$. Using block sizes of $k = 10$ and 100 on a test matrix reduces the number of passes over $\mathbb{P}$ to reach convergence from on the order of 10^5 to 10^3 and 10^2, respectively (Figure 1).

3   Luminal Calcium Release Site Model

In this section, we first describe the Ca^{2+} channel gating model and Ca^{2+} domain compartmental model. We then recast the luminal Ca^{2+} release site model as a discrete biochemical reaction network, using the notation in Section 2.

3.1   Four-State Calcium Channel Gating Model

Many Ca^{2+}-regulated channels have been shown to exhibit both fast Ca^{2+} activation and slower Ca^{2+} inactivation, such as IP3 receptors [10]. The gating of Ca^{2+} channels that are activated and inactivated by local cytosolic Ca^{2+} is represented by a stochastic process with the following state transition diagram,

$$
\begin{array}{ccccc}
\text{(closed)} & \mathcal{C} & \underset{k_a^-}{\overset{k_a^+ c_{cyt}}{\rightleftharpoons}} & \mathcal{O} & \text{(open)} \\
& k_b^+ c_{cyt} \downarrow\uparrow k_b^- & & k_d^- \downarrow\uparrow k_d^+ c_{cyt} & \\
\text{(closed, inactivated)} & \mathcal{CI} & \underset{k_c^-}{\overset{k_c^+ c_{cyt}}{\rightleftharpoons}} & \mathcal{I} & \text{(inactivated)}
\end{array}
\tag{6}
$$

where $k_i^+ c_{cyt}$ and k_i^- are transition rates with units of time^{-1}, c_{cyt} is the local cytosolic $[Ca^{2+}]$, and k_i^+ is an association rate constant with units of concentration^{-1}·time^{-1}, for $i \in \{a, b, c, d\}$. The channel is open when the activation site is Ca^{2+}-bound and the inactivation site is not bound (Figure 2).

In the absence of ion channel gating fluctuations, i.e., for a large number of ion channels $\mathcal{N}_c$, then the dynamics of the fraction of channels in the four states is given by the following system of ordinary differential equations,

$$
\frac{df_{\mathcal{C}}}{dt} = k_a^- f_{\mathcal{O}} + k_b^- f_{\mathcal{CI}} - (k_a^+ + k_b^+) c_{cyt} f_{\mathcal{C}}
\tag{7a}
$$

$$
\frac{df_{\mathcal{CI}}}{dt} = k_b^+ c_{cyt} f_{\mathcal{C}} + k_c^- f_{\mathcal{I}} - (k_b^- + k_c^+ c_{cyt}) f_{\mathcal{CI}}
\tag{7b}
$$

$$
\frac{df_{\mathcal{I}}}{dt} = k_c^+ c_{cyt} f_{\mathcal{CI}} + k_d^+ c_{cyt} f_{\mathcal{O}} - (k_c^- + k_d^-) f_{\mathcal{I}}
\tag{7c}
$$

$$
\frac{df_{\mathcal{O}}}{dt} = k_a^+ c_{cyt} f_{\mathcal{C}} + k_d^- f_{\mathcal{I}} - (k_a^- + k_d^+ c_{cyt}) f_{\mathcal{O}},
\tag{7d}
$$

where $f_{\mathcal{C}}$, $f_{\mathcal{CI}}$, $f_{\mathcal{I}}$, and $f_{\mathcal{O}}$ are the fraction of channels in states $\mathcal{C}$, $\mathcal{CI}$, $\mathcal{I}$, and $\mathcal{O}$, respectively. One of these equations is superfluous, since $f_{\mathcal{C}} + f_{\mathcal{CI}} + f_{\mathcal{I}} + f_{\mathcal{O}} = 1$.

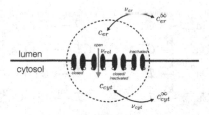

Fig. 2. Illustration of the model components and fluxes in the calcium release unit. Each Ca^{2+} channel has an activation and inactivation binding site. When only the activation site is Ca^{2+}-bound, the channel is open and Ca^{2+} is release at rate v_{rel}. Local cytosolic $[Ca^{2+}]$, c_{cyt}, relaxes to the bulk $[Ca^{2+}]$, c_{cyt}^{∞}, at rate v_{cyt}, and depleted local luminal $[Ca^{2+}]$, c_{er}, refills towards bulk luminal $[Ca^{2+}]$, c_{er}^{∞}, at rate v_{er}.

3.2   Cytosolic and Luminal Domain Compartment Model

Depletion of local Ca^{2+} near the luminal side of the Ca^{2+} channel can alter local Ca^{2+} release events, known as puffs or sparks [10]. Therefore, we consider a four compartment model that accounts for local cytosolic and luminal domains, with $[Ca^{2+}]$ of c_{cyt} and c_{er}, respectively, and bulk cytosolic and luminal compartments, with $[Ca^{2+}]$ of c_{cyt}^{∞} and c_{er}^{∞}, respectively (Figure 2). Assuming local cytosolic $[Ca^{2+}]$, c_{cyt}, relaxes to the bulk $[Ca^{2+}]$, c_{cyt}^{∞}, at rate v_{cyt}, and depleted local luminal $[Ca^{2+}]$, c_{er}, refills towards bulk luminal $[Ca^{2+}]$, c_{er}^{∞}, at rate v_{er}, and in the absence of local cytosolic domain $[Ca^{2+}]$ fluctuations, the dynamics of c_{cyt} and c_{er} are given by the following system of ordinary differential equations,

$$\frac{dc_{cyt}}{dt} = v_{rel} f_O (c_{er} - c_{cyt}) - v_{cyt}(c_{cyt} - c_{cyt}^{\infty}) \tag{8a}$$

$$\frac{dc_{er}}{dt} = \frac{1}{\lambda} \left[-v_{rel} f_O (c_{er} - c_{cyt}) - v_{er}(c_{er} - c_{er}^{\infty}) \right], \tag{8b}$$

where v_{rel} is the luminal Ca^{2+} release flux rate, and $\lambda = \Omega_{er}/\Omega_{cyt}$ is the ratio of the local luminal and cytosolic domain volumes, Ω_{er} and Ω_{cyt}, respectively.

3.3   Stochastic Luminal Calcium Release Site Model

The stochastic luminal calcium release site model that corresponds to the channel gating and compartment models, Eqs. 7-8, respectively, and accounts for fluctuations in both channel gating and local cytosolic $[Ca^{2+}]$ can be described by the following biochemical reaction network consisting of $N = 5$ species and $M = 10$ reactions,

$$\mathcal{C} + \text{Ca}_{cyt}^{2+} \xrightarrow{\bar{k}_a^+} \mathcal{O}, \qquad\qquad \mathcal{O} \xrightarrow{\bar{k}_a^-} \mathcal{C} + \text{Ca}_{cyt}^{2+},$$

$$\mathcal{C} + \text{Ca}_{cyt}^{2+} \xrightarrow{\bar{k}_b^+} \mathcal{CI}, \qquad\qquad \mathcal{CI} \xrightarrow{\bar{k}_b^-} \mathcal{C} + \text{Ca}_{cyt}^{2+},$$

$$\mathcal{CI} + \text{Ca}_{cyt}^{2+} \xrightarrow{\bar{k}_c^+} \mathcal{I}, \qquad\qquad \mathcal{I} \xrightarrow{\bar{k}_c^-} \mathcal{CI} + \text{Ca}_{cyt}^{2+}, \qquad (9)$$

$$\mathcal{O} + \text{Ca}_{cyt}^{2+} \xrightarrow{\bar{k}_d^+} \mathcal{I}, \qquad\qquad \mathcal{I} \xrightarrow{\bar{k}_d^-} \mathcal{O} + \text{Ca}_{cyt}^{2+},$$

$$\varnothing \xrightarrow{\alpha} \text{Ca}_{cyt}^{2+}, \quad \text{Ca}_{cyt}^{2+} \xrightarrow{\beta} \varnothing,$$

where reaction rates are given by $\bar{k}_i^+ = k_i^+/\Omega_{cyt}$, $\alpha(\boldsymbol{x}) = \Omega_{cyt}(v_{cyt}c_{cyt}^\infty + v_{rel}f_\mathcal{O}(\boldsymbol{x})$ $c_{er}(\boldsymbol{x}))$, and $\beta(\boldsymbol{x}) = v_{cyt} + v_{rel}f_\mathcal{O}(\boldsymbol{x})$. The copy numbers of channels in each state are (arbitrarily) defined as $X_1 = \mathcal{C}$, $X_2 = \mathcal{CI}$, $X_3 = \mathcal{I}$, and $X_4 = \mathcal{O}$, such that $f_\mathcal{O}(\boldsymbol{x}) = x_4/\mathcal{N}_c$. Similarly, the copy number of local cytosolic Ca^{2+} ions is defined as $X_5 = \text{Ca}_{cyt}^{2+}$, such that $c_{cyt}(\boldsymbol{x}) = x_5/\Omega_{cyt}$. Local luminal [Ca^{2+}], c_{er}, is assumed to be in rapid equilibrium, such that $c_{er}(\boldsymbol{x}) = (v_{rel}f_\mathcal{O}(\boldsymbol{x})c_{cyt}(\boldsymbol{x}) + v_{er}c_{er}^\infty)/(v_{rel}f_\mathcal{O}(\boldsymbol{x}) + v_{er})$.

The propensity functions and the net stoichiometric matrix $\mathbb{S}$ for the biochemical reaction network defined by Eq. 9 are given by

$$\begin{aligned}
&\pi_1(\boldsymbol{x}) = \bar{k}_a^+ x_1 x_5, \quad \pi_2(\boldsymbol{x}) = k_a^- x_4, \\
&\pi_3(\boldsymbol{x}) = \bar{k}_b^+ x_1 x_5, \quad \pi_4(\boldsymbol{x}) = k_b^- x_2, \\
&\pi_5(\boldsymbol{x}) = \bar{k}_c^+ x_2 x_5, \quad \pi_6(\boldsymbol{x}) = k_c^- x_3, \\
&\pi_7(\boldsymbol{x}) = \bar{k}_d^+ x_4 x_5, \quad \pi_8(\boldsymbol{x}) = k_d^- x_3, \\
&\pi_9(\boldsymbol{x}) = \alpha(\boldsymbol{x}), \quad \pi_{10}(\boldsymbol{x}) = \beta(\boldsymbol{x}),
\end{aligned}
\qquad
\mathbb{S} = \begin{bmatrix}
-1 & 1 & -1 & 1 & 0 & 0 & 0 & 0 & 0 & 0 \\
0 & 0 & 1 & -1 & -1 & 1 & 0 & 0 & 0 & 0 \\
0 & 0 & 0 & 0 & 1 & -1 & 1 & -1 & 0 & 0 \\
1 & -1 & 0 & 0 & 0 & 0 & -1 & 1 & 0 & 0 \\
-1 & 1 & -1 & 1 & -1 & 1 & -1 & 1 & 1 & -1
\end{bmatrix}.$$

As described in Section 2, we calculate the stationary distribution and stationary statistics, given by the q-th moment of the i-th species,

$$\mu_q^i = \sum_{\boldsymbol{x} \in \mathcal{X}} x_i^q \cdot p_{\boldsymbol{X}}^{ss}(\boldsymbol{x}), \qquad (10)$$

where $\mathcal{X}$ is the enumerated state-space, such that the expected channel open and inactivation probability, $\text{E}[f_\mathcal{O}] = \mu_1^4/\mathcal{N}_c$ and $\text{E}[f_\mathcal{I}] = \mu_1^3/\mathcal{N}_c$, respectively, and expected local cytosolic [Ca^{2+}], $\text{E}[c_{cyt}] = \mu_1^5/\Omega_{cyt}$. We compare these stationary statistics that *account for small system size* with the corresponding steady-state values for local cytosolic [Ca^{2+}] and open and inactivated channel fraction that *neglect fluctuations* and *small system size effects*, c_{cyt}^{ss}, $f_\mathcal{O}^{ss}$, and $f_\mathcal{I}^{ss}$, respectively, found by the steady-state solution of Eqs. 7-8. We also calculate the *spark score*, $\mathcal{S} = \text{Var}[f_\mathcal{O}]/\text{E}[f_\mathcal{O}]$, an index of dispersion for $f_\mathcal{O}$, where the $f_\mathcal{O}$ variance, $\text{Var}[f_\mathcal{O}] = [\mu_2^4 - (\mu_1^4)^2]/\mathcal{N}_c^2$, divided by the expectation $\text{E}[f_\mathcal{O}]$, which takes values between 0 and 1. A larger spark score corresponds to more robust, spontaneous luminal Ca^{2+} release events [11].

3.4   Practical Considerations for Enumerating the State-Space

The size of the enumerated state-space $K = \mathcal{R}_5\mathcal{R}_c$, is the product of total number states for Ca$_{cyt}^{2+}$ copy number, $\mathcal{R}_5$, and total number states for the $\mathcal{N}_c$ ion channels, $\mathcal{R}_c$. Enumerating the state-space is straightforward when there is a natural finite range of values that each species can allow. For example, X_1, X_2, X_3 and X_4 allow values between $0, 1, \ldots, \mathcal{N}_c$. However, because Ca^{2+} influx (reaction 9) is zero-order, in theory, the local cytosolic Ca^{2+} ion maximum is infinite. In practice, we define a maximum value $\mathcal{R}_5$, for which the probability of Ca_{cyt}^{2+} exceeding such a value is negligible. We found that a reasonable value for $\mathcal{R}_5 = 2\rho$, where $\rho = \max([c_{cyt}^{max}\Omega_{cyt}], 50)$, $c_{cyt}^{max} = (v_{cyt}c_{cyt}^{\infty} + v_{rel}c_{er}^{\infty})/(v_{cyt} + v_{rel})$ is the hypothetical maximum value for c_{cyt} that occurs for a fully replete local luminal domain ($c_{er} = c_{er}^{\infty}$) and all channels open ($f_\mathcal{O} = 1$), and $\lceil \cdot \rceil$ is the ceiling function. Assuming channels are identical and experience the same local cytosolic $[Ca^{2+}]$, the number of distinguishable states for $\mathcal{N}_c$ channels, with $\mathcal{N}_s$ states, is given by $\mathcal{R}_c = (\mathcal{N}_c + \mathcal{N}_s - 1)!/\mathcal{N}_c!/(\mathcal{N}_s - 1)!$, where $\mathcal{N}_s = 4$ for the gating model in Eq. 6 [12]. For example, for $\Omega_{cyt} = 10^{-2}\ \mu m^3$, $\mathcal{N}_c = 50$ channels, and standard compartment flux parameters (see Figure 3 legend), $\mathcal{R}_5 = 3012$, $\mathcal{R}_c = 23426$, and the state-space size $K \approx 7.06 \cdot 10^7$.

4   Results

We investigate the influence of small system size on the stationary properties of the luminal Ca^{2+} release site model by varying the local cytosolic domain volume Ω_{cyt} and number of Ca^{2+} channels $\mathcal{N}_c$. We plot the joint and marginal distribution for local cytosolic $[Ca^{2+}]$, c_{cyt}, and the fraction of open channels, $f_\mathcal{O}$, and indicate $E[f_\mathcal{O}]$ and $E[c_{cyt}]$ (blue circle, solid line) and $f_\mathcal{O}^{ss}$ and c_{cyt}^{ss} (red X, dashed line).

When the local cytosolic domain volume is small ($\Omega_{cyt} = 10^{-3}\ \mu m^3$) and has a small number of channels ($\mathcal{N}_c = 10$), the $f_\mathcal{O}$-distribution is Poisson-like, while the c_{cyt}-distribution is bimodal, with one peak corresponding to a low $[Ca^{2+}]$ and zero channels open and a second peak correspond to an elevated $[Ca^{2+}]$ and a few channels open (Figure 3A). Steady-state measures, $f_\mathcal{O}^{ss}$ and c_{cyt}^{ss}, that neglect fluctuations correspond closely with the second peak, which illustrates that Ca^{2+} ion and gating fluctuations lead to a subpopulation of channels that are not open, which in turn reduces $E[f_\mathcal{O}]$ and $E[c_{cyt}]$. Further analysis of the joint distribution reveals that most of these channels are in the inactivated states, $\mathcal{I}$ and $\mathcal{CI}$ (not shown). In a local cytosolic domain of larger volume ($\Omega_{cyt} = 10^{-2}\ \mu m^3$, Figure 3D), the two peaks in the c_{cyt}-distribution are narrower (due to smaller $[Ca^{2+}]$ fluctuations). As a consequence of smaller $[Ca^{2+}]$ fluctuations, Ca^{2+}-activation events, $\mathcal{C} \to \mathcal{O}$ and $\mathcal{CI} \to \mathcal{I}$, are less likely, and probability in the stationary distribution shifts such that the closed states, $\mathcal{C}$ and $\mathcal{CI}$, are more likely. As such, $E[f_\mathcal{O}]$ is reduced, and $E[c_{cyt}]$ is reduced as a consequence. However, variability in channel gating slightly increases, such that the spark score $\mathcal{S}$ increases.

As the number of channels in the domain increases ($\mathcal{N}_c = 25$, Figure 3B, E), the $f_\mathcal{O}$-distribution is more Gaussian-like. For a small domain volume (Figure 3B), the c_{cyt}-distribution is bimodal, as in the domain with fewer channels; however, probability has shifted primarily from the low $[Ca^{2+}]$ level to a higher $[Ca^{2+}]$

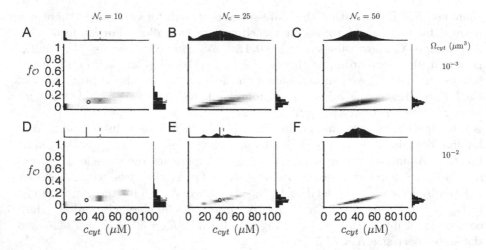

Fig. 3. Luminal Ca^{2+} release site model stationary distribution. For the parameters in each panel, the stationary joint and marginal distribution for local cytosolic $[Ca^{2+}]$, c_{cyt}, and fraction of open channels, f_O are shown. The expected values for f_O and c_{cyt}, $E[f_O]$ and $E[c_{cyt}]$ (blue circle, solid line), respectively, and and steady-state values in the large system limit, f_O^{ss} and c_{cyt}^{ss} (red X, dashed line), respectively, are indicated. Parameters: $\mathcal{N}_c$ = 10 (A, D), 25 (B, E), or 50 (C, F). Ω_{cyt} = 10^{-3} (A-C) or 10^{-2} (D-F) μm^3, Channel gating [13]: $k_a^+ = k_c^+ = 1$ μM^{-1} s^{-1}, $k_a^- = k_c^- = 1$ s^{-1}, $k_b^+ = k_d^+ =$ 0.01 μM^{-1} s^{-1}, $k_b^- = k_d^- = 0.05$ s^{-1}. Compartment fluxes and bulk concentrations [14]: $c_{cyt}^\infty = 0.1$ μM, $c_{er}^\infty = 500$ μM, $v_{cyt} = 10$ s^{-1}, $v_{er} = 10$ s^{-1}, $v_{rel} = 10$ s^{-1}.

level, i.e., $E[c_{cyt}]$ is approaching c_{cyt}^{ss}, as expected for a larger system size. In a larger volume domain, the c_{cyt}-distribution is multimodal, with small peaks corresponding to a distinct number of open channels, including a large peak for zero open channels (Figure 3E).

As the number of ion channels increases further ($\mathcal{N}_c$ = 50), the joint distribution approaches a multivariate Gaussian distribution, with a clear positive correlation between c_{cyt} and f_O (Figure 3C). The expected values for f_O and c_{cyt}, $E[f_O]$ and $E[c_{cyt}]$, respectively, approach the steady-state measures that neglect fluctuations, f_O^{ss} and c_{cyt}^{ss}, respectively. In a local cytosolic domain of larger volume, the c_{cyt}-distribution has a reduced variance, and the correlation between f_O and c_{cyt} is increased (Figure 3F).

In summary, over a wide range of physiological values for $\mathcal{N}_c$ and Ω_{cyt}, we can observe that as $\mathcal{N}_c$ increases, the f_O-distribution transitions from Poisson-like to Gaussian-like, and the f_O variance, $Var[f_O]$ decreases such that the spark score S also decreases. For small domain volumes Ω_{cyt}, the c_{cyt}-distribution transitions from a bimodal to Gaussian distribution as $\mathcal{N}_c$ increases, whereas for a larger Ω_{cyt}, the c_{cyt}-distribution transitions from bimodal, to multimodal, to Gaussian. Over this transition, the variance of c_{cyt} initially increases and then decreases (not shown). Further, in general, f_O and c_{cyt} are more closely correlated for small $\mathcal{N}_c$ and larger Ω_{cyt}.

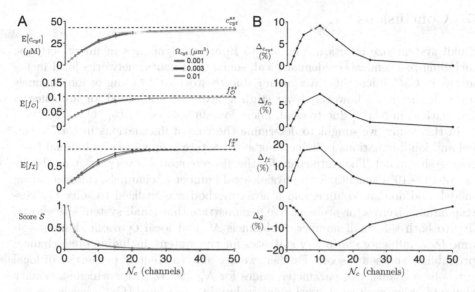

Fig. 4. Stationary statistics for the luminal Ca^{2+} release site model. (A) The expected value for local cytoslic $[Ca^{2+}]$, and fraction of open and inactivated channels, $E[c_{cyt}]$, $E[f_O]$, and $E[f_I]$, respectively, and spark score S are shown as functions of the number of channels N_c, for different local cytosolic domain volume Ω_{cyt}. (B) The small system deviation Δ_z (Eq. 11), for $x \in \{E[c_{cyt}], E[f_O], E[f_I], S\}$ is shown as a function of N_c. Parameters as in Figure 3.

In Figure 4A, we plot $E[c_{cyt}]$, $E[f_O]$, $E[f_I]$, and S as functions of N_c, for different values of Ω_{cyt} (solid, colored lines). We found that S decreases as N_c increases, i.e., spontaneous sparks are less robust in domains with fewer channels. Further, $E[c_{cyt}]$, $E[f_O]$, and $E[f_I]$ all increase as N_c increases and approach the steady-state values that neglects fluctuations, c_{cyt}^{ss}, f_O^{ss}, and f_I^{ss}, respectively (black, dashed). We found that for small number of channels, N_c near $10 - 20$, there is a noticeable difference between these metrics as the cytosolic domain volume increases from $\Omega_{cyt} = 10^{-3}$ (red) to 10^{-2} μm^3 (green). We quantified this deviation, referred to as the small system size deviation in [7],

$$\Delta_z = \frac{E[z] - E[z]_\infty}{E[z]_\infty}, \tag{11}$$

where z is the measurement for the smallest domain volume ($\Omega_{cyt} = 10^{-3}$ μm^3), $E[z]_\infty$ is the measurement for the largest domain volume ($\Omega_{cyt} = 10^{-2}$ μm^3), and $z \in \{c_{cyt}, f_O, f_I, S\}$. Δ_z for the four measurements are biphasic functions of N_c (Figure 4B). For $z \in \{c_{cyt}, f_O, f_I\}$, $\Delta_z > 0$ (positive) and is maximal at $N_c = 10$, i.e., $E[c_{cyt}]$, $E[f_O]$, and $E[f_I]$ all decrease as local cytosolic domain volume Ω_{cyt} increases. $\Delta_S < 0$ (negative) and is minimal at $N_c = 15$, i.e., S increases as Ω_{cyt} increases. The small system size deviation is largest in magnitude for f_I.

5   Conclusions

Small system size effects are known to influence dynamics in many settings, including biochemical, epidemiological, social, and neural networks [6]. Fluctuations in Ca^{2+} microdomain signaling due to stochastic gating of ion channels are well-known [5]; however, fewer studies have also accounted for influence of fluctuations in $[Ca^{2+}]$ due to small microdomain volume [7, 15, 16].

In this study, we sought to determine the role of fluctuations in Ca^{2+} channel and ion fluctuations in influencing steady-state properties of a luminal Ca^{2+} release site model. The state-space for the discrete model is very large, on the order of $10^6 - 10^8$ elements, for a physiological number of channels, channel gating model, and domain volume, and a novel method was utilized to solve the corresponding eigenvector problem. We demonstrate that small system size effects, due to both the small number of channels $\mathcal{N}_c$ and local cytosolic domain volume Ω_{cyt}, influence stationary statistics for the system, including open channel probability and spark score. Further, we are able to identify properties of local cytosolic domains, i.e., parameter values for $\mathcal{N}_c$ and Ω_{cyt}, for which stationary characteristics, such as channel open probability and local $[Ca^{2+}]$ levels, do not agree with the corresponding model that neglect small system size effects. Expected values for c_{cyt}, f_O, f_I, and S were found to have a strong dependence on the number of channels in the domain, $\mathcal{N}_c$. Further, for a given number of channels, in particular, small values near 10-15, these measures deviated as Ω_{cyt} increases, demonstrating that fluctuations in Ca^{2+} ions, in addition to channel gating, also influence system stationary properties. Since local domain, spontaneous Ca^{2+} release events can greatly influence global Ca^{2+} signaling and homeostasis [11, 14], our work suggests that predictive whole-cell models of Ca^{2+} signaling should account for Ca^{2+} ion fluctuations and small system size effects.

In this study, we consider a Ca^{2+} channel gating model that accounts for both Ca^{2+}-dependent activation and inactivation and a Ca^{2+} compartmental model that includes first-order passive exchange between local and bulk domains [10]. Interestingly, we found that small system deviations, $\Delta_{c_{cyt}}$ and Δ_{f_O}, are positive, in contrast with our prior work analyzing a minimal domain and gating model [7], demonstrating that accounting for more physiologically-detailed models of domain compartments and gating is important. In pathological settings, the kinetics of these processes may be altered, leading to more frequent spontaneous Ca^{2+} release events. Further, luminal Ca^{2+} channel gating dynamics may be more complex, including multiple closed, inactivated, and refractory states. Further studies are needed to investigate the influence of small system effects in these settings. However, the general approach presented is independent of model parameters, compartments, or the channel gating model. The stationary statistics of the expansive state-space associated with a pathological or expanded gating model can be similarly analyzed.

Acknowledgments. The authors acknowledge and thank Old Dominion University High Performance Computing for use of the Turing cluster to perform numerical calculations. This work is partially supported by NSF grant 1066471 (YL) and by an ODU Modeling and Simulation Fellowship (HJ). This work used the Extreme Science and Engineering Discovery Environment (XSEDE), which is supported by NSF grant ACI-1053575.

References

1. Keizer, J.: Statistical Thermodynamics of Nonequilibrium Processes. Springer Science & Business Media (August 1987)
2. McQuarrie, D.A.: Kinetics of small systems. I. J. Chem. Phys. 38(2), 433–436 (1963)
3. Berridge, M.J.: Calcium microdomains: organization and function. Cell Calcium 40(5-6), 405–412 (2006)
4. Franzini-Armstrong, C.: Architecture and regulation of the Ca2+ delivery system in muscle cells. Appl. Physiol. Nutr. Metab. 34(3), 323–327 (2009)
5. Greenstein, J.L., Winslow, R.L.: An integrative model of the cardiac ventricular myocyte incorporating local control of Ca2+ release. Biophys. J. 83(6), 2918–2945 (2002)
6. Goutsias, J., Jenkinson, G.: Markovian dynamics on complex reaction networks. Physics Reports 529(2), 199–264 (2013)
7. Weinberg, S.H., Smith, G.D.: Discrete-State Stochastic Models of Calcium-Regulated Calcium Influx and Subspace Dynamics Are Not Well-Approximated by ODEs That Neglect Concentration Fluctuations. Computational and Mathematical Methods in Medicine 2012(12), 1–17 (2012)
8. Stewart, W.J.: Introduction to the Numerical Solution of Markov Chains. Princeton University Press (1994)
9. Golub, G.H., Van Loan, C.F.: Matrix Computations. JHU Press (December 2012)
10. Huertas, M.A., Smith, G.D.: The dynamics of luminal depletion and the stochastic gating of Ca2+-activated Ca2+ channels and release sites. Journal Theor. Biol. 246(2), 332–354 (2007)
11. Wang, X., Weinberg, S.H., Hao, Y., Smith, G.D.: Calcium homeostasis in a local/global whole cell model of permeabilized ventricular myocytes with a Langevin description of stochastic calcium release. Am. J. Physiol. Heart Circ. Phsyiol., 1–62 (November 2014)
12. Deremigio, H., Lamar, M.D., Kemper, P., Smith, G.D.: Markov chain models of coupled calcium channels: Kronecker representations and iterative solution methods. Phys. Biol. 5(3), 036003 (2008)
13. Mazzag, B., Tignanelli, C.J., Smith, G.D.: The effect of residual Ca2+ on the stochastic gating of Ca2+-regulated Ca2+ channel models.. Journal of Theoretical Biology 235(1), 121–150 (2005)
14. Hartman, J.M., Sobie, E.A., Smith, G.D.: Spontaneous Ca2+ sparks and Ca2+ homeostasis in a minimal model of permeabilized ventricular myocytes. Am. J. Physiol. Heart Circ. Physiol. 299(6), H1996–H2008 (2010)

15. Koh, X., Srinivasan, B., Ching, H.S., Levchenko, A.: A 3D Monte Carlo Analysis of the Role of Dyadic Space Geometry in Spark Generation. Biophys. J. 90(6), 1999–2014 (2006)
16. Weinberg, S.H., Smith, G.D.: The Influence of Ca2+ Buffers on Free [Ca2+] Fluctuations and the Effective Volume of Ca2+ Microdomains. Biophys. J. 106(12), 2693–2709 (2014)

Interleaving Global and Local Search for Protein Motion Computation

Kevin Molloy[1] and Amarda Shehu[1,2,3(✉)]

[1] Department of Computer Science,
[2] Department of Bioengineering,
[3] School of Systems Biology, George Mason University, Fairfax, VA 22030, USA
amarda@gmu.edu
http://cs.gmu.edu/~ashehu

Abstract. We propose a novel robotics-inspired algorithm to compute physically-realistic motions connecting thermodynamically-stable and semi-stable structural states in protein molecules. Protein motion computation is a challenging problem due to the high-dimensionality of the search space involved and ruggedness of the potential energy surface underlying the space. To handle the multiple local minima issue, we propose a novel algorithm that is not based on the traditional Molecular Dynamics or Monte Carlo frameworks but instead adapts ideas from robot motion planning. In particular, the algorithm balances computational resources between a global search aimed at obtaining a global view of the network of protein conformations and their connectivity and a detailed local search focused on realizing such connections with physically-realistic models. We present here promising results on a variety of proteins and demonstrate the general utility of the algorithm and its capability to improve the state of the art without employing system-specific insight.

Keywords: Protein motion computation · Conformational path · Roadmap-based algorithm

1 Introduction

Elucidating the detailed motions employed by dynamic protein molecules [1] to switch between different thermodynamically-stable/functional conformations is important to advance our understanding of protein physics and allow drug discovery, protein-based sensor design, and protein engineering [2, 3]. Only computation is capable of providing detailed motions at a microscopic level. However, computational methods are challenged by the size and dimensionality of the protein conformation space, as well as the ruggedness of the underlying protein energy surface. In particular, standard frameworks such as Molecular Dynamics (MD) and Monte Carlo (MC) often get stuck in particular local minima and cannot find conformational paths connecting given functional conformations [4].

Algorithms based on robot motion planning have been proposed over the years, exploiting analogies between protein and robot motions [5]. These algorithms are either limited to small proteins of no more than 100 amino acids when

© Springer International Publishing Switzerland 2015
R. Harrison et al. (Eds.): ISBRA 2015, LNBI 9096, pp. 175–186, 2015.
DOI: 10.1007/978-3-319-19048-8_15

employing no insight on the degrees of freedom (dofs) involved in the motion or heavily employ such insight and in turn have limited applicability [6–16].

Current robotics-inspired methods are tree-based or roadmap-based. Tree-based methods grow a tree search structure in conformation space from a given start to a given goal conformation. The growth of the tree is biased towards the goal. As such, tree-based methods conduct efficient albeit limited sampling of the conformation space. They are limited to finding essentially one path to the goal conformation, a setting known as single-query, and need to be run multiple times to obtain various paths. However, the bias in the growth of the tree causes path correlations among runs. Tree-based methods have successfully been employed to compute motions connecting functional conformations both in small peptides and large proteins of several hundred amino acids [11–16].

Roadmap-based methods can answer multiple queries through graph search algorithms on a constructed graph/roadmap of nearest-neighbor conformations. The conformations are sampled a priori. Such methods have been applied to compute mainly unfolding motions [6–10]. Several challenges limit broad applicability. Sampling conformations in regions of interest is difficult with no a priori knowledge. Once two nearest neighbors are connected with an edge as part of the roadmap construction, the motion represented by that edge needs to be computed or realized through a local search technique known as a local planner. The local planner needs to find intermediate conformations. Doing so is particularly challenging, either because the planner may have to connect vertices of the roadmap far away in conformation space, if the sampling has not been dense, or vertices separated by a high energy barrier. Significant computational time may be spent by local planners to realize all edges in the roadmap before being able to apply simple graph search algorithms to report paths connecting conformations of interest.

We propose here *SPIRAL*, which stands for Stochastic Protein motIon Roadmap ALgorithm. *SPIRAL* is a roadmap-based algorithm that assumes a limited computational budget and spends that budget in a priority-based scheme to realize promising paths. *SPIRAL* balances computational resources between a global search aimed at obtaining a global view of the network of protein conformations and their connectivity and a detailed local search focused on realizing such connections. In particular, *SPIRAL* is an adaptation of the fuzzy probabilistic roadmap method introduced for manipulation planning in robotics [17]. *SPIRAL* is designed to be general and not employ specific insight on where the relevant dofs are. The goal is to provide through *SPIRAL* a first-generation, general algorithm that can be used as a benchmark to further spur research into roadmap-based frameworks for computing protein motions connecting functional conformations arbitrarily far away in conformation space.

2   Methods

SPIRAL consists of two main stages, sampling and roadmap building. The sampling stage generates an ensemble of conformations/samples, Ω, that provide a discrete representation of the conformation space. In roadmap building, the roadmap $G = (V, E)$ first consists of pseudo-edges over nearest neighbors in Ω. A time-limited iterative interplay between a global search and local search/planners converts pseudo-edges residing in a user-specified number (K) of promising paths into tree search structures of actual edges. At the expiration of time or success-ful computation of K paths, the roadmap is augmented with conformations and connections obtained by the local planners. All edge weights in the roadmap are recomputed to reflect energetic difficulty, and the resulting roadmap is queried for a specified number of lowest-cost paths. Various types of analyses can be conducted over these paths, whether in terms of energetic profile or proximity to given functional conformations.

2.1   Sampling Stage

SPIRAL extends the usual setting where two functional conformations are given to an arbitrary number of given conformations. The idea is to accommodate applications where a number $\ell \geq 2$ of stable or semi-stable functional conforma-tions are known from experiment or computation for a protein of interest, and the goal is to map out the connectivity them. Let us refer to these conformations as landmarks. The landmarks are used to initialize Ω. The sampling stage then consists of a cycle of selection and perturbation operators. A selection operator selects a conformation within the current ensemble. Once selected, a perturba-tion operator is then sampled from a set of available ones and applied to the selected conformation to generate a new conformation. The generated confor-mation is checked for energetic feasibility prior to addition to the ensemble Ω. The process repeats until $|\Omega|$ reaches a pre-determined value.

Selection Operator. The selection operator is based on our prior work on tree-based methods for protein motion computation [16] but extended here to deal with an arbitrary number of landmarks. The goal is to promote cover-age of the conformation space enclosed by the landmarks. A progress coordi-nate, $\Delta R(C)_{i,j}$, is defined for each conformation C and a pair of landmarks (C_i, C_j) as in: $\Delta R(C)_{i,j} = \text{lRMSD}(C_i, C) - \text{lRMSD}(C_j, C)$. lRMSD here refers to least root-mean-squared-deviation used to measure the dissimilarity between two conformations after optimal superposition removes differences due to rigid-body motions [18]. The $\Delta R(C)_{i,j}$ coordinate is used to guide sampling towards under-sampled regions. For each pair of landmarks (C_i, C_j), a 1d grid is defined over the range $[-\text{lRMSD}(C_i, C_j) - 2, \text{lRMSD}(C_i, C_j) + 2]$. Each cell in the grid is 1Å wide. All conformations in Ω are projected onto this grid. In this way, each conformation in the growing ensemble Ω has $\binom{\ell}{2}$ projections, one in each of the $\binom{\ell}{2}$ grids. The selection operator proceeds as follows. A pair of landmarks is selected uniformly at random among the $\binom{\ell}{2}$ pairs. This determines the 1d grid, from which a cell is then sampled according to a probability distribution function defined weights w_c associated with cells of a grid. To bias the selection of confor-mations from under-explored regions of the conformation space, $w_c = \frac{1}{(1+\text{ns})*\text{nc}}$, where ns is the number of times the cell has been selected, and nc is the number of conformations projected onto that cell. Once a cell is selected, a conformation from that cell is then selected uniformly at random.

Perturbation Operators. In the absence of any specific insight onto the dofs of relevance, *SPIRAL* employs a set of perturbation operators in order to make moves of different granularities in conformation space in the sampling stage. Each perturbation operator has to satisfy a set of constraints. One of the constraints enforces energetic feasibility of generated conformations. The energy of a conformation C' generated from a selected conformation C, measured through the Rosetta *score3* function, is compared to the energy of C through the Metropolis criterion (*score3* is the backbone-level energy function, as we employ here only backbone-level representations of protein structure). If this fails, C' is not added to the ensemble. If it passes, C' is checked for satisfaction of distance-based constraints. Additional constraints are introduced on the minimum lRMSD ϵ_{min} of C' to any other conformation in the ensemble Ω and the maximum lRMSD δ of C' to the ℓ landmarks. The first constraint prevents redundant conformations from being added to Ω. The second constraint prevents sampling from veering off in regions of the conformation space deemed far from the landmarks to be useful for participating in paths connecting them. While ϵ_{min} is a parameter that can depend on the specific system under investigation (analysis is provided in section3), a reasonable value for δ is 150% of the maximum lRMSD between any pairs of landmarks.

The idea behind making various perturbation operators available to *SPIRAL* is to allow *SPIRAL* to select the perturbation operator deemed most effective based on features of the conformation space and the specific problem at hand. For instance, when the goal is to connect landmarks that reside far way from one another, a perturbation operator capable of making large moves is first desirable. Afterwards, to be able to make connections between such conformations, other perturbation operators capable of making smaller moves may be more effective. We consider here three perturbation operators, detailed below. An optimal weighting scheme that is responsive to emerging features of the search space is difficult to formulate and beyond the scope of the work here. However, we have been able to empirically determine a weighting scheme that is effective on most protein systems studied here.

Molecular Fragment Replacement Operator This operator is inspired from protein structure prediction, where backbone dihedral angles in a bundle/fragment of f consecutive amino acids are replaced altogether with values from a precompiled library. *SPIRAL* employs $f \in \{3, 9\}$ to balance between large ($f = 9$) and small ($f = 3$) moves.

Single Dihedral Replacement Operator This operator modifies a single backbone dihedral angle at a time to allow small moves. Given a selected backbone dihedral angle in a selected conformation, a new value from it is obtained using a normal distribution $\mathcal{N}(\mu, \sigma)$. The angle to perturb is selected uniformly at random. This operator, gaussian sampling, offers the option of biasing the selection of dihedral angles to promote selection of those that differ most between a selected conformation and a landmark, though our application of *SPIRAL* here does not make use of biased gaussian sampling in the sampling stage.

Reactive Temperature Scheme: The energetic constraint that determines whether a conformation C' produced from a perturbation operator applied onto a selected conformation C should be added to Ω is based on the Metropolis criterion. Essentially, a probability $e^{-(E_{C'}-E_C)/(K \cdot T)}$, is measured, where K is the Boltzmann constant, and T is temperature. An arbitrary temperature value is both difficult to justify and obtain constraints-satisfying conformations as Ω grows (if T is low). So, as in previous work on tree-based methods [16], we make use of a reactive temperature scheme but extend it to the multiple-landmark setting here. We maintain a temperature value T_c for each cell c of the 1d grids over the progress coordinate. Each cell's temperature is adjusted every s steps (typical value employed is 25). The temperature of a cell, T_c, is increased if the last s selections of that cell have resulted in no conformations being added to Ω. If conformations are added to Ω more than 60% of the time within a window of s steps, T_c is decreased. Increases and decreases occur over adjacent temperature levels per a proportional cooling scheme that starts with very high temperatures in the 2,000K range and ends with room temperature of 300K.

2.2 Roadmap Building Stage

All conformations in Ω are added to the vertex set V. For each $v \in V$, its k nearest-neighbors are identified, using lRMSD. For each identified neighbor, directional pseudo-edges are added with v. Additional pseudo-edges are added by identifying any vertex $< \epsilon_{max}$ from v that lies in a different connected component from v. Typical values for k and ϵ_{max} are 10 and 5Å, respectively.

The pseudo-edges are assigned a weight to reflect their estimated difficulty of being realizable. At initialization, all pseudo-edges are determined equally difficult with a weight value of 1. A two-layer scheme is then used, which is an iterative interplay between global and local search. The global search, path query, identifies the current most promising/lowest-cost path in the roadmap connecting two given functional conformations. If there are unrealizable edges in the path, these edges are fed to the local search, which launches local planners on unrealized edges, pursuing path realization. The planners are given a limited computational budget, and they report at the end of this budget either a realized edge or a new weight for the unrealized edges. In this iterative interplay between path query and path realization, over time, the pseudo-edges that are most difficult to realize will be assigned high weights and will thus be unlikely to participate in the lowest-cost path pushed to the local planners. This dynamic interplay apportions computational resources in a manner that promotes rapid path discovery. The iterative process continues until a total computational budget is exhausted or a user-specified number of paths is obtained.

Path Query and Path Realization Interplay: A pair of landmarks are selected uniformly at random over the $\ell!$ permutations. The roadmap is then queried for a lowest-cost path, using the assigned pseudo-edge weights. We utilize Yen's K-Shortest path algorithm [19] to identify the lowest non-zero cost path in the graph and allow us to continue obtaining paths after the first path has been successfully realized. Given an identified path, a local planner is assigned to any of the unrealized edges. The planner is given a fixed computational budget, time T. If the local planner succeeds, the pseudo-edge it has realize is assigned a weight of 0 to indicate the pseudo-edge is resolved. If the local planner fails, the pseudo-edge is reweighted as in $w_e = 0.7 \cdot \text{CallsToPlanner} + 0.3 \cdot (\text{ClosestNode} - \text{RequireResolution})^2$. CallsToPlanner tracks the number of times the planner has been requested to work on a particular pseudo-edge, ClosestNode is the node in the tree constructed by the local planner that is closest to the vertex v in the directed pseudo-edge (u, v). For the planner to be successful, it must also generate a path that is within a user-specified lRMSD of the vertex v, so RequireResolution is also employed.

An additional feature of *SPIRAL* is its ability to learn from failures. When a local planner has failed to complete a path more than RefineLimit times, *SPIRAL* augments the graph with conformations identified by the local planner that are otherwise invisible to the global layer. We now proceed to relate details on the local planner and the augmentation procedure.

Local Planner: The local planner is an adaption of the tree-based method proposed in [16]. The adaptation consists of diversifying the types of perturbation operators employed in the expansion of the tree. The local planner selects through a probabilistic scheme shown in section 3 from the menu of perturbation operators described above. While biased gaussian sampling is not used in the sampling stage in *SPIRAL*, it is used by the local planner.

Roadmap Augmentation: Some regions of conformational space may be challenging to connect through local planners. This can be due to high energetic barriers or inadequate sampling. To address this issue, *SPIRAL* makes use of a feedback mechanism to augment the roadmap. When a local planner encounters difficulty realizing a pseudo-edge connecting given conformations p and r more than RefineLimit times (set at 25), the problem of connecting p to r is considered as a mini-version of the entire motion computation problem. The sampling scheme is repeated, essentially treating p and r as start and goal conformations. The perturbation operators described above are used together with a new one based on straight-line interpolation. The produced conformations are then minimized using the Rosetta *relax* protocol. Only the lowest-energy conformation is considered for addition. Conformations obtained from the perturbation operators are checked for satisfaction of the energetic and geometric constraints also used in the sampling stage. The operators are applied under a probabilistic scheme detailed in section 3 until either 25 conformations have been added to the roadmap or a maximum of 2500 attempts to do so have been made.

Roadmap Analysis: Each edge in the roadmap is reweighted to reflect energetic difficulty per the Metropolis criterion. A room temperature value is used for this purpose. The reweighted graph is queried for one or more lowest-cost paths, which are then analyzed in terms of energetic profile or distance within which they come of the goal landmark structures, as related in section 3.

3   Results

3.1   Systems of Study

Table 1 lists the protein systems selected for testing here. These are carefully gathered from published literature to provide comparisons where possible; not many published methods exist, and many of them either focus on few specific systems or are limited by system size. For most of the collected systems, two functional conformations have been extracted from literature (we consider both directions). The final column in Table 1 shows the lRMSD between the start and goal conformations. Neither size nor the lRMSD between functional conformations do by themselves define system difficulty. We have observed that the larger systems that exhibit smaller motions (less than 4.5Å lRMSD) between the start and goal conformations may require the protein chain to partially unfold before returning to a folded state. The process of unfolding a large, compact structure is computationally costly, as effectively an energy barrier needs to be crossed to get out of the compact state. Indeed, many computational studies avoid computing the motions involved in transitions from a closed to an open structural state because of this challenge.

Table 1. Protein systems for evaluation of performance

System	Length	Start ↔ Goal	lRMSD(start, goal)
CVN	101	2ezm ↔ 1l5e	16.01 Å
CaM	140	1cfd ↔ 1cll	10.7 Å
		1cfd ↔ 2f3y	9.9 Å
		1cll ↔ 2f3y	13.44 Å
AdK	214	1ake ↔ 4ake	6.96 Å
LAO	238	1laf ↔ 2lao	4.7 Å
DAP	320	1dap ↔ 3dap	4.3 Å
OMP	370	1omp ↔ 3mbp	3.7 Å
BKA	691	1cb6 ↔ 1bka	6.4 Å

3.2   Implementation Details

SPIRAL is implemented in C++. A hard termination criterion is set with regards to the total number of energy evaluations. The sampling stage is terminated if the total number of energy evaluations exceeds $1,000$ times the requested ensemble size. That is, a maximum of 25 attempts are made to obtain a sample. The roadmap building stage is terminated after $10,000$ iterations of the interplay between path query and path realization. This stage may terminate earlier if $K = 250$ paths are obtained for all $\ell!$ landmarks as a way to control computational cost. The analysis stage reports the 50 lowest-cost paths. In terms of CPU time, the computational time demands of all these three stages in *SPIRAL* spans anywhere from 56 hours on one CPU for protein systems around 100 amino acids long to 300 hours on one CPU for systems around 700 amino acids long.

During sampling, the molecular fragment replacement perturbation operator with $f = 3$ is selected 75% of the time, the operator with $k = 9$ is selected 20% of the time, and the gaussian sampling operator is selected 5% of the time. The reason for this scheme is to make large moves more often than small ones so as to spread out conformations in conformation space during sampling.

During roadmap building, the probabilistic scheme with which local planners and the roadmap augmentation make use of the perturbation operators is different, as shown in Table 2. A local planner can use two different schemes depending on the lRMSD between the two conformation/vertices it is asked to connect by the global layer. These schemes are not fine-tuned; essentially, when the distance is $\leq$ 2.5Å, smaller moves are promoted as opposed to when the distance is > 2.5Å. The reason for basing the decision at 2.5Åis due to prior work on tree-based planners showing that molecular fragment replacement can result in step sizes greater than 2.5Å [16].

Table 2. The perturbation operator set and their weights during roadmap building

Roadmap Building	Perturbation Operator	Prob.
	Molecular Fragment Replacement ($f = 3$)	0.70
Local Planner (> 2.5 Å lRMSD)	Gaussian Sampling ($\mu = 0, \sigma = 15$)	0.15
	Biased Gaussian Sampling ($\mu = 0, \sigma = 15$)	0.15
	Molecular Fragment Replacement ($f = 3$)	0.20
Local Planner ($\leq$ 2.5 Å lRMSD)	Gaussian Sampling ($\mu = 0, \sigma = 15$)	0.40
	Biased Gaussian Sampling ($\mu = 0, \sigma = 15$)	0.40
	Molecular Fragment Replacement($f = 3$)	0.20
Augmentation	Gaussian Sampling ($\mu = 0, \sigma = 15$)	0.40
	Biased Gaussian Sampling ($\mu = 0, \sigma = 15$)	0.40
	Interpolation-based	0.05

The ϵ_{min} parameter controls how close neighboring conformations will be in the roadmap. Intuitively, smaller ϵ_{min} values would produce a better-quality roadmap. Our analysis indicates that this is not the case. Small values of ϵ_{min} (< 1Å) can result in many small cliques being formed in the roadmap around local minima conformations. This is not surprising, particularly for the broad minima that contain the stable and semi-stable landmarks. On these minima, it is rather easy to sample a very large number of conformations nearby a landmark and thus essentially "get stuck" in the same local minimum. Insisting on a minimum distance separation among sampled conformations forces sampling not to provide refinement or exploitation of a particular local minimum but rather explore the breadth of the conformation space. Not insisting on a minimum distance pushes all the work to obtaining intermediate conformations to bridge local minima to the local planners, which is an ineffective use of computational time. The ϵ_{min} parameter is set to 2.0Å for systems where the lRMSD between

Table 3. Column 4 reports smallest distance to goal over all paths obtained by *SPI-RAL*. Columns 5 − 7 shows such distances from tree-based methods. Max Step in column 3 refers to the maximum lRMSD between any two consecutive conformations in the *SPIRAL* path that comes closest to the goal. '-' indicates lack of published data.

System	Start → Goal	Max Step	Dist to Goal (Å)			
			SPIRAL	Tree-based [16]	Cortés[15]	Haspel [11, 12]
CVN	2ezm → 1l5e	1.5	**1.5**	–	2.1	2.1
(101 aa)	1l5e → 2ezm	1.5	1.3	–	–	–
CaM (144 aa)	1cll → 1cfd	3.4	**1.46**	3.35	–	–
	1cfd → 1cll	2.67	**1.12**	3.17	–	–
	1cll → 2f3y	2.77	**1.26**	1.67	–	–
	2f3y → 1cll	3.5	1.12	0.73	–	1.33
	1cfd → 2f3y	3.33	**1.26**	3.5	–	–
	2f3y → 1cfd	3.48	**1.46**	3.2	–	–
AdK	1ake → 4ake	3.0	**1.86**	3.8	2.56	2.2
(214 aa)	4ake → 1ake	3.12	**1.33**	3.6	1.56	–
Lao	2lao → 1laf	2.0	**1.21**	–	1.32	–
(238 aa)	1laf → 2lao	3.2	1.90	–	–	–
DAP	1dap → 3dap	1.42	1.5	–	1.31	–
(320 aa)	3dap → 1dap	1.46	0.92	–	–	–
OMP	1omp → 3mbp	1.04	3.04	–	–	–
(370 aa)	3mbp → 1omp	0.91	3.61	–	–	–
BKA	1bka → 1cb6	3.87	**1.55**	–	2.79	–
(691 aa)	1cb6 → 1bka	3.98	1.69	–	–	–

landmarks is > 6Å, 1.5Å for systems where the lRMSD between landmarks is > 4.5 but ≤ 6Å, and 1.0 for systems where the lRMSD between landmarks is ≤ 4.5Å.

3.3 Comparison of Found Paths with Other Methods

We compare *SPIRAL* to published tree-based methods [11, 12, 15, 16]. These methods make use of specific moves. For instance, our tree-based method in [16] uses molecular fragment replacements with $f = 3$, the method in [15] uses moves over low-frequency modes revealed by normal mode analysis, and the method in [11, 12] considers only backbone dihedral angles whose values change between the given functional conformations. The last two methods consider a low-dimensional search space of no more than 30 dimensions.

We report the closest that any path computed by *SPIRAL* comes to the specified goal conformation and compare such values on all protein systems to those reported in other published work. Columns 4−7 in Table 3 show these values for *SPIRAL* and other published work. Column 3 reports some more details on the path with which *SPIRAL* comes closest to the goal conformation by listing the maximum lRMSD between any two consecutive conformations

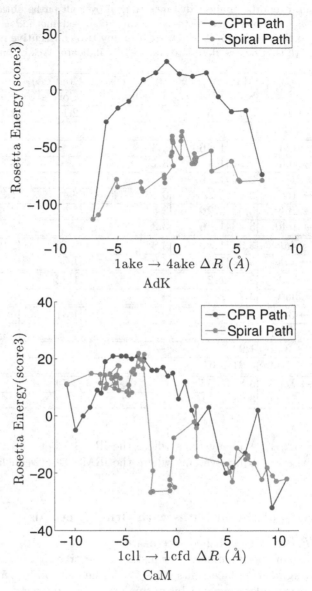

Fig. 1. Energy profiles of conformational paths computed for AdK (top) and of CaM (bottom) by *SPIRAL*(green) and an interpolation-based planner (red)

in the path. *SPIRAL* typically generates paths with conformations closer to the goal conformation than other methods (highlighted in bold where true). A video illustrating the lowest-cost conformational path reported by *SPIRAL*for the CVN protein can be found at http://youtu.be/7P4reYO3k-c.

3.4   Analysis of Energetic Profiles

We show the energetic profile of the lowest-cost path obtained by *SPIRAL* on two selected systems, AdK and CaM. We compare these profiles to those obtained by the interpolation-based planner described in section 2. For this planner, the resolution distance ϵ is set to 1.0 Å, and 50 cycles are performed to obtain a path. This provides a fair comparison, given that we also analyze 50 paths obtained after the analysis stage in *SPIRAL* and report here the lowest-cost one. Figure 1 shows that on proteins, such as AdK, where the distance between the start and goal conformations is large, paths provided by the interpolation-based planner tend to have higher energies than those provided by *SPIRAL*. On systems, such as CaM, where the start-to-goal distance is smaller, an interpolation-based planner can perform comparably to *SPIRAL*.

4   Conclusions

This paper has proposed *SPIRAL*, a novel protein motion computation algorithm capable of handling proteins of various sizes and settings where distances among functional conformations of interest can exceed 16Å. The algorithm is inspired by frameworks used in robot motion planning as opposed to MD- or MC-based frameworks. The main reason for doing so is to address the limited sampling in MD- or MC-based frameworks, particularly when motions involve disparate time and length scales.

SPIRAL exploits no particular information on any protein at hand. It is expected that tunings of the probabilistic scheme or employment of additional perturbation operators and moves based on specific system insight will improve performance. Future work will consider such directions, but the current need of the community is for a powerful, general, baseline method for the purpose of benchmarking.

The results shown here suggest *SPIRAL* produces good-quality paths and can be employed both to extract information on protein motions, possible long-lived intermediate conformations in such motions, as well as to advance algorithmic work in motion computation frameworks. In particular, the inherent prioritization scheme in *SPIRAL* allows the sampling of both low-cost paths and high-cost paths, provided enough computational budget is allocated. The latter paths may highlight possible local unfolding involved in protein motions connecting functional conformations. An executable of *SPIRAL* can be provided to researchers upon demand.

Acknowledgments. Experiments were run on ARGO, a research computing cluster provided by the Office of Research Computing at George Mason University, VA (URL: http://orc.gmu.edu). Funding for this work is provided in part by the National Science Foundation (Grant No. 1440581 and CAREER Award No. 1144106).

References

1. Jenzler-Wildman, K., Kern, D.: Dynamic personalities of proteins. Nature 450, 964–972 (2007)
2. Wong, C.F., McCammon, J.A.: Protein simulation and drug design. Adv. Protein Chem. 66, 87–121 (2003)
3. Merkx, M., Golynskiy, M.V., Lindenburg, L.H., Vinkenborg, J.L.: Rational design of FRET sensor proteins based on mutually exclusive domain interactions. Biochem. Soc. Trans. 41, 128–134 (2013)
4. Amaro, R.E., Bansai, M.: Editorial overview: Theory and simulation: Tools for solving the insolvable. Curr. Opinion Struct. Biol. 25, 4–5 (2014)
5. Singh, A.P., Latombe, J.C., Brutlag, D.L.: A motion planning approach to flexible ligand binding. In: Schneider, R., Bork, P., Brutlag, D.L., Glasgow, J.I., Mewes, H.W., Zimmer, R. (eds.) Proc. Int. Conf. Intell. Sys. Mol. Biol (ISMB), vol. 7, pp. 252–261. AAAI, Heidelberg (1999)
6. Amato, N.M., Dill, K.A., Song, G.: Using motion planning to map protein folding landscapes and analyze folding kinetics of known native structures. J. Comp. Biol. 10, 239–255 (2002)
7. Thomas, S., Song, G., Amato, N.: Protein folding by motion planning. Physical Biology S148–S155 (2005)
8. Thomas, S., Tang, X., Tapia, L., Amato, N.M.: Simulating protein motions with rigidity analysis. J. Comput. Biol. 14, 839–855 (2007)
9. Tapia, L., Tang, X., Thomas, S., Amato, N.: Kinetics analysis methods for approximate folding landscapes. Bioinformatics 23, i539–i548 (2007)
10. Tapia, L., Thomas, S., Amato, N.: A motion planning approach to studying molecular motions. Communications in Information Systems 10, 53–68 (2010)
11. Haspel, N., Moll, M., Baker, M.L., Chiu, W., Kavraki, L.E.: Tracing conformational changes in proteins. BMC Struct. Biol. 10, S1 (2010)
12. Luo, D., Haspel, N.: Multi-resolution rigidity-based sampling of protein conformational paths. In: Proc. of ACM-BCB (ACM International Conference on Bioinformatics and Computational Biology), CSBW (Computational Structural Bioinformatics Workshop), pp. 787–793 (2013)
13. Cortés, J., Simeon, T., de Angulo, R., Guieysse, D., Remaud-Simeon, M., Tran, V.: A path planning approach for computing large-amplitude motions of flexible molecules. Bioinformatics 21, 116–125 (2005)
14. Jaillet, L., Corcho, F.J., Perez, J.J., Cortés, J.: Randomized tree construction algorithm to explore energy landscapes. J. Comput. Chem. 32, 3464–3474 (2011)
15. Al-Bluwi, I., Vaisset, M., Siméon, T., Cortés, J.: Modeling protein conformational transitions by a combination of coarse-grained normal mode analysis and robotics-inspired methods. BMC Structural Biology 13, S8 (2013)
16. Molloy, K., Shehu, A.: Elucidating the ensemble of functionally-relevant transitions in protein systems with a robotics-inspired method. BMC Struct. Biol. 13, S8 (2013)
17. Nielsen, C., Kavraki, L.: A two level fuzzy prm for manipulation planning. In: Proceedings of the 2000 IEEE/RSJ International Conference on Intelligent Robots and Systems, IROS 2000, vol. 3, pp. 1716–1721 (2000)
18. McLachlan, A.D.: A mathematical procedure for superimposing atomic coordinates of proteins. Acta Crystallogr. A. 26, 656–657 (1972)
19. Yen, J.Y.: Finding the k shortest loop less paths in a network. Management Science 17, 712–716 (1971)

On the Complexity of Duplication-Transfer-Loss Reconciliation with Non-binary Gene Trees

Misagh Kordi[1] and Mukul S. Bansal[1,2(✉)]

[1] Department of Computer Science and Engineering,
University of Connecticut, Storrs, USA
[2] Institute for Systems Genomics, University of Connecticut, Storrs, USA
misagh.kordi@uconn.edu, mukul@engr.uconn.edu

Abstract. Duplication-Transfer-Loss (DTL) reconciliation has emerged as a powerful technique for studying gene family evolution in the presence of horizontal gene transfer. DTL reconciliation takes as input a gene family phylogeny and the corresponding species phylogeny, and reconciles the two by postulating speciation, gene duplication, horizontal gene transfer, and gene loss events. Efficient algorithms exist for finding optimal DTL reconciliations when the gene tree is binary. However, gene trees are frequently non-binary. With such non-binary gene trees, the reconciliation problem seeks to find a binary resolution of the gene tree that minimizes the reconciliation cost. Given the prevalence of non-binary gene trees, many efficient algorithms have been developed for this problem in the context of the simpler Duplication-Loss (DL) reconciliation model. Yet, no efficient algorithms exist for DTL reconciliation with non-binary gene trees and the complexity of the problem remains unknown. In this work, we resolve this open question by showing that the problem is, in fact, NP-hard. Our reduction applies to both the dated and undated formulations of DTL reconciliation. By resolving this long-standing open problem, this work will spur the development of both exact and heuristic algorithms for this important problem.

1 Introduction

Duplication-Transfer-Loss (DTL) reconciliation is one of the most powerful techniques for studying gene and genome evolution in microbes and other non-microbial species engaged in horizontal gene transfer. DTL reconciliation accounts for the role of gene duplication, gene loss, and horizontal gene transfer in shaping gene families and can infer these evolutionary events through the systematic comparison and reconciliation of gene trees and species trees. Specifically, given a gene tree and a species tree, DTL reconciliation shows the evolution of the gene tree inside the species tree, and explicitly infers duplication, transfer, and loss events. Accurate knowledge of gene family evolution has many uses in biology, including inference of orthologs, paralogs and xenologs for functional genomic studies, e.g., [1, 2], reconstruction of ancestral gene content, e.g., [3,4], and accurate gene tree and species tree construction, e.g., [2,5–7], and the DTL reconciliation problem has therefore been widely studied, e.g., [4,8–15].

DTL reconciliation is typically formulated using a parsimony framework where each evolutionary event is assigned a cost and the goal is to find a reconciliation with minimum

© Springer International Publishing Switzerland 2015
R. Harrison et al. (Eds.): ISBRA 2015, LNBI 9096, pp. 187–198, 2015.
DOI: 10.1007/978-3-319-19048-8_16

total cost. The resulting optimization problem is called the *DTL-reconciliation prob-lem*. DTL-reconciliations can sometimes be *time-inconsistent*; i.e, the inferred transfers may induce contradictory constraints on the dates for the internal nodes of the species tree. The problem of finding an optimal *time-consistent* reconciliation is known to be NP-hard [10, 16]. Thus, in practice, the goal is to find an optimal (not necessarily time-consistent) DTL-reconciliation [4,10,11,13,15] and this problem can be solved in $O(mn)$ time [11], where m and n denote the number of nodes in the gene tree and species tree, respectively. Interestingly, the problem of finding an optimal time-consistent reconcili-ation actually becomes efficiently solvable [9, 17] in $O(mn^2)$ time if the species tree is fully dated. Thus, these two efficiently solvable formulations, regular and dated, are the two standard formulations of the DTL-reconciliation problem.

Both these formulations of the DTL-reconciliation problem assume that the input gene tree and species tree are binary. However, gene trees are frequently non-binary in practice. This is due to the fact that there is often insufficient information in the under-lying gene sequences to fully resolve gene tree topologies. When the input consists of a non-binary gene tree, the reconciliation problem seeks to find a binary resolution of the gene tree that minimizes the reconciliation cost. Given the prevalence of non-binary gene trees, many efficient algorithms have been developed for this problem in the con-text of the simpler Duplication-Loss (DL) reconciliation model [5, 18–20], with the most efficient of these algorithms having an optimal $O(m + n)$ time complexity [20]. However, the DTL reconciliation model is more general and significantly more complex than the DL reconciliation model. Consequently, no efficient algorithms exist for DTL reconciliation with non-binary gene trees and the complexity of the problem remains unknown. As a result, DTL reconciliation is currently inapplicable to non-binary gene trees, significantly reducing its utility in practice.

In this work, we settle this open problem by proving that the DTL-reconciliation problem on non-binary gene trees is, in fact, NP-hard. Our proof is based on a reduction from the minimum 3-set cover problem and applies to both formulations of the DTL-reconciliation problem. An especially desirable feature of our reduction is that it implies NP-hardness for biologically relevant settings of the event cost parameters, showing that the problem is difficult even for biologically meaningful scenarios. The uncertainty about the complexity of DTL-reconciliation for non-binary gene trees has prevented the development of any algorithms, exact or heuristic, for the problem. By settling this question, our work will spur the development of both exact (better than brute-force) and efficient approximation and heuristic algorithms for this important problem.

We develop our NP-hardness proof in the context of the regular (undated) DTL-reconciliation formulation, and revisit dated DTL-reconciliation later in Section 4. The next section introduces basic definitions and preliminaries, and we present the NP-hardness proof for the optimal gene tree resolution problem in Section 3. Concluding remarks appear in Section 5. In the interest of brevity, proofs for all Lemmas are de-ferred to the full version of this paper.

2   Definitions and Preliminaries

We follow the basic definitions and notation from [11]. Given a tree T, we denote its node, edge, and leaf sets by $V(T)$, $E(T)$, and $Le(T)$ respectively. If T is rooted, the root node of T is denoted by $rt(T)$, the parent of a node $v \in V(T)$ by $pa_T(v)$, its set of children by $Ch_T(v)$, and the (maximal) subtree of T rooted at v by $T(v)$. The set of *internal nodes* of T, denoted $I(T)$, is defined to be $V(T) \setminus Le(T)$. We define $\leq_T$ to be the partial order on $V(T)$ where $x \leq_T y$ if y is a node on the path between $rt(T)$ and x. The partial order $\geq_T$ is defined analogously, i.e., $x \geq_T y$ if x is a node on the path between $rt(T)$ and y. We say that y is an *ancestor* of x, or that x is a *descendant* of y, if $x \leq_T y$ (note that, under this definition, every node is a descendant as well as ancestor of itself). We say that x and y are *incomparable* if neither $x \leq_T y$ nor $y \leq_T x$. Given a non-empty subset $L \subseteq Le(T)$, we denote by $lca_T(L)$ the last common ancestor (LCA) of all the leaves in L in tree T. Throughout this work, the *term* tree refers to rooted trees. A tree is *binary* if all of its internal nodes have exactly two children, and *non-binary* otherwise. We say that a tree T' is a *binary resolution* of T if T' is binary and T can be obtained from T' by contracting one or more edges. We denote by $\mathcal{BR}(T)$ the set of all binary resolutions of a non-binary tree T.

Gene trees may be either binary or non-binary while the species tree is always assumed to be binary. Throughout this work, we denote the gene tree and species tree under consideration by G and S, respectively. If G is restricted to be binary we refer to it as G^B and as G^N if it is restricted to be non-binary. We assume that each leaf of the gene tree is labeled with the species from which that gene was sampled. This labeling defines a *leaf-mapping* $\mathcal{L}_{G,S} \colon Le(G) \rightarrow Le(S)$ that maps a leaf node $g \in Le(G)$ to that unique leaf node $s \in Le(S)$ which has the same label as g. Note that gene trees may have more than one gene sampled from the same species. We will implicitly assume that the species tree contains all the species represented in the gene tree.

2.1   Reconciliation and DTL-scenarios

A binary gene tree can be reconciled with a species tree by mapping the gene tree into the species tree. Next, we define what constitutes a valid reconciliation; specifically, we define a Duplication-Transfer-Loss scenario (DTL-scenario) [10, 11] for G^B and S that characterizes the mappings of G^B into S that constitute a biologically valid reconciliation. Essentially, DTL-scenarios map each gene tree node to a unique species tree node in a consistent way that respects the immediate temporal constraints implied by the species tree, and designate each gene tree node as representing either a speciation, duplication, or transfer event.

Definition 1 (DTL-scenario). *A DTL-scenario for G^B and S is a seven-tuple $\langle \mathcal{L}, \mathcal{M}, \Sigma, \Delta, \Theta, \Xi, \tau \rangle$, where $\mathcal{L} \colon Le(G^B) \rightarrow Le(S)$ represents the leaf-mapping from G^B to S, $\mathcal{M} \colon V(G^B) \rightarrow V(S)$ maps each node of G^B to a node of S, the sets Σ, Δ, and Θ partition $I(G^B)$ into speciation, duplication, and transfer nodes respectively, Ξ is a subset of gene tree edges that represent transfer edges, and $\tau \colon \Theta \rightarrow V(S)$ specifies the recipient species for each transfer event, subject to the following constraints:*

1. *If $g \in Le(G^B)$, then $\mathcal{M}(g) = \mathcal{L}(g)$.*
2. *If $g \in I(G^B)$ and g' and g'' denote the children of g, then,*
   (a) $\mathcal{M}(g) \not\leq_S \mathcal{M}(g')$ and $\mathcal{M}(g) \not\leq_S \mathcal{M}(g'')$,
   (b) At least one of $\mathcal{M}(g')$ and $\mathcal{M}(g'')$ is a descendant of $\mathcal{M}(g)$.

3. *Given any edge* $(g, g') \in E(G^B)$, $(g, g') \in \Xi$ *if and only if* $\mathcal{M}(g)$ *and* $\mathcal{M}(g')$ *are incomparable.*
4. *If* $g \in I(G^B)$ *and* g' *and* g'' *denote the children of* g, *then,*
   (a) $g \in \Sigma$ *only if* $\mathcal{M}(g) = lca(\mathcal{M}(g'), \mathcal{M}(g''))$ *and* $\mathcal{M}(g')$ *and* $\mathcal{M}(g'')$ *are incomparable,*
   (b) $g \in \Delta$ *only if* $\mathcal{M}(g) \geq_S lca(\mathcal{M}(g'), \mathcal{M}(g''))$,
   (c) $g \in \Theta$ *if and only if either* $(g, g') \in \Xi$ *or* $(g, g'') \in \Xi$.
   (d) *If* $g \in \Theta$ *and* $(g, g') \in \Xi$, *then* $\mathcal{M}(g)$ *and* $\tau(g)$ *must be incomparable, and* $\mathcal{M}(g')$ *must be a descendant of* $\tau(g)$, *i.e.,* $\mathcal{M}(g') \leq_S \tau(g)$.

DTL-scenarios correspond naturally to reconciliations and it is straightforward to infer the reconciliation of G^B and S implied by any DTL-scenario. Figure 1 shows an example of a DTL-scenario. Given a DTL-scenario α, one can directly count the minimum number of gene losses, $Loss_\alpha$, in the corresponding reconciliation. For brevity, we refer the reader to [11] for further details on how to count losses in DTL-scenarios.

Let P_Δ, P_Θ, and P_{loss} denote the non-negative costs associated with duplication, transfer, and loss events, respectively. The reconciliation cost of a DTL-scenario is defined as follows.

Definition 2 (Reconciliation cost of a DTL-scenario). *Given a DTL-scenario* $\alpha = \langle \mathcal{L}, \mathcal{M}, \Sigma, \Delta, \Theta, \Xi, \tau \rangle$ *for* G^B *and* S, *the* reconciliation cost *associated with* α *is given by* $\mathcal{R}_\alpha = P_\Delta \cdot |\Delta| + P_\Theta \cdot |\Theta| + P_{loss} \cdot Loss_\alpha$.

A most parsimonious reconciliation is one that has minimum reconciliation cost.

Definition 3 (Most Parsimonious Reconciliation (MPR)). *Given* G^B *and* S, *along with* P_Δ, P_Θ, *and* P_{loss}, *a* most parsimonious reconciliation (MPR) *for* G^B *and* S *is a DTL-scenario with minimum reconciliation cost.*

2.2 Optimal Gene Tree Resolution

Non-binary gene trees cannot be directly reconciled against a species tree. Thus, given a non-binary gene tree G^N, the problem is to find a binary resolution of G^N whose MPR with S has the smallest reconciliation cost. An example of a non-binary gene tree and a binary resolution is shown in Figure 1.

Problem 1 (Optimal Gene Tree Resolution (OGTR)). *Given* G^N *and* S, *along with* P_Δ, P_Θ, *and* P_{loss}, *the* Optimal Gene Tree Resolution (OGTR) *problem is to find a binary resolution* G^B *of* G^N *such that the MPR of* G^B *and* S *has the smallest reconciliation cost among all* $G^B \in \mathcal{BR}(G^N)$.

3 NP-hardness of the OGTR Problem

We claim that the OGTR problem is NP-hard; specifically, that the corresponding decision problem is NP-Complete. The decision version of the OTGR problem is as follows:

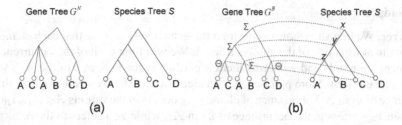

Fig. 1. DTL reconciliation and OGTR problem. Part (a) shows a non-binary gene tree G^N and binary species tree S. Part (b) shows a DTL reconciliation between a possible binary resolution G^B of G^N and species tree S. The dotted arcs show the mapping $\mathcal{M}$ (with the leaf mapping being specified by the leaf labels on the gene tree), and the label at each internal node of G^B specifies the type of event represented by that node. This reconciliation invokes two transfer events.

Problem 2 (Decision-OGTR (D-OGTR)).

Instance: G^N and S, event costs P_Δ, P_Θ, and P_{loss}, and a non-negative integer l.
Question: Does there exist a $G^B \in \mathcal{BR}(G^N)$ such that the MPR of G^B and S has reconciliation cost at most l?

Theorem 1. *The D-OGTR problem is NP-Complete.*

The D-OGTR problem is clearly in NP. In the remainder of this section we will show that the D-OGTR problem is NP-hard using a poly-time reduction from the decision version of the NP-hard *minimum 3-set cover* problem [21].

3.1 Reduction from Minimum 3-set Cover

The decision version of minimum 3-set cover can be stated as follows.

Problem 3 (Minimum 3-Set Cover (M3SC)).

Instance: *Given a set of n elements $U=\{u_1, u_2, \ldots, u_n\}$, a set $A=\{A_1, A_2, \ldots, A_m\}$ of m subsets of U such that $|A_i| = 3$ for each $1 \le i \le m$, and a nonnegative integer $k \le m$.*
Question: *Is there a subset of A of size at most k whose union is U?*

We point out that the M3SC problem as defined above is a slight variation of the traditional minimum 3-set cover problem: In our formulation the subsets of U in A are restricted to have *exactly* three elements each while the traditional formulation allows for the subsets to have *less than or equal to* three elements [21]. However, it is easy to establish that the NP-Completeness of the traditional version directly implies the NP-Completeness of our formulation of the M3SC problem. We will also assume, without any loss of generality, that each element u_i appears in at least two subsets from A.

Consider an instance ϕ of the M3SC problem with $U = \{u_1, u_2, \ldots, u_n\}$, $A = \{A_1, A_2, \ldots, A_m\}$, and k given. We now show how to transform ϕ into an instance λ of the D-OGTR problem by constructing G^N and S and setting the three event costs in such a way that there exists a YES answer to the M3SC instance ϕ if and only if there exists a YES answer to the D-OGTR instance λ with $l = 10k + 39m - 12n$.

3.2 Gadget

Gene Tree. We first show how to construct the gene tree G^N. Note that each element of U occurs in at least two of the subsets from A. We will treat each of the occurrences of an element separately and will order them according to the indices p of the A_p's which contain that element. More precisely, for an element $u_i \in U$, we denote by $x_{i,j}$ the jth occurrence of u_i in A. For instance, if element u_5 occurs in the subsets A_2, A_4, A_{10}, and A_{25}, then $x_{5,2}$ refers to the occurrence of u_5 in A_4, while $x_{5,4}$ refers to the occurrence of u_5 in A_{25}.

Let c_i denote the cardinality of the set $\{A_p: u_i \in A_p, \text{ for } 1 \le p \le m\}$. Then, $x_{i,j}$ is well defined as long as $1 \le i \le n$ and $1 \le j \le c_i$. Each $x_{i,j}$ will correspond to exactly four leaves, $x_{i,j,1}, x_{i,j,2}, x_{i,j,3}$, and $x_{i,j,4}$ in the gene tree G^N. In addition, the leaf set of G^N also contains a special node labeled *start*, provided for orientation.

Thus, $Le(G^N) = \{x_{i,j,1}, x_{i,j,2}, x_{i,j,3}, x_{i,j,4}: 1 \le i \le n \text{ and } 1 \le j \le c_i\} \cup \{start\}$. The overall structure of G^N is shown in Figure 2(a). As shown, the root node of the gene tree is unresolved and has $3m + 3n + 1$ children consisting of (i) the *start* node, (ii) the $\sum_{i=1}^{n} c_i = 3m$ leaf nodes, collectively called *blue* nodes, and (iii) the $3n$ internal nodes labeled g_i, g_i', and g_i'', for each $1 \le i \le n$. These internal nodes represent the n elements in U and the subtrees rooted at those nodes have the structure shown in Figure 2(a). Note that the number of children for each of the internal nodes labeled g_i, g_i', and g_i'', for $1 \le i \le n$, is c_i. These nodes may thus be either binary or non-binary. The leaves labeled $x_{i,j,3}$ appear in the node g_i', those labeled $x_{i,j,4}$ appear in g_i'', and those labeled $x_{i,j,1}$ or $x_{i,j,2}$ appear in g_i. The $x_{i,j,1}$'s also appear in the collection of blue nodes and thus appear twice in the gene tree. Note, also, that all the children of a node g_i, for $1 \le i \le n$, are themselves internal nodes and are labeled $y_{i,j}$, where $1 \le j \le c_i$.

Species Tree. Next, we show how to construct the species tree S. The tree S is binary and consists of m subtrees whose root nodes are labeled $s_1, \ldots s_m$, each corresponding to a subset from A, connected together through a backbone tree as shown in Figure 2(b). The exact structure of this backbone tree is unimportant, as long as each s_i is sufficiently separated from the roots of the rest of the subtrees. For concreteness, we will assume that this backbone consists of a "caterpillar" tree as shown Figure 2(b), and that $9m$ extraneous leaves (not present in the gene tree) have been added to this backbone as shown in the figure to ensure that each pair of subtrees is sufficiently separated.

Recall that we use $x_{i,j}$ to denote the jth occurrence of u_i in A. Assuming that $u_i \in A_p$ and that $x_{i,j}$ refers to the occurrence of u_i in A_p, we define $f(i,p)$ to be j. In other words, if the jth occurrence of an element u_i is in the subset A_p, then we assign $f(i,p)$ to be j. Each S_i corresponds to the subset A_i and has the structure depicted in Figure 2(b). In particular, if A_i contains the three elements u_a, u_b, and u_c, then S_i contains the 12 leaves labeled $x_{a,f(a,i),j}, x_{b,f(b,i),j}$, and $x_{c,f(c,i),j}$, for $1 \le j \le 4$.

Event Costs. We assign the following event costs for problem instance λ: $P_\Delta = 2$, $P_\Theta = 4$, and $P_{loss} = 1$.

Note that the D-OGTR instance λ can be constructed in time polynomial in m and n.

Claim 1. *There exists a YES answer to the M3SC instance ϕ if and only if there exists a YES answer to the D-OGTR instance λ with $l = 10k + 39m - 12n$.*

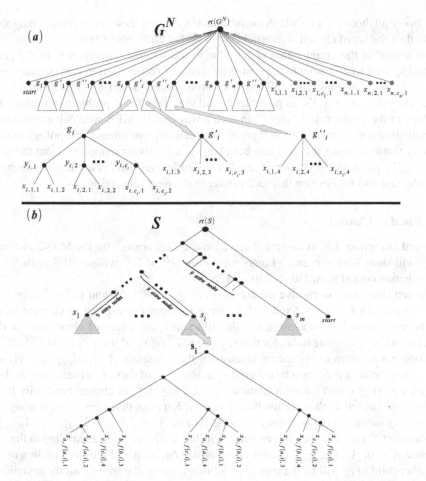

Fig. 2. Construction of non-binary gene tree and species tree. (a) Structure of the non-binary gene tree G^N. (b) Structure of the species tree S.

The remainder of this section is devoted to proving this claim which, in turn, would complete our proof for Theorem 1. We begin by explaining the main idea of the reduction and describing the association between the instances ϕ and λ, and then prove the forward and reverse directions of the claim.

3.3 Key Insight

The main idea behind our reduction can be explained as follows: In the gene tree G^N, subtrees $G^N(g_i)$, $G^N(g_i')$ and $G^N(g_i'')$ correspond to the element u_i, for each $1 \leq i \leq n$, while in the species tree the subtree $S(s_j)$ corresponds to the subset A_j, for each $1 \leq j \leq m$. Let G^B be any binary resolution of G^N. It can be shown that in any MPR of any optimal binary resolution G^B of G^N the following must hold: For each $i \in \{1, \ldots, n\}$, g_i (along with g_i' and g_i'') must map to an $S(s_j)$ for which $u_i \in A_j$. Under these restrictions on the mappings, observe that if we were to solve the OGTR problem on G^N and S and then choose all those A_j's for which the subtree $S(s_j)$ has at least one of the g_i's mapping into it, then the set of chosen A_j's would cover all the elements of U.

The source of the optimization is that, due to the specific construction of the gene tree and species tree, it is more expensive (in terms of reconciliation cost) to use more $S(s_j)$'s for the mapping. Thus, all the g_i's (along with g_i''s and g_i'''s) must map to as few of the subtrees, $S(s_j)$'s, as possible. Recall that the OGTR problem optimizes the topology of the binary resolution G^B in such a way that its MPR with S has minimum reconciliation cost. Thus, the OGTR problem effectively optimizes the topology of G^B in a way that minimizes the total number of $S(s_j)$'s receiving mappings from the g_i's, g_i''s, or g_i'''s, yielding a set cover of smallest possible size. This is the key idea behind our reduction and we develop this idea further in the next subsection.

3.4   Proof of Claim 1

Forward Direction. Let us assume that we have a YES answer for the M3SC instance ϕ. We will show how to create a binary resolution G^B of G^N whose MPR with S has reconciliation cost at most $10k + 39m - 12n$.

We first show how to resolve the subtrees $G^N(g_i)$, $G^N(g_i')$, and $G^N(g_i'')$, for $1 \leq i \leq n$. Recall that, for any fixed i, these three subtrees correspond to element u_i of U. The $y_{i,j}$'s in $G^N(g_i)$ correspond to the different occurrences of element u_i in the subsets from A. The same holds for the $x_{i,j,3}$'s in $G^N(g_i')$ and the $x_{i,j,4}$'s in $G^N(g_i'')$.

Suppose a solution to instance ϕ consists of the k subsets $A_{r(1)}, A_{r(2)}, \ldots, A_{r(k)}$. Since every element in U must be covered by at least one of these k subsets, we can designate a *covering subset* for each element $u_i \in U$, $1 \leq i \leq n$, chosen arbitrarily from among those subsets in the solution that contain u. Suppose that element u_i is assigned the covering subset A_j (so we must have $u_i \in A_j$ and $A_j \in \{A_{r(1)}, A_{r(2)}, \ldots, A_{r(k)}\}$). The subtree $G^N(g_i)$ will then be resolved as follows: The $y_{i,j}$ corresponding to the occurrence of u_i in A_j, i.e., $y_{i,f(i,j)}$, will be separated out as one of the two children of g_i. The other child of g_i will be the root of an arbitrary caterpillar tree on all the remaining $y_{i,j}$'s in $G^N(g_i)$. This is depicted in Figure 3(d). The subtrees $G^N(g_i')$ and $G^N(g_i'')$ are resolved similarly, except that in $G^N(g_i')$ the leaf node $x_{i,f(i,j),3}$ is separated out and in $G^N(g_i'')$ the leaf node $x_{i,f(i,j),4}$ is separated out. Thus, the resolution of $G^N(g_i)$, $G^N(g_i')$, and $G^N(g_i'')$ is done based on the assigned covering subset of element u_i. This is repeated for all i, where $1 \leq i \leq n$.

Next, we show how to resolve the root node of G^N to obtain G^B. The *start* node will become an outgroup to the rest of G^B. The backbone of the rest of G^B consists of an arbitrary caterpillar tree on k "leaf" nodes as shown in Figure 3(a). These k nodes are labeled $h_{r(1)}, \ldots h_{r(k)}$ and are the root nodes of k subtrees. Each of the k subtrees corresponds to one of the subsets $A_{r(1)}, A_{r(2)}, \ldots, A_{r(k)}$. In particular, subtree $G^B(h_{r(i)})$, for $1 \leq i \leq k$ corresponds to the subset $A_{r(i)}$. Each of the blue nodes and the subtrees rooted at the g_i's, g_i''s, and g_i'''s, for $1 \leq i \leq n$ will be included in one of these k subtrees. Specifically, the subtree $G^B(h_{r(j)})$ will include all those g_i's, g_i''s, and g_i'''s for which the covering subset of the corresponding u_i is $A_{r(j)}$. Since there may be 0, 1, 2, or 3 i's for which the covering subset of u_i is $A_{r(j)}$, the sizes of different $G^B(h_{r(j)})$ subtrees may vary. The structure of $G^B(h_{r(j)})$ when there are 3 $i's$ is depicted in Figure 3(b). The structure of $G^B(h_{r(j)})$ when there are only 1 or 2 such

$i's$ is similar and is the induced subtree, on the relevant i's, of the full subtree for all 3 i's. As shown in the figure, note that each subtree $G^B(h_{r(j)})$ also includes exactly three blue nodes, corresponding to the three elements in $A_{r(j)}$. These three blue nodes are included even for cases where there are fewer than 3 i's. Thus, when there are 0 such i's, which can happen when the size of the minimum set cover for instance ϕ is less than k, the subtree $G^B(h_{r(j)})$ consists of the three blue nodes.

This results in the assignment of all g_i's, g_i''s, and g_i'''s, for $1 \leq i \leq n$ to one of the subtrees $G^B(h_{r(j)})$, for $1 \leq j \leq k$. As discussed above, $3k$ out of the $3m$ blue nodes also get assigned in this process. The remaining $3m - 3k$ of the blue nodes are organized into an arbitrary caterpillar tree and added to the subtree $G^B(h_{r(k)})$ as shown in Figure 3(c).

This finishes our description of G^B. The following two lemmas imply the forward direction of Claim 1. The next lemma follows from the construction of G^B above.

Lemma 1. *Gene tree G^B is a binary resolution of G^N.*

It is not difficult to construct a DTL-scenario for G^B and S with cost exactly $10k + 39m - 12n$, yielding the following lemma.

Lemma 2. *Any MPR of G^B with S has reconciliation cost at most $10k + 39m - 12n$.*

Reverse Direction. Conversely, let us assume that we have a YES answer for the OGTR instance λ with $l = 10k + 39m - 12n$. We will show that there exists a solution of size at most k for the set cover instance ϕ. We first characterize the structure of optimal resolutions and their most parsimonious reconciliations.

Lemma 3. *For any optimal binary resolution G^B of G^N there exists an MPR of G^B with S such that:*

1. *For any $i \in \{1, \ldots, n\}$, g_i, g_i' and g_i'' map to the same subtree $S(s_j)$, where j is such that $u_i \in A_j$.*
2. *If there is a subtree $S(s_j)$ for which at least one of the nodes of G^B labeled g_i, g_i', or g_i'', for any $i \in \{1, \ldots, n\}$, maps to a node in $S(s_j)$, then there exists an $i \in \{1, \ldots, n\}$ such that g_i, g_i' and g_i'' all map to $S(s_j)$.*
3. *If g_i maps to a node in subtree $S(s_j)$, then g_i, g_i', g_i'', and the three blue nodes corresponding to the elements in A_j are arranged in such a way that the subtree of G^B connecting these six nodes does not contain any transfer nodes.*
4. *If two nodes, say a and b map to different subtrees $S(s_j)$, for $1 \leq j \leq m$, then the path connecting them in G^B must contain at least one transfer event.*

Lemma 4. *For any optimal binary resolution G^B of G^N, all MPRs of G^B with S must be such that:*

1. *Each $G^B(g_i)$, $G^B(g_i')$ and $G^B(g_i'')$, for $1 \leq i \leq n$, has exactly $(c_i - 1)$ transfer nodes, no duplications, and invokes no losses.*
2. *Each blue node that maps to an $S(s_j)$, $1 \leq j \leq m$, to which none of the g_i's map must be the recipient of a transfer edge.*

The next lemma implies the reverse direction and is based on the two lemmas above.

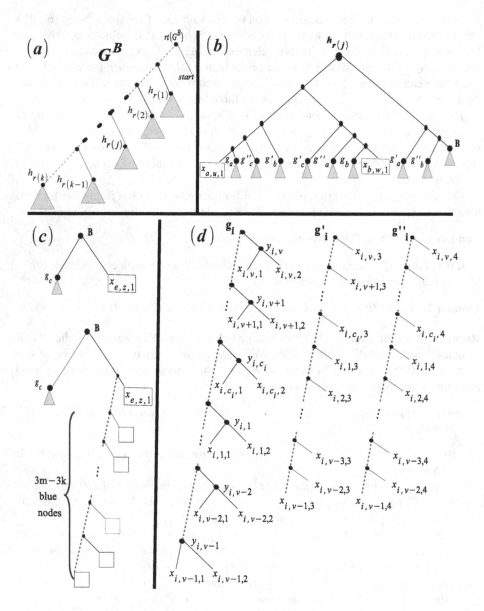

Fig. 3. Resolution of G^N into G^B. (a) The structure of the backbone of the gene tree G^B. (b) Structure of the subtree $h_{r(j)}$ for any $j \in \{1, \ldots, k\}$. (c) The two possible structures of the subtree with root B in $h_{r(j)}$. For any $j \in \{1, \ldots, k\}$, this subtree is as shown at the top of part (c) while, for $j = k$, it is as shown at the bottom and includes all the "remaining" $3m - 3k$ blue nodes. (d) The resolution of the g_i's, g'_i's, g''_i's. In the figure, u_a, u_b, and u_c represent the three elements in $A_{r(j)}$, with $u = f(a, r(j))$, $w = f(b, r(j))$, and $z = f(c, r(j))$. In part (d), if the covering subset of element u_i is A_p, then v represents $f(i, p)$. The labels inside the blue boxes represent blue nodes.

Lemma 5. *If there exists a binary resolution of G^N such that its MPR with S has reconciliation cost at most $10k + 39m - 12n$, then there exists a solution of size at most k for the M3SC instance ϕ.*

4 Extension to Dated DTL Reconciliation

An alternative model of DTL reconciliation has been proposed when the internal nodes of the species tree can be fully ordered in time [9]. We refer to this model as the *Dated-DTL* reconciliation model. Dated-DTL reconciliation makes use of the total order on the species nodes to ensure that the reconstructed optimal reconciliation is time-consistent. A key feature of this model is that it subdivides the species tree into different *time slices* [9] and then restricts transfer events to only occur within the same time slice.

We show how to assign divergence times to each node of the species tree. Observe that all subtrees $S(s_i)$, for each $i \in \{1 \ldots m\}$, have identical structure. All nodes at the same level in each $S(s_i)$ are assigned the same divergence time across all the subtrees. The rest of the nodes in S may be assigned arbitrary divergence times respecting the topology of S. It can be shown that there exists an optimal resolution of the gene tree for which an MPR exists that only invokes transfer events that respect the timing constraints of this dated species tree as required by the dated-DTL reconciliation model. This implies that, for our gadget, any optimal resolution of the gene tree under the undated DTL reconciliation model has the same minimum reconciliation cost as the dated-DTL reconciliation model.

Theorem 2. *The OGTR problem under the dated-DTL reconciliation model is NP-hard.*

5 Conclusion

In this work, we have shown that the OGTR problem, i.e., the problem of reconciling non-binary gene trees with binary species trees under the DTL reconciliation model, is NP-hard. Our reduction applies to both the undated and dated formulations of DTL-reconciliation and, furthermore, shows that the problem is NP-hard even for a biologically meaningful event cost assignment of 1, 2, and 4 for losses, duplications, and transfers, respectively. The uncertainty about its complexity has prevented the development of algorithms for the OGTR problem. This work will lead to the development of effective exact, approximate, and heuristic algorithms for this problem, making it possible to apply the powerful DTL reconciliation framework to non-binary gene trees. Interesting open problems include determining if efficient algorithms exist for the special case when the degree of each gene tree node is bounded above by a constant, and investigating the approximability of the dated and undated OGTR problems.

Funding: This work was supported in part by startup funds from the University of Connecticut to MSB.

References

1. Koonin, E.V.: Orthologs, paralogs, and evolutionary genomics. Annual Review of Genetics 39(1), 309–338 (2005)

2. Vilella, A.J., Severin, J., Ureta-Vidal, A., Heng, L., Durbin, R., Birney, E.: Ensemblcompara genetrees: Complete, duplication-aware phylogenetic trees in vertebrates. Genome Research 19(2), 327–335 (2009)
3. Chen, K., Durand, D., Farach-Colton, M.: Notung: dating gene duplications using gene family trees. In: RECOMB, pp. 96–106 (2000)
4. David, L.A., Alm, E.J.: Rapid evolutionary innovation during an archaean genetic expansion. Nature 469, 93–96 (2011)
5. Durand, D., Halldórsson, B.V., Vernot, B.: A hybrid micro-macroevolutionary approach to gene tree reconstruction. J. Comput. Biol. 13(2), 320–335 (2006)
6. Burleigh, J.G., Bansal, M.S., Eulenstein, O., Hartmann, S., Wehe, A., Vision, T.J.: Genome-scale phylogenetics: Inferring the plant tree of life from 18,896 gene trees. Syst. Biol. 60(2), 117–125 (2011)
7. Scornavacca, C., Jacox, E., Szöllosi, G.J.: Joint amalgamation of most parsimonious reconciled gene trees. Bioinformatics (in press)
8. Gorbunov, K.Y., Liubetskii, V.A.: Reconstructing genes evolution along a species tree. Molekuliarnaia Biologiia 43(5), 946–958 (2009)
9. Doyon, J.-P., Scornavacca, C., Gorbunov, K.Y., Szöllősi, G.J., Ranwez, V., Berry, V.: An efficient algorithm for gene/species trees parsimonious reconciliation with losses, duplications and transfers. In: Tannier, E. (ed.) RECOMB-CG 2010. LNCS, vol. 6398, pp. 93–108. Springer, Heidelberg (2010)
10. Tofigh, A., Hallett, M.T., Lagergren, J.: Simultaneous identification of duplications and lateral gene transfers. IEEE/ACM Trans. Comput. Biology Bioinform. 8(2), 517–535 (2011)
11. Bansal, M.S., Alm, E.J., Kellis, M.: Efficient algorithms for the reconciliation problem with gene duplication, horizontal transfer and loss. Bioinformatics 28(12), 283–291 (2012)
12. Stolzer, M., Lai, H., Xu, M., Sathaye, D., Vernot, B., Durand, D.: Inferring duplications, losses, transfers and incomplete lineage sorting with nonbinary species trees. Bioinformatics 28(18), 409–415 (2012)
13. Bansal, M.S., Alm, E.J., Kellis, M.: Reconciliation revisited: Handling multiple optima when reconciling with duplication, transfer, and loss. J. Comput. Biol. 20(10), 738–754 (2013)
14. Scornavacca, C., Paprotny, W., Berry, V., Ranwez, V.: Representing a set of reconciliations in a compact way. J. Bioinform. Comput. Biol. 11(02), 1250025 (2013)
15. Libeskind-Hadas, R., Wu, Y.C., Bansal, M.S., Kellis, M.: Pareto-optimal phylogenetic tree reconciliation. Bioinformatics 30(12), i87–i95 (2014)
16. Ovadia, Y., Fielder, D., Conow, C., Libeskind-Hadas, R.: The cophylogeny reconstruction problem is NP-complete. J. Comput. Biol. 18(1), 59–65 (2011)
17. Libeskind-Hadas, R., Charleston, M.: On the computational complexity of the reticulate cophylogeny reconstruction problem. J. Comput. Biol. 16, 105–117 (2009)
18. Chang, W.-C., Eulenstein, O.: Reconciling gene trees with apparent polytomies. In: Chen, D.Z., Lee, D.T. (eds.) COCOON 2006. LNCS, vol. 4112, pp. 235–244. Springer, Heidelberg (2006)
19. Lafond, M., Swenson, K.M., El-Mabrouk, N.: An optimal reconciliation algorithm for gene trees with polytomies. In: Raphael, B., Tang, J. (eds.) WABI 2012. LNCS, vol. 7534, pp. 106–122. Springer, Heidelberg (2012)
20. Zheng, Y., Zhang, L.: Reconciliation with non-binary gene trees revisited. In: Sharan, R. (ed.) RECOMB 2014. LNCS, vol. 8394, pp. 418–432. Springer, Heidelberg (2014)
21. Karp, R.M.: Reducibility among combinatorial problems. In: Proceedings of a Symposium on the Complexity of Computer Computations, held March 20-22, 1972, at the IBM Thomas J. Watson Research Center, Yorktown Heights, New York, pp. 85–103 (1972)

On the Near-Linear Correlation
of the Eigenvalues Across BLOSUM Matrices

Jin Li[1], Yen Kaow Ng[2], Xingwu Liu[3], and Shuai Cheng Li[1(✉)]

[1] Department of Computer Science, City University of Hong Kong, Hong Kong,
China
{jinli28,shuaicli}@cityu.edu.hk
[2] Department of Computer Science,
Faculty of Information and Communication Technology,
Universiti Tunku Abdul Rahman, Kampar, Malaysia
ykng@utar.edu.my
[3] Institute of Computing Technology, Chinese Academy of Sciences,
Beijing, China
liuxingwu@ict.ac.cn

Abstract. The BLOSUM matrices estimate the likelihood for one amino
acid to be substituted with another, and are commonly used in sequence
alignments. Each BLOSUM matrix is associated with a parameter x—the
matrix elements are computed based on the diversity among sequences
of no more than $x\%$ similar. In an earlier work, Song *et al.* observed a
property in the BLOSUM matrices—eigendecompositions of the matrices produce nearly identical sets of eigenvectors. Furthermore, for each
eigenvector, a nearly linear trend is observed in all its eigenvalues. This
property allowed Song *et al.* to devise an iterative alignment and matrix
selection process to produce more accurate matrices. In this paper, we
investigate the reasons behind this property of the BLOSUM matrices.
Using this knowledge, we analyze the situations under which the property
holds, and hence clarify the extent of the earlier method's validity.

1 Introduction

In a protein sequence alignment, each amino acid in one sequence is matched
to an amino acid (or to a gap) in the other sequence. The likeliness that the
two sequences are related (under the alignment) is often evaluated through the
likeliness for amino acids in the matched pairs to appear in the place of each
other. It is hence very important to accurately assess the likeliness for these
amino acid substitutions. Several standards have emerged for this purpose, such
as BLOSUM [11], PAM [3], and GONNET [9]. Each of these gives a family
of substitution matrices of 20×20 elements; each element is a score for the
transition between two amino acids.

We consider the BLOSUM matrices in this paper. Each matrix in the BLO-
SUM family is distinguished by a parameter x—the transitional probabilities of
that matrix are calculated from the diversity among the sequences of no more

© Springer International Publishing Switzerland 2015
R. Harrison et al. (Eds.): ISBRA 2015, LNBI 9096, pp. 199–210, 2015.
DOI: 10.1007/978-3-319-19048-8_17

than $x\%$ similarity. The matrix of a parameter x is denoted BLOSUMx, e.g. the matrix where $x = 50$ is written BLOSUM50. The BLOSUM62 matrix is the most common, and is the substitution matrix used by default in the popular sequence comparison tool called BLAST [1]. Fifteen other BLOSUM matrices are in common use, namely, the matrices for x=30, 35, ..., 100. Which of these matrices to use depends very much on the situation; matrices of lower x are better at aligning more distant sequences while matrices of higher x are better for aligning more closely related sequences [16].

Many methods have been proposed to automate the process of finding the optimal BLOSUMx [5,6,2,8,13,4,10,7,18,12,15]. Many of these methods employ machine learning methods which are trained on sequence examples. Recently, Song *et al.* proposed a method which is based completely on the BLOSUMx matrices—eliminating the need for training examples [17].

The method is based on a property they observed of the BLOSUM matrices, namely, that in general, the eigenvectors obtained from eigendecompositions of the BLOSUM matrices are nearly identical, and that when all the eigenvalues (from the BLOSUM matrices of different x) that correspond to the same (or nearly identical) eigenvector are plotted against x, a near-linearly increasing correlation is observed. This is shown in Figure 1, where we eigendecompose the BLOSUMx matrices for x=30, 35, ..., 100, (and x=62,) and then, for each eigenvector obtained, we plot its corresponding eigenvalues against x.

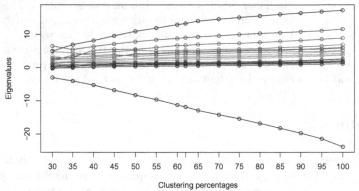

Fig. 1. Linearity between eigenvalues and clustering percentages

What causes this observation is unknown, and could be related to earlier studies on the eigenvalues of other score matrices [14]. Our task in this paper is to investigate a sufficient condition under which this near-linearly increasing trend holds. From this, we examine the extent to which the phenomena can be expected. One of the immediate implications of this is that we now have a better understanding of when the method by Song *et al.* is applicable.

2 Preliminaries

We first give the necessary background, as well as the notations in this paper.

2.1 The Construction of BLOSUM Matrices

The BLOSUM matrices are constructed from the blocks of sequences from the BLOCKS database. Each block in the database is an ungapped multiple sequence alignment. The sequences in different blocks are considered unrelated.

Each BLOSUM matrix is associated with a parameter x, and is constructed as follows. First, sequences are clustered such that the sequences between different clusters are no more than $x\%$ identical. Let f_i denote the number of times an amino acid A_i appears in the sequences of all the clusters, over the number of all amino acid occurrences. We consider all possible pairs of sequences, each from a different cluster within a same block. For every such pair, we count the total number of times that one amino acid, A_i, is aligned with another, A_j; this count is weighted by the size of the clusters from which the sequences are drawn to prevent over-representation of sequences in large clusters. The sum of these counts from all possible pairs is computed, and divided by the total number of pairs of aligned amino acids from all possible pairs of sequences; the resultant value is denoted $f_{i,j}$. (Note that $f_{i,j} = f_{j,i}$.) Finally, we calculate $s_{i,j}$, an estimate of the log-odds ratio of finding A_i interchanged with A_j, as

$$s_{i,j} = \begin{cases} \log_2 \frac{f_{i,i}}{f_i f_i} & \text{if } i = j, \\ \log_2 \frac{f_{i,j}}{2 f_i f_j} & \text{if } i \neq j. \end{cases}$$

We are now prepared to define the BLOSUMx matrix. For simplicity we refer to the BLOSUMx matrix by simply B_x. The (i, j)th element of B_x is set to the integer that is closest to $\frac{1}{0.347} s_{i,j}$. Using this procedure, Heinikoff and Heinikoff computed the matrices for B_{30}, B_{35}, ..., B_{100}, as well as B_{62} [11]. We refer to these matrices which they published implicitly in this work.

We let J_x denote the matrix of the counts $f_{i,j}$ obtained under parameter x. That is,

$$J_x = \begin{pmatrix} f_{1,1} & \frac{f_{1,2}}{2} & \cdots & \frac{f_{1,20}}{2} \\ \frac{f_{2,1}}{2} & f_{2,2} & \cdots & \frac{f_{2,20}}{2} \\ & & \ddots & \\ \frac{f_{20,1}}{2} & \frac{f_{20,2}}{2} & \cdots & f_{20,20} \end{pmatrix}$$

For a matrix M, we denote its (i, j)th element by $M(i, j)$. The ith element of a vertor v is written $v(i)$. We denote the eigenvectors from a eigendecomposition of M as u_M^1, u_M^2, ..., u_M^{20}, and denote the corresponding eigenvalues λ_M^1, λ_M^2, ..., λ_M^{20}.

We now create the notations to formalize our observation. Our first observation is that the eigendecomposition of B_{30}, B_{35}, ..., B_{100} resulted in very similar eigenvectors. For simplicity we assume that there is an ideal set of eigenvectors shared commonly. We denote these eigenvectors as $\mathbf{u}_B^1$, $\mathbf{u}_B^2$, ..., $\mathbf{u}_B^{20}$. For each eigenvector $\mathbf{u}_B^i$, we denote its corresponding eigenvalue from B_x as $\lambda_{B_x}^i$. Hence, the observation stated in the introduction can be restated as:

Observation 1. In general, one can observe a nearly linear trend in the values $\lambda_{B_{30}}^i$, $\lambda_{B_{35}}^i$, ..., $\lambda_{B_{100}}^i$, for each individual i from 1, 2, ..., to 20.

   This observation, in fact, extends to the joint probability matrices, J_x. That is, the same observation holds when one replaces the B_x in the paragraphs above to J_x, giving:

Observation 2. In general, one can observe a nearly linear trend in the values $\lambda_{J_{30}}^i$, $\lambda_{J_{35}}^i$, ..., $\lambda_{J_{100}}^i$, for each i from 1, 2, ..., to 20.

3   A Sufficient Property For Nearly Linear Eigenvalue Changes

We shall now show a link between the properties above, which is concerned with the eigenvalues, and a more immediate and intuitive property of the matrices. This property is with regard to the elements of the joint probability matrices that is, $J_x(i, j)$. The property can be stated as follows:

Observation 3. In general, for each i and j, one can observe a nearly linear trend in the values $J_{30}(i, j)$, $J_{35}(i, j)$, ..., $J_{100}(i, j)$.

   To illustrate this property, we show this trend in the diagonal elements $J_{30}(i, i)$, $J_{35}(i, i)$, ..., $J_{100}(i, i)$ for each individual i from 1, 2, ..., to 20, in Figure 2.

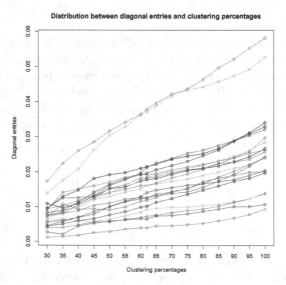

Fig. 2. Diagonal entries of J_x

   In this section we shall show, through both simulations and mathematical argument, that this property of J_x gives rise to Observations 1 and 2.

3.1   Simulation

We first verify the validity of the claim through simulations. We generate pseudo-joint probability matrices J'_{30}, J'_{35}, ..., J'_{100}, with the only assumptions that

1. Each J_x is symmetric and its elements sum up to one,
2. The diagonal elements of J_x increase linearly with respect to x. More precisely, we assume for the ith diagonal elements of J'_{30}, J'_{35}, ..., J'_{100}, there is a constant δ_i, such that $J_{x+5}(i,i) = J_x(i,i) + \delta_i$,
3. The non-diagonal elements of the matrices are multiplied by a (same) constant factor as we proceed from J'_{30}, J'_{35}, to J'_{100}.

Generating Pseudo-joint Probability Matrices. We first generate a base matrix J'_{100} of sum 1. The remaining J'_{30}, J'_{35}, ..., J'_{95}, are then generated from J'_{100}, as follows.

The diagonal elements of J'_{100-5k}, $k \in \{1, \ldots 14\}$ is computed as

$$J'_{100-5k}(i,i) = J'_{100}(i,i) - k\delta_i, \quad \text{for } i \in \{1, \ldots, 20\}. \tag{1}$$

We now consider the non-diagonal elements of J'_x. As mentioned, the sum of the elements of each matrix is 1, that is, $\sum_{i,j\in\{1,\ldots,20\}} J'_x(i,j) = 1$. Let $d_{100} = \sum_{i=1}^{20} J'_{100}(i,i)$. Then,

- the non-diagonal elements in J'_{100} have a total probability of $(1 - d_{100})$, while
- the diagonal elements in J'_{100-5k} have a total probability of $(d_{100} - k\delta)$, where

$$\delta = \sum_{i=1}^{20} \delta_i.$$

As mentioned, we assume that each non-diagonal element of J'_{100-5k} is of a value s_k times its original value in $J'_{100}(i,j)$, that is, $J'_{100-5k}(i,j) = s_k J'_{100}(i,j)$. To find s_k, we solve

$$(d_{100} - k\delta) + s_k \sum_{i,j,i\neq j} J'_{100}(i,j) = 1$$

$$(d_{100} - k\delta) + s_k(1 - d_{100}) = 1$$

to obtain

$$s_k = 1 + \frac{k\delta}{1 - d_{100}}. \tag{2}$$

Our joint probability matrices J'_{30}, J'_{35}, ..., J'_{95} are generated from J'_{100} and $\delta_1, \ldots, \delta_{20}$, by Eqn. (1) and (2), which combines into

$$J'_{100-5k}(i,j) = \begin{cases} J'_{100}(i,i) - k\delta_i & i = j \\ J'_{100}(i,j) + \frac{k\delta}{1-d}J'_{100}(i,j) & i \neq j \end{cases} \tag{3}$$

Generating Pseudo-BLOSUM Matrices. Our pseudo-BLOSUM matrices B'_{30}, B'_{35}, ..., B'_{100}, are generated straightforwardly from the pseudo-joint probability matrices, as

$$B'_x(i,j) = \log_2 \frac{J'_x(i,j)}{P_x(i)P_x(j)},$$

where $P_x(i)$ are the marginal probabilities of the pseudo-joint probability matrix J'_x, that is,

$$P_x(i) = \sum_{j=1}^{20} J'_x(i,j).$$

Simulation Results. We examine the pseudo-joint matrices and pseudo-BLOSUM matrices separately. We perform eigendecomposition of the matrices to produce, for each matrix, a set of eigenvectors. To match the eigenvectors from a matrix to those from another, we construct a bipartite graph where every eigenvector is used as a vertex. A weighted edge is placed between every two vertices; its weight is computed as the dot product of its corresponding eigenvectors. Then, a maximum weighted bipartite matching of the graph is used as the match between the eigenvectors. Finally, the eigenvalues for each group of matching eigenvectors are collected and plotted in Figure 3.

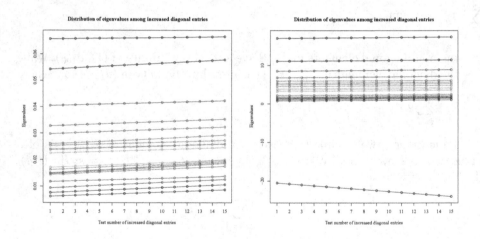

Fig. 3. (a) Eigenvalues from the pseudo-joint probability matrices, (b) Eigenvalues from the pseudo-BLOSUM matrices

3.2   Theoretical Justification

We will now provide mathematical reasons for the observed relationship.

Our aim is to show that linearly increasing elements in the joint probability matrices J_{30}, J_{35}, ..., J_{100} imply linearly increasing $\lambda^i_{J_{30}}$, $\lambda^i_{J_{35}}$, ..., $\lambda^i_{J_{100}}$, as well as linearly increasing $\lambda^i_{B_{30}}$, $\lambda^i_{B_{35}}$, ..., $\lambda^i_{B_{100}}$, for each i from 1, 2, ..., to 20.

Linear Trend in the Eigenvalues of J_x. We start with our assumptions, that

1. Each J_x is real and symmetric. (Here, we do not need the elements to sum up to one.)
2. For each i and j, all the (i, j)th elements from the matrices J_x are linearly correlated. More precisely, we can write

$$J_{100-5k} = J_{100} + k\Delta$$

for a symmetric matrix

$$\Delta = \begin{bmatrix} \delta_{1,1} & \delta_{1,2} & \cdots & \delta_{1,20} \\ \delta_{2,1} & \delta_{2,2} & \cdots & \delta_{2,20} \\ \vdots & \vdots & & \vdots \\ \delta_{20,1} & \delta_{20,2} & \cdots & \delta_{20,20} \end{bmatrix}.$$

However, to simplify the analysis, we also assume that

3. The eigendecomposition of each of J_{30}, J_{35}, $\ldots$, J_{100} results in perfectly identical sets of orthogonal eigenvectors, namely, $\mathbf{u}_J^1$, $\mathbf{u}_J^2$, $\ldots$, $\mathbf{u}_J^{20}$. (Since J_x is real and symmetric, achieving orthogonality is not a problem.)

We start with the eigendecomposition of J_x,

$$J_x = \mathbf{U}\Lambda_x\mathbf{U}^T$$

where $\mathbf{U} = \begin{bmatrix} \mathbf{u}_J^1 & \mathbf{u}_J^2 & \cdots & \mathbf{u}_J^{20} \end{bmatrix}$, and $\Lambda_x = \mathrm{diag}(\lambda_{J_x}^1, \lambda_{J_x}^2, \ldots, \lambda_{J_x}^{20})$. Left-multiplying $\mathbf{U}^T$ and right-multiplying $\mathbf{U}$ at both sides,

$$\Lambda_x = \mathbf{U}^T J_x \mathbf{U}$$
$$\Lambda_{100-5k} = \mathbf{U}^T \left(J_{100} + k\Delta\right)\mathbf{U} = \mathbf{U}^T J_{100}\mathbf{U} + k\mathbf{U}^T\Delta\mathbf{U}$$
$$= \Lambda_{100} + k\mathbf{U}^T\Delta\mathbf{U}.$$

Hence $\lambda_{J_{100-5k}}^i = \lambda_0(i) + k\,\mathrm{trace}(\Delta^T(\mathbf{u}_J^i)(\mathbf{u}_J^i)^T)$, which shows a linear relationship between λ_x^i and x.

We now proceed to show the same result for the BLOSUM matrices.

Linear Trend in the Eigenvalues of B_x. Our analysis for the B_x requires more assumptions of J_x. Besides the earlier conditions, we furthermore require that:

1. The change in each element of J_x is small. More precisely, for all x and all pairs (i, j)

$$\frac{|\, J_x(i, j) - J_{100}(i, j)\,|}{J_{100}(i, j)} \ll 1$$

2. The change in marginal probability of all J_x from J_{100} is small. More precisely, for all x and all i,

$$\frac{|\sum_j J_x(i,j) - \sum_j J_{100}(i,j)|}{\sum_j J_{100}(i,j)} \ll 1$$

These conditions are required of J_x. For B_x, we only assume that the eigendecompositions of each of $B_{30}, B_{35}, \ldots, B_{100}$ results in the same set of orthogonal eigenvectors, namely, $\mathbf{u}_B^1, \mathbf{u}_B^2, \ldots, \mathbf{u}_B^{20}$.

Our proof will show that B_x has the same properties as required in Section 3.2 for J_x, and hence, by the same argument therein, B_x has linearly correlated eigenvalues.

For a fixed i and j,

$$B_x(i,j) = \log_2 \frac{J_x(i,j)}{P_x(i)P_x(j)} = \log_2 \frac{J_x(i,j)}{\left(\sum_t J_x(i,t)\right)\left(\sum_t J_x(j,t)\right)}$$

$$B_{100-5k}(i,j) = \log_2 \frac{J_{100}(i,j) + k\delta_{i,j}}{\left(\sum_t (J_{100}(i,t) + k\delta_{i,t})\right)\left(\sum_t (J_{100}(j,t) + k\delta_{j,t})\right)}$$

$$= \log_2 \frac{J_{100}(i,j) + k\delta_{i,j}}{(r_{100}(i) + km(i))(r_{100}(j) + km(j))}$$

in which, we have let $r_{100}(z) = \sum_t J_{100}(z,t)$, and $m(z) = \sum_t \delta_{z,t}$. Note that $m(z)$ is the change in the marginal probability at the zth row. Subsequently,

$$B_{100-5k}(i,j) = \log_2(J_{100}(i,j) + k\delta_{i,j})$$
$$- \log_2(r_{100}(i) + km(i)) - \log_2(r_{100}(j) + km(j))$$

$$= \log_2\left(J_{100}(i,j)(1 + k\frac{\delta_{i,j}}{J_{100}(i,j)})\right)$$
$$- \log_2\left(r_{100}(i)(1 + k\frac{m(i)}{r_{100}(i)})\right) - \log_2\left(r_{100}(j)(1 + k\frac{m(i)}{r_{100}(j)})\right)$$

$$= \log_2\frac{J_{100}(i,j)}{r_{100}(i)r_{100}(j)} + \log_2(1 + k\frac{\delta_{i,j}}{J_{100}(i,j)})$$
$$- \log_2(1 + k\frac{m(i)}{r_{100}(i)}) - \log_2(1 + k\frac{m(j)}{r_{100}(j)})$$

$$= B_{100}(i,j) + \log_2(1 + k\frac{\delta_{i,j}}{J_{100}(i,j)})$$
$$- \log_2(1 + k\frac{m(i)}{r_{100}(i)}) - \log_2(1 + k\frac{m(j)}{r_{100}(j)})$$

Let $x = \dfrac{k\delta_{i,j}}{J_{100}(i,j)}$, and $y(z) = \dfrac{km(z)}{r_{100}(z)}$, and this becomes

$$B_{100-5k}(i,j) = B_{100}(i,j) + \log_2(1+x) - \log_2(1 + y(i)) - \log_2(1 + y(j)) \quad (4)$$

Due to our assumptions, $x = \dfrac{k\delta_{i,j}}{J_{100}(i,j)} \ll 1$, and $y(z) = \dfrac{km(z)}{r_{100}(z)} \ll 1$. This allows us to use Taylor expansion to expand Eqn. (4) into

$$
\begin{aligned}
B_{100-5k}(i,j) &= B_{100}(i,j) + \frac{1}{\ln 2}(x + o(x^2)) \\
&\quad -(y(i) + o(y(i)^2)) - (y(j) + o(y(j)^2))) \\
&\approx B_{100}(i,j) - \frac{1}{\ln 2}(y(i) + y(j) - x) \\
&= B_{100}(i,j) - \frac{k}{\ln 2}\left(\frac{m(i)}{r_{100}(i)} + \frac{m(j)}{r_{100}(j)} - \frac{\delta_{i,j}}{J_{100}(i,j)}\right).
\end{aligned}
$$

Or more intuitively,

$$
B_{100-5k}(i,j) \approx B_{100}(i,j) - k\left(\frac{1}{\ln 2}\right)\left(\frac{m(i)}{r_{100}(j)} + \frac{m(i)}{r_{100}(j)} - \frac{\delta_{i,j}}{J_{100}(i,j)}\right).
$$

With this, and with the fact that each B_x computed from J_x is real and symmetric, we have arrived at the same condition for $J_{30}, J_{35}, \ldots, J_{100}$ assumed at the beginning of Section 3.2, which was sufficient to show a linear trend in their eigenvalues. The exact same method can be applied here to demonstrate the same for B_x to give us the desired result.

4   Achievability of Condition

The result that we have just obtained indicates that, for a set of substitution matrices to have the near-linearly increasing eigenvalues stated, it suffices that the joint probability matrices have incremental, and preferrably constantly increasing elements.

In this section we shall investigate how demanding this condition is. Let $g(x)$ denote the number of aligned sequence pairs at level x (that is, the sequences that are $100 * x\%$ identical). We are interested in the dependency of J_x on x for different forms of $g(x)$. Particularly, we want to know the forms of $g(x)$ under which the elements of J_x increase linearly with respect to x.

We assume that $g(x)$ is monotonic, and we furthermore assume that $g(x)$ is continuous (in order to obtain integrals). Then, the diagonal elements of J_x can be evaluated as

$$
J_x(i,i) \propto \frac{\int_0^x y\, g(y)dy}{\int_0^x g(y)dy}
$$

and non-diagonal elements of J_x can be evaluated as

$$
J_x(i,j) \propto \frac{\int_0^x (1-y)\, g(y)dy}{\int_0^x g(y)dy}.
$$

We consider the following two growth behavior for $g(x)$:

Case (1). $g(x) = ax^n$, for $n > 0, a > 0$,
Case (2). $g(x) = a(e^{bx} - 1)$ for $a > 0, b > 0$.

For Case (1), the integrals are straight-forward, and evaluates to

$$J_x(i, i) \propto \frac{\int_0^x y^{n+1}dy}{\int_0^x y^n dy} = \frac{n+1}{n+2}x,$$

and

$$J_x(i, j) = 1 - \frac{\int_0^x y^{n+1}dy}{\int_0^x y^n dy} = 1 - \frac{n+1}{n+2}x,$$

which show that the elements of J_x correlate linearly with x.

Case (2) presents a more involved integral:

$$\begin{aligned}
J_x(i, i) &\propto \frac{\int_0^x y(e^{by} - 1)dy}{\int_0^x (e^{by} - 1)dy} \\
&= \frac{xe^{bx} - \frac{1}{b}e^{bx} + \frac{1}{b} - \frac{bx^2}{2}}{e^{bx} - bx - 1} \\
&= x + \frac{(-1)x(e^{bx} - bx - 1) + xe^{bx} - \frac{1}{b}e^{bx} + \frac{1}{b} - \frac{bx^2}{2}}{e^{bx} - bx - 1} \\
&= x + \frac{-\frac{1}{b}e^{bx} + \frac{1}{b} + \frac{bx^2}{2} + x}{e^{bx} - bx - 1} \\
&= x - \frac{1}{b} + \frac{\frac{bx^2}{2}}{e^{bx} - bx - 1}
\end{aligned}$$

We are interested in whether $x - \frac{1}{b} + \frac{\frac{bx^2}{2}}{e^{bx}-bx-1}$ is nearly linear for $0 \leq x \leq 1$. Since $b > 0$, this expression monotonically increases with x. We can hence deduce its range to $\left[0, 1 - \frac{1}{b} + \frac{\frac{b}{2}}{e^b-b-1}\right]$ for $0 \leq x \leq 1$. We compare the expression, denoted $S(b, x)$, to the straight line L_b which passes through $(0, 0)$ and $(1, 1 - \frac{1}{b} + \frac{\frac{b}{2}}{e^b-b-1})$, namely

$$L_b(x) = \left(1 - \frac{1}{b} + \frac{\frac{b}{2}}{e^b - b - 1}\right) x.$$

Let the discrepancy between $S(b, x)$ and $L_b(x)$ be

$$d(b) = \frac{\max_{0 < x < 1} |S(b, x) - L_b(x)|}{S(b, 1) - S(b, 0)}$$

Figure 4(a) shows how this discrepancy varies with respect to b. The curve indicates that $d(b)$ reaches a maximum of 0.0504 at $b = 7.83$. Figure 4(b) shows how $S(b, x)$ varies with x at $b = 7.83$; similar trends are observed for other values of b.

We hence conclude that in Case (2) as well, the elements of J_x correlate nearly linearly with x.

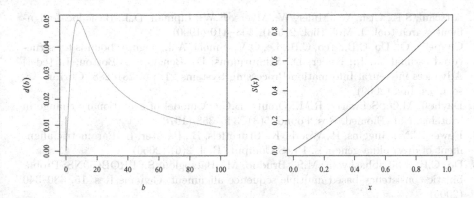

Fig. 4. (a) $d(b)$ at different b; (b) $S(b, x)$ vs x at $b = 7.83$

Finally, we examine the growth behavior of $g(x)$ in the actual data. We show this function in Figure 5, together with an exponential curve $f(x) = a(e^{bx} - bx - 1)$ approximated through nonlinear least-squares regression; the parameters $a = 74.1418, b = 8.1846, \text{erf} = 1.013e + 09$.

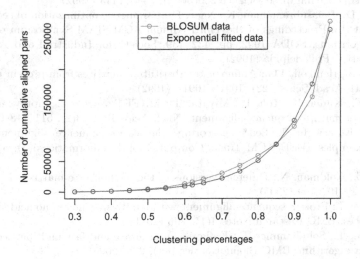

Fig. 5. Cumulative aligned pairs of BLOSUM clustering trees

Acknowledgments. This research was supported by Hong Kong GRF Grant No. 9041901 (CityU 118413), and partially supported by National Natural Science Foundation of China (61173009).

References

1. Altschul, S.F., Gish, W., Miller, W., Myers, E.W., Lipman, D.J.: Basic local alignment search tool. J. Mol. Biol. 215(3), 403–410 (1990)
2. Chapelle, O., Do, C.B., Teo, C.H., Le, Q.V., Smola, A.J.: Tighter bounds for structured estimation. In: Koller, D., Schuurmans, D., Bengio, Y., Bottou, L. (eds.) Advances in Neural Information Processing Systems 21, pp. 281–288. Curran Associates, Inc. (2009)
3. Dayhoff, M.O., Schwartz, R.M., Orcutt, B.C.: A model of evolutionary change in proteins. Nat. Biomed. Res. Found. 5(3), 345–358 (1978)
4. Dewey, C.N., Huggins, P., Woods, K., Sturmfels, B., Pachter, L.: Parametric alignment of drosophila genomes. PLoS Comput. Biol. 2(6) (2006)
5. Do, C.B., Mahabhashyam, M.S., Brudno, M., Batzoglou, S.: PROBCONS: Probabilistic consistency-based multiple sequence alignment. Genome Res. 15, 330–340 (2005)
6. Do, C.B., Woods, D.A., Batzoglou, S.: CONTRAfold: RNA secondary structure prediction without physics-based models. In: ISMB (Supplement of Bioinformatics), vol. 22, pp. 90–98 (2006)
7. Edgar, R.C.: Optimizing substitution matrix choice and gap parameters for sequence alignment. BMC Bioinformatics 10(396) (2009)
8. Flannick, J., Novak, A., Do, C.B., Srinivasan, B.S., Batzoglou, S.: Automatic parameter learning for multiple local network alignment. J. Comput. Biol. 16(8), 1001–1022 (2006)
9. Gaston, H., Gonnet, M., Cohen, A., Benner, S.: Exhaustive matching of the entire protein sequence database. Science 256(5062), 1443–1445 (1992)
10. Gusfield, D., Balasubramanian, K., Naor, D.: Parametric optimization of sequence alignment. In: Proceedings of the Third Annual ACM-SIAM Symposium on Discrete Algorithms, SODA 1992, pp. 432–439. Society for Industrial and Applied Mathematics, Philadelphia (1992)
11. Henikoff, S., Henikoff, J.G.: Amino acid substitution matrices from protein blocks. Proc. Natl. Acad. Sci. 89(22), 10915–10919 (1992)
12. Katoh, K., Kuma, K., Toh, H., Miyata, T.: MAFFT version 5: improvement in accuracy of multiple sequence alignment. Nucl. Acids Res. 33(2), 511–518 (2005)
13. Kim, E., Kececioglu, J.: Learning scoring schemes for sequence alignment from partial examples. IEEE/ACM Trans. Comput. Biol. Bioinformatics 5(4), 546–556 (2008)
14. Kosial, C., Goldman, N.: Different versions of the dayhoff rate matrix. Mol. Biol. Evol. 22(2), 193–199 (2005)
15. Kuznetsov, I.: Protein sequence alignment with family-specific amino acid similarity matrices. BMC Research Notes 4(1), 296 (2011)
16. Lassmann, T., Sonnhammer, E.: Kalign – an accurate and fast multiple sequence alignment algorithm. BMC Bioinformatics 6(1), 298 (2005)
17. Song, D., Chen, J., Chen, G., Li, N., Li, J., Fan, J., Bu, D., Li, S.C.: Parameterized blosum matrices for protein alignment. IEEE/ACM Trans. Comput. Biol. Bioinformatics PP(99), 1 (2014)
18. Wang, H.-C., Susko, E., Roger, A.J.: An amino acid substitution-selection model adjusts residue fitness to improve phylogenetic estimation. Mol. Biol. Evol. 31(4), 779–792 (2014)

Predicting RNA Secondary Structures: One-grammar-fits-all Solution

Menglu Li[1], Micheal Cheng[2], Yongtao Ye[1], Wk Hon[2], Hf Ting[1],
Tw Lam[1], Cy Tang[2,3], Thomas Wong[4], and Sm Yiu[1(✉)]

[1] Department of Computer Science, The University of Hong Kong, Hong Kong, China
[2] Department of Computer Science, National Tsinghua University, Hsinchu City, Taiwan
[3] Department of Computer Science, Providence University, Taichung City, Taiwan
[4] CSIRO Ecosystem Sciences, Canberra, Australia
smyiu@cs.hku.hk

Abstract. RNA secondary structures are known to be important in many biological processes. Many available programs have been developed for RNA secondary structure prediction. Based on our knowledge, however, there still exist secondary structures of known RNA sequences which cannot be covered by these algorithms. In this paper, we provide an efficient algorithm that can handle all RNA secondary structures found in Rfam database. We designed a new stochastic context-free grammar named Rectangle Tree Grammar (RTG) which significantly expands the classes of structures that can be modelled. Our algorithm runs in $O(n^6)$ time and the accuracy is reasonably high, with average PPV and sensitivity over **75%**. In addition, the structures that RTG predicts are very similar to the real ones.

1 Introduction

Secondary structures of RNA molecules play important roles in their functionalities [1, 2]. Many methods have been proposed to predict RNA secondary structures. Although the majority of RNAs have simple secondary structures, pseudoknots (base pairs crossing each other) are found in almost all classes of RNAs. Pseudoknots are known to be involved in biological functions such as stimulating ribosomal frameshifting [3, 4]. The existence of pseudoknots make the secondary structure prediction an NP-hard problem, in general [5, 6]. Existing algorithms attempt to solve the problem by considering a restricted set of pseudoknots [7–18]. Not all existing pseudoknots can be modelled. In terms of prediction accuracy, CentroidAlifold[9] generalized a centroid estimator that maximizes the expected accuracy of structure prediction. Tabei, Yasuo and Kiryu[16] proposed a fast multiple sequence alignment method named MXScarna in which the optimal structure that maximized a heuristic scoring function was found during the group alignments of stem component sequences. RNAaliFold[17] precomputed alignments using a combination of free-energy and a covariation measures, whilst TurboFold[18] utilized an iterative probabilistic method to predict secondary structures for multiple RNA sequences.

© Springer International Publishing Switzerland 2015
R. Harrison et al. (Eds.): ISBRA 2015, LNBI 9096, pp. 211–222, 2015.
DOI: 10.1007/978-3-319-19048-8_18

Despite of so many algorithms to predict RNA secondary structures, there exist secondary structures of known RNAs in Rfam [19] that cannot be covered by existing efficient algorithms[1]. Figure 1 shows such an example.

Fig. 1. A structure in Rfam which cannot be handled by existing efficient algorithms

In this paper, we proposed a grammar-based machine learning method to predict secondary structures for all RNA sequences in Rfam. Enlightened by [20], we designed a new stochastic context-free grammar called Rectangular Tree Grammar (RTG), which can model all possible secondary structures of known RNA sequences in the Rfam database. Each structure can be generated by a unique operation path, that is, the only sequence of operations that yields this sequence. A set of paths is obtained using some real RNA sequences with known structures. Rule transition probabilities and base emission probabilities are calculated based on this set. In order to determine the unknown secondary structure of a RNA sequence, dynamic programming is adopted to generate the most probable structure. This procedure takes $O(n^6)$ time, where n is the length of input RNA sequence.

The proposed approach was evaluated using several sets of sequences with one containing pseudoknot-free structures and the others with different types of pseudoknots. We compared the performance of RTG with popular prediction algorithms including gfold[7], CentroidAlifold[9], pknotsRG[21], NUPACK[22], MXScarna[16], RNAaliFold[17] and TurboFold[18]. The experimental results have shown that our approach outperforms others substantially with high PPV and sensitivity, especially on highly-pseudoknotted sequences.

2   Method

2.1   RNA Secondary Structure Definitions

Let $S = s_1 s_2 \ldots s_n$ be an RNA sequence of length n. $M_{x,y}$ is the set of base pairs in the range $[x, y]$, $M_{x,y} = \{(i, j) | x \leq i < j \leq y, (s_i, s_j)$ is a base pair$\}$.

Banding: The secondary structure of $s_x \ldots s_y$ is a *banding* if it satisfies the following conditions:
(i) for any $i, j, k, l \in [x, y], i \neq k, j \neq l$, if $(i, j) \in M_{x,y}$ and $(k, l) \in M_{x,y}$, then $i < k < l < j$ or $k < i < j < l$.
(ii) $(x, y) \in M_{x,y}$.

---

[1] We only consider algorithms which run in $O(n^6)$ time.

Gapped Banding: The secondary structure of $s_x \ldots s_y \cup s_p \ldots s_q$ is a *gapped banding* if it satisfies the following conditions:
(i) By cutting out the gap $s_{y+1} \ldots s_{p-1}$, the secondary structure over this new sequence $s_x \ldots s_y s_p \ldots s_q$ is a banding.
(ii) $\forall (i,j) \in M_{[x,y] \cup [p,q]}$, (i,j) is across the gap.

Regular Structure: A structure is a regular structure if no base pair crossing exists, that is, the secondary structure of $s_x \ldots s_y$ is a *regular structure* if $\not\exists i,j,k,l \in [x,y]$ such that $(i,j) \in M_{x,y}, (k,l) \in M_{x,y}$, and $i < k < j < l$.

Standard Pseudoknot of Degree k: A structure is a *standard pseudoknot* of degree k ($k \geq 3$) if it is either a *simple standard pseudoknot* of degree k or a *gapped standard pseudoknot* of degree k.

For any $1 \leq w \leq k-1$, let $H_w = \{(i,j) \in M_{x,y} | x_{w-1} \leq i < x_w \leq j < x_{w+1}\}$. We allow $j = x_k$ for H_{k-1} to resolve the boundary case.

The secondary structure of $s_x \ldots s_y$ is a *simple standard pseudoknot* of degree k ($k \geq 3$) if there exists a set of $x_1, x_2, \ldots, x_{k-1}$ that satisfies the following conditions (Figure 2):
(i) $x = x_0 < x_1 < x_2 < \ldots < x_{k-1} < x_k = y$.
(ii) $\forall w \in [1, k-1]$, H_w is a gapped banding.
(iii) $\forall (i,j) \in M_{x,y}$, $\exists w$ such that $(i,j) \in H_w$.
(iv) $\forall w \in [1, k-1]$, if $(i,j) \in H_w, (k,l) \in H_{w+1}$, then $i < k < j < l$.
(v) there does not exist two base pairs $(i,j) \in H_w, (k,l) \in H_v, v - w \geq 2$, such that $i < k < j < l$.

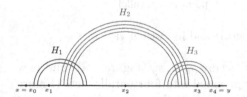
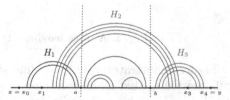

Fig. 2. A simple standard pseudoknot of degree 4

Fig. 3. A gapped standard pseudoknot of degree 4, where $s_a \ldots s_b$ forms a regular structure

The secondary structure of $s_x \ldots s_y$ is a *gapped standard pseudoknot* of degree k ($k \geq 3$) if there exists a, b such that $s_{a+1} \ldots s_{b-1}$ is a structure defined above and $s_x \ldots s_a \cup s_b \ldots s_y$ satisfies the following conditions (Figure 3):
(i) By cutting out the gap $s_{a+1} \ldots s_{b-1}$, the secondary structure over this new sequence $s_x \ldots s_a s_b \ldots s_y$ is a standard pseudoknot.
(ii) if $(i,j) \in M_{[x,a] \cup [b,y]}$ is across the gap, then $\exists w$ such that $(i,j) \in H_w$. Moreover, $\forall (k,l) \in H_w$ is across the gap.

Based on our analysis to Rfam database, we focus on all standard pseudoknots of degree k ($k \geq 3$) in this paper.

Three Banding Structure: The secondary structure of $s_x \ldots s_y$ is a *three banding structure* if we can find x_1, x_2, x_3 such that all the following conditions are satisfied.

(i) $x \le x_1 \le x_2 \le x_3 \le y$.

(ii) $\forall (i,j) \in M_{[x,y]}$, it must belong to one of the sets $L_{12}, L_{23}, L_{34}, L_{14}$ as defined below.

(iii) for any two pairs $(i,j) \in L_{ab}$ and $(k,l) \in L_{ab}$, then $i < k < l < j$ or $k < i < j < l$. where $L_{ab} = L_{12}, L_{23}, L_{34}$ or L_{14}.

Let $L_{12} = \{(i,j) | x \le i \le x_1 \le j \le x_2\}$, $L_{23} = \{(i,j) | x_1 \le i \le x_2 \le j \le x_3\}$, $L_{34} = \{(i,j) | x_2 \le i \le x_3 \le j \le y\}$, $L_{14} = \{(i,j) | x \le i \le x_1, x_3 \le j \le y\}$.

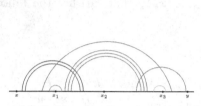

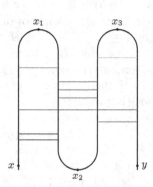

(a) A three banding structure.

(b) Twist this three banding structure so that all base pairs are parallel.

Fig. 4. A three banding structure and its twisted view

Figure 4 illustrates a three banding structure $s_x \ldots s_y$ (Figure 4(a)) and how it is twisted so that all its base pairs are grouped into four parallel sets (Figure 4(b)), i.e., L_{12} (blue and cyan pairs), L_{23} (red pairs), L_{34} (green and lime pairs) and L_{14} (magenta pairs).

k-Crossing Structure: $s_x s_{x+1} \ldots s_y$ is a *k-crossing structure* ($k \ge 3$) if it is either a *simple k-crossing structure* or a *gapped k-crossing structure*. Intuitively, in a k-crossing structure, there exist k gapped bandings where any two of them crosses each other.

For any ($1 \le w \le k$), let $H_w = \{(i,j) \in M_{x,y} | x_{w-1} \le i < x_w, x_{w-1+k} \le j < x_{w+k}\}$. We allow $j = x_{2k}$ for H_k to resolve the boundary case. Let $C_w (1 \le w \le 2k) = \{(i,j) \in M_{x,y} | x_{w-1} \le i < j < x_w\}$. $j = x_j$ is allowed for C_{2k}. A *crossing set* is defined as $CH_w = H_w \cup C_w \cup C_{w+k} (1 \le w \le k)$.

The secondary structure of $s_x \ldots s_y$ is a *simple k-crossing structure* ($k \ge 3$) if there exist $x_0, x_1, \ldots, x_{2k}$ that satisfy the following conditions:

(i) $x = x_0 < x_1 < \ldots < x_{2k-1} < x_{2k} = y$.

(ii) $\forall \ (i,j) \in M_{x,y}, \exists w$ such that $(i,j) \in CH_w$.

(iii) $\forall w \in [1,k], CH_w$ is a regular structure, a standard pseudoknot or a three banding structure.

The secondary structure of $s_x \ldots s_y$ is a *gapped k-crossing structure* if and only if there exists a, b such that $s_{a+1} \ldots s_{b-1}$ is a defined structure and $s_x \ldots s_a \cup s_b \ldots s_y$ satisfies the following conditions:

(i) By cutting out the gap $s_{a+1} \ldots s_{b-1}$, the secondary structure over this new sequence $s_x \ldots s_a s_b \ldots s_y$ is a k-crossing structure.

(ii) $\forall w \in [1, k], \forall (i, j) \in H_w, (i, j)$ is across the gap $s_{a+1} \ldots s_{b-1}$.

(iii) $\forall (i, j) \in M_{[x,a] \cup [b,y]}$ is across the gap, $\exists w \in [1, k]$ such that $(i, j) \in H_w$ and $\not\exists w \in [1, k]$ such that $(i, j) \in C_w$.

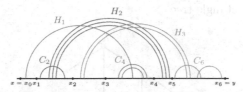

Fig. 5. A 3-crossing structure

Figure 5 depicts a 3-crossing structure. Each color denotes a crossing set, i.e., CH_1 (cyan), CH_2 (red) and CH_3 (green).

Recall the example in Figure 1. The difficulty of this structure lies on two mixed substructures called 3-crossing and standard pseudoknots for which none of the existing algorithms can model (the green basepairs form a standard pseudoknot; the green, blue, and the red basepairs form a 3-crossing structure). As a matter of fact, among all classes of Rfam structures we defined below, only gfold[7] can generate some extremely simple 3-crossing structures with $CH_w = H_w$. None of the aforementioned algorithms can generate k-crossing structures ($k \geq 4$).

2.2 Rectangle Tree and Complete Tree

We have observed that the classic grammar-based algorithm, Simple Linear Tree Adjoining Grammar[23], is incapable of predicting some highly-pseudoknotted structures (e.g k-crossing structures). To predict these structures, we introduce a new grammar called *Rectangle Tree Grammar*(RTG).

Let V be a finite set of alphabets and Σ be a set of terminal alphabets where $\Sigma \subset V$. Let γ be a *tree* over V such that

(i) each internal node must be labeled with a nonterminal symbol.

(ii) each leaf node can be labeled with a nonterminal or terminal symbol.

(iii) each internal node can have any number of children.

(iv) each edge can be labeled red or black.

$Y(\gamma)$ (ie. *yield* of tree) is defined as breadth-first search output of γ where all the nonterminal symbols are ignored.

A tree is *rectangle* if it satisfies all the conditions below:

(i) all the internal nodes should be labeled with nonterminal symbols.

(ii) there is only one leaf labeled with nonterminal symbol. This node is called N_4. The path from the root to N_4 is called the *backbone*.

(iii) there is only one *red edge* that defines the *insertion point* of this tree which is along the backbone.

(iv) considering the red edge $N_2 - N_3$, N_3 is the only child of N_2.

(v) the path from root to N_2 is the longest path in upper tree, the path from N_3 to N_4 is the longest path in bottom tree.

According to the definition above, a rectangle tree can be divided into two parts by splitting through red edge, the yield of upper tree is γ_U, the yield of bottom tree rooted at N_3 is γ_B. $Y(\gamma) = \gamma_U\gamma_B$, the position between γ_U and γ_B is called an *insertion point* (where other structures can be inserted in). Figure 6 is an example of a rectangle tree.

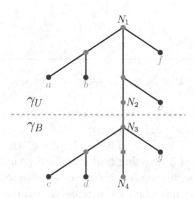

Fig. 6. A rectangle tree. Orange nodes and blue are labeled with nonterminal and terminal symbols, respectively. Its yield is $fabe, gcd$. The comma represents its insertion point.

A tree is *complete* if it satisfies all the following conditions:

(i) only one leaf (labeled as N_4) is labeled with nonterminal symbol.

(ii) there is no red edge, i.e., no more base pairs will be added.

(iii) the path from root to N_4 is the longest path in the tree.

By labeling the red edge in Figure 6, the rectangle tree becomes a complete tree. The yield is $fabegcd$. To predict the secondary structure of an RNA sequence, we compute the most probable rectangle tree whose yield is exactly the given sequence.

2.3   Grammar States

A rectangle tree or a complete tree has a unique state. As shown in Table 1, a *state* corresponds to the secondary structure represented by this tree. The first seven states are for rectangle trees. The remaining three are for complete trees.

2.4   Tree Operations

There are multiple ways to add bases into a tree. A *tree operation* defines how a single base, a base pair or another tree are allowed to be added. In this section,

Table 1. Grammar states and the corresponding secondary structures of RNA sequence $s_i \ldots s_k \cup s_l \ldots s_j$ or $s_i \ldots s_j$

State	Structure Description
B	banding or gapped banding
$B3$	the structure is three banding, insertion point is the insertion point of the second banding.
BL	the structure is standard pseudoknot of degree $k(k \geq 3)$, insertion point is the insert point of the rightmost banding.
BR	the structure is standard pseudoknot of degree $k(k \geq 3)$, insertion point is the insert point of the leftmost banding.
BLR	the structure is standard pseudoknot of degree $k(k \geq 4)$, insertion point can be insertion point of any banding except the leftmost and rightmost one.
G	2-crossing structure, after another Cr operation, it will transit to state H.
H	k-crossing structure$(k \geq 3)$.
CPP	both s_i and s_j are paired bases.
CPS	s_i is a paired base, s_j is a single base.
CSP	s_i is a single base, s_j is a paired base.
CSS	both s_i and s_j are single bases.

we introduce tree operations from state to state so that it is clear why each operation is needed. For simplicity, we use S_1 to denote γ_U (the yield of upper tree) and S_2 to denote γ_B (the yield of bottom tree).

Gapped Banding (State B). A gapped banding is divided by insertion point into two parts: S_1 and S_2. Basically, base pairs and single bases of a banding is allowed to be added from outmost inwards. For rectangle trees, single bases can only be added at the end of S_1 (at N_2) or at the beginning of S_2 (at N_3). To obtain a gapped banding, the following tree operations are designed:

- $L23$: add a base pair X into the tree, where the head and tail of X are added to the end of S_1 and the beginning of S_2, respectively.
- $Ls2$: add a single base to the end of S_1.
- $Ls3$: add a single base to the beginning of S_2.

Three Banding (State $B3$). A basic idea to generate a three banding structure is to add gapped bandings in the twisted structure in a top-down manner. Besides $L23, Ls2$ and $Ls3$, there are three more legal operations to add a gapped banding (or a base pair) X:

- $L12$: add the head of X to the beginning of S_1; tail of X to the end of S_1.
- $L34$: add the head of X to the beginning of S_2; tail of X to the end of S_2.
- $L14$: add the head of X to the beginning of S_1; tail of X to the end of S_2.

Standard Pseudoknot of Degree k (State BL). To generate standard pseudoknot of degree $k(k \geq 3)$, we designed operation LR (Figure 7). Operation LR inserts the upper tree of α above N_1 of β and its bottom tree above N_2. At the same time, the insertion point is updated to insertion point of β. As a result, base pairs across upper tree and bottom tree in α and β would cross. Moreover, the update of insertion point prevents base pairs in α from crossing base pairs in subsequent trees adjoined with LR later. After $(k-2)$ LR operations, standard pseudoknot of degree k is generated.

k-Crossing (State G and H). In k-crossing structures, without considering embedded substructures, all base pairs can be grouped into different crossing sets $CH_1 \ldots CH_k$, where CH_w ($\forall w \in [1, k]$) is a regular structure (state B), a standard pseudoknot (state BL, BR or BLR) or a three banding structure (state $B3$). Standard pseudoknots of state BL and BR have their insertion point within the leftmost and rightmost banding, respectively. Otherwise, if the insertion point comes from neither the leftmost nor the rightmost banding, this standard pseudoknot is in state BLR.

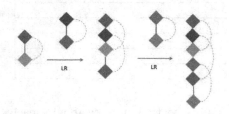 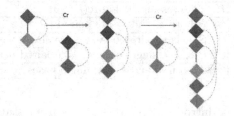

Fig. 7. After 2 LR operations over 3 gapped bandings, a standard pseudoknot of degree 4 is generated

Fig. 8. A 3-crossing can be generated by 2 Cr operations

Operation LL is designed for state BR and BLR. An LL operation on rectangle tree ϵ with ζ inserts the upper tree of ζ above N_3 of ϵ and its bottom tree under N_4. Then by operating LR on $\alpha_1 LR \ldots LR\alpha_i$ with $\epsilon_1 LL \ldots LL\epsilon_j$, standard pseudoknot with insertion point in its $(i+1)$th banding is generated.

After generation of all the crossing sets, we designed the operation Cr to link them up. As is shown in Figure 8, operation Cr on rectangle tree α with β inserts upper tree of β under N_2 of α and bottom tree under N_4. So base pairs between upper tree and bottom tree in α and β would cross. After $(k-1)$ Cr operations, k-crossing can be generated.

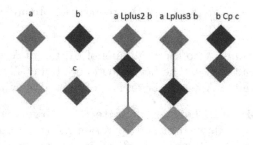

Fig. 9. $Lplus2$, $Lplus3$ and Cp for embedding and concatenation. a is a rectangle tree, b and c are complete trees

Embedding and Concatenation (State CPP). By applying operation Lm to label the red edge of a rectangle tree to black, a complete tree is generated. Embedding operations $Lplus2$ and $Lplus3$ insert a complete tree to N_2 and N_3 of a complete tree, respectively. The concatenation operation Cp can concatenate two complete trees. The above three operations are also explained in Figure 9. Note that if a rectangle tree α embeds (using $Lplus2$ or $Lplus3$) some complete trees, it becomes a new rectangle tree α'. And the state of α' remains the same as that of α.

Single Bases at Both Ends (State CPS, CSP and CSS). As required by RTG, gapped bandings are always generated at first. Afterwards, applying proper tree operations as defined above, these gapped bandings compose a more complicated structure. When no base pairs are to be inserted, Lm alters this rectangle tree (representing this complicated structure) to a complete tree of state CPP. Note that the first base s_i and the last base s_j must have been boundary bases of gapped bandings. When there are single bases at either end of an RNA sequence, operation $Ls1$ and $Ls4$ are used to add single bases to the beginning of S_1 and the end of S_2, respectively.

2.5 Grammar

A *RTG grammar rule* clarifies whether a specific operation is applicable to rectangle trees (in CPP, CPS, CSP or CSS state) or complete trees (in any other state). All the rules are tabulated in Table 2. In the table, α is a single base. (α,β) is a base pair. $(b1,b2)$ and $(b3,b4)$ are rectangle trees, where comma denotes their insertion points. (c), $(c1)$, and $(c2)$ represent complete trees.

After applying RTG grammar rules, the state of the predicted structure transits into another. All valid transitions defined by the grammar rules will be given in the full paper. For the dynamic programming algorithm for structure prediction and the parameter training, we follow the standard techniques (details will be given in the full paper).

Table 2. RTG rules

Operation	Input	output
$Ls2\_\alpha$	$(b1,b2)*(\alpha)$	$(s1\alpha,s2)$
$Ls3\_\alpha$	$(b1,b2)*(\alpha)$	$(b1,\alpha b2)$
$L12$	$(b1,b2)*(b3,b4)$	$(b3b1b4,b2)$
$L23\_\alpha\_\beta$	$(b1,b2)*(\alpha,\beta)$	$(b1\alpha,\beta b2)$
$L34$	$(b1,b2)*(b3,b4)$	$(b1,b3b2b4)$
$L14$	$(b1,b2)*(b3,b4)$	$(b3b1,b2b4)$
LL	$(b1,b2)*(b3,b4)$	$(b3,b1b4b2)$
LR	$(b1,b2)*(b3,b4)$	$(b1b3b2,b4)$
$Lplus2$	$(b1,b2)*(c)$	$(b1c,b2)$
$Lplus3$	$(b1,b2)*(c)$	$(b1,cb2)$
Lm	$(b1,b2)$	$(b1b2)$
Cr	$(b1,b2)*(b3,b4)$	$(b1b3,b2b4)$
Cp	$(c1)*(c2)$	$(c1c2)$
$Ls1$	$(c1)*(\alpha)$	$(\alpha c1)$
$Ls4$	$(c1)*(\alpha)$	$(c1\alpha)$

3   Experiments

A total of 564 RNA sequences from 44 families were extracted from Rfam database for our experiments. All these families were classified into three sets D1, D2 and D3. D1 consists of regular structures (15 families). D2 contains standard pseudoknots of degree ≥ 3 (27 families). D3[2]comprises a set of 3-crossing structures (2 families). We carried out a 10-fold cross-validation on D1, D2 and D3 datasets separately. More specifically, take D1 as an example. In each round of validation, a total of 334 sequences in D1 were randomly partitioned into ten equal-size subsets. Out of these ten subsets, one subset was retained to test the model, while the other nine subsets were used to train this model. To eliminate variability, 10 rounds were performed using different partitions. The performance evaluated below is based on the average among 10 rounds. We compared the performance of our RTG method with seven popular softwares. For softwares that take multiple sequences as inputs, like TurboFold, CentroidAlifold and RNAalifold, we provided them with each family of sequences as an input. The performance was measured using positive predictive value (PPV) and sensitivity defined below. PPV $= \frac{\alpha}{\gamma}$ and sensitivity $= \frac{\alpha}{\beta}$, where α is the number of correctly reported base pairs, β is the total number of reported base pairs, and γ is the total number of base pairs in the Rfam.

Table 3. PPV and sensitivity of RTG and seven other softwares on $D1$, $D2$ and $D3$

Dataset	Software	PPV(%)	Sensitivity(%)	Software	PPV(%)	Sensitivity(%)
D1		62.21	28.97		54.5	24.6
D2	pknotsRG[21]	71.72	65.92	gfold[7] [3]	67.35	53.28
D3		19.78	10.13		11.00	6.70
D1		51.41	24.36		93.53	36.73
D2	NUPACK[22]	74.24	62.63	CentroidAlifold[9]	50.24	43.71
D3		37.52	18.88		24.89	12.38
D1		75.54	38.76		77.69	45.60
D2	MXScarna[16]	48.01	52.50	RNAalifold[17]	43.98	51.77
D3		13.73	7.30		23.40	24.88
D1		75.46	34.01		**80.22**	**62.81**
D2	TurboFold[18]	55.09	42.07	**RTG**	**80.56**	**75.09**
D3		20.80	10.74		**71.95**	**71.36**

Table 3 summarized the comparison of secondary structure prediction for RTG and seven other state-of-the-art programs. Our RTG program often outperforms other programs in terms of PPV and sensitivity. The experiment has revealed that 3-crossing dataset is hard to predict for other programs, which is consistent with our analysis of previous algorithms. However, the prediction of RTG program is accurate to a certain extent.

---

[2] There are only two families in Rfam with this complicated structures and one of the families (RF02032) is too long that our server does not have enough memory to handle it, we only extracted the 3-crossing structure (without considering embedded substructure) to run.

Apart from RTG, NUPACK[22] and RNAalifold[17] performed best in estimating the secondary structure for 3-crossing dataset. The performance regarding this dataset is further illustrated in Figure 10, which presents the predicted
structure of NUPACK, RNAalifold and RTG over AE005174-2 as well as the
trusted annotation in Rfam. The underlined parentheses($<->$, $A-a$ and
$B-b$) denotes the correctly predicted base pairs.

Fig. 10. An detailed comparison for predicting the structure of AE005174-2(RF00140)
in Rfam

Evidently, RTG behaved the best with PPV = 87.5% and sensitivity = 70.0%.
The PPV and sensitivity of RNAalifold were 56.2% and 45.0%, respectively. NU-
PACK reached even lower PPV and sensitivity. In addition to its high accuracy
evaluated using PPV and sensitivity, RTG predicted a structure much more similar to the ground truth. RTG thought the secondary structure of AE005174-2
is a 3-crossing. Furthermore, it almost pointed out all the bandings correctly.
Even for pairs denoted by $B-b$, the pairing position was very close. However,
NUPACK and RNAalifold predicted it as regular structures, which was way far
from its real structure.

References

1. Ten Dam, E., Pleij, K., Draper, D.: Structural and functional aspects of rna pseudoknots. Biochemistry 31(47), 11665–11676 (1992)
2. Lee, K., Varma, S., SantaLucia Jr., J., Cunningham, P.R.: In vivo determination
   of rna structure-function relationships: analysis of the 790 loop in ribosomal rna.
   Journal of Molecular Biology 269(5), 732–743 (1997)
3. Brierley, I., Digard, P., Inglis, S.C.: Characterization of an efficient coronavirus
   ribosomal frameshifting signal: requirement for an rna pseudoknot. Cell 57(4),
   537–547 (1989)
4. Giedroc, D.P., Theimer, C.A., Nixon, P.: Structure, stability and function of rna
   pseudoknots involved in stimulating ribosomal frameshifting. J. Mol. Biol. 298,
   167–185 (2000)
5. Lyngsø, R.B., Pedersen, C.N.S.: RNA pseudoknot prediction in energy-based models. Journal of Computational Biology 7(3-4), 409–427 (2004)

---

[3] There are some families gfold cannot run, so its PPV and sensitivity does not include
these families.

6. Lyngsø, R.B.: Complexity of pseudoknot prediction in simple models. In: Díaz, J., Karhumäki, J., Lepistö, A., Sannella, D. (eds.) ICALP 2004. LNCS, vol. 3142, pp. 919–931. Springer, Heidelberg (2004)
7. Reidys, C.M., Huang, W.D., Andersen, F., Penner, J.E., Stadler, R.C., Nebel, P.F., Topology, M.E.: prediction of rna pseudoknots. Bioinformatics 27(8), 1076–1085 (2011)
8. Ren, J., Rastegari, B., Condon, A., Hoos., H.H.: Hotknots: Heuristic prediction of rna secondary structures including pseudoknots. RNA 11, 1494–1504 (2005)
9. Hamada, M., Kiryu, H., Sato, K., Mituyama, T., Asai., K.: Predictions of RNA secondary structure using generalized centroid estimators. Bioinformatics 25(4), 465–473 (2009)
10. Zakov, S., Goldberg, Y., Elhadad, M., Ziv-Ukelson., M.: Rich parameterization improves rna structure prediction. Journal of Computational Biology 18(11), 1525–1542 (2011)
11. Bindewald, E., Shapiro., T.K.B.: Cylofold: secondary structure prediction including pseudoknots. Nucleic Acids Research suppl.(W), 368–387 (2010)
12. Akutsu., T.: Dynamic programming algorithms for rna secondary structure prediction with pseudoknots. Discrete Applied Mathematics 104, 45–62 (2000)
13. Chen, H., Condon, A., Jabbari, H.: An o(n(5)) algorithm for mfe prediction of kissing hairpins and 4-chains in nucleic acids. Discrete Applied Mathematics 16(6), 803–815 (2009)
14. Dirks, R., Pierce, N.: A partition function algorithm for nucleic acid secondary structure including pseudoknots. J. Comput. Chem. 24(13), 1664–1677 (2003)
15. Rivas, E., Eddy, S.R.: A dynamic programming algorithm for rna structure prediction including pseudoknots. J. Mol. Biol. 285, 2053–2068 (1999)
16. Tabei, Y., Kiryu, H., Kin, T., Asai, K.: A fast structural multiple alignment method for long rna sequences. BMC Bioinformatics 9(1), 33 (2008)
17. Bernhart, S.H., Hofacker, I.L., Will, S., Gruber, A.R., Stadler, P.F.: Rnaalifold: improved consensus structure prediction for rna alignments. BMC Bioinformatics 9(1), 474 (2008)
18. Harmanci, A.O., Sharma, G., Mathews, D.H.: Turbofold: iterative probabilistic estimation of secondary structures for multiple rna sequences. BMC Bioinformatics 12(1), 108 (2011)
19. Griffiths-Jones, S., Moxon, S., Marshall, M., Khanna, A., Eddy, S.R., Bateman, A.: Rfam: annotating non-coding rnas in complete genomes. Nucleic Acids Research 33(suppl. 1), D121–D124 (2005)
20. Cai, L., Malmberg, R.L., Wu, Y.: Stochastic modeling of rna pseudoknotted structures: a grammatical approach. Bioinformatics 19(suppl. 1), i66–i73 (2003)
21. Reeder, J., Steffen, P., Giegerich, R.: pknotsrg: Rna pseudoknot folding including near-optimal structures and sliding windows. Nucl. Acids Res. 35, 320–324 (2007)
22. Zadeh, J.N., Steenberg, C.D., Bois, J.S., Wolfe, B.R., Pierce, M.B., Khan, A.R., Dirks, R.M., Pierce, N.A.: Nupack: Analysis and design of nucleic acid systems. Journal of Computational Chemistry 32, 170–173 (2011)
23. Uemura, Y., Hasegawa, A., Kobayashi, S., Yokomori, T.: Tree adjoining grammars for rna structure prediction. Theoretical Computer Science 210(2), 277–303 (1999)

An Approach for Matching Mixture MS/MS Spectra with a Pair of Peptide Sequences in a Protein Database

Yi Liu[1(✉)], Weiping Sun[1], Gilles Lajoie[3], Bin Ma[2], and Kaizhong Zhang[1]

[1] Department of Computer Science, The University of Western Ontario,
London, Ontario, N6A 5B7, Canada
{yliu766,wsun62,kzhang}@csd.uwo.ca

[2] David R. Cheriton School of Computer Science, University of Waterloo, Waterloo,
Ontario, N2L 3G1, Canada
binma@uwaterloo.ca

[3] Department of Biochemistry, The University of Western Ontario,
London, Ontario, N6A 5B8, Canada
glajoie@uwo.ca

Abstract. The large amount of data collected in an MS experiment requires effective computational approaches for the automated analysis of those data. However, one specific challenge exists now is the unsatisfactory analysis of mixture spectra by the traditional computational approaches. Mixture spectra are observed quite frequently in mass spectrometry experiment which result from the concurrent fragmentation of multiple precursors, therefore effective approaches for characterizing those non-canonical spectra are highly desired. In this manuscript, we proposed an approach for matching mixture tandem mass spectra with a pair of peptide sequences acquired from the protein sequence database by incorporating a special *de novo* assisted filtration. The preliminary experimental results demonstrated the efficiency of the integrated filtration strategy in reducing examination space and verified the effectiveness of the proposed matching method.

Keywords: Mass spectrometry · Mixture spectra · Filtration strategy · Protein database

1 Introduction

The application of mass spectrometry on the characterization of large biopolymers has obtained remarkable achievements over the past decade [1]. Now mass spectrometry has become a standard choice for the high-throughput identification and quantification of proteins. In the predominant "bottom-up" proteomics method, the proteins are first proteolyzed into peptides mixtures which are then separated by the reversed-phase liquid chromatography. After ionization using ESI or MALDI, the targeted peptides are sequenced by the intensity dependent selection and gas phase fragmentation. Subsequently, the fragments are

© Springer International Publishing Switzerland 2015
R. Harrison et al. (Eds.): ISBRA 2015, LNBI 9096, pp. 223–234, 2015.
DOI: 10.1007/978-3-319-19048-8_19

detected and measured by mass spectrometers in very high speed which usually generates a large amount of MS/MS spectra in one single run. Computational approaches are necessary for the automated interpretation of the large volume of proteomics data collected from mass spectrometers. Software tools developed for this purpose can be generally divided into two categories: *de novo* sequencing and database search. *De novo* sequencing techniques directly reconstructs the peptide sequence from each spectrum, while in contrast a database search method scores each spectrum against peptides in a protein sequence database. Typical *de novo* sequencing software tools include PEAKS [2], PepNovo [3], and Lutefisk [4], meanwhile there are also many software packages available for database search purpose, including Mascot [5], SEQUEST [6], PEAKS DB [7], X!Tandem [8], and OMSSA [9].

Even with all those rapid progresses, there remain several challenging areas unsolved. One significant challenge is that in a typical tandem mass spectrometry experiment, the ratio of the collected spectra being successfully identified is rather low. One specific factor that contributes to the situation of low identification rate is the frequent occurrence of concurrent fragmentation of multiple peptides in one MS/MS spectrum. Traditional computational approaches have difficulties in characterizing the resulting mixture spectra because they always take the assumption that each MS/MS spectrum is generated from a single precursor. Some preliminary literatures have reported that the frequency of multiple peptides co-fragmented simultaneously can be quite phenomenal. Hoopmann et al. in [10] investigated the frequency of mixture spectra observed in datasets collected from a high resolution LTQ-Orbitrap mass spectrometer and estimated that 11% of MS/MS spectra are chimeras, with an additional 29% of MS/MS spectra with parent iosotope distribution inconsistent with peptide analysis. Houel et al. in [11] also conducted an assessment on the occurrence of chimera spectra in shotgun proteomics under the Data-Dependent Acquisition(DDA) mode, indicating that the percentage of chimeras may reach as high as 50% of total spectra collected. The frequently observed mixture spectra in proteomics experiment necessitate the developing of new computational approaches for interpreting those non-canonical spectra. In addition, mixture spectra have also been explored by alternative experimental configurations to circumvent the limitations of current mass spectrometry data acquisition strategies [12,13,14,15].

There have been several preliminary literatures appeared to address such an issue by utilizing the mixture spectra to identify more peptides. Zhang et al. in [16] proposed a method to identify the co-eluting peptides from mixture MS/MS spectra. The database search engine, ProbIDtree, proceeds in an iterative manner to identify multiple peptides from one spectrum by database search. During the process, ions assigned to a tentative peptide are subtracted from the spectrum, and the remaining spectrum is then used to match with another peptide sequence. The software then calculated an adjusted probability score to select the best matched peptides from all the tentative identifications. Wang et al. in [17] introduced M-SPLIT, an MS/MS spectra library search software and demonstrated its potential to identiy peptides from mixture spectra by matching the acquired spectra against a spectral library. This approach is limited in application because that it can only be used to search against a library comprising of the previously observed and confidently interpreted spectra. Recently, Wang et al. in [18] designed another method, a database search tool, MixDB to correlate mixture spectra with a pair of peptide sequences filtered from a protein database by using a specifically designed scoring function. This method explicitly formulates the occurrence of co-sequenced peptides in the same spectrum, and provide a strategy to report the FDR (False Discovery Rate) for the computational results validation.

Despite of those advances above, the research on characterizing mixture spectra is still far from satisfactory. Methods based on spectral library search will lose effectiveness when the target peptides have not been observed or identified before, meanwhile the research of *de novo* sequencing of mixture spectra identification is currently in its very primitive phase which suffers the problem of low accuracy in computation, more research will be necessary before it become practical for real application. Database search is traditionally recognized as the mainstream approach for peptide identification with mass spectrometry. However when considering two precursors in a single spectra, the number of possible candidate peptide pairs that fall into the required mass error bound is always enormous [18]. Efficient filtration strategy is highly desired to avoid the huge computational overload of searching the spectrum against all possible pairs in the protein database. Researchers lately have showed an increasing interest in combining the *de novo* sequencing results with database search to sufficiently interpret MS/MS spectra [7,19]. In those research, the *de novo* sequencing results that contain only partially correct sequences are used to reduce the examination space. In this paper, we will formulate the mixture spectra identification problem formally, and propose an approach for matching mixture MS/MS spectra with a pair of peptides from a protein database by incorporating a special filtration strategy assisted with the preliminary *de novo* sequencing results. Experimental results demonstrated that when equipped with such filtration process, the correct matches can be found by only considering a minuscule fraction of all possible pairs.

2   Notations and Problem Formulation

Assume that a mixture spectrum $\mathcal{M}$ is generated by the co-fragmentation of peptides $P1$ and $P2$, and $\mathcal{M}$ can be represented by a peak list $\mathcal{M} = \{(x_i, h_i)|i = 1, 2, ..., n\}$. Each element (x_i, h_i) represents a peak in the spectrum, in which x_i is the m/z value and h_i is the intensity of the peak. Meanwhile we use two molecular weight MW_1 and MW_2 to denote the precursor mass values of the two peptides that satisfy $\|MW_1 - MW_2\| \leq \Delta$, and Δ is a small value predefined by the width of the mass spectrometer selection window.[1] More formally, we formulated a mixture spectrum $\mathcal{M}$ as $\mathcal{M} = A + \alpha B$, where A and B are the

---

[1] In the typical Data-dependent Acquisition (DDA) mode, the selection window in the fragmentation is usually only a few Daltons wide.

MS/MS spectra from the co-sequenced peptides P_1 and P_2 respectively, and the mixture coefficient α represents the relative abundance of A and B when sequenced. Without losing generality, we assume that both A and B are scaled to the same magnitude, and A always corresponds to the peptide with higher abundance level, therefore we have $0 \leq \alpha \leq 1$. In addition, we use Σ to denote the alphabet of 20 different types of amino acids. For an amino acid $a \in \Sigma$, we use $\|a\|$ to symbolize the mass of the amino acid residue. Let $P = a_1 a_2 ... a_k$ be the string of amino acids, we define the residue mass of the peptide as $\|P\| = \Sigma_{1 \leq j \leq k} \|a_i\|$ and the actual mass of the peptide as $\|P\| + \|H_2O\|$.

Similar to the identification by comparing the single-peptide MS/MS spectra against all possible peptides from a protein database, our objective is to identify mixture spectra by comparison against all possible pairs of peptides in a given protein sequence database. Thus, the MIXTURE SPECTRA DATABASE SEARCH PROBLEM can be formulated as follows: Given a mixture spectrum $\mathcal{M}$, two precursor mass values MW_1 and MW_2, a predefined error bound δ, and a protein database D, we want to find a coefficient α and a pair of peptides P_1 and P_2 from D that maximize the value under a specific scoring function $H(\mathcal{M}, A + \alpha B)$, such that $\|\|P_1\| + \|H_2O\| - MW_1\| \leq \delta$, $\|\|P_2\| + \|H_2O\| - MW_2\| \leq \delta$.[2] The scoring function $H(\mathcal{M}, A + \alpha B)$ measures how much the peptide pair (P_1, P_2) matches with the mixture spectrum $\mathcal{M}$, and A and B are the theoretical spectra predicted from their peptide sequence P_1 and P_2 respectively, and the constant α in the scoring function indicates the relative abundance of the co-sequenced precursors.

We use the *normalized dot product* of two real-value vectors to measure the spectral similarity in this research. When all the spectra are scaled to Norm 1, the *normalized dot product* will simply reduce to calculate the *cosine* value between two unit vectors. Such measurement scheme considers no special requirements regarding the real peak intensity values, and the spectral similarity is measured based on the shape of the two spectra only. More specifically, we define the following way to convert each spectrum to a real-value vector. Assume that a spectrum $\mathcal{S}$ can be transformed to a real-value vector $V_{\mathcal{S}} = s_1, s_2, ... s_n$, in which each element s_i corresponds to the total intensity of peaks falling into the ith mass bin. The value s_i is calculated as follows:

$$s_i = \sum_{(x_j, h_j) \in \mathcal{S}, x_j \in [(i-0.5)\delta, (i+0.5)\delta]} h_j \tag{1}$$

The bin size δ in the equation above is chosen according to the resolution of the instruments.[3] After the conversion, each vector is normalized to unit vector

---

[2] The molecular mass MW_1 or MW_2 is the neutral mass of the precursor without extra protons attached. We can obtain this value directly from the m/z and charge state z reported by the instruments.

[3] The bin size in Equation 1, and the error bound in the PROBLEM FORMULATION are consistent with each other, both are predetermined by the resolution of the experimental configurations, therefore we use the same symbol δ to denote them.

with each element in the vector divided by its Euclidean Norm, thus the scoring function can be rewritten in the following way:

$$H(\mathcal{M}, A + \alpha B) = \frac{V_{\mathcal{M}} \cdot (V_A + \alpha V_B)}{\|V_A + \alpha V_B\|} \qquad (2)$$

In the equation above, the vector $V_{\mathcal{M}}$, V_A and V_B are all unit vectors. The $\|V_A + \alpha V_B\|$ in the denominator indicates the Euclidean Norm(or Euclidean Distance) of the new vector $V_A + \alpha V_B$, which is the linear combination of two separate unit vectors V_A and V_B.

3   Main Method

Even though the proposed MIXTURE SPECTRA DATABASE SEARCH PROBLEM is formulated in a very simple form, the direct implementation will suffer a major computational disadvantage when considering multiple precursors in one single spectrum. The number of the possible candidate peptide pairs that fall into the required mass error range will be very large. The computational burden for scoring each spectrum against all the possible peptide pairs will make it impractical for applying to large datasets. Moreover, the quadratic explosion in search space will also dramatically increase the chances of false-positive identifications. Under such circumstance, efficient filtration strategy is highly necessary for reducing the search space before scoring and ranking all the candidate peptide pairs.

3.1   Filtration of Database Peptides

In our most recent research, we formulated the problem of peptide *de novo* sequencing from mixture MS/MS spectra mathematically, and proposed a dynamic algorithm to report candidate pairs for each of the query spectrum [20,21]. The algorithm has the ability to provide partially correct, yet useful peptide pairs for a given mixture spectrum. We will be utilizing these incomplete results to screen the peptide pairs acquired from the protein database.

Assume that for some mixture spectrum $\mathcal{M}$, the *de novo* algorithm will output the top-ranked peptide pairs in the following list:

$$L_m = \{(A_1, B_1), (A_2, B_2), ..., (A_m, B_m)\}$$

in which, the subscript m indicates the number of reported pairs, and it can be adjusted if required. Each element (A_i, B_i) in the list contains two individual peptides generated for the two different precursors in the query mixture spectrum respectively.

For each of the molecular mass values MW_1 and MW_2, we filtered the whole protein database to find all the theoretically digested peptide sequences that satisfy the required mass tolerance. We use the following list to include all the tentative peptide sequences after filtered by precursor mass value MW_1:

$$L_n^1 = (R_1, R_2, ..., R_n)$$

in which, each element R_i satisfy the requirement of mass tolerance δ such that $|\|R_i\| + \|H_2O\| - MW_1| \leq \delta$, and n is the number of the filtered sequences. Similarly we put all the possible peptide sequences for the second molecular mass value MW_2 in the following list:

$$L_k^2 = (Q_1, Q_2, ..., Q_k)$$

where for each element Q_j, the inequality $|\|Q_j\| + \|H_2O\| - MW_2| \leq \delta$ also holds.

Assuming that we intend to filter the peptide sequence list L_n^1: Firstly, for each element R_i, we compare it with all of its counterpart sequences in list L_m, that is to compare R_i with each A_j in the *de novo* candidate pairs. In the comparison of two sequences R_i and A_j, we use a special alignment algorithm which takes linear time to count the number of common amino acids $N_c^{(i,j)}$ between them. An example to illustrate the comparison is shown in Fig. 1. Secondly, we calculate an

de novo sequence A_j: [W P] LV [N] YGTR
|| |||
database peptide R_i: [G LI] LV [G G] Y.GTR

Fig. 1. Comparing a *de novo* sequence with a database peptide. The alignment ensures that the mass of the aligned block(letters wrapped by brackets) is equal for both sequences. Although in this example the masses of [WP] and [GLI] are slightly different, we allow a tiny error tolerance δ exist, therefore we treat them as equal in comparison. The number of common amino acids here is $N_c = 6$.

initial score for each R_i according to a *Triplet* $\mathcal{T}^i = (l_{max}^i, l_{sum}^i, m_{num}^i)$ obtained during the comparison in the previous step. In the *Triplet*, the notation l_{max}^i represents the largest $N_c^{(i,j)}$ obtained when comparing the target peptide R_i with some sequence A_j from the *de novo* list L_m. And the notation l_{sum} is calculated as: $l_{sum}^i = \sum_{1 \leq j \leq n, N_c^{(i,j)} \geq 3} N_c^{(i,j)}$ which represents the summation over all the $N_c^{(i,j)}$ larger than or equal to 3. And the notation m_{num}^i denotes how many sequences in the *de novo* list L_m has common amino acids $N_c^{(i,j)} \geq 3$ when aligning to the current peptide R_i. The filtration scoring function we chosen in the research is as follows:[4]

$$S_{ini}(R^i) = \log(l_{sum}^i) * l_{max}^i \tag{3}$$

Thirdly, we rank all the peptides in R_i based on the filtration score. The score calculated for each peptide indicates its likelihood of being a correct precursor, at

---

[4] The filtration scoring function S_{ini} is chosen empirically and can be adjusted. Each of the three values contained in the *Triplet* indicates on some level how much the related peptide should be considered(or correct). Different scoring functions are evaluated based on the *Triplet*, however we have observed similar performance in search space reduction after the filtration procedure. Another acceptable scoring function we have evaluated is $S_{ini}(R^i) = \log\left(\frac{l_{sum}^i}{m_{num}^i}\right) * l_{max}^i$.

least on some level. Similar operations can be carried out for the peptide list L_k^2. After scoring and ranking both L_n^1 and L_k^2, we will select a portion of peptides from each list to form the candidate pairs which will subsequently be matched with the query mixture spectrum. The algorithm for this part is described in Algorithm 1.

Algorithm 1. Filtration: Scoring and Ranking Database Peptides

INPUT: Given mixture spectrum $\mathcal{M}$, the list of *de novo* sequence pairs: $L_m = \{(A_1, B_1), (A_2, B_2), ..., (A_m, B_m)\}$, the database peptide list $L_n^1 = (R_1, R_2, ..., R_n)$ for MW_1, and the database peptide list $L_k^2 = (Q_1, Q_2, ..., Q_k)$ for MW_2.

OUTPUT: Both lists L_n^1 and L_k^2 sorted according to the scoring function S_{ini}

1: Inializing *Triplet* array $\mathcal{T}^1[1, ..., n]$ and $\mathcal{T}^2[1, ..., k]$.
2: Inializing the *Initial Score* array $S_{ini}^1[1, ..., n]$ and $S_{ini}^2[1, ..., k]$.
3: **for** i from 1 to n **do**
4:     **for** j from 1 to m **do**
5:         Calculate $N_c^{(i,j)}$ between database peptide R_i and *de novo* sequence A_j
6:         **if** $N_c^{(i,j)} \geq l_{max}^i$ **then**
7:             $l_{max}^i = N_c^{(i,j)}$
8:         **if** $N_c^{(i,j)} \geq 3$ **then**
9:             $m_{num}^i = m_{num}^i + 1$
10:            $l_{sum}^i = l_{sum}^i + N_c^{(i,j)}$
11:    Calculate the initial score $S_{ini}^1[i]$ for R_i based on Equation 3
12: Similar operations(line 3 to line 11)are carried out for L_k^2 to obtain $\mathcal{T}^2$ and S_{ini}^2.
13: Sort both lists L_n^1 and L_k^2 in decreasing order according to the filtration score.

In line 5 of Algorithm 1, given the fact that both peptide R_i and sequence A_j have very limited length, the time complexity for finding the common amino acids between them can be regarded as a constant. Without losing generosity, assume $n > k$, thus the overall complexity for Algorithm 1 is $O(mn + n \log n)$. m in the complexity denotes the number of candidate pairs reported by the *de novo* procedure, it is adjustable according to real requirements of balancing results accuracy and computational speed. In line 8 of Algorithm 1, we only count the case that the number of common amino acids are larger than 3. because larger number indicates more confidently that those common letters are true matching evidence rather than random hits between the aligned sequences.

After the database peptides are sorted, we calculate the ratio between the first-ranked and second-ranked sequences in each database peptide lists. We use ratio $r = \frac{R_{1st}}{R_{2ed}}$ to determine how many peptides in each list should go through further examination. A relatively larger value of r strongly suggests that the top-ranked peptide has a great chance of being one of the co-fragmented precursors. If $r - 1 \geq \beta$ in which β is a threshold value satisfies $\beta \geq 0$, we only have to take out the very few top-ranked peptides. The threshold we use in this research is $\beta = 0.3$, in case that $r - 1 \geq \beta$ we only select the top $\lfloor \log n \rfloor$ peptides out of the list, otherwise if $r - 1 < \beta$ we will take out all the peptides in the list which has

$m_{num}^i > 0.$[5] Peptides with $m_{num}^i > 0$ means that there is at least one *de novo* sequence that has 3 or more common amino acids with the current database peptide. Thus in case that we can't rely on the ratio r to reduce the list, we will consider all the peptides with such *de novo* matching evidence. This also helps to cut down the number of peptides to be considered in the next step. After that, we will pair up the peptides selected from different lists to form peptide pairs and score each peptide pair against the query mixture spectrum based on Equation 2 to report the best matched pair.

3.2   Estimation of Mixture Coefficient

In the query mixture spectrum $\mathcal{M} = A + \alpha B$, we assume that α is unknown before the identification. It is necessary to give a reasonable estimation of coefficient α prior to scoring the query mixture spectrum against the target candidate pair, because a biased mixture coefficient will compromise the accuracy of calculating the *normalized dot product* between $\mathcal{M}$ and the correct peptides. We use a similar method as [17] to estimate the mixture coefficient. Assume that α' denotes the estimated value of the mixture coefficient, we obtain the optimal value of α' such that the *cosine* similarity between $V_{\mathcal{M}}$ and $V_A + \alpha' V_B$ is maximized. Because vectors $V_{\mathcal{M}}$, V_A and V_B are all normalized to unit vectors, then $V_{\mathcal{M}}^2 = V_A^2 = V_B^2 = 1$. We rewrote the mixture spectra scoring function in Formula 2 as the following function with respect variable α:

$$f(\alpha) = \frac{V_{\mathcal{M}} \cdot (V_A + \alpha V_B)}{\sqrt{1 + 2\alpha V_A \cdot V_B + \alpha^2}}$$

in which $f(\alpha) = \cos \Theta$, and Θ is the angle between the vectors $V_{\mathcal{M}}$ and $V_A + \alpha V_B$.

The function $f(\alpha)$ will have a maximum value at some value α'. In order to achieve this, the first derivative of $f(\alpha)$ with respect to α will be zero at this specific α', meanwhile the second derivative of $f(\alpha)$ at the corresponding α' is negative. To simplify the following derivations, we denote $V_{\mathcal{M}} \cdot V_A = x$, $V_{\mathcal{M}} \cdot V_B = y$ and $V_A \cdot V_B = z$. The first derivative of function $f(\alpha)$ is calculated as:

$$f'(\alpha) = \frac{y\sqrt{1 + 2\alpha z + \alpha^2} - (x + \alpha y)(\alpha + z)(1 + 2\alpha z + \alpha^2)^{-\frac{1}{2}}}{1 + 2\alpha z + \alpha^2}$$

In which, the denominator will always be greater than zero, therefore to make $f'(\alpha) = 0$ is equivalent to let the numerator be zero, then we will have:

$$y\sqrt{1 + 2\alpha' z + \alpha'^2} - (x + \alpha' y)(\alpha' + z)(1 + 2\alpha' z + \alpha'^2)^{-\frac{1}{2}} = 0$$

$$y(1 + 2\alpha' z + \alpha'^2) - (x + \alpha' y)(\alpha' + z) = 0$$

$$y + 2\alpha' yz + \alpha'^2 y - \alpha' x - xz - \alpha'^2 y - \alpha' yz = 0$$

$$y - xz + \alpha' yz - \alpha' x = 0$$

---

[5] The threshold β is also an empirical value in the research. Base on our preliminary experiment on a dataset of limited size, we found that $\beta = 0.3$ is a threshold value large enough to distinguish the first-ranked peptide from all other followers.

From the induction above, we will obtain the following formula to calculate the estimation of mixture coefficient:

$$\alpha' = \frac{y - xz}{x - yz} = \frac{V_{\mathcal{M}} \cdot V_B - (V_{\mathcal{M}} \cdot V_A)(V_A \cdot V_B)}{V_{\mathcal{M}} \cdot V_A - (V_{\mathcal{M}} \cdot V_B)(V_A \cdot V_B)} \qquad (4)$$

Furthermore, to make sure that at this specific point α' the original function $f(\alpha)$ obtains the maximum value, we also need to guarantee that the second derivative of $f(\alpha)$ is negative at α'. We know that $f(\alpha) = \cos\Theta$ will always be positive, thus $\Theta \in (0, \frac{\pi}{2})$, the second derivative of this function is $-\cos\Theta$. In the domain $\Theta \in (0, \frac{\pi}{2})$, it will ways be negative value for $-\cos\Theta$. Thus, we can conclude that the value α' calculated by Equation 4 will make the original function $f(\alpha)$ achieve the maximum value.

3.3   Algorithm for Scoring Peptide Pairs

Given an query mixture spectrum $\mathcal{M}$, our proposed method will firstly conduct a preliminary *de novo* procedure to report a list of sequence pairs, and then those *de novo* results possessing only limited accuracy will be used to filtrate the database peptides acquired from a protein sequence database. After filtration of the database peptides, it is expected that the number of potential peptide sequences that require a more rigorous examination will be reduced. For each peptide sequence in the shortened lists, we use a similar method as [6] to predict its corresponding theoretical spectrum. All the fragment types considered in the theoretical spectrum are denoted in $\Pi = \{y, b, a, c, x, z, y^*, y^o, b^*, b^o\}$. The theoretical spectrum is furthermore converted to a spectrum vector using

---

Algorithm 2. Scoring Peptide Pairs against Query Mixture Spectrum

INPUT: The query mixture spectrum $\mathcal{M}$ and shortened database peptide lists after filtration: $L_x^1 = (R_1, R_2, ..., R_x)$ for precursor molecular value MW_1, and $L_y^i = (Q_1, Q_2, ..., Q_y)$ for precursor molecular value MW_2.
OUTPUT: The best matched peptide pair (R_i, Q_j) and its matching score.
1: Initializing an variable $Score_{max} = 0$, and two indices x_{max} and y_{max}
2: Convert $\mathcal{M}$ to vector and normalize to unit vector $V_{\mathcal{M}}$.
3: **for** i from 1 to x **do**
4:     Predict theoretical spectrum S_{R_i} from sequence R_i
5:     Convert S_{R_i} and normalize to unit vector V_{R_i}
6:     **for** j from 1 to y **do**
7:         Predict theoretical Spectrum S_{Q_j} from sequence Q_j
8:         Convert S_{Q_j} and normalize to unit vectors V_{Q_j}
9:         Estimate coefficient α' based on Equation 4
10:        Calculate the *cosine* value using $\cos = \frac{V_{\mathcal{M}} \cdot (V_{R_i} + \alpha' V_{Q_j})}{\|V_{R_i} + \alpha' V_{Q_j}\|}$
11:        **if** $\cos > Score_{max}$ **then**
12:            $Score_{max} = \cos$
13:            $x_{max} = i$ and $y_{max} = j$
14: Output peptide pair $(R_{x_{max}}, Q_{y_{max}})$ with its matching score $Score_{max}$

Equation 1 and subsequently normalized to a unit vector. Sequences from two individual lists are paired up to constitute the tentative peptide pairs. Then for each of the tentative peptide pairs, we will score it against the query mixture spectrum based on the scoring function in Equation 2, and the best matched pair will be outputted in the final step. The following Algorithm 2 describes the general outline of seeking the best matched pair among all the tentative pairs:

The time consumed in Line 2, 5, 8 for converting a vector to its corresponding unit vector depends on the number of bins considered in the spectra conversion, therefore the complexity is $O(\frac{MW}{\delta})$. The overhead in Line 10 for calculating the *normalized dot product* between $V_{\mathcal{M}}$ and the linear combined vector $V_{R_i} + \alpha' V_{Q_j}$ also relies on the number of bins(or dimensions) in the vector, thus the complexity for this part is the same as above. Meanwhile, we have nested loops iterative with i and j, therefore the integrated complexity for Algorithm 2 is $O(x \times y \times \frac{MW}{\delta})$, in which x and y are denoted in Algorithm 2 representing the size of the shortened database peptide lists respectively.

4   Experimental Result and Discussion

To verify the efficiency of the proposed method, we use the published software package MSPLIT [17] on a tryptic yeast databaset released on PRIDE [22] repository to obtain some real mixture spectra. The MSPLIT software is run in the target/decoy strategy on the yeast dataset collected from an Ion Trap mass spectrometer. In our experiment, we further fitered those reported mixture spectra according to our current requirements. We only select those spectra in which both precursor peptides in the mixture spectra have charge 2, and contain no Post-translational Modifications. Furthermore, we use the renowned database search software PEAKS DB [7] to re-confirm those mixture spectra reported by MSPLIT. Each individual peptide is searched against the protein database with the other peptide sequence removed from the database to eliminate the potential interference between two peptides. Only those mixture spectra in which both peptides can be confirmed by PEAKS DB are kept. In total, we obtained 7 distinct mixture spectra. We implemented a software prototype based on our proposed method and searched those mixture spectra against a yeast protein sequence database. The theoretically enzymatic peptides acquired from the protein sequence database contain both fully-tryptic and semi-tryptic peptide sequences, and also contain the peptide sequences with one missing cleavage. The mass error tolerance considered throughout the experiment is $\pm 0.1 Da$. Our software can successfully identify all the 14 different peptides contained in those mixture spectra above. The experimental results are listed in Table 1.

From Table 1, we can clearly see that the proposed filtration strategy can effectively reduced the number of candidate peptide pairs to be examined. For most of the entries, the reduction ratio f is less than 1‰. Even with the worst case in entry 4, the proposed method can still exclude more than two thirds of the sequence pairs acquired from the protein sequence database after filtration. Another point worth noticing is that in the real mixture spectra, we don't know

Table 1. Preliminary experiment results on a dataset containing 7 mixture spectra. The columns N_b and N_u represent the number of database peptides before and after the filtration procedure for each mass value respectively. The column f shows the ratio between the number of candidate sequence pairs after filtration and the number of all possible peptide pairs acquired directly from the protein database. The column α' is the estimated mixture coefficient. The column $\cos\Theta$ is the score(*normalized dot product*) calculated based on Equation 2.

m/z	Peptides	N_b	N_a	f	α'	$\cos\Theta$
553.321	GLILVGGYGTR	1853	7	0.5831‰	0.827	0.327
553.7795	GPPGVFEFEK	1924	297			
419.724	ASIASSFR	1823	7	0.0204‰	1.101	0.444
420.2585	IAGLNPVR	1319	7			
506.78	FHLGNLGVR	1709	7	0.0156‰	0.917	0.365
507.2820	KFPVFYGR	1837	7			
600.2756	GYSTGYTGHTR	976	6	0.0292‰	1.149	0.321
600.340	DAGTIAGLNVLR	1476	7			
462.706	DNEIDYR	1239	474	267.23‰	1.347	0.343
463.3077	IVAALPTIK	743	519			
521.253	YSDFEKPR	1434	7	0.0182‰	0.996	0.302
521.7932	GAIAAAHYIR	1880	7			
675.364	GKPFFQELDIR	1560	360	1.0147‰	1.847	0.388
675.8437	ANLGFFQSVDPR	1592	7			

which precursor have the higher abundance, thus in estimated value for α' can be either larger than 1 or smaller than 1. The correctness of the identification results demonstrated the effectiveness of the proposed method for matching a mixture spectrum with a pair of database peptides by incorporating a preliminary *de novo* filtration. In our future work, we will evaluate the performance of our proposed method on datasets of different size and different acquisition methods, also we will develop a method for the validation of the identification results.

Acknowledgments. KZ was partially supported by an NSERC Discovery Grant and a Discovery Accelerator Supplements Grant.

Reference

1. Abersold, R., Mann, M.: Mass Spectrometry-based Proteomics. Nature 422(6928), 198–207 (2003)
2. Ma, B., Zhang, K., Hendrie, C., et al.: PEAKS: Powerful Software for Peptide De Novo Sequencing by Tandem Mass Spectrometry. Rapid Commun. Mass Spectrom. 17(20), 2337–2342 (2003)
3. Frank, A., Pevzner, P.: PepNovo: De Novo Peptide Sequencing via Probabilistic Networking Modeling. Anal. Chem. 73(11), 2594–2604 (2001)
4. Taylor, J.A., Johnson, R.S.: Sequence Database Searches via De Novo Peptide Sequencing by Tandem Mass Spectrometry. Rapid Commun. Mass Spectrom. 11(9), 1067–1075 (1997)

5. Perkins, D.N., Pappin, D.J.C., Creasy, D.M., et al.: Probability-based Protein Identification by Searching Sequence Database using Mass Spectrometry Data. Electrophoresis 20(18), 3551–3567 (1999)

6. Eng, J.K., Mccormack, A.L., Yates, J.R.: An Approach to Correlate Tandem Mass Spectral Data of Peptides with Amino Acid Seqeuences in a Protein Database. J. Am. Soc. Mass Spectrom. 5(11), 976–989 (1994)

7. Zhang, J., Xin, L., Shan, B., et al.: PEAKS DB: De Novo Sequencing Assisted Database Search for Sensitive and Accurate Peptide Identification. Mol. Cell. Proteomics 11(4), M111-010587 (2012)

8. Graig, R., Beavis, R.C.: TANDEM: Matching Proteins with Tandem Mass Spectra. Bioinformatics 20(9), 1466–1467 (2004)

9. Geer, L.Y., Markey, S.P., Kowalak, J.A.: Open Mass Spectrometry Search Algorithm. J. Proteome Res. 3(5), 958–964 (2004)

10. Hoopmann, M.R., Finney, G.L.: High-speed Data Reduction, Feature Detection, and MS/MS Spectrum Quality Assessment of Shotugn Proteomics Datasets Using High-resolution Mass Spectrometry. Anal. Chem. 79(15), 5620–5632 (2007)

11. Houel, S., Abernathy, K., Renganathan, K.: Quantifying the Impact of Chimera MS/MS Spectra on Peptide Identification in Large-Scale Proteomics Studies. J. Proteome Res. 9(8), 4152–4160 (2010)

12. Venable, J.D., Dong, M.Q., Wohlschlegel, J., et al.: Automated Approach for Quantitative Analysis of Complex Peptide Mixtures from Tandem Mass Spectra. Nat. Methods 1(1), 39–45 (2004)

13. Chakraborty, A.B., Berger, S.J., Gebler, J.C.: Use of an Integrated MS-multiplexed MS/MS Data Acquisition strategy for High-coverage Peptide Mapping Studies. Rapid Commun. Mass Spectrom. 21(5), 730–733 (2007)

14. Hakansson, K., et al.: Combined Electron Capture and Infrared Multiphoton Dissociation for Multistage MS/MS in a Fourier Transform Ion Cyclotron Resonance Mass Spectrometer. Anal. Chem. 75(13), 3256–3262 (2003)

15. Geiger, T., et al.: Proteomics on an Orbitrap Benchtop Mass Spectrometer using All-ion Fragmentation. Mol. Cell. Proteomics 9(10), 2252–2261 (2010)

16. Zhang, N., et al.: ProbIDtree: An Automated Software Program Capable of Identifying Multiple Peptides from a Single Collision-induced Dissociation Spectrum Collected by a Tandem Mass Spectrometer. Proteomics 5(16), 4096–4106 (2005)

17. Wang, J., Perez-Santiago, J., Katz, J.E., et al.: Peptide Identification from Mixture Tandem Mass Spectra. Mol. Cell. Proteomics 9(7), 1476–1485 (2010)

18. Wang, J., Bourne, P.E., Bandeira, N.: Peptide Identification by Database Search of Mixture Tandem Mass Spectra. Mol. Cell. Proteomics 10(12), M111-010017 (2011)

19. Frank, A.M., Savitski, M.M., Nielsen, M.L.: De Novo Peptide Sequencing and Identification with Precision Mass Spectrometry. J. Proteome Res. 6(1), 114–123 (2007)

20. Liu, Y., Ma, B., Zhang, K., Lajoie, G.: An Effective Algorithm for Peptide de novo Sequencing from Mixture MS/MS Spectra. In: Basu, M., Pan, Y., Wang, J. (eds.) ISBRA 2014. LNCS, vol. 8492, pp. 126–137. Springer, Heidelberg (2014)

21. Liu, Y., et al.: An Approach for Peptide Identification by De Novo Sequencing of Mixture Spectra. IEEE/ACM Trans. Comput. Biol. Bioinform (2015), doi: 10.1109/TCBB, 2407401

22. Vizcaino, J.A., Cote, R.G., et al.: The Proteomics Identification(PRIDE) Database and Associated Tools: Status in, Nucleic Acids Res. 41(D1), D1063–D1069 (2013)

Diploid Alignments and Haplotyping

Veli Mäkinen[✉] and Daniel Valenzuela

Helsinki Institute for Information Technology HIIT
Department of Computer Science
University of Helsinki, Helsinki, Finland
{vmakinen,dvalenzu}@cs.helsinki.fi

Abstract. Sequence alignments have been studied for decades under the simplified model of a consensus sequence representing a chromosome. A natural question is if there is some more accurate notion of alignment for diploid (and in general, polyploid) organisms. We have developed such a notion in our recent work, but unfortunately the computational complexity remains open for such a diploid pair-wise alignment; only a trivial exponential algorithm is known that goes over all possible diploid alignments. In this paper, we shed some light on the complexity of diploid alignments by showing that a haplotyping version, involving three diploid inputs, is polynomial time solvable.

1 Introduction and Related Work

There are myriads of variants of pair-wise sequence alignments trying to capture various biological sequence features, such as mutation biases, repeats (DNA), splicing (RNA), alternative codons (protein) [4,5], but the fundamental feature of a genome of a higher organism being diploid or even polyploid has remained largely unexplored in alignment literature. The closest come some fairly recent approaches in progressive multiple alignment that model a multiple alignment profile as a *labeled directed acyclic graph* (labeled DAG) [7,8]. These works define the alignment of two such labeled DAGs A and B as the problem of finding a path P^A through A and a path P^B through B such that the optimal alignment score of P^A and P^B is maximized. Since a pair-wise alignment models a diploid chromosome pair accurately, giving the synchronization of their haploid sequences, the labeled DAG alignment could be applied to model diploid alignment. However, the caveat is that this approach takes only partial information into account from the diploids, not their full content. It was shown in [9] how to modify the approach into a *covering* version that takes full content of diploids into account. Unfortunately, the computational complexity of this accurate model of diploid alignment remains open.

In this paper, we shed some light on the complexity of diploid alignment by showing that a haplotyping version, involving three diploid inputs, is polynomial time solvable. In addition to the theoretical interest, the haplotyping version may also be of practical value as a complementary technique to haplotype

---

Partially supported by Academy of Finland under grant 284598 (CoECGR).

R. Harrison et al. (Eds.): ISBRA 2015, LNBI 9096, pp. 235–246, 2015.
DOI: 10.1007/978-3-319-19048-8_20

assembly. We give some proof-of-concept simulation results that show excellent performance on realistic input scenarios on an implementation of the approach.

In what follows, we fix our mindset on the haplotyping problem to fix the terminology and to motivate our study also from the practical point of view. Then we formalize the notion of diploid alignments and show how this formalization can be extended to modeling the haplotyping problem.

1.1 Genotyping and Haplotyping

In *diploid organism* a pair of haplotype DNA sequences forms a chromosome pair, where one haplotype is inherited from the mother and one from the father. Each inherited haplotype sequence is a mixture (*recombination*) of the two haplotype sequences forming the corresponding diploid chromosome pair in the parent.

Genotyping consists in the discovery of variants in specific positions in the genome of an individual with respect to a *consensus genome* of the species. After genotyping of child, mother, and father, one can reason which variants came from the mother, which came from the father, and which are new. The inherited variants are called *germ-line variants* and the new *de novo variants*. A *homozygous variant* is inherited from both mother and father, and a *heterozygous variant* is inherited only from one of them.

Haplotyping consists in the assignment of heterozygous variants to the correct *phase*, that is, to a haplotype inherited from the mother or to a haplotype inherited from the father. The importance of this process is not just in revealing the inheritance pattern, but also in understanding the function of each haplotype; after all, genes and other functional units are residing in haplotypes, and the function always depends on the exact sequence content. Genotype information is just enough to argue about the effect of a single mutation, while haplotype information gives the full power of reasoning about the combined effect of a set of mutations.

The state of the art is that genotyping is nowadays a rather routinely conducted process, when studying e.g. human individuals in the hunt of disease causing mutations. It can be conducted by high-throughput sequencing of individual DNA, aligning the sequencing reads to the consensus genome, and analyzing the read alignments for variants supported by many reads. *Single-nucleotide polymorphisms (SNPs)* affecting a single genome position can be revealed with high accuracy, but larger indels and structural variants are much more hard to identify [11].

Given a set of predicted heterozygous variants, haplotyping is still a challenging task, and it is often solved using statistical methods [1,10,2]. Recent advances in pseudo-polynomial algorithms for *haplotype assembly* are however making large-scale haplotyping feasible [12]. In haplotype assembly the j-th read is reduced to a sequence R^j from alphabet $\{*, 0, 1\}$ with $R^j[i] = *$ denoting that the read does not overlap i-th heterozygous variant, with $R^j[i] = 0$ denoting that the read overlaps the i-th heterozygous variant but does not support it, and with $R^j[i] = 1$ denoting that the read overlaps the i-th heterozygous variant and supports it. The task is to assign each read to one of the haplotypes such that minimal flipping of bits inside reads is required to make them uniform with their chosen consensus haplotype pattern. The approach works if the read length is long enough to contain many variants.

The reads can be separated into small independent blocks (i.e. such that there are no shared variants among different blocks) and then solve the problem for each block independently. Some blocks have been identified as particularly difficult to phase [3].

There are also more tailored approaches to haplotyping that combine computational methods with problem-targeted sequencing technology [13].

In this paper, we propose a haplotyping algorithm that works directly at the DNA sequence level, thus differing from previous approaches. Our method is independent of the sequencing technology. We do not require any specific information about reads, so the variants might have been obtained using different methods.

The purpose of our algorithm is to haplotype *complex genome regions*, that may contain long and possibly overlapping variants that are difficult to capture by the haplotype assembly framework. We assume that one has identified complex regions of the child genome and predicted variants from those regions as well as from the same regions in the mother and father genomes. Such data can be produced e.g. by targeted high-coverage sequencing followed by variant prediction. Our haplotyping algorithm takes $O(n^3)$ time, where n is the length of the genome region in question. This approach fits a haplotyping project, where a computationally light approach can be applied on easy-to-haplotype regions, and a more computational heavy approach can be applied on the identified complex regions.

1.2   Alignment of Diploid Individuals

Recently, new alignment models that are designed for diploid organisms, incorporating the possibility of recombination, were introduced [9]. In this section we briefly present these models and some basic definitions.

A *pair-wise alignment* (or simply an *alignment*, when it is clear from the context) of sequences A and B is a pair of sequences (S^A, S^B) such that S^A is a supersequence of A, S^B is a supersequence of B, $|S^A| = |S^B| = n$ is the length of the alignment, and all positions which are not part of the subsequence A (respectively B) in S^A (respectively S^B), contain the *gap* symbol $'-'$, which is not present in the original sequences.

Given a similarity function $s(a, b)$ that assets the similarity between two characters, the similarity of a pair-wise alignment is simply defined as

$$S(S^A, S^B) = \sum_{i=1}^{n} s(S^A[i], S^B[i]).$$

The similarity of two sequences is then defined as

$$S(A, B) = max\{\sum_{1=1}^{n} s(S^A[i], S^B[i]) : (S^A, S^B) \text{ is an alignment of } A \text{ and } B\}.$$

An alignment that achieves that value is called an *optimal alignment*.

We say that $(S^{A'}, S^{B'})$ is a *recombination* of an alignment (S^A, S^B) if both alignments have the same length n and there exists a binary string P (for *phase*) such that $S^{A'}[i] = S^A[i]$ and $S^{B'}[i] = S^B[i]$ if $P[i] = 0$, and $S^{A'}[i] = S^B[i]$ and $S^{B'}[i] = S^A[i]$ if $P[i] = 1$. We say that the characters are *swapped* in the positions in which $P[i] = 1$.

We denote this *recombination relation* by $(S^{A'}, S^{B'})\Re(S^A, S^B)$.

Diploid to Diploid Similarity[9][1]: Given two pair-wise alignments (S^A, S^B) and (S^X, S^Y) the diploid to diploid similarity is the sum of the optimal similarity scores by components, given by the best possible recombination of both individuals. More formally:

$$S_{d-d}((A, B), (X, Y)) = \max\{S(A', X') + S(B', Y') :$$
$$(S^{A'}, S^{B'})\Re(S^A, S^B) \wedge (S^{X'}, S^{Y'})\Re(S^X, S^Y))\}$$

Neither algorithms nor complexity bounds were provided for the similarity above. However, a simpler version where one of the individuals is considered as a diploid and the other as a pair of haploids was developed:

Pair of Haploids to Diploid Similarity[9]: Given a pair-wise alignment (S^A, S^B) and two sequences X and Y, the pair of haploids to diploid distance is defined as the sum of the optimal similarity scores by components, given by the best possible recombination of the diploid individual. More formally:

$$S_{d-hh}((A, B), (X, Y)) = \max\{S(A', X) + S(B', Y) :$$
$$(S^{A'}, S^{B'})\Re(S^A, S^B)\}$$

The best algorithm for this similarity measure runs in $O(n^3)$ time and requires $O(n^2)$ memory. A modified version of the above was also presented, which can be computed in $O(nk)$ time and $O(n)$ memory, where k is the resulting edit distance (considering the analogous problem with min instead of max and costs instead of scores).

2   Haplotype Sequences via Alignment

The measures covered in the previous section intent to measure the similarity between individuals in which heterozygous and homozygous variations are known, but there is no knowledge about the correct phasing of the variations.

As such, they are not useful for haplotype phasing, as the evolutionary recombination pattern is unique to the individual. To take the evolutionary context into account, we need to extend the measures to mother-father-child trios, as it is considered next.

---

[1] The original paper considers the distance measure instead of the similarity, but these are computationally equivalent.

2.1   The Similarity Model

Let us consider three pair-wise alignments, $(S^{M_1}, S^{M_2}), (S^{F_1}, S^{F_2})$ and (S^{C_1}, S^{C_2}) of length L_M, L_F and L_C respectively. Those represent the diploid sequences of the mother, father and child, and we call the three of them a mother-father-child (m-f-c) trio.

We define the *haplotyping similarity of an m-f-c trio*, H(m-f-c), as the maximum pair-wise similarity between one of the sequences of the mother and one of the sequences of the child, plus the pair-wise similarity between the other child sequence and one of the father sequences, *assuming that none of the diploids is phased correctly*. This means that we need to allow free recombination in each pair-wise alignment, to let the model discover the real phase. More formally:

$$H(\text{m-f-c}) = \max\{S(M_1', C_1') + S(F_1', C_2') : (S^{M_1}, S^{M_2})\Re(S^{M_1}, S^{M_2})$$
$$\wedge\,(S^{F_1}, S^{F_2})\Re(S^{F_1}, S^{F_2}) \wedge (S^{C_1}, S^{C_2})\Re(S^{C_1}, S^{C_2})\}$$

Figure 1 shows an optimal alignment of a m-f-c trio.

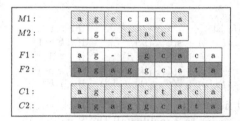
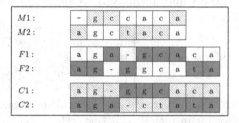

Fig. 1.   On the left we show how the pair-wise alignments would be if we knew the correct phasing of the three individuals. On the right, the same individuals are presented, but the haplotype phasing is not known a priori. We assume a similarity function that scores 1 for equal characters, and -1 for indels and mismatches. The colored recombinations show the sequences M_1' (hatched blue), F_1' (green) from the recombination that gives the optimal alignment. The haplotyping similarity of the trio is $S(\text{AGCTACA}, \text{AGCTACA}) + S(\text{AGAGGCATA}, \text{AGAGGCATA}) = 7 + 9 = 16$. The binary strings associated to the recombinations are $P_M = 1001110$, $P_F = 110100011$ and $P_C = 000111000$. Note that the latter corresponds to the predicted phase for the child genome. It is also important to note that none of the binary strings are signaling evolutionary recombinations, but they are signaling phasing errors in the input data.

Notice that this similarity measure is an extension of the diploid to pair of haploids similarity, but it is easier than the diploid to diploid similarity measure. The feasibility of the solution comes from the fact that only the child genome needs to be *covered* by the alignment: Sequences induced by $S^{C_1'}$ and $S^{C_2'}$ are both aligned in the H(m-f-c) definition, whereas only sequences induced by $S^{M_1'}$ and $S^{F_1'}$ are aligned from the recombined father and mother sequences, respectively. The difficulty of the diploid to diploid measure lies in the requirement of covering both recombined inputs, which appears difficult to capture at least by dynamic programming. For the m-f-c trio case, we can luckily extend the cubic solution of the diploid to pair of haploids similarity.

2.2 Dynamic Programming Algorithm

In this section we present our algorithm to compute the m-f-c similarity. We propose a dynamic programming formulation that computes values $H_{i,j,k,m,f,c}$ with $i \in \{1, L_M\}$, $j \in \{1, L_F\}$, $k \in \{1, L_C\}$, $m \in \{1, 2\}$, $f \in \{1, 2\}$ and $c \in \{1, 2\}$. The value stored in $H_{i,j,k,m,f,c}$ stands for the similarity score between $(S^{M_1}[1, i], S^{M_2}[1, i])$, $(S^{F_1}[1, j], S^{F_2}[1, j])$, and $(S^{C_1}[1, k], S^{C_2}[1, k])$, with the additional constrain that the last character of the mother alignment is swapped if and only if $m = 1$, the last character of the father alignment is swapped if and only if $f = 1$ and the last character of the child alignment is swapped if and only if $c = 1$.

We first consider the particular case when the input alignments contain no gaps. That is, $S^{C_1} = C1$, $S^{C_2} = C_2$, $S^{F_1} = F_1$, etc. It is possible to compute those values recursively as follows:

$$H_{i,j,k,m,f,c} = \max \begin{cases} H_{i-1,j,k,*,f,c} & + s('-', M_m[i]) & \text{if } i > 1 \\ H_{i-1,j,k-1,*,f,*} & + s(C_c[k], M_m[i]) + s(C_{c\oplus1}[k], '-') & \text{if } i, k > 1 \\ H_{i,j-1,k,m,*,c} & + s('-', F_f[j]) & \text{if } j > 1 \\ H_{i,j-1,k-1,m,*,*} & + s(C_c[k], '-') + s(C_{c\oplus1}[k], F_f[j]) & \text{if } j, k > 1 \\ H_{i-1,j-1,k-1,*,*,*} & + s(C_c[k], M_m[i]) + s(C_{c\oplus1}[k], F_f[j]) & \text{if } i, j, k > 1 \end{cases}$$

The recurrence uses several short-hand notations as follows. With $H_{i,j,k,*,*,*}$ we mean $\max_{\{m,f,c\} \in \{1,2\}^3} \{H_{i,j,k,m,f,c}\}$ in order to consider all the 8 valid subproblems where the previous last characters could have been swapped or not (and analogously when only one or two $*$ symbols are present). With $c \oplus 1$ we mean 2 if $c = 1$ and 1 otherwise.

The first and third cases correspond to the scenarios where the last character of the mother (respectively, of the father) is not aligned with any character of the child, and therefore a gap symbol is inserted. The second and fourth cases corresponds to the scenarios where one of the last characters of the child is aligned with one of the last characters of the mother (respectively, of the father), and the other character of the child is not aligned, therefore, a gap is inserted. The fifth case is the scenario where the last character of one of the child sequences is aligned with one of the last sequences of the mother, and the last character of the other child sequence is aligned with the last character of one of the sequences of the father.

The correctness of the algorithm can be seen as a generalization of the classic dynamic programming algorithm: Firstly, all the possibilities of alignment among the last characters of the input sequences are considered. For each of those case, it remains the subproblem of the m-f-c alignment where the characters that had just been aligned (with another character or with a gap) are removed. For the alignment to be optimal, it is required that the subproblem is solved optimally too, and therefore, the recursion holds true.

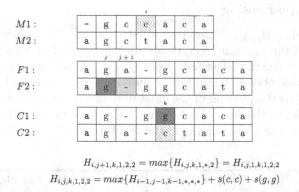

$$H_{i,j+1,k,1,2,2} = max\{H_{i,j,k,1,*,2}\} = H_{i,j,1,k,1,2,2}$$
$$H_{i,j,k,1,2,2} = max\{H_{i-1,j-1,k-1,*,*,*}\} + s(c,c) + s(g,g)$$

Fig. 2. Example showing two steps of the dynamic programming algorithm. First for the computation of $H_{i,j+1,k,1,2,2}$ we highlight the characters that need to be considered. As the $j + 1$ character of the father sequence that is being considered is a gap, the recursion returns the previous value of j, keeping all the parameters constant, except for the sequence of the father that can be considered (line 10 of Algorithm 1.) For the computation of $H_{i,j,k,1,2,2}$ the previous values indicated by lines 6,7,11,12, and 14 needs to be considered. Those correspond to all possible combinations of alignments between the highlighted characters (allowing some of them to be ignored, but not all of them).

Notice that we are allowing free recombinations when a path change its value in either m, f, or c indexes. It is straightforward to include a penalty in those changes of indexes, as it was proposed in [9], however, we decided to stay free of such penalties for reasons that are discussed in Section 3.

It remains to consider the scenarios where the input sequences do have gaps. Observe that the gaps in the input alignments need to be ignored without any cost in order to model the similarity measure correctly; the gaps in the input sequences are just required for keeping the positions of the two haplotypes of the diploid synchronized so that recombination can be modeled. Algorithm 1 shows our pseudo-code to handle this: If for a given configuration the last character of the mother is a gap, we can immediately resort to the value computed for the position $i - 1$. This just ignores that gap character. The case for the father is handled analogously. When the gap character comes in one of the child sequences, it is handled implicitly by the recursion, given that $s('-','-') = 0$ allows the gap character from the child to be ignored through the second and fourth cases of the recurrence. Figure 2 simulates one step of the computation.

The straightforward implementation as in Algorithm 1 would require $O(n^3)$ time and memory, as we need to retrieve the phasing. We implemented the check-point method [14], a flexible variant of Hirschberg's algorithm [6] that allows our algorithm to run in $O(n^3)$ time and $O(n^2)$ memory. Still, in order to obtain a scalable method, it is possible to apply some heuristics [9].

2.3   Haplotype Phasing

Once all the values $H_{i,j,k,m,f,c}$ are computed, it is enough to trace back the path that originated the optimal score to obtain the optimal alignment. We notice that if we collect the three last indexes m, f and c from the path we will obtain the recombination binary strings for the mother, father, and child. In particular, the latter gives us the phasing of the child diploid that maximizes the similarity of the m-f-c trio. In the example of Figure 1 the binary strings are P_M, P_F and P_C; the last one being the phasing of the child diploid.

Algorithm 1. Haplotyping similarity of an m-f-c trio. The algorithm corresponds to the dynamic programming implementation of the recurrence presented in Section 2.2, modified to handle the gaps in the input sequences properly.

```
 1:  function HAPLOIDSIMILARITY(M₁, M₂, F₁, F₂, C₁, C₂)
 2:
 3:      H[0, 0, 0, *, *, *] ← 0
 4:      SetGlobal(H,M1,M2,F1,F2,C1,C2)
 5:      for i ← 0 to Lₘ do
 6:          for j ← 0 to Lₘ do
 7:              for k ← 0 to Lₘ do
 8:                  for m ← 1 to 2 do
 9:                      for f ← 1 to 2 do
10:                          for c ← 1 to 2 do
11:                              H[i, j, k, m, f, c] ← HValue(i,k,k,m,f,c)
12:      return max H[Lₘ, L_f, L_c, *, *, *]
```

$$
\begin{aligned}
&1:\ \textbf{function } \text{HVALUE}(i, j, k, m, f, c)\\
&2:\quad value \leftarrow -\inf\\
&3:\quad \textbf{if } i > 0 \textbf{ then}\\
&4:\quad\quad \textbf{if } M_m[i] =' -' \textbf{ then}\\
&5:\quad\quad\quad \textbf{return } \max\{H[i-1, j, k, *, f, c]\}\\
&6:\quad\quad value \leftarrow \max\{value, H[i-1, j, k, *, f, c] + s('-', M_m[i])\}\\
&7:\quad\quad value \leftarrow \max\{value, H[i-1, j, k-1, *, f, *] + s(C_c[k], M_m[i]) + s(C_{c\oplus 1}[k], '-')\}\\
&8:\quad \textbf{if } j > 1 \textbf{ then}\\
&9:\quad\quad \textbf{if } F_f[i] =' -' \textbf{ then}\\
&10:\quad\quad\quad \textbf{return } \max\{H[i, j-1, k, m, *, c]\}\\
&11:\quad\quad value \leftarrow \max\{value, H[i, j-1, k, m, *, c] + s('-', F_f[j])\}\\
&12:\quad\quad value \leftarrow \max\{value, H[i, j-1, k-1, m, *, c] + s(C_c[k], '-') + s(C_{c\oplus 1}[k], F_f[j])\}\\
&13:\quad \textbf{if } j > 1\ \&\ > 1 \textbf{ then}\\
&14:\quad\quad value \leftarrow \max\{value, H[i-1, j-1, k-1, m, *, c] + s(C_c[k], M_m[i]) + s(C_{c\oplus 1}[k], F_f[j])\}\\
&15:\quad \textbf{return } value
\end{aligned}
$$

---

3   From Variants to Unphased Diploid Genome

Our phasing algorithm assumes the inputs as pair-wise alignments representing unphased diploid genomes. These can be constructed as follows. After sequencing the target region (or whole genome) on each individual involved, one can align the reads and analyse the variants [11]. All the predicted homozygous variants on an individual can be *applied* to the consensus genome to produce a base S of a pair-wise alignment; with applying we mean that the content of consensus is replaced by the variants. Then the heterozygous variants can be greedily applied to S from left to right such that if a variant overlaps a previously applied variant, it is not applied. This forms the non-gapped content T of the top row of a pair-wise alignment. The remaining set of heterozygous variants are applied to S to produce the non-gapped content B of the bottom row of a pair-wise alignment. Finally, enough gaps are added such that T and B are synchronized according to their origin in S. This process produces an unphased representation of a diploid genome. It is important to notice that our algorithm is invariant to the phasing of the input variants: In Figure 1 it is shown two different inputs that correspond to the same variants; in the left the input is already correctly phased for the trio, and in the right the variants are incorrectly phased, and the result in both scenarios is the same. This is possible because we allow free recombination in each of the diploids, thus giving an equal opportunity for each variant to be phased either way.

We shall consider in Sect. 5 the case when the overlap depth is higher, so that some variants are left after constructing B.

4   Experimental Results

We implemented our method fully in $C++$, and the source code is freely available[2]. We ran our experiments in a computer node with 2 Intel Xeon E5540 2.53GHz processors, 32GB of RAM. The operating system was Ubuntu 12.04.4. Our code was compiled with gcc 4.6.4, optimization option $-O3$.

To study the difficult areas to phase we simulated our father-mother-child trios directly without adding the variant analysis step of Sect. 3. In this way we could control the amount the mutation ratio, the type of variations, and the measurement/predictions errors that are present in the input data.

To simulate the mother sequences, we started from two identical copies of a sample from human chromosome 21. We inserted different types of variations, and then we simulated a recombination of that pair-wise alignment to obtain the child chromosome that is inherited from the mother. For the recombination process we choose recombination points at random. We did analogously to simulate the father, and the chromosome that the child inherits from the father.

The variations planted on the parents were as follow: *point mutations* consisted of SNPs and single nucleotide deletions. Then we introduced *long indels* (larger than 50 base pairs). In the case of insertions, those consisted of random base pairs. We also inserted *short tandem repeats*[15] consisting of insertions of length between 20 and 60 repeating a sequence between 2 and 6 base pairs. After the recombination has been simulated to generate the child sequences, we introduced *de novo point mutations*. (We also considered *short indels* but as the results turned out to be analogous to long indels, we omitted them from the results reported here for the lack of space.)

In addition to all the previous parameters, we also introduced *random errors* over all the sequences at the end, to take into account errors during the

---

[2] http://www.cs.helsinki.fi/u/dvalenzu/code/haplotyping/

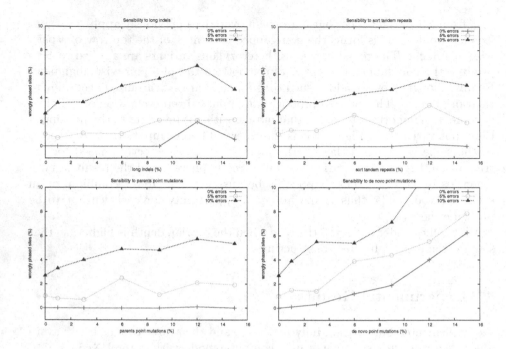

Fig. 3. We show the percentage of incorrectly phased positions versus the mutation ratio. For every graphic we include three levels of random noise. The two first plots (above) show the behavior against long indels and short tandem repeats. The third plot (below left) shows the behavior against point mutations in the parents and the fourth plot (below right) shows de novo variants.

sequencing of the sequences and during variant calling to discover the genotype patterns that constitute the inputs of our algorithm.

We studied the quality of our phasing algorithm by measuring how sensible it was with respect to each of the parameters. For that sake, we made our simulations from a 1000 bp sequence and we measured the percentage of positions that were incorrectly phased. For each different type of variation, the ratio ranged between 0 and 15%. Regarding to the error ratios we considered three scenarios: a very optimistic one, were the genotyping was done without errors (0%), a moderated scenario where error ratio is 5%, and a pessimistic scenario where the error ratio is 15%. The time used for haplotyping each simulated trio was less than a hour. The results are shown in Figure 3.

5   Discussion

Our case study and experimentation on the haplotyping problem show that our method can provide a complementary technique to perform haplotype phasing in complex genome sequences that are too difficult for haplotype assembly modeling. Our experiments are so far showing the proof-of-concept, and several aspects need to be taken into account in order to apply the method on the real data setting of Sect. 3. As the motivation for our approach is complex genome regions, it should be observed that just the detection of variants in such areas is challenging. It can happen that variant predictions overlap such that at some positions a diploid genome is not enough to cover all variants [16]. For the process in Sect. 3, this means that after constructing T and B for the content of an unphased diploid genome, there are still some heterozygous variants to be applied. Consider continuing the process further to create a *multiple alignment* with some small number c of sequences for child genome, m for mother genome, and f for father genome. Our dynamic programming approach can be extended to this scenario, by considering all $\binom{c}{2}$ pairs of rows from child multiple alignment at each column to be aligned to m possible rows in mother multiple alignment at column i and f possible rows in father multiple alignment at column j. We plan to implement this scheme so as to compare our approach to other haplotyping methods.

Our liberal model of allowing crossover at every position can be made more restrictive by exploiting the connection to labeled DAGs we already discussed in Sect. 1: Consider the alignment visualization in Fig. 1. This can be viewed as a labeled DAG, by interpreting each cell as a vertex, and drawing an arc from bottom cell to its neighbor on the right and to its neighbor on its top right, and symmetrically for top cells. Each vertex has then two outgoing arcs except for the two last vertices. With some existing local haplotype information, some crossovers can be forbidden by removing non-horizontal arcs. The problem to be solved becomes that of finding two paths C_1 and C_2 through the child DAG, a path M_1 through mother DAG, and a path F_1 through father DAG, such that $S(C_1, M_1) + S(C_2, F_1)$ is maximized. Extending our dynamic programming approach to this generalization is left as future work, but we think this is feasible. On a more direct extension, it is straightforward to include a penalty cost for each recombination in our equations in order to avoid overfitting.

Finally, the main objective of this study is to illustrate that sequence alignments can be extended to take the full content of diploid chromosome representations into account, and that meaningful alignment problems under this model can be stated and solved in polynomial time. Given the unknown complexity of the very basic diploid alignment problem on two diploid inputs and the connection to covering problems on labeled DAGs, the current study is probably only scratching the surface of a prominent subarea of research.

References

1. Browning, S.R., Browning, B.L.: Haplotype phasing: existing methods and new developments. Nature Reviews Genetics 12(10), 703–714 (2011)
2. Chen, W., Li, B., Zeng, Z., Sanna, S., Sidore, C., Busonero, F., Kang, H.M., Li, Y., Abecasis, G.R.: Genotype calling and haplotyping in parent-offspring trios. Genome Research 23(1), 142–151 (2013)

3. Zhi-Zhong Chen, Fei Deng, and Lusheng Wang. Exact algorithms for haplotype assembly from whole-genome sequence data. Bioinformatics, btt349 (2013)
4. Durbin, R., Eddy, S.R., Krogh, A., Mitchison, G.: Biological Sequence Analysis: Probabilistic Models of Proteins and Nucleic Acids. Cambridge University Press (1998)
5. Gusfield, D.: Algorithms on Strings, Trees and Sequences: Computer Science and Computational Biology. Cambridge University Press (1997)
6. Hirschberg, D.S.: A linear space algorithm for computing maximal common subsequences. Communications of the ACM 18(6), 341–343 (1975)
7. Lee, C., Grasso, C., Sharlow, M.F.: Multiple sequence alignment using partial order graphs. Bioinformatics 18(3), 452–464 (2002)
8. Löytynoja, A., Vilella, A.J., Goldman, N.: Accurate extension of multiple sequence alignments using a phylogeny-aware graph algorithm. Bioinformatics 28(13), 1684–1691 (2012)
9. Mäkinen, V., Valenzuela, D.: Recombination-aware alignment of diploid individuals. BMC Genomics 15(suppl. 6), S15 (2014)
10. Marchini, J., Cutler, D., Patterson, N., Stephens, M., Eskin, E., Halperin, E., Lin, S., Qin, Z.S., Munro, H.M., Abecasis, G.R., et al.: A comparison of phasing algorithms for trios and unrelated individuals. The American Journal of Human Genetics 78(3), 437–450 (2006)
11. Pabinger, S., Dander, A., Fischer, M., Snajder, R., Sperk, M., Efremova, M., Krabichler, B., Speicher, M.R., Zschocke, J., Trajanoski, Z.: A survey of tools for variant analysis of next-generation genome sequencing data. Briefings in Bioinformatics 15(2), 256–278 (2014)
12. Patterson, M., Marschall, T., Pisanti, N., van Iersel, L., Stougie, L., Klau, G.W., Schönhuth, A.: WHATSHAP: Haplotype assembly for future-generation sequencing reads. In: Sharan, R. (ed.) RECOMB 2014. LNCS, vol. 8394, pp. 237–249. Springer, Heidelberg (2014)
13. Peters, B.A., Kermani, B.G., Sparks, A.B., Alferov, O., Hong, P., Alexeev, A., Jiang, Y., Dahl, F., Tang, T., Haas, J., et al.: Accurate whole-genome sequencing and haplotyping from 10 to 20 human cells. Nature 487(7406), 190–195 (2012)
14. Powell, D.R., Allison, L., Dix, T.I.: A versatile divide and conquer technique for optimal string alignment. Information Processing Letters 70(3), 127–139 (1999)
15. Weber, J.L., Wong, C.: Mutation of human short tandem repeats. Human molecular genetics 2(8), 1123–1128 (1993)
16. Wittler, R.: Unraveling overlapping deletions by agglomerative clustering. BMC Genomics 14(S-1), S12 (2013)

Structural Comparative Analysis of Ecto- NTPDase Models from *S. Mansoni* and *H. Sapiens*

Vinicius Schmitz Nunes[1(✉)], Eveline Gomes Vasconcelos[2],
Priscila Faria-Pinto[2], Carlos Cristiano H. Borges[1], and Priscila V.S.Z. Capriles[1]

[1] Instituto de Ciências Exatas, Universidade Federal de Juiz de Fora, Juiz de Fora,
MG, Brazil
[2] Instituto de Ciências Biológicas, Universidade Federal de Juiz de Fora,
Juiz de Fora, MG, Brazil
vschmitzn@gmail.com, eveline.vasconcelos@ufjf.edu.br,
priscila.faria@ufjf.edu.br, cchborges@ice.ufjf.br,
priscila.capriles@ufjf.edu.br

Abstract. The control of extracellular nucleoside concentrations by Extracellular Nucleoside Triphosphate Diphosphohydrolases (Ecto-NTPDases) is essential in the regulation of the purinergic signalling and also in immune response. In mammals, four isoforms of Ecto-NTPDases have been described (NTPDase1-3 and NTPDase8). The isoform 1 of human Ecto-NTPDase (HsNTPDase1 or CD39) is expressed in endothelial cells of veins and arteries. An Ecto-NTPDase have been identified in the tegument of adult worms of *Schistosoma mansoni* (SmNTPDase1), and it was located on the outer surface of parasite's tegument. Due to the location of the SmATPDase1, it was proposed that these Ecto-NTPDase participate in the evasion of the host immune system by parasite. These assumptions reinforce the importance of researching the SmATPDase1 as a drug target candidate for the schistosomiasis treatment. In this work, we propose the three-dimensional structure model of the enzymes SmATPDase1 and CD39 using comparative modeling. The results show similarities between these proteins, especially in the active site region, and become necessary to search for alternative binding site of drugs aiming new therapies for schistosomiasis.

Keywords: ATP diphosphohydrolases · *Schistosoma mansoni* · Three-dimensional models · Structural comparison

1 Introduction

Schistosomiasis is a neglected disease caused by helminthes of the genus *Schistosoma* [3]. In 2009, there were 239 million people infected by the parasite in the world [13]. This disease occurs predominantly in tropical countries of Africa, Southeast Asia, South America, Middle East and the Caribbean [11]. In Brazil, the specie responsible for schistosomiasis is *Schistosoma mansoni*, with an estimate of 4-6 million people infected [12].

© Springer International Publishing Switzerland 2015
R. Harrison et al. (Eds.): ISBRA 2015, LNBI 9096, pp. 247–259, 2015.
DOI: 10.1007/978-3-319-19048-8_21

In humans, the adult parasite lives in blood vessels [17], and the main form of treatment is the praziquantel administration, though the literature has reported cases of parasite resistance to this drug [6]. Consequently, the Schistosomiasis Research Agenda advises about the importance to detect proteins that could be considered as new putative drug targets [3,24].

ATP diphosphohydrolases (EC 3.6.1.5), also known as NTPDases (Nucleoside Triphosphate Diphosphohydrolases) or apyrases, are enzymes that hydrolyze nucleoside di- and triphosphates in the presence of bivalent ions. A characteristic shared between these enzymes is the occurrence of five conserved regions called ACR (Apyrase Conserved Regions) [18,14,19].

Two isoforms of *S. mansoni* ATP diphosphohydrolase (SmATPDase1 and 2) were previously described both of molecular weight around 63 kDa and localized on the external surface of tegument from adult schistosome [21,22]. They are also present in other life cycle stages of the parasite, as eggs and miracidia [8].

Molecular studies showed that SmATPDase1 is possibly anchored to the outer surface of adult worms through two transmembrane (TM) domains [5], being the second most expressed protein in the tegument [1]. Results [4] showed that in the suppression of enzymes in the tegument of adult worms, involved in the catalysis of ATP and ADP, SmATPDase1 was the only that presented a reduction in the ability to hydrolyze extracellular ATP and ADP.

Due to the catalytic properties and location of SmATPDase1, and the importance of di- and triphosphate nucleotides in the activation of haemostatic and immune system cells, it was suggested that this isoform is involved in the regulation of the concentrations of nucleotides surrounding the parasite contributing with its evasion [21,22,5,15]. The authors point out the SmATPDase1 as a drug target candidate for the schistosomiasis treatment.

In mammals, eight isoforms (NTPDases1-8) have been described in numerous cells and tissues. They are divided into three groups: (i) called Ecto-NTPDases, this group is formed by NTPDase1-3 and 8 that can be found on the outer surface of the cell and anchored to the plasma membrane by two TM domains, (ii) group of intracellular forms (NTPDase5 and 6) that can be secreted after heterologous expression, (iii) group consisted by NTPDase4 and 7, forms associated with membrane of organelles [18,14,19].

Ecto-NTPDases of mammalian cells act in the regulation of extracellular concentrations of nucleotides, which are important chemical signals stimulating P2 type purinergic receptors, in many physiological processes such as blood coagulation, cell proliferation, inflammation, and immune response [18,19]. The human NTPDase1 (HsNTPDase1 or CD39) is expressed in endothelial cells of veins and arteries, some immune cells such as B cells and dendritic cells [18,19].

In this work, we propose the first three-dimensional (3D) structure models for related proteins SmATPDase1 and CD39, and a comparative structural analysis performed aiming to find differences between them. This work can help the further investigations about structure-based drug design studies for schistosomiasis treatment, aiming an inhibitor of SmATPDase1 that does not interfere with the CD39, causing less or none side effects during treatment in humans.

2 Methodology

2.1 Template Searching

We ran BlastP on the PDB website (using the default parameters and as cutoff e-value≤ 0.001), aligning deposited proteins and target sequences (SmATPDase1 and CD39). Templates for 3D model construction were selected based on amino acid sequence identity, similarity and the presence of ligands. We could not identify templates (neither fragments) for TM domains of both enzymes. To solve this problem, we constructed TM regions applying structural restrictions during comparative modeling.

2.2 Detection of Transmembrane Domain

The prediction of TM domains was performed using the programs Phobius, TMHMM and TMpred. In order to verify the consensus between target sequences and the secondary structure (SS) from templates, we used the programs PSIPREP, Jpred3 and JUFO, using default parameters.

TMhit and TMhhcp programs were used to predict transmembrane helices contacts between TM1 and TM2. All contacts were restricted to a distance of 8Å between carbons-α and a standard deviation of 0.5Å. Significative contacts were defined with score $s \geq 0.5$ for TMhhcp and probability $p \geq 0.8$ for TMhit.

2.3 Three-Dimensional Protein Model Construction

Multiple sequence alignment between the amino acid sequences of protein targets and templates were performed using ClustalΩ program. For generation of the models we used the Modeller program, with applying restrictions of SS and contact between TM1 and TM2.

2.4 SAS, Volume and Electrostatic Profile

We evaluated overall and catalytic site differences between the electrostatic profile, volume and solvent accessible surface area (SASA) of SmATPDase1 and CD39. The analyses of the overall volume and SASA were performed using the plugin VolArea in the VMD program and CASTp program was used for the active sites. The calculation and analyses of the electrostatic potentials were performed using the programs PDB2PQR v1.9 and APBS v1.4.1.

3 Results and Discussion

3.1 NTPDases's Phylogeny of Mammals and Parasites

Figure 1 presents phylogenetic relationships between NTPDase of mammals and parasites. The phylogeny were made in ClustalΩ program, using the neighbor-joining method, and the Bootstrap was calculated in MEGA v6.0 program. Relationships between NTPDases of these organisms may provide clues about the function of NTPDase of parasites not shown yet [19]. It is possible to see that the mammalian Ecto-NTPDases and parasite NTPDases are in separate clades.

Additionally, it is interesting to note that SmATPDase1 and NTPDase from *Plasmodium falciparum* (PfNTPDase) have a higher phylogenetic relationship with the mammalian Ecto-NTPDase than with NTPDase of parasites.

The phylogenetic relationship becomes more interesting, taking into account: (i) the occurrence of two TM in SmATPDase1 and mammalian Ecto-NTPDases, in special CD39, (ii) the location on the cell surface, anchored on the outer surface of plasma membrane through TM, (iii) because the location, has been proposed a probable involvement of SmATPDase1 inhibition of platelet aggregation [21].

With respect to *P. falciparum*, was demonstrated that platelets are able to kill the parasite within the erythrocyte and this effect was abrogated in the presence of platelet inhibitors, including ADPase [16]. The presence of a putative NTP-Dase gene in the *P. falciparum* genome and the prediction of two TM domains, suggest that this enzyme is present in the parasite surface acting against platelet activation [19].

3.2   Templates Selection

The structure of mammalian Ecto-NTPDase is characterized by an extracellular domain (ECD) where is located the ACR and the active site, and two TM domains located near the Nter and Cter [10,18,14]. Recently, the crystallographic structure of the ECD domain of both NTPDase1 and NTPDase2 of *Rattus norvegicus* (RnNTPDase1 and RnNTPDase2) were published [26,27].

We select as templates the structure of RnNTPDase1 (3ZX3) [27] and Rn-NTPDase2 (3CJA) [26], this last one due to the presence of ATP analogue. Sequence identity (and similarity) between SmATPDase1 (UniProtID: Q7YTA4) and 3ZX3 and 3CJA were 37%(53%) and 33%(49%), respectively. Whereas for the CD39 (UniProtID: P49961) these values were 74%(85%) and 46%(62%). In this work, the structure 3CJA was used as reference in the identification of catalytic residues, ligand and water molecules in the active site.

3.3   Prediction of Transmembrane Domains

The analyses with TM, SS and contact helix predictors confirmed the presence of two TM in both models. The results of predictors can be seen in Table 1. With respect to contacts between TMs were used the following pairs of contacts: (i) SmATPDase1 - 40-533, 43-520, 43-524, 45-526 and 53-522, (ii) CD39 - 22-488 and 36-477.

It has been shown that the mobility of TM regulates the active site of CD39 [10]. The removal of one or both TM reduced by 90% the enzymatic activity, indicating that the enzyme activity depends on both transmembrane domains [9]. Both helices display a high degree of rotational movement, and this movement is regulated by the binding of substrate [9,10].

Da'dara and coworkers (2014) demonstrated that the lack of both TM domains compromises the ATPase and ADPase activities of SmATPDase1. They concluded that TM domains help to maintain the protein in an enzymatically favorable conformation [4].

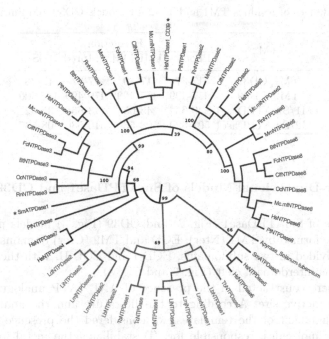

Fig. 1. Phylogenetic tree constructed with amino acid sequences of mammalian and parasites NTPDases. (green) Ecto-NTPDases of mammals, (red) NTPDases of parasites, (brown) intracellular human isoforms, (purple) secreted isoforms of humans and *S. mansoni*, and apyrase of *Solanum tuberosum*. The SmATPDase1 and CD39 (HsNTPDase1) are in asterisk. The phylogeny were made in ClustalΩ program, and edited in the FigTree program. The Bootstrap values are marked in black. The NCBI access number of each sequence used is following listed: Rattus norvegicus (RnNTPDase1: PDBID 3ZX3, RnNTPDase2: PDBID 3CJA, RnNTPDase3: NP_835207.1, RnNTPDase8: NP_001028737.1); *Mus musculus* (MmNTPDase1: NP_033978.1, MmNTPDase2: NP_033979.2, MmNTPDase8: AAH31143.2); *Bos taurus* (BtNTPDase1: XP_005225502.1, BtNTPDase2: DAA24062.1, BtNTPDase3: XP_005222451.1, BtNTPDase8: NP_001071395.1); *Oryctolagus cuniculus* (OcNTPDase3: XP_002713133.1, OcNTPDase3: XP_002723105.1); *Felis catus* (FcNTPDase1: XP_006938097.1, FcNTPDase3: XP_003992264.1, FcNTPDase8: XP_003996175.1); *Canis lupus familiaris* (ClfNTPDase1: XP_005637614.1, ClfNTPDase2: XP_548362.1, ClfNTPDase3: XP_542723.3, ClfNTPDase8: XP_003435393.1); *Pan troglodytes* (PtNTPDase1: XP_003312746.1, PtNTPDase3: XP_001135726.1, PtNTPDase8: XP_001140109.1); *Macaca mulatta* (Mc.mlNTPDase1: XP_001100060.1, Mc.mlNTPDase2: XP_001117802.2, Mc.mlNTPDase3: NP_001253677.1, Mc.mlNTPDase8: XP_001087078.1); *Homo sapiens* (CD39 or HsNTPDase1: NP_001767.3, HsNTPDase2: NP_982293.1, HsNTPDase3: NP_001239.2, HsNTPDase4: NP_004892.1, HsNTPDase5: NP_001240.1, HsNTPDase6: NP_001238.2, HsNTPDase7: NP_065087.1, HsNTPDase8: NP_001028285.1); *Plasmodium falciparum* (PfNTPDase: XP_001348471.2); *Trypanosoma cruzi* (TcNTPDase: AAS75599.1); *Trypanosoma brucei* (TcNTPDase1: XP_847211.1, TcNTPDase2: XP_845817.1); *Leishmania major* (LmjNTPDase1: XP_001681917.1, LmjNTPDase2: XP_001681345.1); *Leishmania infantum* (LiNTPDase1: XP_001464341.1, LiNTPDase2: XP_001463665.1); *Leishmania braziliensis* (LbNTPDase1: XP_001562178.1, LbNTPDase2: XP_001562788.1); *Leishmania donovani* (LdNTPDase1: CBZ32820.1, LdNTPDase2: CBZ32136.1); *Leishmania mexicana* (LmxNTPDase1: CBZ25018.1, LmxNTPDase2: CBZ24328.1); *Solanum tuberosum* (Apyrase: P80595) and *Schistosoma mansoni* (SmATPDase1: AAP94734.1, SmNTPDase2: ABI79456.1).

Table 1. Ranges of residues TM1 and TM2 for models CD39 and SmATPDase1

TM	Prediction			
	Phobius	TMHMM	TMpred	SS
TM1/CD39	17-37	17-39	17-37	12-38
TM2/CD39	477-499	477-499	479-499	477-500
TM1/SmATPDase1	40-65	43-65	43-65	35-62
TM2/SmATPDase1	507-529	508-530	510-530	508-537

3.4 Three-Dimensional Models of SmATPDase1 and CD39

3D structure of SmATPDase1 (Fig. 2) and CD39 (Fig. 3) models proposed in this work are formed by TM1 (Nter), ECD and TM2 (Cter) domains. The ECD domain is divided in two subdomains, ECD-I and ECD-II, with the active site located in the interdomain cleft (Fig. 2 and Fig. 3).

Models were constructed in the presence of ANP (ATP analogue) and the Ca^{+2} in the active site. According to the literature and the analysis of the structural alignment of the templates, we considered the presence of six conserved water molecules, responsible for: (i) stabilizing the metal ion, and (ii) nucleophilic attack [2,23,26,28]. The six water molecules (O503, O504, O505, O506, O507, O508) were selected from 3CJA.

According Zebisch and Sträter (2008), there are twenty residues in the catalytic site of RnNTPDases2 which interact with the ligand. From these, eight residues (D45, T122, A123, E165, D201, S206, Q208 and W436) interact with six conserved water molecules cited above, and four residues (R245, A347, Y350 and R394) that perform hydrogen bonds with the ligand allowing the docking into the catalytic site [26].

Based on Zebisch and coworkers [26,27], it was possible to identify the likely catalytic residues of SmATPDase1 and CD39. In both models, all residues involved in catalytic activity are in ACR (Fig. 2 and Fig. 3). Studies have shown that mutation of these residues implies a loss of catalytic ability of the enzyme [20,7,25]. In SmATPDase1, the catalytic residues are D78 (ACR1), T154 and A155 (both ACR2), E201 (ACR3), D232, S237 and Q239 (ACR4) and W483 (ACR5) (Fig. 4). In CD39, they are D54 (ACR1), T131 and A132 (ACR2), E174 (ACR3), D213, S218 and Q220 (ACR4) and W450 (ACR5) (Fig. 4). Additionaly, there are two residues that form the stacking of nitrogenous base in SmATPDase1 (Y397 and F441), and CD39 (F365 and Y408).

The total volume calculated for CD39 was $89,788 \mathring{A}^3$ and for SmATPDase1 was $96,329 \mathring{A}^3$. The volume of active site was $1,273.91 \mathring{A}^3$ to SmATPDase1 and $1,234.17 \mathring{A}^3$ to CD39. Regarding the SASA of the active site of both models the values calculated were $716.37 \mathring{A}^2$ to SmATPDase1 and $789.57 \mathring{A}^2$ to CD39.

The electrostatic profile of SmATPDase1 is more positively charged than CD39 (Fig. 5). This profile can also be observed in the active site (Fig. 6), although around the D438 is more negative. Despite having a more negative

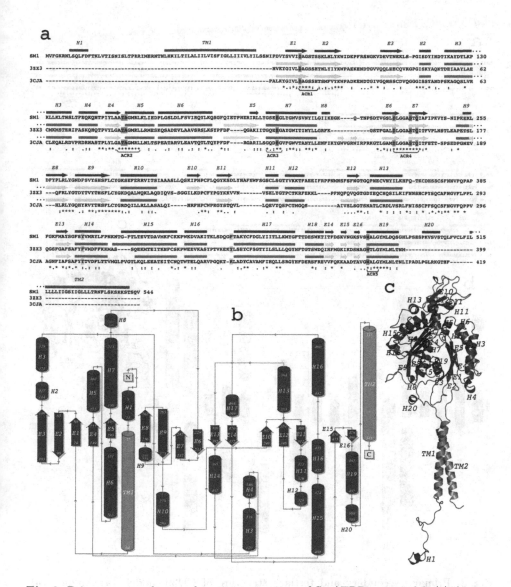

Fig. 2. Primary, secondary and tertiary structure of SmATPDase1 model. (a) Amino acid sequences alignment of SmATPDase1 and templates 3ZX3 and 3CJA. Secondary structures are represented by rectangles (helices) and arrows (sheets). Catalytic residues are shown in gray and residues that form the base-stacking of nitrogenous base of nucleotide in beige. (b) Topology diagram of SmATPDase1 model. Sheets (E) and helices (H) are numbered starting from the Nter to Cter. The diagram was generated using the PDBsum program.(c) 3D model of SmATPDase1. The structure is formed by TM1, TM2 and ECD domains. The last is divided into two subdomains: ECD-I (E1-E5, H2-H6, E15-E16 e H19) and ECD-II (E6-E14 e H7-H18). The composition of subdomains ECD-I and ECD-II is based on the work of Zebisch and Sträter (2008).

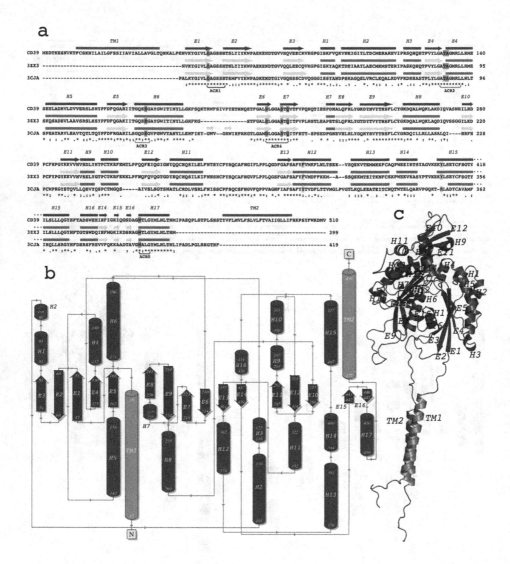

Fig. 3. Primary, secondary and tertiary structure of CD39 model. (a) Amino acid sequences alignment of CD39 and templates 3ZX3 and 3CJA. Secondary structures are represented by rectangles (helices) and arrows (sheets). Catalytic residues are shown in gray and residues that form the base-stacking of nitrogenous base of nucleotide in beige. (b) Topology diagram of SmATPDase1 model. Sheets (E) and helices (H) are numbered starting from the Nter to Cter. The diagram was generated using the PDBsum program.(c) 3D model of CD39. The structure is formed by TM1, TM2 and ECD domaind. The last is divided into two subdomains: ECD-I (E1-E5, H1-H5, H17 and E15-E16) and ECD-II (E6-E14 and H6-H16). The composition of subdomains ECD-I and ECD-II is based on the work of Zebisch and Sträter (2008).

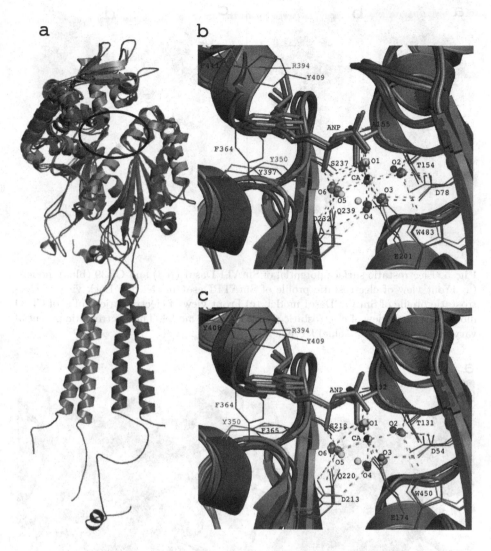

Fig. 4. The 3D structure and active site of SmATPDase1 and CD39. (a) Structural alignment of SmATPDase1 model (red) and CD39 (blue), with templates 3ZX3 (purple) and 3CJA (green). (b) Structural alignment of SmATPDase1 model and templates in the active site region. (c) Structural alignment of CD39 model and templates in the active site region. In sticks is presented the ligand (ANP), catalytic residues (black), conserved waters (O1, O2, O3, O4, O5 and O6) between the structures 3CJ1 (light red), 3CJA7 (yellow), 3CJ9 (orange), 3CJA (green) and 3ZX3 (purple), and the calcium ion from structures 3CJ9 (black), 3CJA (light green), SmATPDase1 (brown) and CD39 (under the calcium ion of 3CJA). The dashed lines show the contacts between the catalytic residues, the six conserved water molecules and the calcium ion. It can also observe the residues that form stacking with adenine in SmATPDase1 (Y397 and F441) and CD39 (F365 and Y408).

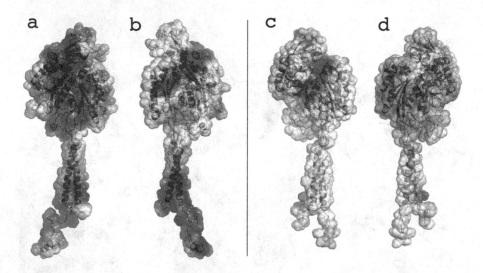

Fig. 5. Electrostatic surface potential of SmATPDase1 (red) and CD39 (blue) models. (a) Front view of electrostatic profile of SmATPDase1 model. (b) Back view of electrostatic profile of SmATPDase1 model. (c) Front view of electrostatic profile of CD39 model. (d) Back view of electrostatic profile of CD39 model. The electrostatic potential vary from -5 (red) to 5 (blue) kT/e.

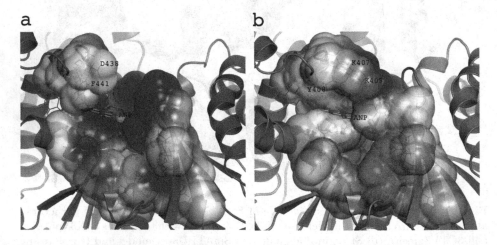

Fig. 6. Electrostatic potential profile of SmATPDase1 and CD39 active sites. (a) Electrostatic profile of active site of SmATPDase1 model (red). (b) Electrostatic profile of active site of CD39 model (blue). 3D structures are presented in Cartoon, ligand (ANP) in Licorice, Ca^{2+} and water molecules are presented in Spheres. The electrostatic potential vary from -3 (red) to 3 (blue) kT/e.

profile, the active site of CD39 (Fig. 6) shows a more positive electrostatic profile around the lysines K405 and K407 next to the Y408.

According Zebisch and coworkers [27], the binding site of the polyoxometallates in RnNTPDase1 is located between the ECD-I and ECD-II subdomains and close to the residues Y409 (stacked with nucleotide nitrogenous base), K406 and K407 [27]. This region would present a positive electrostatic potential [27], probably by the presence of these two lysines. They suggested that K406 and K407 are involved in electrostatic interactions with polyoxometallates [27]. Interestingly, the residue K405 (CD39) corresponds to residue K406 in RnNTPDase1 (3ZX3) and residue K392 in RnNTPDase2 (3CJA), and it is not observerd in SmATPDase1 (D438).

4   Conclusions

This paper presented 3D structure models of the isoforms 1 of NTPDases of *S. mansoni* (SmATPDase1) and human (CD39). Both models were generated by comparative modeling based on templates 3ZX3 and 3CJA. The proposed models for SmATPDase1 and CD39 follow the pattern of Ecto-NTPDases of mammals already described in the literature, namely the presence of a extracellular domain (ECD) and two transmembrane domains (TM1 and TM2) whereby the enzyme is anchored to the plasma membrane of cells.

As the templates used in the present study had only the ECD domain, a specific approach was required for modeling TM1 and TM2 based on consensus predictors of secondary structure, TM domains and contact. Both models showed several regions with high structural conservation in relation to templates, especially in regions that correspond to ACR. It was also possible to identify the likely residues of the active site involved in catalytic activity.

Besides the high structural conservation, in could observed significative electrostatic differences, mainly in the active sites which present some residue substitutions that could induce local changes in the electrostatic profiles. Further investigations about alternative druggable pockets should be also performed.

We expect that these analyses could help future structure-based drug design studies for schistosomiasis treatment, aiming an inhibitor of SmATPDase1 that not interfere with the structure of CD39, causing less or none side effects during treatment in humans.

Acknowledgments. This work was supported by CAPES.

References

1. Castro-Borges, W., et al.: Abundance of tegument surface proteins in the human blood fluke Schistosoma mansoni determined by QconCAT proteomics. Journal of Proteomics 74, 1519–1533 (2011)
2. Cohn, M., Meek, G.A.: The mechanism of hydrolysis of adenosine di- and triphosphate catalysed by potato apyrase. Biochemical Journal 66, 128–130 (1956)
3. Colley, D.G., Secor, W.E.: A schistosomiasis research agenda. PLOS Neglected Tropical Diseases 1, e32 (2007)

4. Da'dara, A.A., et al.: Schistosome tegumental ecto-apyrase (SmATPDase1) degrades exogenous pro-inflammatory and pro-thrombotic nucleotides. PeerJ 2, e316 (2014)
5. DeMarco, R., et al.: Molecular characterization and immunolocalization of *Schistosoma mansoni* ATP-diphosphohydrolase. Biochemical and Biophysical Research Communications 307(4), 831–838 (2003)
6. Doenhoff, M.J., et al.: Praziquantel: mechanisms of action, resistance and new derivatives for schistosomiasis. Current Opinion in Infectious Diseases 21, 659–667 (2008)
7. Drosopoulos, J.H., et al.: Site-directed mutagenesis of human endothelial cell ecto-ADPase/soluble CD39: requirement of glutamate 174 and serine 218 for enzyme activity and inhibition of platelet recruitment. Biochemistry 39, 6936–6943 (2000)
8. Faria-Pinto, P., et al.: ATP diphosphohydrolase from *Schistosoma mansoni* egg: characterization and immunocytochemical localization of a new antigen. Parasitology 129, 51–57 (2004)
9. Grinthal, A., Guidotti, G.: Dynamic motions of CD39 transmembrane domains regulate and are regulated by the enzymatic active site. Biochemistry 43, 13849–13858 (2004)
10. Grinthal, A., Guidotti, G.: CD39, NTPDase 1, is attached to the plasma membrane by two transmembrane domains. Why? Purinergic Signalling 2, 391–398 (2006)
11. Gryseels, B., et al.: Seminar Human schistosomiasis. Lancet 368, 1106–1118 (2006)
12. Guiguet Leal, D.A., et al.: Acute schistosomiasis in brazilian traveler: the importance of tourism in the epidemiology of neglected parasitic diseases. Case Reports in Infectious Diseases, 2012:650929 (2012)
13. King, C.H., et al.: Utility of repeated praziquantel dosing in the treatment of schistosomiasis in high-risk communities in Africa: a systematic review. PLOS Neglected Tropical Diseases 5, e1321 (2011)
14. Knowles, A.F.: The GDA1_CD39 superfamily: NTPDases with diverse functions. Purinergic Signalling 7(1), 21–45 (2011)
15. Levano-Garcia, J., et al.: Characterization of *Schistosoma mansoni* ATPDase2 gene, a novel apyrase family member. Biochemical and Biophysical Research Communications 352(2), 384–389 (2007)
16. McMorran, B.J., et al.: Platelets kill intraerythrocytic malarial parasites and mediate survival to infection. Science 323, 797–800 (2009)
17. Pearce, E.J., MacDonald, A.S.: The immunobiology of schistosomiasis. Nature Reviews Immunology 2, 499–511 (2002)
18. Robson, S.C., et al.: The E-NTPDase family of ectonucleotidases: Structure function relationships and pathophysiological significance. Purinergic Signalling 2(2), 409–430 (2006)
19. Sansom, F.M.: The role of the NTPDase enzyme family in parasites: what do we know, and where to from here? Parasitology 139(8), 963–980 (2012)
20. Smith, T.M., Kirley, T.L.: Site-directed mutagenesis of a human brain ecto-apyrase: evidence that the E-type ATPases are related to the actin/heat shock 70/sugar kinase superfamily. Biochemistry 38, 321–328 (1999)
21. Vasconcelos, E.G., et al.: Characterization and localization of an ATP-diphosphohydrolase on the external surface of the tegument of *Schistosoma mansoni*. Molecular and Biochemical Parasitology 58, 205–214 (1993)
22. Vasconcelos, E.G., et al.: Partial purification and immunohistochemical localization of ATP diphosphohydrolase from *Schistosoma mansoni*. Immunological cross-reactivities with potato apyrase and *Toxoplasma gondii* nucleoside triphosphate hydrolase. Journal of Biological Chemistry 271(36), 22139–22145 (1996)

23. Vorobiev, S., et al.: The structure of nonvertebrate actin: implications for the ATP hydrolytic mechanism. PNAS 100(10), 5760–5765 (2003)
24. Wilson, R.A.: Virulence factors of schistosomes. Microbes Infect. 14, 1442–1450 (2012)
25. Yang, F., et al.: Site-directed mutagenesis of human nucleoside triphosphate diphosphohydrolase 3: the importance of residues in the apyrase conserved regions. Biochemistry 40, 3943–3950 (2001)
26. Zebisch, M., Sträter, N.: Structural insight into signal conversion and inactivation by NTPDase2 in purinergic signaling. PNAS 105(19), 6882–6887 (2008)
27. Zebisch, M., et al.: Crystallographic evidence for a domain motion in rat nucleoside triphosphate diphosphohydrolase (NTPDase) 1. Journal of Molecular Biology 415(2), 288–306 (2012)
28. Zebisch, M., et al.: Crystallographic Snapshots along the Reaction Pathway of Nucleoside Triphosphate Diphosphohydrolases. Structure 21(8), 1460–1475 (2013)

Assessing the Robustness of Parsimonious Predictions for Gene Neighborhoods from Reconciled Phylogenies: Supplementary Material

Ashok Rajaraman[1(✉)], Cedric Chauve[1], and Yann Ponty[1,2,3]

[1] Department of Mathematics, Simon Fraser University, Burnaby, Canada
{arajaram,cedric.chauve}@sfu.ca
[2] Pacific Institute for Mathematical Sciences, CNRS UMI3069, Vancouver, Canada
[3] CNRS/LIX, Ecole Polytechnique, Palaiseau, France
ponty@lri.fr

Abstract. The availability of many assembled genomes opens the way to study the evolution of syntenic character within a phylogenetic context. The DeCo algorithm, recently introduced by Bérard *et al.*, computes parsimonious evolutionary scenarios for gene adjacencies, from pairs of reconciled gene trees. Following the approach pioneered by Sturmfels and Pachter, we describe how to modify the DeCo dynamic programming algorithm to identify classes of cost schemes that generate similar parsimonious evolutionary scenarios for gene adjacencies. We also describe how to assess the robustness, again to changes of the cost scheme, of the presence or absence of specific ancestral gene adjacencies in parsimonious evolutionary scenarios. We apply our method to six thousands mammalian gene families, and show that computing the robustness to changes of cost schemes provides interesting insights on the DeCo model.

1 Introduction

Reconstructing evolutionary histories of genomic characters along a given species phylogeny is a long-standing problem in computational biology. This problem has been studied for several types of genomic characters (DNA sequences and gene content for example), for which efficient algorithms exist to compute parsimonious evolutionary scenarios. Recently, Bérard *et al.* [2] extended the corpus of such results to syntenic characters. They defined a model for the evolution of gene adjacencies within a species phylogeny, together with an efficient dynamic programming (DP) algorithm, called DeCo, to compute parsimonious evolutionary histories that minimize the total cost of gene adjacencies gain and break, for a given cost scheme associating a cost to each of these two events. Reconstructing evolutionary scenarios for syntenic characters is an important step towards more comprehensive models of genome evolution, going beyond classical sequence/ content frameworks, as it implicitly integrates genome rearrangements [5]. Application of such methods include the study of genome rearrangement rates and the reconstruction of ancestral gene order. Moreover, DeCo

© Springer International Publishing Switzerland 2015
R. Harrison et al. (Eds.): ISBRA 2015, LNBI 9096, pp. 260–271, 2015.
DOI: 10.1007/978-3-319-19048-8_22

is the only existing tractable model that considers the evolution of gene adjacencies within a general phylogenetic framework; so far other tractable models of genome rearrangements accounting for a given species phylogeny are either limited to single-copy genes and ignore gene-specific events [3,18], assume restrictions on the gene duplication events, such as considering only whole-genome duplication (see [7] and references there), or require a dated species phylogeny [11].

The evolutionary events considered by DeCo, gene adjacency gain and break caused by genome rearrangement, are rare evolutionary events compared to gene-family specific events. It is then important to assess the robustness of inferences made by DeCo, whether it is of a parsimony cost or of an individual feature such as the presence of a specific ancestral adjacency. We recently explored an approach that considers the set of all possible evolutionary scenarios under a Boltzmann probability distribution for a fixed cost scheme [6]. A second approach consists of assessing how robust features of evolutionary scenarios are to changes in the cost associated to evolutionary events (the cost scheme). Such approaches have recently been considered for the gene tree reconciliation problem and have been shown to significantly improve the results obtained from purely parsimonious approaches [1,10]. This relates to the general problem of deciding the precise cost to assign to evolutionary events in evolutionary models, a recurring question in the context of parsimony-based approaches in phylogenetics.

This motivates the precise questions tackled in this work. First, how robust is a parsimonious evolutionary scenario to a change of the costs associated to adjacency gains and breaks? Similarly, how robust is an inferred parsimonious gene adjacency to a change in these costs? We address this problem using a methodology that has been formalized into a rigorous algebraic framework by Pachter and Sturmfels [15,14,13], that we refer to as the *polytope approach*. Its main features, summarized in Fig. 1 for assessing the robustness of evolutionary scenarios, are (1) associating each evolutionary scenario to a *signature*, a vector of two integers (g, b) where g is the number of adjacency gains and b the number of adjacency breaks; and (2) partitioning the space of cost schemes into convex regions such that, for all the cost schemes within a region, all optimal solutions obtained with such cost schemes have the same signature. This partition can be computed by an algorithm that is a direct translation of the DP algorithm into a polytope framework. Furthermore, the same framework can be extended to assess the robustness of inferred parsimonious ancestral adjacencies.

2   Preliminary: Models and Problems

A *phylogeny* is a rooted tree which describes the evolutionary relationships of a set of elements (species, genes, ...) represented by its nodes: internal nodes correspond to ancestral elements, leaves to extant elements, and edges represent direct descents between parents and children. For a node v of a phylogeny, we denote by $s(v)$ the species it belongs to. For a tree T and a node x of T, we denote by $T(x)$ the subtree rooted at x. If x is an internal node, we assume it has either one child, denoted by $a(x)$, or two children, denoted by $a(x)$ and $b(x)$.

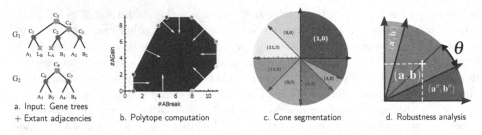

a. Input: Gene trees
+ Extant adjacencies

b. Polytope computation

c. Cone segmentation

d. Robustness analysis

Fig. 1. Outline of our method for assessing the robustness of an evolutionary scenario: Starting from two reconciled gene trees and a set of extant adjacencies (a.), the polytope of parsimonious signatures is computed (b.). Its normal vectors define a segmentation of the space of cost schemes into cones (c.), each associated with a signature. Here, the positive quadrant is fully covered by a single cone, meaning that the parsimonious prediction does not depend on the precise cost scheme. In general (d.), the robustness of a prediction (here, obtained using the $(1,1)$ scheme) to perturbations of the scheme can be measured as the smallest angle θ such that a cost scheme at angular distance θ no longer predicts the signature (a, b).

Species Tree and Reconciled Gene Trees. A species tree S is a binary tree that describes the evolution of a set of species from a common ancestor through the mechanism of *speciation*. A reconciled gene tree is a binary tree that describes the evolution of a set of genes, called a *gene family*, within a given species tree S, through the evolutionary mechanisms of speciation, *gene duplication* and *gene loss*. Therefore, each leaf of a gene tree G represents either a gene loss or an an extant gene, while each internal node represents an ancestral gene. In a reconciled gene tree, we associate every ancestral gene (an internal node g) to an evolutionary event $e(g)$ that leads to the creation of the two children $a(g)$ and $b(g)$: $e(g)$ is a speciation (denoted by Spec) if the species pair $\{s(a(g)), s(b(g))\}$ is equal to the species pair $\{a(s(g)), b(s(g))\}$, $s(a(g)) \neq s(b(g))$, or a gene duplication (GDup) if $s(a(g)) = s(b(g)) = s(g)$. If g is a leaf, then $e(g)$, as stated before, indicates either a gene loss (GLoss) or an extant gene (Extant), in which case $e(g)$ is not an evolutionary event *stricto sensu*. A *pre-speciation* ancestral gene is an internal node g such that $e(g) =$ Spec. See Fig. 2 for an illustration.

Adjacency Trees and Forests. We consider now that we are given two reconciled gene trees G_1 and G_2, representing two gene families evolving within a species tree S. A *gene adjacency* is a pair of genes (one from G_1 and one from G_2) that appear consecutively along a chromosome, for a given species, ancestral or extant. Gene adjacencies evolve within a species tree S through the evolutionary events of speciation, gene duplication, gene loss (these three events are modeled in the reconciled gene trees), and *adjacency duplication* (ADup), *adjacency loss* (ALoss) and *adjacency break* (ABreak), that are adjacency-specific events.

Following the model introduced in [2], we represent such an evolutionary history using an *adjacency forest*, composed of *adjacency trees*. An adjacency tree

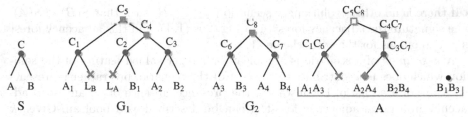

Fig. 2. A species tree S, with two extant species A and B and an ancestral species C. Two reconciled gene trees G_1 and G_2, with four extant genes in genome A, four extant genes in genome B and three ancestral genes in genome C. The set of extant gene adjacencies is (A_1A_3, B_1B_3, B_2B_4). An adjacency forest A composed of two adjacency trees. Blue dots represent speciation nodes. Leaves are extant species/genes/adjacencies, except the one labeled by a red cross (gene loss) or a red diamond (adjacency breaks). Green squares are (gene or adjacency) duplication nodes. Gene labels refer to the species they belong to. Every node of the adjacency tree is labeled by a gene adjacency. Figure adapted from [2].

represents the evolution of an ancestral gene adjacency (located at the root of the tree) through the following events: (1) The duplication of an adjacency $\{g_1, g_2\}$, where g_1 and g_2 are respectively genes from G_1 and G_2 such that $s(g_1) = s(g_2)$, follows from the simultaneous duplication of both its genes g_1 and g_2 (so $e(g_1) = e(g_2) = \mathsf{GDup}$), resulting in the creation of two distinct adjacencies each belonging to $\{a(g_1), b(g_1)\} \times \{a(g_2), b(g_2)\}$; (2)The loss of an adjacency, which can occur due to several events, such as the loss of exactly one of its genes (gene loss, GLoss), the loss of both its genes (adjacency loss, ALoss) or a genome rearrangement that breaks the contiguity between the two genes (adjacency break, ABreak); (3) The creation/gain of an adjacency (denoted by AGain), for example due to a genome rearrangement, that results in the creation of a new adjacency tree whose root is the newly created adjacency.

With this model, one can model the evolution of two gene families along a species phylogeny by a triple (G_1, G_2, A): G_1 and G_2 are reconciled gene trees representing the evolution of these families in terms of gene-specific events and A is an adjacency forest consistent with G_1 and G_2. Similar to species trees and reconciled gene trees, internal nodes of an adjacency tree are associated to ancestral adjacencies, while leaves are associated to extant adjacencies or lost adjacencies (due to a gene loss, adjacency loss or adjacency break), and are labeled by evolutionary events. The label $e(v)$ of an internal node v of an adjacency forest A belongs to $\{\mathsf{Spec}, \mathsf{GDup}, \mathsf{ADup}\}$, while the label $e(v)$ of a leaf belongs to $\{\mathsf{Extant}, \mathsf{GLoss}, \mathsf{ALoss}, \mathsf{ABreak}\}$, as shown in Fig. 2.

Signatures, Descriptors and Parsimonious Scenarios. The *signature* of an adjacency forest A is an ordered pair of integers $\sigma(A) = (g_A, b_A)$ where g_A (resp. b_A) is the number of adjacency gains (resp. adjacency breaks) in A. A *cost scheme* is a pair $\mathbf{x} = (x_0, x_1)$ of non-negative real numbers, where x_0 is the cost of an adjacency gain and x_1 the cost of an adjacency break. The *cost* of an adjacency forest A for a given cost scheme $\mathbf{x}$ is the number $S(A) = x_0 \times g_A + x_1 \times b_A$. The adjacency forest A in an evolutionary scenario (G_1, G_2, A) is *parsimonious for*

x if there is no other evolutionary scenario (G_1, G_2, B) such that $S(B) < S(A)$. The signature the adjacency forest A in Fig. 2 is $(1, 1)$, and this adjacency forest is parsimonious for the cost scheme $(1, 1)$.

A *descriptor* of a scenario is a boolean or integer valued feature of the solution which does not contribute to the cost of the scenario, but rather represents a feature of a scenario. For instance, the presence/absence of an ancestral adjacency in a given adjacency forest A can be described as a boolean. Given k descriptors $a_1, \ldots, a_k$, we define an *extended signature* of a scenario A as a tuple $\sigma_{a_1,\ldots,a_k}(A) = (g, b, s_{a_1}, \ldots, s_{a_k})$, where g, b are the numbers of adjacency gains and breaks in A respectively, and s_{a_i} is the value of the descriptor a_i for A.

The DeCo Algorithm. Bérard *et al.* [2] showed that, given a pair of reconciled gene trees G_1 and G_2, a list L of extant gene adjacencies, and a cost scheme **x**, one can use a DP algorithm to compute an evolutionary scenario (G_1, G_2, A), where A is a parsimonious adjacency forest such that L is exactly the set of leaves of A labeled Extant. The DeCo algorithm computes, for every pair of nodes g_1 (from G_1) and g_2 (from G_2) such that $s(g_1) = s(g_2)$, two quantities $c_1(g_1, g_2)$ and $c_0(g_1, g_2)$, that correspond respectively to the cost of a parsimonious adjacency forest for the pairs of subtrees $G(g_1)$ and $G(g_2)$, under the hypothesis that g_1 and g_2 form (for c_1) or do not form (for c_0) an ancestral adjacency. As usual in dynamic programming along a species tree, the cost of a parsimonious adjacency forest for G_1 and G_2 is given by $\min(c_1(r_1, r_2), c_0(r_1, r_2))$ where r_1 is the root of G_1 and r_2 the root of G_2. In [6], we recently generalized DeCo into a DP algorithm DeClone that allows one to explore the space of all possible adjacency evolutionary scenarios for a given cost scheme.

Robustness Problems. The first problem we are interested in is the *signature robustness problem*. A signature $\sigma = (g, b)$ is parsimonious for a cost scheme **x** if there exists at least one adjacency forest A that is parsimonious for **x** and has signature $\sigma(A) = \sigma$. The robustness of the signature σ is defined as the difference between **x** and the closest cost scheme for which σ is no longer parsimonious. To measure this difference, we rely on a geometric representation of a cost scheme. Assuming that a cost scheme $\mathbf{x} = (x_0, x_1) \in \mathbb{R}^2$ provides sufficient information to evaluate the cost of an adjacency forest, the predictions under such a model remain unchanged upon multiplying **x** by any positive number, allowing us to assume that $\|\mathbf{x}\| = 1$ without loss of generality. So $\mathbf{x} = (x_0, x_1)$ can be summarized as an angle θ (expressed in radians), and the difference between two cost schemes is indicated by their associated angular distance.

However, signatures only provide a quantitative summary of the evolutionary events described by a parsimonious adjacency forest. In particular, signatures discard any information about predicted sets of ancestral adjacencies. We address the robustness of inferred parsimonious adjacencies through the *parsimonious adjacency robustness problem*. Let $a = (g_1, g_2)$ be an ancestral adjacency featured in a parsimonious adjacency forest for a cost scheme **x**. We say that a is parsimonious for a cost scheme **y** if a belongs to every adjacency forest that is parsimonious for **y**. The robustness of a is defined as the angular distance from **x** to the closest cost scheme **y** for which a is no longer parsimonious.

3   Methods

If the signature for a given adjacency forest A is given by the vector $\sigma(A) = (g, b)$, and the cost scheme is given by the vector $\mathbf{x} = (x_0, x_1)$, then the parsimony cost of DeCo can be written as the inner product $\langle \mathbf{x}, \sigma(A) \rangle = g \times x_0 + b \times x_1$. DeCo computes the following quantity for a pair of gene trees G_1 and G_2.

$$c(G_1, G_2) = \min_{A \in \mathcal{F}(G_1, G_2)} \langle \mathbf{x}, \sigma(A) \rangle, \tag{1}$$

where $\mathcal{F}(G_1, G_2)$ denotes the set of all possible adjacency forests that can be constructed from G_1 and G_2, irrespective of the cost scheme.

For a given adjacency forest A, we will consider a single descriptor a, indicating the presence or absence of an ancestral adjacency $a = (g_1, g_2) \in G_1 \times G_2$ in A, where $s_a = 1$ if it is present in A, and 0 otherwise. Since, by definition, a descriptor does not contribute to the cost, when considering the robustness of specific adjacencies, we will consider cost schemes of the form $\mathbf{x} = (x_0, x_1, 0)$, and DeCo will compute Eq. (1) as usual.

For a given cost scheme $\mathbf{x}$, two adjacency forests A_1 and A_2 such that $\sigma(A_1) = \sigma(A_2)$ will have the same associated cost. We can thus define an equivalence class in $\mathcal{F}(G_1, G_2)$ based on the signatures. However, for a given potential ancestral adjacency $a = (g_1, g_2) \in G_1 \times G_2$, the adjacency forests in this equivalence class may have different extended signatures, differing only in the last coordinate. Thus, there may be two adjacency forests A_1 and A_2 with extended signatures $(g, b, 1)$ and $(g, b, 0)$ respectively, and they will have the same cost for all cost schemes. Evolutionary scenarios with the same extended signature also naturally form an equivalence class in $\mathcal{F}(G_1, G_2)$.

Convex Polytopes from Signatures. Let us denote the set of signatures of all scenarios in $\mathcal{F}(G_1, G_2)$ by $\sigma(\mathcal{F}(G_1, G_2))$, and the set of extended signatures for a given adjacency a by $\sigma_a(\mathcal{F}(G_1, G_2))$. Each of these is a point in $\mathbb{R}^d$, where $d = 2$ for signatures and $d = 3$ for extended signatures. In order to explore the parameter space of parsimonious solutions to DeCo, we use these sets of points to construct a *convex polytope* in $\mathbb{R}^d$. A convex polytope is simply the set of all convex combinations of points in a given set, in this case the set of signatures or extended signatures [15]. Thus, for each pair of gene trees G_1, G_2 and a list of extant adjacencies, we can theoretically construct a convex polytope in $\mathbb{R}^2$ by taking the convex combinations of all signatures in $\sigma(\mathcal{F}(G_1, G_2))$. This definition generalizes to a convex polytope in $\mathbb{R}^3$ when extended signatures $\sigma_a(\mathcal{F}(G_1, G_2))$ are considered for some ancestral adjacency a. Viewing the set of evolutionary scenarios as a polytope allows us to deduce some useful properties:

1. Any (resp. extended) signature that is parsimonious for some cost scheme $\mathbf{x}$ lies on the surface of the polytope;
2. If a (resp. extended) signature is parsimonious for two cost schemes $\mathbf{x}$ and $\mathbf{x}'$, then it is also parsimonious for any cost scheme *in between* (i.e. for any convex combination of $\mathbf{x}$ and $\mathbf{x}'$).

Traditionally, a polytope is represented as a set of inequations, which is inappropriate for our intended application. Therefore, we adopt a slighty modified representation, and denote the polytope of $\mathcal{F}(G_1, G_2)$ as the list of signatures that are represented within $\mathcal{F}(G_1, G_2)$ and lie on the convex hull of the polytope.

A *vertex* in a polytope is a signature (resp. extended signature) which is parsimonious for some cost scheme. The domain of parsimony of a vertex $\mathbf{v}$ is the set of cost schemes for which $\mathbf{v}$ is parsimonious. From Property 2, the domain of parsimony for a vertex $\mathbf{v}$ is a *cone* in $\mathbb{R}^d$, formally defined as:

$$Cone\,(\mathbf{v}) = \left\{ \mathbf{x} \in \mathbb{R}^d : \langle \mathbf{x}, \mathbf{v} \rangle \leq \langle \mathbf{x}, \mathbf{w} \rangle \; \forall \; \mathbf{w} \in P \right\}. \tag{2}$$

The set of cones associated with the vertices of a polytope form a partition of the cost schemes space [15], which allows us to assess the effect of perturbing the cost scheme on the optimal solution of DeCo for this cost scheme.

Computing the Polytope. Building on earlier work on parametric sequence alignment [8], Pachter and Sturmfels [14,15] described the concept of *polytope propagation*, based on the observation that the polytope of a DP (minimization) scheme can be computed through an algebraic substitution. Accordingly, any point that lies strictly within the polytope is suboptimal for any cost scheme, and can be safely discarded by a procedure that repeatedly computes the *convex hull* $H(P)$ of the (intermediates) polytopes produced by the modified DP scheme. In the context of the DeCo DP scheme, the precise modifications are:

1. Any occurrence of the $+$ operator is replaced by $\oplus$, the (convex) *Minkowski sum* operator, defined for P_1, P_2 two polytopes as

$$P_1 \oplus P_2 = H(\{p_1 + p_2 \mid (p_1, p_2) \in P_1 \times P_2\});$$

2. Any occurrence of the min operator is replaced by $\uplus$, the *convex union* operator, defined for P_1, P_2 two polytopes as

$$P_1 \uplus P_2 = H(P_1 \cup P_2);$$

3. Any occurrence of an *adjacency gain* cost is replaced by the vector $(1, 0)$ (resp. $(1, 0, 0)$ for extended signatures);
4. Any occurrence of an *adjacency break* cost is replaced by the vector $(0, 1)$ (resp. $(0, 1, 0)$ for extended signatures);
5. (Extended signatures only) An event that corresponds to the prediction of a fixed ancestral adjacency a in a scenario is replaced by the vector $(0, 0, 1)$;

By making this substitution, we can efficiently compute the polytope associated with two input gene trees G_1 and G_2, having sizes n_1 and n_2 respectively, through $O(n_1 \times n_2)$ executions of the convex hull procedure. In place of the integers used by the original minimization approach, intermediate convex polytopes are now processed by individual operations, and stored in the DP tables, so the overall time and space complexities of the algorithm critically depend on the size of the polytopes, i.e. its number of vertices. Pachter and Sturmfels proved that, in general, the number of vertices on the surface of the polytope is $O\left(n^{d-1}\right)$, where d is the number of dimensions, and n is the size of the DP table. In our case, the number of vertices in the 2D polytope associated with simple signatures is in $O(n_1 \times n_2)$. This upper bound also holds for extended signatures, as the third coordinate is a boolean, and the resulting 3D polytope is in fact the union of two

2D polytopes. The total cost of computing the polytope is therefore bounded by $O\left(n_1^2 \times n_2^2 \times \log(n_1 \times n_2)\right)$, e.g. using Chan's convex hull algorithm [4]. As for the computation of the cones, let us note that the cone of a vertex v in a given polytope P is fully delimited by a set of vectors, which can be computed from P as the normal vectors, pointing towards the center of mass of P, of each of the facets in which v appears. This computation can be performed as a postprocessing using simple linear algebra, and its complexity will remain largely dominated by that of the DP-fuelled polytope computation.

Assessing Signature and Adjacency Robustness. The cones associated with the polytope of a given instance cover all the real-valued cost schemes, including those associating negative costs to events. These later cost schemes are not valid, and so, we only consider cones which contain at least one positive cost scheme. Given a fixed cost scheme **y**, the vertex associated to the cone containing this cost scheme corresponds to the signature of all parsimonious scenarios for this cost scheme. In order to assess the robustness of this signature, we can calculate the smallest angular perturbation needed to move from **y** to a cost scheme whose parsimonious scenarios do not have this signature. This is simply the angular distance from **y** to the nearest boundary of the cone which contains it. Using this methods, we assign a numerical value to the robustness of the signatures of parsimonious scenarios on a number of instances for a particular cost scheme.

In the case of extended signatures $\sigma_a\left(\mathcal{F}(G_1, G_2)\right)$ for an adjacency a, the polytope is 3-dimensional. The cones associated with the vertices, as defined algebraically, now partition $\mathbb{R}^3$, the set of cost schemes (x_0, x_1, x_2), where x_2 indicates the cost of a distinguished adjacency. Since the third coordinate is a descriptor, it does not contribute to the cost scheme, and we therefore restrict our analysis to the $\mathbb{R}^+ \times \mathbb{R}^+ \times \{0\}$ subset of the cost scheme space. Precisely, we take the intersection of the plane $x_2 = 0$ with each cone associated to a vertex (g, b, s_a), and obtain the region in which the extended signature (g, b, s_a) is parsimonious. This region is a 2D cone.

However, the cost of an extended signature is independent of the entry in its last coordinate, and there may exist two different extended signatures $(g, b, 0)$ and $(g, b, 1)$, both parsimonious for all the cost schemes found in the 2D cone. It is also possible for adjacent cones to have different signatures, yet feature a given adjacency. The robustness of a given adjacency a is computed from the cones using a greedy algorithm which, starting from the cone containing **x**, explores the adjacent cones in both directions (clockwise/counter-clockwise) until it finds one that no longer predicts a, i.e. is associated with at least one signature $(g', b', 0)$.

4   Results

We considered $5,039$ reconciled gene trees and $50,389$ extant gene adjacencies, forming $6,074$ `DeCo` instances, with genes taken from 36 extant mammalian genomes from the Ensembl database in 2012. In [2], this data was analyzed with `DeCo`, using the cost scheme $(1, 1)$. These adjacency forests defined $96,482$

ancestral adjacencies (adjacencies between two pre-speciation genes from the same ancestral species), covering 112,188 ancestral genes.

We first considered all 6,074 instances, and computed for each signature the robustness of the parsimonious signature obtained with the cost scheme $(1,1)$. We observe (Fig. 3(A)) that for more than half of the instances, the parsimonious signature is robust to a change of cost scheme, as the associated cone is the complete first quadrant of the real plane. On the other hand, for 945 instances the parsimonious signature for the cost scheme $(1,1)$ is not robust to any change in the cost scheme; these cases correspond to interesting instances where the cost scheme $(1,1)$ lies at the border of two cones, meaning that two parsimonious signatures exist for the cost scheme $(1,1)$, and any small change of cost scheme tips the balance towards one of these two signatures. More generally, we observe (Fig. 3(A)) an extreme robustness of parsimonious signatures: there is a $\sim 80\%$ overlap between the sets of signatures that are parsimonious for any (positive) cost scheme, and for the $(1,1)$ cost scheme. This observation supports the notion of a sparsely-populated search space for attainable signatures. In this vision, signatures are generally isolated, making it difficult to trade adjacency gains for breaks (or vice-versa) in order to challenge the $(1,1)$-parsimonious prediction. We hypothesize that such a phenomenon is essentially combinatorial, as extra adjacency gains typically lead, through duplications to more subsequent adjacency breaks.

Next, to evaluate the stability of the total number of evolutionary events inferred by parsimonious adjacency forests, we recorded two counts of evolutionary events for each instance: the number of syntenic events (adjacencies gains and breaks) of the parsimonious signature (called the *parsimonious syntenic events count*), and the maximum number of syntenic events taken over all signatures that are parsimonious for some cost scheme (called the *maximum syntenic events count*). We observe that the average parsimonious (resp. maximum) syntenic events count is 1.25 (resp. 1.66). This shows a strong robustness of the (low) number of syntenic events to changes in the cost scheme.

We then considered the robustness of individual ancestral adjacencies. Using the variant DeClone of DeCo that explores the set of all evolutionary scenarios [6], we extracted, for each instance, the set of ancestral adjacencies that belong to all parsimonious solutions for the cost scheme $(1,1)$, and computed their robustness as defined in the previous sections. This set of ancestral adjacencies contains 87,019 adjacencies covering 106,903 ancestral genes. The robustness of these adjacencies is summarized in Fig. 3(B, left and center columns). It is interesting to observe that few adjacencies have a low robustness, while, conversely, a large majority of the universally parsimonious adjacencies are completely robust to a change of cost scheme (97,593 out of 106,639). This suggests that the DeCo model of parsimonious adjacency forests is robust, and infers highly supported ancestral adjacencies, which is reasonable given the relative sparsity of genome rearrangements in evolution compared to smaller scale evolutionary events.

Besides the notions of robustness, an indirect validation criterion used to assess the quality of an adjacency forest is the limited presence of *syntenic conflicts*.

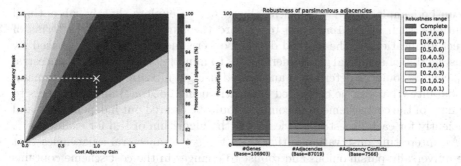

Fig. 3. (A) Average robustness of signatures predicted using the $(1,1)$ cost scheme. At each point (x,y), the colour indicates the proportion of signatures that are parsimonious, and therefore predicted, for the $(1,1)$ cost scheme, and remain parsimonious for the (x,y) cost scheme. (B) Universally parsimonious adjacencies and syntenic conflicts. (Left) Percentage of ancestral genes present in universally parsimonious adjacencies per level of minimum robustness of the adjacencies, expressed in radians. (Center) Percentage of universally parsimonious adjacencies per level of minimum robustness. (Right) Percentage of conserved conflicting adjacencies per level of minimum robustness.

An ancestral gene is said to participate in a syntenic conflict if it belongs to three or more ancestral adjacencies, as a gene can only be adjacent to at most two neighboring genes along a chromosome. An ancestral adjacency participates in a syntenic conflict if it contains a gene that does. Among the ancestral adjacencies inferred by DeCo, $16,039$ participate in syntenic conflicts, covering $5,817$ ancestral genes. This represents a significant level of syntenic conflict and a significant issue in using DeCo to reconstruct ancestral gene orders. It was observed that selecting universally parsimonious ancestral adjacencies, as done in the previous analysis, significantly reduced the number of syntenic conflicts, as almost all discarded ancestral adjacencies participated in syntenic conflicts. Considering syntenic conflicts, we observe (Fig. 3(B, right column) a positive result, i.e. that filtering by robustness results in a significant decrease of the ratio of conflicting adjacencies. However, even with robust universally parsimonious ancestral adjacencies, one can observe a significant number of adjacencies participating in syntenic conflicts. We discuss these observations in the next section.

5 Discussion and Conclusion

From an application point of view, the ability to exhaustively explore the parameter space leads to the observation that, on the considered instances, the DeCo model is extremely robust. Even taking parsimonious signatures that maximize the number of evolutionary syntenic events (i.e. considering cost schemes that lead to the maximum number of events) results in an average increase of roughly 33% events (1.25 to 1.66), and stays very low, much lower than gene specific events such as gene duplications (average of 3.38 event per reconciled gene tree). This is consistent with the fact that for rare evolutionary events such as genome rearrangements, a parsimony approach is relevant, especially when it can be complemented by efficient algorithms to explore slightly sub-optimal solutions, such as DeClone, and to explore the parameter space. In terms of direct applications of the method developed here and in [6], gene-tree based reconstruction of ancestral gene orders comes to mind [5]; more precisely, ancestral adjacencies could be determined and scored using a mixture of their Boltzmann probability (that can be computed efficiently using DeClone) and robustness to changes of the cost scheme, and conflicts could be cleared out independently and efficiently for each ancestral species using the algorithm of [12] for example.

An interesting observation is that even the set of ancestral adjacencies that are universally-parsimonious and robust to changes in the cost scheme contains a significant number of adjacencies participating in syntenic conflict. We conjecture that the main reason for syntenic conflicts is in the presence of a significant number of erroneous reconciled gene trees. This is supported by the observation that the ancestral species with the highest number of syntenic conflict are also species for which the reconciliation with the mammalian species tree resulted in a significantly larger number of genes than expected (data not shown). This points clearly to errors in either gene tree reconstruction or in the reconciliation with the mammalian species phylogeny, which tends to assign wrong gene duplications in some specific species, resulting an inflation of the number of genes, especially toward the more ancient species [9]. It would be interesting to see if the information about highly suported conflicting adjacencies can be used in reconciled gene tree correction.

From a methodological point of view, we considered here extended signatures for a single ancestral adjacency at a time. It would be natural to extend this concept to the more general case of several ancestral adjacencies considered at once. We conjecture that this case can be addressed without an increase in the asymptotic complexity of computing the polytope; this problem will be considered in the full version of the present work. Next, there exists another way to explore the parameter space of a dynamic programming phylogenetic algorithm. It consists of computing the *Pareto-front* of the input instance [10,16], rather than optimal signatures for classes of cost schemes. A signature v is said to be *Pareto-optimal* if there is no other signature whose entries are equal or smaller than the corresponding entries in v, and is strictly smaller at at least one coordinate. The Pareto-front is the set of all Pareto-optimal signatures, and can be efficiently computed by dynamic programming [17,16,10]. The Pareto-front differs from the approach we describe in the present work in several aspects. An advantage of the Pareto-front is that it is a notion irrespective of the type of cost function being used. This contrasts with the polytope propagation technique, which requires that the cost function be a linear combination of its terms. However, so far, the Pareto-approach has only been used to define a partition of the parameter space when the cost function is restricted to be linear/affine, and it remains to investigate the difference with the polytope approach in this case.

Acknowledgements. The authors wish to express their gratitude towards Anatoliy Tomilov (Ural Federal University) for contributing the open source C++ 11 implementation of the `QuickHull` algorithm used for the convex hull computations within `DeClone`. This research was supported by an NSERC Discovery Grant to C.C. and an SFU Michael Stevenson scholarship to A.R.

References

1. Bansal, M.S., Alm, E.J., Kellis, M.: Reconciliation revisited: Handling multiple optima when reconciling with duplication, transfer, and loss. J. Comput. Biol. 20(10), 738–754 (2013)
2. Bérard, S., Gallien, C., Boussau, B., Szöllosi, G.J., Daubin, V., Tannier, E.: Evolution of gene neighborhoods within reconciled phylogenies. Bioinformatics 28(18), 382–388 (2012)
3. Biller, P., Feijão, P., Meidanis, J.: Rearrangement-based phylogeny using the single-cut-or-join operation. IEEE/ACM Trans. Comput. Biol. Bioinf. 10(1), 122–134 (2013)
4. Chan, T.M.: Optimal output-sensitive convex hull algorithms in two and three dimensions. Discrete Comput. Geom. 16, 361–368 (1996)
5. Chauve, C., El-Mabrouk, N., Gueguen, L., Semeria, M., Tannier, E.: Duplication, rearrangement and reconciliation: A follow-up 13 years later. Models and Algorithms for Genome Evolution, 47–62 (2013)
6. Chauve, C., Ponty, Y., Zanetti, J.P.P.: Evolution of genes neighborhood within reconciled phylogenies: An ensemble approach. In: Campos, S. (ed.) BSB 2014. LNCS, vol. 8826, pp. 49–56. Springer, Heidelberg (2014)
7. Gagnon, Y., Blanchette, M., El-Mabrouk, N.: A flexible ancestral genome reconstruction method based on gapped adjacencies. BMC Bioinformatics 13(S-19), S4 (2012)
8. Gusfield, D., Balasubramanian, K., Naor, D.: Parametric optimization of sequence alignment. Algorithmica 12(4/5), 312–326 (1994)
9. Hahn, M.W.: Bias in phylogenetic tree reconciliation methods: implications for vertebrate genome evolution. Genome Biol. 8, R141 (2007)
10. Libeskind-Hadas, R., Wu, Y., Bansal, M.S., Kellis, M.: Pareto-optimal phylogenetic tree reconciliation. Bioinformatics 30(12), 87–95 (2014)
11. Ma, J., Ratan, A., Raney, B.J., Suh, B.B., Zhang, L., Miller, W., Haussler, D.: DUPCAR: reconstructing contiguous ancestral regions with duplications. J. Comput. Biol. 15(8), 1007–1027 (2008)
12. Manuch, J., Patterson, M., Wittler, R., Chauve, C., Tannier, E.: Linearization of ancestral multichromosomal genomes. BMC Bioinformatics 13(S-19), S11 (2012)
13. Pachter, L., Sturmfels, B.: Parametric inference for biological sequence analysis. Proc. Natl. Acad. Sci. USA 101(46), 16138–16143 (2004)
14. Pachter, L., Sturmfels, B.: Tropical geometry of statistical models. Proc. Natl. Acad. Sci. USA 101(46), 16132–16137 (2004)
15. Pachter, L., Sturmfels, B. (eds.): Algebraic Statistics for Computational Biology. Cambridge University Press (2005)
16. Saule, C., Giegerich, R.: Observations on the feasibility of exact Pareto optimization. In: Proceedings of CMSR 2014, pp. 43–56 (2014)
17. Schnattinger, T., Schöning, U., Kestler, H.A.: Structural RNA alignment by multi-objective optimization. Bioinformatics 29(13), 1607–1613 (2013)
18. Tannier, E., Zheng, C., Sankoff, D.: Multichromosomal median and halving problems under different genomic distances. BMC Bioinformatics, 10 (2009)

Sorting Signed Circular Permutations
by Super Short Reversals

Gustavo Rodrigues Galvão[1][(✉)], Christian Baudet[2,3], and Zanoni Dias[1]

[1] Institute of Computing, University of Campinas, Campinas, Brazil
{ggalvao,zanoni}@ic.unicamp.br
[2] Laboratoire Biométrie et Biologie Evolutive, Université de Lyon, Université Lyon 1, CNRS, Villeurbanne, UMR5558, France
[3] Inria Grenoble - Rhône-Alpes, Erable Team, Villeurbanne Cedex, France
christian.baudet@inria.fr

Abstract. We consider the problem of sorting a circular permutation by reversals of length at most 2, a problem that finds application in comparative genomics. Polynomial-time solutions for the unsigned version of this problem are known, but the signed version remained open. In this paper, we present the first polynomial-time solution for the signed version of this problem. Moreover, we perform an experiment for inferring distances and phylogenies for published *Yersinia* genomes and compare the results with the phylogenies presented in previous works.

1 Introduction

Distance-based methods form one of the three large groups of methods to infer phylogenetic trees from sequence data [8, Chapter 5]. Such methods proceed in two steps. First, the evolutionary distance is computed for every sequence pair and this information is stored in a matrix of pairwise distances. Then, a phylogenetic tree is constructed from this matrix using a specific algorithm, such as *Neighbor-Joining* [9]. Note that, in order to complete the first step, we need some method to estimate the evolutionary distance between a sequence pair. Assuming the sequence data correspond to complete genomes, we can resort to the genome rearrangement approach [4] in order to estimate the evolutionary distance.

In genome rearrangements, one estimates the evolutionary distance between two genomes by finding the rearrangement distance between them, which is the length of the shortest sequence of rearrangement events that transforms one genome into the other. Assuming genomes consist of a single chromosome, share the same set of genes, and contain no duplicated genes, we can represent them as permutations of integers, where each integer corresponds to a gene. If, besides the order, the orientation of the genes is also regarded, then each integer has a sign, + or −, and the permutation is called a signed permutation (similarly, we also refer to a permutation as an unsigned permutation when its elements do not have signs). Moreover, if the genomes are circular, then the permutations are also circular; otherwise, they are linear.

© Springer International Publishing Switzerland 2015
R. Harrison et al. (Eds.): ISBRA 2015, LNBI 9096, pp. 272–283, 2015.
DOI: 10.1007/978-3-319-19048-8_23

A number of publications address the problem of finding the rearrangement distance between two permutations, which can be equivalently stated as a problem of sorting a permutation into the identity permutation (for a detailed survey, the reader is referred to the book of Fertin *et al.* [4]). This problem varies according to the rearrangement events allowed to sort a permutation. Reversals are the most common rearrangement event observed in genomes. They are responsible for reversing the order and orientation of a sequence of genes within a genome. Although the problem of sorting a permutation by reversals is a well-studied problem, most of the works concerning it do not take into account the length of the reversals (*i.e.* the number of genes affected by it). Since it has been observed that short reversals are prevalent in the evolution of some species [1, 2, 7, 10], recent efforts have been made to address this issue [3, 5].

In this paper, we add to those efforts and present a polynomial-time solution for the problem of sorting a signed circular permutation by super short reversals, that is, reversals which affect at most 2 elements (genes) of a permutation (genome). This solution closes a gap in the literature since polynomial-time solutions are known for the problem of sorting an unsigned circular permutation [3, 6], for the problem of sorting an unsigned linear permutation [6], and for the problem of sorting a signed linear permutation [5]. Moreover, we reproduce the experiment performed by Egri-Nagy *et. al.* [3] to infer distances and phylogenies for published *Yersinia* genomes, but this time we consider the orientation of the genes (they have ignored it in order to treat the permutations as unsigned).

The rest of this paper is organized as follows. Section 2 succinctly presents the solution developed by Jerrum [6] for the problem of sorting by cyclic super short reversals. Section 3 builds upon the previous section and presents the solution for the problem of sorting by signed cyclic super short reversals. Section 4 briefly explains how we can use the solutions described in Sect(s). 2 and 3 to solve the problem of sorting a (signed) circular permutation by super short reversals. Section 5 presents experimental results performed on *Yersinia pestis* data. Finally, Sect. 6 concludes the paper.

2 Sorting by Cyclic Super Short Reversals

A *permutation* π is a bijection of $\{1, 2, \ldots, n\}$ onto itself. A classical notation used in combinatorics for denoting a permutation π is the two-row notation

$$\pi = \begin{pmatrix} 1 & 2 & \ldots & n \\ \pi_1 & \pi_2 & \ldots & \pi_n \end{pmatrix},$$

$\pi_i \in \{1, 2, \ldots, n\}$ for $1 \leq i \leq n$. This notation indicates that $\pi(1) = \pi_1$, $\pi(2) = \pi_2$, $\ldots$, $\pi(n) = \pi_n$. The notation used in genome rearrangement literature, which is the one we will adopt, is the one-row notation $\pi = (\pi_1 \ \pi_2 \ \ldots \ \pi_n)$. We say that π has size n. The set of all permutations of size n is S_n.

A *cyclic reversal* $\rho(i, j)$ is an operation that transforms a permutation $\pi = (\pi_1 \ \pi_2 \ \ldots \ \pi_{i-1} \ \pi_i \ \pi_{i+1} \ \ldots \ \pi_{j-1} \ \pi_j \ \pi_{j+1} \ \ldots \ \pi_n)$ into the permutation $\pi \cdot \rho(i, j) = (\pi_1 \ \pi_2 \ \ldots \ \pi_{i-1} \ \underline{\pi_j \ \pi_{j-1} \ \ldots \ \pi_{i+1} \ \pi_i} \ \pi_{j+1} \ \ldots \ \pi_n)$ if $1 \leq i < j \leq n$ and transforms a permutation $\pi = (\pi_1 \; \pi_2 \ldots \pi_i \; \pi_{i+1} \ldots \pi_{j-1} \; \pi_j \; \pi_{j+1} \ldots \pi_n)$ into the permutation $\pi \cdot \rho(i, j) = (\pi_j \; \pi_{j+1} \ldots \pi_n \; \pi_{i+1} \ldots \pi_{j-1} \; \pi_i \; \pi_{i-1} \ldots \pi_1)$ if $1 \le j < i \le n$. The cyclic reversal $\rho(i, j)$ is called a *cyclic k-reversal* if $k \equiv j - i + 1 \pmod{n}$. It is called *super short* if $k = 2$.

The problem of sorting by cyclic super short reversals consists in finding the minimum number of cyclic super short reversals that transform a permutation $\pi \in S_n$ into $\iota_n = (1 \; 2 \ldots n)$. This number is referred to as the *cyclic super short reversal distance* of permutation π and it is denoted by $d(\pi)$.

Let $S(\pi_i, \pi_j)$ denote the act of switching the positions of the elements π_i and π_j in a permutation π. Note that the cyclic 2-reversal $\rho(i, j)$ can be alternatively denoted by $S(\pi_i, \pi_j)$. Given a sequence S of cyclic super short reversals and a permutation $\pi \in S_n$, let $R_S(\pi_i)$ be the number of cyclic 2-reversals of the type $S(\pi_i, \pi_j)$ and let $L_S(\pi_i)$ be the number of cyclic 2-reversals of the type $S(\pi_k, \pi_i)$. In other words, $R_S(\pi_i)$ denotes the number of times a cyclic 2-reversal moves the element π_i to the right and $L_S(\pi_i)$ denotes the number of times a cyclic 2-reversal moves the element π_i to the left. We define the *net displacement* of an element π_i with respect to S as $d_S(\pi_i) = R_S(\pi_i) - L_S(\pi_i)$. The *displacement vector* of π with respect to S is defined as $d_S(\pi) = (d_S(\pi_1), d_S(\pi_2), \ldots, d_S(\pi_n))$.

Lemma 1. *Let $S = \rho_1, \rho_2, \ldots, \rho_t$ be a sequence of cyclic super short reversals that sorts a permutation $\pi \in S_n$. Then, we have that*

$$\sum_{i=1}^{n} d_S(\pi_i) = 0, \tag{1}$$

$$\pi_i - d_S(\pi_i) \equiv i \pmod{n}. \tag{2}$$

Proof. Let L_S be the number of times a cyclic super short reversal of S moves an element to the left and let R_S be the number of times a cyclic super short reversal of S moves an element to the right. Then, $L_S = R_S$ because a cyclic super reversal always moves two elements, one for each direction. Therefore, we have that $\sum_{i=1}^{n} d_S(\pi_i) = \sum_{i=1}^{n} (R_S(\pi_i) - L_S(\pi_i)) = R_S - L_S = 0$ and equation 1 follows. The equation 2 follows from the fact that, once the permutation is sorted, all of its elements must be in the correct position. $\square$

Note that, in one hand, we can think of a sequence of cyclic super short reversals as specifying a displacement vector. On the other hand, we can also think of a displacement vector as specifying a sequence of cyclic super short reversals. Let $x = (x_1, x_2, \ldots, x_n) \in Z^n$ be a vector and $\pi \in S_n$ be a permutation. We say that x is a *valid vector* for π if $\sum_i x_i = 0$ and $\pi_i - x_i \equiv i \pmod{n}$. Given a vector $x = (x_1, x_2, \ldots, x_n) \in Z^n$ and two distinct integers $i, j \in \{1, 2, \ldots, n\}$, let $r = i - j$ and $s = (i + x_i) - (j + x_j)$. The *crossing number* of i and j with respect to x is defined by

$$c_{ij}(x) = \begin{cases} |\{k \in [r, s] : k \equiv 0 \pmod{n}\}| & \text{if } r \le s, \\ -|\{k \in [s, r] : k \equiv 0 \pmod{n}\}| & \text{if } r > s. \end{cases}$$

The crossing number of x is defined by $C(x) = \frac{1}{2}\sum_{i,j}|c_{ij}(x)|$. Intuitively, if S is a sequence of cyclic super short reversals that sorts a permutation π and $d_S(\pi) = x$, then $c_{ij}(x)$ measures the number of times the elements π_i and π_j must "cross", that is, the number of cyclic 2-reversals of type $S(\pi_i, \pi_j)$ minus the number of cyclic 2-reversals of type $S(\pi_j, \pi_i)$. Using the notion of crossing number, Jerrum [6] was able to prove the following fundamental lemma.

Lemma 2 (Jerrum [6]). *Let S be a minimum-length sequence of cyclic super short reversals that sorts a permutation $\pi \in S_n$ and let $x \in Z^n$ be a valid vector for π. If $d_S(\pi) = x$, then $d(\pi) = C(x)$.*

The Lemma 2 allows the problem of sorting a permutation π by cyclic super short reversals to be recast as the optimisation problem of finding a valid vector $x \in Z^n$ for π with minimum crossing number. More specifically, as Jerrum [6] pointed out, this problem can formulated as the integer program:

$$\text{Minimize } C(x) \text{ over } Z^n$$
$$\text{subject to } \sum_i x_i = 0, \ \pi_i - x_i \equiv i \pmod{n}.$$

Although solving an integer program is NP-hard in the general case, Jerrum [6] presented a polynomial-time algorithm for solving this one.

Firstly, Jerrum [6] introduced a transformation $T_{ij} : Z^n \to Z^n$ defined as follows. For any vector $x \in Z^n$, the result, $x' = T_{ij}(x)$, of applying T_{ij} to x is given by $x'_k = x_k$ for $k \notin \{i,j\}$, $x'_i = x_i - n$, and $x'_j = x_j + n$. Lemma 3 shows what is the effect of this transformation on the crossing number of a vector.

Lemma 3. *Let x and x' be two vectors over Z^n such that $x' = T_{ij}(x)$. Then, $C(x') - C(x) = 2(n + x_j - x_i)$.*

Proof. The proof of this lemma is given by Jerrum [6, Theorem 3.9]. We note, however, that he mistakenly wrote that $C(x') - C(x) = 4(n + x_j - x_i)$. In other words, he forgot to divide the result by 2. This division is necessary because the crossing number of a vector is the half of the sum of the crossing numbers of its indices.   □

Let $\max(x)$ and $\min(x)$ respectively denote the maximum and minimum component values of a vector $x \in Z^n$. The transformation T_{ij} is said to *contract* x iff $x_i = \max(x)$, $x_j = \min(x)$ and $x_i - x_j \geq n$. Moreover, T_{ij} is said to *strictly contract* x iff, in addition, the final inequality is strict. The algorithm proposed by Jerrum [6] starts with a feasible solution to the integer program and performs a sequence of strictly contracting transformations which decrease the value of the crossing number. When no further strictly contracting transformation can be performed, the solution is guaranteed to be optimal. This is because, as showed by Jerrum [6], any two local optimum solutions (*i.e* solutions which admit no strictly contracting transformation) can be brought into agreement with each other via a sequence of contracting transformations. The detailed algorithm is given below (Algorithm 1).

Regarding the time complexity of Algorithm 1, we have that line 1 and the for loop of lines 2-4 take $O(n)$ time. Jerrum [6] observed that none of the variables x_i

---

Data: A permutation $\pi \in S_n$.
Result: Number of cyclic super short reversals applied for sorting π.

1 Let x be a n dimension vector
2 **for** $k = 1$ **to** n **do**
3 | $x_k \leftarrow \pi_k - k$
4 **end**
5 **while** $\max(x) - \min(x) > n$ **do**
6 | Let i,j be two integers such that $x_i = \max(x)$ and $x_j = \min(x)$
7 | $x_i \leftarrow x_i - n$
8 | $x_j \leftarrow x_j + n$
9 **end**
10 **return** $C(x)$

Algorithm 1. Algorithm for sorting by cyclic super short reversals

changes value more than once, therefore the while loop iterates only $O(n)$ times. As the lines 6-8 take $O(n)$ time, the while loop takes $O(n^2)$ time to execute. Since we can compute the value of $C(x)$ in $O(n^2)$ time, the overall complexity of the algorithm is $O(n^2)$.

Note that, in this section, we have focused on the problem of computing the cyclic super short reversal distance of a permutation rather than finding the minimum number of cyclic super short reversals that sorts it. As Jerrum [6] remarked, his proofs are constructive and directly imply algorithms for finding the sequence of cyclic super short reversals.

3   Sorting by Signed Cyclic Super Short Reversals

A *signed permutation* π is a bijection of $\{-n, \ldots, -2, -1, 1, 2, \ldots, n\}$ onto itself that satisfies $\pi(-i) = -\pi(i)$ for all $i \in \{1, 2, \ldots, n\}$. The two-row notation for a signed permutation is

$$\pi = \begin{pmatrix} -n & \ldots & -2 & -1 & 1 & 2 & \ldots & n \\ -\pi_n & \ldots & -\pi_2 & -\pi_1 & \pi_1 & \pi_2 & \ldots & \pi_n \end{pmatrix},$$

$\pi_i \in \{1, 2, \ldots, n\}$ for $1 \leq i \leq n$. The notation used in genome rearrangement literature, which is the one we will adopt, is the one-row notation $\pi = (\pi_1\ \pi_2\ \ldots\ \pi_n)$. Note that we drop the mapping of the negative elements since $\pi(-i) = -\pi(i)$ for all $i \in \{1, 2, \ldots, n\}$. By abuse of notation, we say that π has size n. The set of all signed permutations of size n is $S_n^{\pm}$.

A *signed cyclic reversal* $\rho(i, j)$ is an operation that transforms a signed permutation $\pi = (\pi_1\ \pi_2\ \ldots\ \pi_{i-1}\ \underline{\pi_i\ \pi_{i+1}\ \ldots\ \pi_{j-1}\ \pi_j}\ \pi_{j+1}\ \ldots\ \pi_n)$ into the signed permutation $\pi \cdot \rho(i, j) = (\pi_1\ \pi_2\ \ldots\ \pi_{i-1}\ \underline{-\pi_j\ -\pi_{j-1}\ \ldots\ -\pi_{i+1}\ -\pi_i}\ \pi_{j+1}\ \ldots\ \pi_n)$ if $1 \leq i \leq j \leq n$ and transforms a signed permutation $\pi = (\underline{\pi_1\ \pi_2\ \ldots\ \pi_i\ \pi_{i+1}\ \ldots\ \pi_{j-1}}\ \pi_j\ \pi_{j+1}\ \ldots\ \pi_n)$ into the signed permutation $\pi \cdot \rho(i, j) = (\underline{-\pi_j\ -\pi_{j+1}\ \ldots\ -\pi_n}\ \pi_{i+1}\ \ldots\ \pi_{j-1}\ \underline{-\pi_i\ -\pi_{i-1}\ \ldots\ -\pi_1})$ if $1 \leq j < i \leq n$. The signed cyclic reversal

$\rho(i, j)$ is called a *signed cyclic k-reversal* if $k = j - i + 1 \pmod{n}$. It is called *super short* if $k \leq 2$.

The problem of sorting by signed cyclic super short reversals consists in finding the minimum number of signed cyclic super short reversals that transform a permutation $\pi \in S_n^{\pm}$ into ι_n. This number is referred to as the *signed cyclic super short reversal distance* of permutation π and it is denoted by $d^{\pm}(\pi)$.

Let $S(|\pi_i|, |\pi_j|)$ denote the act of switching the positions and flipping the signs of the elements π_i and π_j in a signed permutation π. Note that the signed cyclic 2-reversal $\rho(i, j)$ can be alternatively denoted by $S(|\pi_i|, |\pi_j|)$. Given a sequence S of cyclic signed super short reversals and a signed permutation $\pi \in S_n^{\pm}$, let $R_S(\pi_i)$ be the number of signed cyclic 2-reversals of the type $S(|\pi_i|, |\pi_j|)$ and let $L_S(\pi_i)$ be the number of signed cyclic 2-reversals of the type $S(|\pi_k|, |\pi_i|)$. We define the *net displacement* of an element π_i with respect to S as $d_S(\pi_i) = R_S(\pi_i) - L_S(\pi_i)$. The *displacement vector* of π with respect to S is defined as $d_S(\pi) = (d_S(\pi_1), d_S(\pi_2), \ldots, d_S(\pi_n))$. The following lemma is the signed analog of Lemma 1. We omit the proof because it is the same as of the proof of Lemma 1.

Lemma 4. *Let $S = \rho_1, \rho_2, \ldots, \rho_t$ be a sequence of signed cyclic super short reversals that sorts a signed permutation $\pi \in S_n^{\pm}$. Then, we have that*

$$\sum_{i=1}^{n} d_S(\pi_i) = 0, \tag{3}$$

$$|\pi_i| - d_S(\pi_i) \equiv i \pmod{n}. \tag{4}$$

Let $x \in Z^n$ be a vector and $\pi \in S_n^{\pm}$ be a signed permutation. We say that x is a *valid vector* for π if $\sum_i x_i = 0$ and $|\pi_i| - x_i \equiv i \pmod{n}$. Given a valid vector x for the signed permutation π, we define the set $podd(\pi, x)$ as $podd(\pi, x) = \{i : \pi_i > 0 \text{ and } |x_i| \text{ is odd}\}$ and we define the set $neven(\pi, x)$ as $neven(\pi, x) = \{i : \pi_i < 0 \text{ and } |x_i| \text{ is even}\}$. Moreover, let $U(\pi, x)$ denote the union of these sets, that is, $U(\pi, x) = podd(\pi, x) \cup neven(\pi, x)$. The following lemma is the signed analog of Lemma 2.

Lemma 5. *Let S be a minimum-length sequence of signed cyclic super short reversals that sorts a signed permutation $\pi \in S_n^{\pm}$ and let $x \in Z^n$ be a valid vector for π. If $d_S(\pi) = x$, then $d^{\pm}(\pi) = C(x) + |U(\pi, x)|$.*

Proof. Note that the sequence S can be decomposed into two distinct subsequences S_1 and S_2 such that S_1 is formed by the signed cyclic 1-reversals of S and S_2 is formed by the signed cyclic 2-reversals of S. Moreover, we can assume without loss of generality that the signed cyclic reversals of subsequence S_2 are applied first. We argue that $|S_1| = |U(\pi, x)|$ regardless the size of S_2. To see this, suppose that we apply a signed cyclic 2-reversal $\rho(i, j)$ of S_2 in π, obtaining a signed permutation π'. Moreover, let S' be the resulting sequence after we remove $\rho(i, j)$ from S. We have that $d_{S'}(\pi'_k) = d_S(\pi_k)$ for $k \notin \{i,j\}$, $d_{S'}(\pi'_i) = d_S(\pi_i) - 1$, and $d_{S'}(\pi'_j) = d_S(\pi_j) + 1$. Then, assuming the vector $x' \in Z^n$ is equal to $d_{S'}(\pi')$, we can conclude that $U(\pi', x') = U(\pi, x)$ because $\rho(i, j)$ has changed both the parities of $|x_i|$ and $|x_j|$ and the signs of π_i and π_j. Since $|S_1| = |U(\pi, x)|$ regardless the size of S_2 and we know from Lemma 2 that $|S_2| \geq C(x)$, we can conclude that $|S_2| = C(x)$, therefore the lemma follows.   □

The Lemma 5 allows the problem of sorting a signed permutation π by signed cyclic super short reversals to be recast as the optimisation problem of finding a valid vector $x \in Z^n$ for π which minimizes the sum $C(x) + |U(\pi, x)|$. The next theorem shows how to solve this problem in polynomial time.

Theorem 1. *Let $\pi \in S_n^{\pm}$ be a signed permutation. Then, we can find a valid vector $x \in Z^n$ which minimizes the sum $C(x) + |U(\pi, x)|$ in polynomial time.*

Proof. We divide our analysis into two cases:

i) n is even. In this case, we have that the value of $|U(\pi, x)|$ is the same for any feasible solution x. This is because, in order to be a feasible solution, a vector x has to satisfy the restriction $|\pi_i| - x_i \equiv i \pmod{n}$. This means that x_i is congruent modulo n with $a = |\pi_i| - i$ and belongs to the equivalent class $\{\ldots, a - 2n, a - n, a, a + n, a + 2n, \ldots\}$. Since n is even, the parities of the absolute values of the elements in this equivalence class are the same, therefore the value of $|U(\pi, x)|$ is the same for any feasible solution x. It follows that we can only minimize the value of $C(x)$ and this can be done by performing successive strictly contracting transformations.

ii) n is odd. In this case, it is possible to minimize the values of $|U(\pi, x)|$ and $C(x)$. Firstly, we argue that minimizing $C(x)$ leads to a feasible solution x'' such that $C(x'') + |U(\pi, x'')|$ is at least as low as $C(x') + |U(\pi, x')|$, where x' can be any feasible solution such that $C(x')$ is not minimum. To see this, let x' be a feasible solution such that $C(x')$ is not minimum. Then, we can perform a sequence of strictly contracting transformations which decrease the value of $C(x)$. When no further strictly contracting transformation can be performed, we obtain a solution x'' such that $C(x'')$ is minimum. On one hand, we know from Lemma 3 that each strictly contracting transformation T_{ij} decreases $C(x)$ by at least 2 units. On the other hand, since n is odd, its possible that the parities of $|x_i|$ and $|x_j|$ have been changed in such a way that the value of $|U(\pi, x)|$ increases by 2 units. Therefore, in the worst case, each strictly contracting transformation does not change the value of $C(x) + |U(\pi, x)|$, so $C(x') + |U(\pi, x')| \geq C(x'') + |U(\pi, x'')|$. Now, we argue that, if there exists more than one feasible solution x such that $C(x)$ is minimum, then it is still may be possible to minimize the value of $|U(\pi, x)|$.

Jerrum [6, Theorem 3.9] proved that if there is more than one feasible solution such that $C(x)$ is minimum, then each of these solutions can be brought into agreement with each other via a sequence of contracting transformations. Note that a contracting transformation T_{ij} does not change the value of $C(x)$, but it can change the value of $|U(\pi, x)|$ because n is odd and the parities of $|x_i|$ and $|x_j|$ change when T_{ij} is performed. This means that, among all feasible solutions such that $C(x)$ is minimum, some of them have minimum $|U(\pi, x)|$ and these solutions are optimal. Therefore, we can obtain an optimal solution by first obtaining a feasible solution with minimum $C(x)$ (this can be done by performing successive strictly contracting transformations) and then we can apply on it every possible contracting transformation T_{ij} which decreases the value of $|U(\pi, x)|$. □

The proof of Theorem 1 directly implies an exact algorithm for sorting by signed cyclic super short reversals. Such an algorithm is described below (Algorithm 2). Regarding its time complexity, we know from previous section that lines 1-9 take $O(n^2)$ time. Since lines 13-23 take $O(1)$ time, we can conclude that the nested for loops take $O(n^2)$ times to execute. Finally, we can compute $C(x) + |U(\pi, x)|$ in $O(n^2)$, therefore the overall complexity of Algorithm 2 is $O(n^2)$.

Data: A permutation $\pi \in S_n^{\pm}$.
Result: Number of signed cyclic super short reversals applied for sorting π.

1 Let x be a n dimension vector
2 **for** $k = 1$ **to** n **do**
3 $\quad\big|\quad x_k \leftarrow |\pi_k| - k$
4 **end**
5 **while** $\max(x) - \min(x) > n$ **do**
6 $\quad\big|\quad$ Let i,j be two integers such that $x_i = \max(x)$ and $x_j = \min(x)$
7 $\quad\big|\quad x_i \leftarrow x_i - n$
8 $\quad\big|\quad x_j \leftarrow x_j + n$
9 **end**
10 **if** n is odd **then**
11 $\quad\big|\quad$ **for** $i = 1$ **to** $n - 1$ **do**
12 $\quad\quad\big|\quad$ **for** $j = i + 1$ **to** n **do**
13 $\quad\quad\quad\big|\quad$ **if** $x_i > x_j$ **then**
14 $\quad\quad\quad\quad\big|\quad min \leftarrow j$
15 $\quad\quad\quad\quad\big|\quad max \leftarrow i$
16 $\quad\quad\quad\big|\quad$ **else**
17 $\quad\quad\quad\quad\big|\quad min \leftarrow i$
18 $\quad\quad\quad\quad\big|\quad max \leftarrow j$
19 $\quad\quad\quad\big|\quad$ **end**
20 $\quad\quad\quad\big|\quad$ **if** $x_{max} - x_{min} = n$ and $min \in U(\pi, x)$ and $max \in U(\pi, x)$ **then**
21 $\quad\quad\quad\quad\big|\quad x_{max} \leftarrow x_i - n$
22 $\quad\quad\quad\quad\big|\quad x_{min} \leftarrow x_j + n$
23 $\quad\quad\quad\big|\quad$ **end**
24 $\quad\quad\big|\quad$ **end**
25 $\quad\big|\quad$ **end**
26 **end**
27 **return** $C(x) + |U(\pi, x)|$

Algorithm 2. Algorithm for sorting by signed cyclic super short reversals

Note that, in this section, we have focused on the problem of computing the signed cyclic super short reversal distance of a signed permutation rather than finding the minimum number of signed cyclic super short reversals that sorts it. We remark that the proofs are constructive and directly imply algorithms for finding the sequence of signed cyclic super short reversals.

4   Sorting Circular Permutations

In this section, we briefly explain how we can use the solution for the problem of sorting by (signed) cyclic super short reversals to solve the problem of sorting a (signed) circular permutation by super short reversals. This explanation is based on Sect. 2.3 of the work of Egri-Nagy *et al.* [3] and on Sect. 2.5 of the book of Fertin *et al.* [4], where one can find more details.

Note that a circular permutation can be "unrolled" to produce a linear permutation, such as defined in the two previous sections. This process can produce n different linear permutations, one for each possible rotation of the circular permutation. Moreover, since a circular permutation represents a circular chromosome, which lives in three dimension, it can also be "turned over" before being unrolled. This means that, for each possible rotation of the circular permutation, we can first turn it over and then unroll it, producing a linear permutation. Again, this process can produce n different linear permutations. The n linear permutations produced in the first process are different from the n linear permutations produced in the second process, thus both processes can produce a total of $2n$ different linear permutations. Each of these $2n$ linear permutations represents a different viewpoint from which to observe the circular permutation, therefore they are all equivalent.

The discussion of the previous paragraph leads us to conclude that, in order to sort a (signed) circular permutation by super short reversals, we can sort each of the $2n$ equivalent (signed) linear permutations by (signed) cyclic super short reversals, generating $2n$ different sorting sequences. Then, we can take the sequence of minimum length as the sorting sequence for the (signed) circular permutation and the *super short reversal distance* of the (signed) circular permutation is the length of this sequence. Note that this procedure takes $O(n^3)$ time because we have to execute Algorithm 1 or Algorithm 2 $O(n)$ times.

5   Experimental Results and Discussion

We implemented the procedure described in the previous section for computing the super short reversal distance of a signed circular permutation and we reproduced the experiment performed by Egri-Nagy *et. al.* [3] for inferring distances and phylogenies for published *Yersinia* genomes. In fact, we performed the same experiment, except that we considered the orientation of the genes rather than ignoring it and we considered that each permutation has 78 elements rather than 79^1. More specifically, we obtained from Darling *et al.* [2] the signed circular permutations which represent eight *Yersinia* genomes. Then, we computed the super short reversal distance between every pair of signed circular permutation and this information was stored in a matrix of pairwise distances (Table 1). Finally, a phylogenetic tree was constructed from this matrix using *Neighbor-Joining* [9] method. The resulting phylogeny is shown in Fig. 1.

Table 1. Matrix of the super short reversal distances among the signed circular permutations which represent the *Yersinia* genomes. The names of the species were abbreviated so that YPK refers to *Y. pestis Kim*, YPA to *Y. pestis Antiqua*, YPM to *Y. pestis Microtus 91001*, YPC to *Y. pestis CO92*, YPN to *Y. pestis Nepal516*, YPP to *Y. pestis Pestoides F 15-70*, YT1 to *Y. pseudotuberculosis IP31758*, and YT2 to *Y. pseudotuberculosis IP32953*.

	YPK	YPA	YPM	YPC	YPN	YPP	YT1	YT2
YPK	0	243	752	205	338	533	764	760
YPA	243	0	772	352	279	510	724	773
YPM	752	772	0	728	747	643	361	385
YPC	205	352	728	0	381	656	776	760
YPN	338	279	747	381	0	547	617	624
YPP	533	510	643	656	547	0	434	457
YT1	764	724	361	776	617	434	0	189
YT2	760	773	385	760	624	457	189	0

Considering the pair of *Y. pseudotuberculosis* as outgroup, the obtained phylogeny shows that *Y. pestis Microtus 91001* was the first to diverge. It was followed then by the divergences of *Y. pestis Pestoides F 15-70*, *Y. pestis Nepal516*, *Y. pestis Antiqua* and the final divergence of *Y. pestis Kim* and *Y. pestis CO92*. This result is different of the one obtained by Egri-Nagy *et. al.* [3] which used super short reversal distance between unsigned permutations. On their results, the divergence of *Y. pestis Nepal516* happened before the divergence of *Y. pestis CO92* which occurred previous to the divergence of *Y. pestis Kim* and *Y. pestis Antiqua*.

In our work and in the work of Egri-Nagy *et. al.* [3], the use of super short reversals resulted on topologies which are different from the one of Darling *et al.* [2], which considered inversions of any size. The first difference observed on the result of Darling *et al.* [2] is that *Y. pestis Pestoides F 15-70* diverged before *Y. pestis Microtus 91001*. The second difference shows that *Y. pestis Nepal516* is sibling of *Y. pestis Kim*, that *Y. pestis CO92* is sibling of *Y. pestis Antiqua*

---

[1] In their article, Darling *et al.* [2] state that they could identify 78 conserved segments (or blocks) using Mauve, but they provided permutations with elements ranging from 0 to 78. In a personal communication, Darling confirmed that there are actually 78 blocks, with 0 and 78 being part of the same block. Nevertheless, we performed another experiment, this time considering the permutations have 79 elements. Although the distances were greater, the topology of the tree was the same.

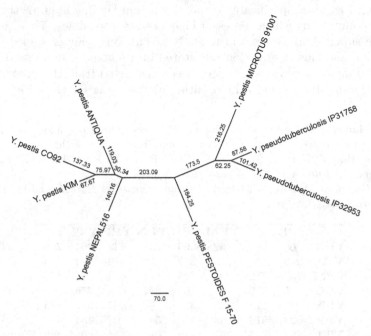

Fig. 1. Phylogeny of the *Yersinia* genomes based on the super short reversal distance of the signed circular permutations.

and that these four bacteria have a common ancestor that is descendant of *Y. pestis Microtus 91001.*

If we look to the branch lengths of the two trees obtained with super short reversal distances and we compare with the branch lengths of the topology obtained by Darling *et al.* [2], we can see that our results are more consistent than the one obtained by Egri-Nagy *et al.* [3]. For instance, on our results the distance between the two *Y. pseudotuberculosis* is smaller than the one observed between the pair *Y. pestis Kim* and *Y. pestis Antiqua*, what agrees with the configuration obtained by Darling *et al.* [2].

6   Conclusions

In this paper, we presented a polynomial-time solution for the problem of sorting a signed circular permutation by super short reversals. From a theoretical perspective, this solution is important because it closes a gap in the literature. From a biological perspective, it is important because signed permutations constitute a more adequate model for genomes. Moreover, we performed an experiment to infer distances and phylogenies for published *Yersinia* genomes and compared the results with the phylogenies presented in previous works [2,3]. Our obtained topology is similar to the one obtained by Egri-Nagy *et. al.* [3]. However, the distances calculated with our algorithm are more consistent with the topology obtained by Darling *et al.* [2]. Some theoretical questions remain open (for instance, the diameter of the super short reversal distance for signed permutations), and we intend to address them in our future research.

Acknowledgments. GRG acknowledges the support by the FAPESP and CAPES under grant #2014/04718-6, São Paulo Research Foundation (FAPESP). CB acknowledges the support by the French Project ANR MIRI BLAN08-1335497 and by the ERC Advanced Grant SISYPHE (FP7/2007-2013)/ERC grant agreement no. [247073]10. ZD acknowledges the support by the CNPq under grants 306730/2012-0, 477692/2012-5, and 483370/2013-4. Finally, the authors thank the Center for Computational Engineering and Sciences at Unicamp for financial support through the FAPESP/CEPID Grant 2013/08293-7.

References

1. Dalevi, D.A., Eriksen, N., Eriksson, K., Andersson, S.G.E.: Measuring genome divergence in bacteria: A case study using chlamydian data. Journal of Molecular Evolution 55(1), 24–36 (2002)
2. Darling, A.E., Miklós, I., Ragan, M.A.: Dynamics of genome rearrangement in bacterial populations. PLoS Genetics 4(7), e1000128 (2008)
3. Egri-Nagy, A., Gebhardt, V., Tanaka, M.M., Francis, A.R.: Group-theoretic models of the inversion process in bacterial genomes. Journal of Mathematical Biology 69(1), 243–265 (2014)
4. Fertin, G., Labarre, A., Rusu, I., Tannier, E., Vialette, S.: Combinatorics of Genome Rearrangements. The MIT Press, Cambridge (2009)
5. Galvão, G.R., Lee, O., Dias, Z.: Sorting signed permutations by short operations. Algorithms for Molecular Biology 10(12) (2015)
6. Jerrum, M.R.: The complexity of finding minimum-length generator sequences. Theoretical Computer Science 36, 265–289 (1985)
7. Lefebvre, J.F., El-Mabrouk, N., Tillier, E., Sankoff, D.: Detection and validation of single gene inversions. Bioinformatics 19(suppl 1), i190–i196 (2003)
8. Lemey, P., Salemi, M., Vandamme, A.: The Phylogenetic Handbook: A Practical Approach to Phylogenetic Analysis and Hypothesis Testing. Cambridge University Press, New York (2009)
9. Saitou, N., Nei, M.: The neighbor-joining method: a new method for reconstructing phylogenetic trees. Molecular Biology and Evolution 4(1), 406–425 (1987)
10. Seoighe, C., Federspiel, N., Jones, T., Hansen, N., Bivolarovic, V., Surzycki, R., Tamse, R., Komp, C., Huizar, L., Davis, R.W., Scherer, S., Tait, E., Shaw, D.J., Harris, D., Murphy, L., Oliver, K., Taylor, K., Rajandream, M.A., Barrell, B.G., Wolfe, K.H.: Prevalence of small inversions in yeast gene order evolution. Proceedings of the National Academy of Sciences of the United States of America 97(26), 14433–14437 (2000)

Multiple Alignment of Structures Using Center Of ProTeins

Kaushik Roy, Satish Chandra Panigrahi[✉], and Asish Mukhopadhyay

School of Computer Science
University of Windsor
401 Sunset Avenue
Windsor, ON N9B 3P4, Canada
{roy113,panigra,asishm}@uwindsor.ca

Abstract. Multiple Structure Alignment (MStA) is a fundamental problem in Computational Biology and proteomics in particular. Therefore, a number of algorithms have been proposed, chief among which are MUSTANG, POSA, MultiProt, CE-MC. In this paper we propose a new algorithm MASCOT that imitates the center-star algorithm for multiple sequence alignment. We report the root mean square deviations (RMSD) and execution times for a large number of alignments and discuss their significance. We also include a comparison of the execution times of MASCOT with the well-known and widely-used algorithm MUSTANG.

1 Introduction

Multiple Structure Alignment (MStA) is a fundamental problem in Computational Biology, and proteomics in particular, in view of the applications to homology modelling, molecular replacement, finding conserved structures in protein families to name a few. Thus it has been the focus of extensive research by computational biologists. The multiple alignment problem is more challenging than pairwise alignment even for sequences, and we resort to heuristics to find as best an approximation as possible, in polynomial time. In view of the rapid growth of the Protein Data Bank [1], the need for fast, robust and reliable MStA algorithms can hardly be overstated.

Based on the technique used, MStA algorithms fall into roughly one of four categories. MUSTANG [2], msTali [3], and CE-MC [4] imitates the progressive alignment approach used for multiple sequence alignment. Despite their successes, these algorithms suffer from the disadvantages inherent to this technique, such as not guaranteeing convergence to the global optimum. Algorithms based on other approaches, [5] [6], often outperform the progressive ones, both in speed and accuracy. A second approach is to optimize a consensus structure, sometimes with several iterations, and report a common core of the input proteins. The goal is to find out a structurally conserved subset of residues among the proteins to gain some insight into their evolutionary origins. However, such cores

This research is supported by an NSERC Discovery Grant.

are often pseudo-structures that, although geometrically interesting, may not have any biological significance. MATT [5], MultiProt [7], Mass [8], Mapsci [9], and Smolign [10] belong to this category of MStA algorithms. Ye and Godzik's graph-based POSA [6] takes a totally different path by representing a protein as a directed acyclic graph (DAG) of residues, connected in the order following the backbone. POSA then creates a combined non-planar, multi-dimensional DAG, taking hinge rotation into account, to come up with residue equivalences among the input proteins. Though POSA is novel in incorporating the flexibility of protein structures, it is known to completely miss motifs on TIM-barrel and helix-bundle proteins [10], and incur a higher cost of alignment than for example, MATT or Smolign [10]. The pivot-based approach selects one of the input molecules, 'closest' to all the other proteins as the pivot. The remaining proteins are then iteratively aligned to the pivot either in a bottom-up [11], or in a top-down manner [12] to come up with residue-residue correspondences that are later used to minimize some objective function and derive a score as a similarity measure. Some of the few published algorithms in this category, are Mistral [13], [11], and [12]. Our approach is an adaptation of the center-star method for producing a multiple sequence alignment (MSA) to the MStA problem.

In this paper we introduce a new algorithm, **MASCOT** (acronym for **M**ultiple **A**lignment of **S**tructures using **C**enter **O**f pro**T**eins), for aligning two or more protein structures. Our algorithm takes advantage of the linear structure of the protein polypeptide chain, while judiciously preserving the secondary structure elements (SSEs). The justification for this approach is that SSEs are fundamental components of protein structures, serving as a well-preserved scaffold. As a result, SSEs are evolutionarily remarkably conserved while mutations affect the loops, thus modifying functionality. For example, the substrate specificity of different serine proteases is governed by the conformation of the binding loops [14]. Further, representing protein structures by their SSEs has been successfully used on several previous occasions for the pairwise alignment problem ([15], [16], [17]; [18], [19]). Our goal is to develop a fast algorithm that uses these sequences of SSEs and produces a multiple alignment with high accuracy. Thus we have designed MASCOT as a hybrid algorithm that uses a center protein, obtained by minimizing a sum pairwise-distances, to drive a layout that identifies a set of residues from each protein that are similar. We then proceed to find an optimal correspondence among the backbone carbon atoms of these molecular structures, using inter-residue Euclidean distance threshold, and report the *centerRMSD* (*cRMSD*)of the structures aligned in space as a measure of similarity. We also include a comparison of the execution times of MASCOT with the well-known and widely-used algorithm MUSTANG.

2   Method

2.1   Input Data Set

Protein structures are stored as PDB files in the Protein Data Bank [1], which currently contains more than 99000 structures and is growing rapidly. The data can come from standard protein databases available, or from a local repository. Either way, the first step is to retrieve the correct data and suitably preprocess it for the algorithm to proceed. This is a tedious but important step as we might have to align thousands of macromolecules, and we must then ensure that all the requisite molecules are fetched and ready for the next step. For example, an input set could be 1AOR:A 7ACN 2ACT 1TTQ:B etc. As we can see, with inputs having such varied descriptions, in order to supply the right data to the algorithm, the preprocessing step is an important and crucial one.

2.2   Representing the Proteins

Once we have the correct input set, we take each protein in turn and represent it in a way that makes further processing convenient, while retaining all vital information. The following observations are factored into this key step towards obtaining a robust and fast multiple alignment algorithm.

(a) Too simple a representation could potentially miss crucial structural and functional information, whereas too complex a representation will demand innovative methods at every turn; Thus while a primary sequence is the simplest possible representation, it is well-known that they do not necessarily determine functionality. At the other extreme, POSA uses partial order graphs, while mulPBA uses PROFIT elements.
(b) Distinguish between equal and unequal length proteins; for equal length proteins we can represent the proteins by their coordinates and apply, for example, the algorithm of Panigrahi et. al [20]. Otherwise, we have available Janardan's method [12] that handles unequal length proteins by representing a protein as a set of vectors using gap vectors for a gap alignment.
(c) Speed of performance is crucial.

Thus, to strike a balance between complexity and functionality, we represent the proteins by motif sequences based on DSSP-program [21] output, assigning each residue to one of eight possible structural motifs (see Table 1). Simply put, we use the linear structure of the protein to stretch it out into a straight line, keeping the SSEs intact. For example, a DSSP representation of the protein SEA CUCUMBER CAUDINA (1HLM) is as below.

...HHHHGGGGZZIIIIITTHHHHHHHTTSSI...

This approach, we believe, captures the biological essence of the problem.

The main advantage of this is that we are now free to use a host of pattern matching algorithms that can compute an optimal alignment given any two such sequences. We exploit this observation in the next step.

2.3   Pairwise Global Alignment

First, we obtain the global alignment of every pair of the N DSSP sequences, corresponding to the N input proteins. For this we apply the Needleman Wunsch algorithm [22] with appropriate affine gap opening penalty, and the scoring matrix in Table 2.

Table 1. The DSSP code

Symbol	Motif
H	Alpha Helix
B	Beta bridge
G	Helix 3
E	Beta strand
T	Turn
S	Bend
I	Helix 5
Z	No motif

Table 2. Scoring Matrix

–	H	B	G	E	T	S	I	Z
H	1	0	1	0	0	0	1	0
B	0	1	0	1	0	0	0	0
G	1	0	1	0	0	0	1	0
E	0	1	0	1	0	0	0	0
T	0	0	0	0	1	1	0	0
S	0	0	0	0	1	1	0	0
I	1	0	1	0	0	0	1	0
Z	0	0	0	0	0	0	0	1

These pairwise alignments produce a primitive picture of SSE-SSE alignments. For example, pairwise alignments of the DSSP-based sequences of the globins 1DM1, 1MBC, 1MBA generate the following output.

```
1DM1 ..ZZZHHHHHHHHHHHHHHHHHTHHHHHHHHHHHHHHHHHSGGG-...
1MBA ..ZZZHHHHHHHHHHHHHHHHHHT-HHHHHHHHHHHHHHHHZGGG...

1DM1 ..ZZZHHHHHHHHHHHHHHHHHHTHHHHHHHHHHHHHHHH-SGGG...
1MBC ..ZZZHHHHHHHHHHHHHHHHGGGHHHHHHHHHHHHHHHHHZTHHH...

1MBC ..ZZZHHHHHHHHHHHHHHHHGGGHHHHHHHHHHHHHHHHHZTHH-H...
1MBA ..ZZZHHHHHHHHHHHHHHHHHHHTHHHHHHHHHHHHHHHHZGGGGG...
```

We can see that the helices are properly aligned against one another. These alignments are saved in a list and referred to when needed.

2.4 Center Protein

As in the center star method [23] for multiple sequence alignment, an $N \times N$ symmetric matrix is created (see Table 3) whose entries are edit distances between aligned pairs in the list mentioned above.

From this matrix we find the sum-of-pairs score (SP-score) and the center protein using the following equation.

$$P_c = min_i \sum_{j=1}^{N} EditDistance(P_i, P_j) \tag{1}$$

A protein having the minimum SP-score is chosen as the center protein, P_c, with respect to which other proteins are aligned. An edit-distance matrix for the globins 1DM1, 1MBC, 1MBA is shown in Table 4 from which it is easy to see that the globin 1DM1 with SP-score of 41 can be chsoen as the center protein, P_c.

Table 3. A sample edit-distance matrix

−	P_1	P_2	...	P_N
P_1	0	10	...	20
P_2	10	0	...	30
⋮	...	...	...	⋮
P_N	20	30	⋯	0

Table 4. Edit-distance matrix for the globins 1DM1, 1MBC, 1MBA

	1DM1	1MBC	1MBA
1DM1	0	37	4
1MBC	34	0	35
1MBA	7	35	0

Many MStA algorithms calculate a consensus structure in place of a real protein in the hope of finding a common core. MUSTANG, MultiProt, Janaradan [12], all attempt to obtain a template structure to drive their alignment process. However, it should be noted that a consensus structure, while geometrically and, perhaps, computationally convenient, may turn out to be a pseudo-structure, bereft of any biological significance. Thus we have chosen to work with an actual protein to drive the alignment.

2.5   Correspondence Matrix

Formally, a correspondence matrix is an $N \times l$ matrix with respect to a center protein P_c in the set of input proteins, $P = \{P_1, P_2, \ldots, P_N\}$ such that

$$max(|P_1|, |P_2|, \ldots, |P_N|) \leq l \leq |P_1| + |P_2| + \cdots + |P_N|$$

with the following properties:

(a) The ith row contains the DSSP-based sequence of protein P_i, with gaps.
(b) No column consists entirely of gaps
(c) Gives a good idea of residue equivalences to work with, should we have to apply rigid body superposition.

In this step, all alignment pairs between P_c and every other protein are retrieved from the saved list, and merged sequentially, using the following algorithm:

A sample correspondence matrix for the globin family is shown in Fig. 1. We can clearly see how the SSEs of all the proteins are aligned together in a column-wise fashion.

At this point we have identified conserved regions across all the proteins, but not aligned them in any way. Janaradan's method [12] reaches a similar result, creating a correspondence matrix by carefully manipulating vectors.

Algorithm. CORRESPONDENCE MATRIX

Require: Protein DSSP sequences S_1, S_2 upto S_N
Ensure: MSA of Sequences S_1 to S_N

    for all $i = 1...N - 1$ **do**
        use the alignment pair (S_c, S_i) and MSA$(S_c, S_1, S_2, \ldots, S_{i-1})$ to obtain
MSA$(S_c, S_1, S_2, \ldots, S_i)$ following the 'once a gap, always a gap' rule

---

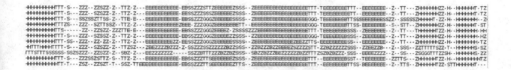

Fig. 1. A correspondence matrix [Notice there are no columns with gaps in all rows.]

The result is an MSA of DSSP-based sequences S_i for proteins P_i, $1 \leq i \leq N$. The output of this step is used to identify as many residue equivalences as possible given the raw protein structures. To actually align them in 3D space we feed this output to the next step.

2.6 Rigid Body Superposition

To generate the spatial alignment of a set of input proteins we have to apply the proper translation and rotation so that the distance between the alpha carbon atoms of equivalent residues is below a threshold value. For this we need a set of equivalences, and a reference frame against which the rigid body superposition is to take place. We have both: the correspondence matrix from phase 2 gives us the residue-residue equivalences, and our chosen center protein is the reference frame. So, if the correspondence is as in Table 5, some annotated equivalences between the center protein and the other protein are (1,3) (2,4) (4,5) (5,6) and (6,8).

For each protein P_i we apply Kabsch's method [24] to superpose the structures in space with respect to the center protein P_c. For example, the spatial alignment of the globins 1DM1, 1MBC, 1MBA (see Figs. 2, 3, 4) generated by our algorithm is shown in Fig. 5.

Residue no.			1	2	3	4	5		6
Center protein	-	-	H	H	T	I	E	-	G
Other protein	S	S	H	H	-	G	E	E	I
Residue no.	1	2	3	4		5	6	7	8

Table 5. Identifying equivalences

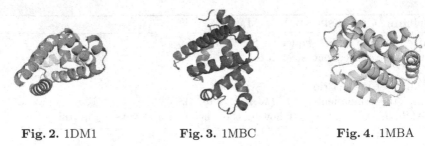

Fig. 2. 1DM1   **Fig. 3.** 1MBC   **Fig. 4.** 1MBA

Fig. 5. Alignment of 1DM1, 1MBC, and 1MBA

2.7   Dynamic Programming and Scoring

Once the proteins are spatially aligned by rigid body superposition, equivalent residues have been brought close to each other. We increase the number of equivalences between a pair of proteins by calculating the Euclidean distance between every pair of alpha carbon atoms, and declare the ones that fall within a threshold value(5Å) as equivalent pairs.

Finally, we use the following formula to derive the *centerRMSD* (*cRMSD*) that represents the quality of the alignment.

$$\frac{1}{N-1} \sum_{i=1,i\neq c}^{N} RMSD(P_i, P_c)$$

A good alignment is one where the score is typically less than half the threshold value(2.5Å). However, difficult alignments having biological relevance can exceed this value by about 1.5Å. The *cRMSD* value of the alignment in Fig. 5 is 0.44Å.

2.8   Pseudocode

A pseudo-code version of our algorithm is described below.

Algorithm. MASCOT

Require: Protein $pdbids : (pdbid_1, pdbid_2, \cdots, pdbid_N)$
Ensure: Multiple alignment of proteins with files created for $pdbids_{1...N}$

▷ Phase 1

1: Extract protein structures into $P = \{P_1, P_2, \ldots, P_N\}$
2: Represent P by corresponding DSSP-based sequences $S = \{S_1, S_2, \ldots, S_N\}$ consisting of DSSP-defined SSE motifs
3: Perform pairwise global alignment of every (S_i, S_j) pair, using custom similarity matrix

▷ Phase 2

4: Create an edit-distance matrix that stores the distances between every (P_i, P_j) using a custom scoring function
5: Choose the protein(sequence) with index c having minimum SP-score as the center protein(sequence) $P_c(S_c)$
6: Create an MSA of S w.r.t S_c using the center-star approach

▷ Phase 3

7: Treat all alignments of symbols with non-gaps as residue-residue equivalences of the pair (P_i, P_j)
8: Apply Kabsch's method to every (P_i, P_c) pair to obtain $(trans_i, rot_i)$ for this pair
9: Use $(trans_i, rot_i)$ from Step 8 to transform and place P_i in space, with P_c being brought to origin first, to produce output pdb files

---

3 Results and Discussion

MASCOT was implemented in Python 2.7.5 using packages from Bio-python 2.0 on an Alienware laptop-Intel corei7 CPU, 3 GHz and 8 GB RAM, running under Fedora(64 bit). A large variety of experiments were conducted, among which a representative set of 7 results have been presented here. Due to space constraints, the results of another set of six experiments have been relegated to the arXiv report CoRR abs/1412.8093 (2014) [25]. Note that T_T and T_G represents the time taken right from giving the input to producing the output files for MASCOT and MUSTANG respectively.

3.1 Globins

Globins are some of the most rigorously studied proteins by the MStA community. The globin family has long been known from studies of approximately 150-residue proteins such as vertebrate myoglobins and haemoglobins. The following globins have been aligned using MASCOT:

Table 6. The table below shows the globins used in this section

Name	PDB ids	Count	T_T	T_G
Set 1	1HHO:A 2DHB:A 2DHB:B 1HHO:B 1MBD 1DLW 1DLY 1ECO 1IDR:A 2LH7	10	23s	29s
Set 2	1MBC 1MBA 1DM1 1HLM 2LHB 2FAL 1HBG 1FLP 1ECA 1ASH	10	24s	26s
Set 3	5MBN 1ECO 2HBG 2LH3 2LHB 4HHB:B 4HHB:A	7	13s	13s
Set 4	1ASH 1ECA 1GDJ 1HLM 1MBA 1BAB:A 1EW6:A 1H97:A 1ITH:A 1SCT:A 1DLW:A 1FLP 1HBG 1LHS 1MBC 1DM1 2LHB 2FAL 1HBG 1FLP	20	1m 38s	1m 47s

Fig. 6. Set 1   **Fig. 7.** Set 2   **Fig. 8.** Set 3   **Fig. 9.** Set 4

Set 1 is used by [13], and [10] to show how their algorithms align globins. The *cRMSD* for this superposition is 2.765Å. Set 2, taken from [12], has been aligned with an *cRMSD* of 2.39Å. Set 3 is [7]'s test data with *cRMSD* 2.41Å. Set 4 is a custom assortment of 20 globins created from [26] and [9]. The purpose is to see how well they are aligned visually and with how much *cRMSD*. As one can see, the helices and the hinges are placed within the threshold distance as much as possible, with *cRMSD* 2.038Å.

3.2   Serpins

Serpins play an important role in the biological world. For instance, thyroxine-hinding globulin is a serpine which transports hormones to various parts of the body, and Maspin is a serpine which controls gene expression of certain tumors [27]. The name Serpin stands for Serine Protease Inhibitors. The following serpins have been aligned using MASCOT:

Table 7. The table below shows the serpins used in this section

Name	PDB ids	Count	T_T	T_G
Set 5	7API:A 8API:A 1HLE:A 1OVA:A 2ACH:A 9API:A 1PSI 1ATU 1KCT 1ATH:A 1ATT:A 1ANT:L 2ANT:L	13	3m 33s	4m 14s

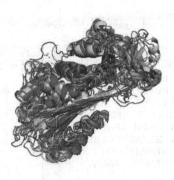

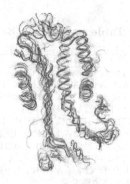

Fig. 10. Set 5   **Fig. 11.** Set 5 LIPR

The serpins in set 5 is the same one used by [7] and is said to be quite difficult owing to their large size and motif distribution. Unlike [7] we do not attempt to find a common core. Instead, we perform a global alignment over the length of the proteins. Fig. 10 shows how the beta sheets, hinges, and helices are aligned together in spite of the difficulty. Also some non-alignable parts have been correctly identified and left out. The *cRMSD* for this alignment is 2.99Å. Fig. 11 is a low intensity PyMol rendition (LIPR) of the same alignment viewed from another angle. It uses a ribbon representation to condense the output and show most of the aligned portions of the proteins. The pictures suggest that all these serpins share functionality and purpose, within the body. We can club all these proteins into a single family, and keep adding to it as and when such high similarities are found.

Fig. 12. Set 6   **Fig. 13.** Set 6 LIPR   **Fig. 14.** Set 7

3.3   Barrels

The eight-stranded TIM-barrel is found in a lot of enzymes, but the evolutionary history of this family has been the subject of rigorous debate. The ancestry of this family is still a mystery. Aligning TIM-barrel proteins will allow us to add to this ever-expanding family. The proteins aligned in this category are as follows:

MASS [8] has used the 66 molecules in set 6 to show how it aligns proteins with barrels. MASCOT produces an *cRMSD* of 3.4Å for this alignment. Fig. 12

Table 8. The table below shows the barrels used in this section

Name	PDB ids	Count	T_T	T_G
Set 6	1A49:A 1A49:B 1A49:C 1A49:D 1A49:E 1A49:F 1A49:G 1A49:H 1A5U:A 1A5U:B 1A5U:C 1A5U:D 1A5U:E 1A5U:F 1A5U:G 1A5U:H 1AQF:A 1AQF:B 1AQF:C 1AQF:D 1AQF:E 1AQF:F 1AQF:G 1AQF:H 1F3X:A 1F3X:B 1F3X:C 1F3X:D 1F3X:E 1F3X:F 1F3X:G 1F3X:H 1PKN 1F3W:A 1F3W:B 1F3W:C 1F3W:D 1F3W:E 1F3W:F 1F3W:G 1F3W:H 1PKM 1PKL:A 1PKL:B 1PKL:C 1PKL:D 1PKL:E 1PKL:F 1PKL:G6 1PKL:H 1A3W:A 1A3W:B 1A3X:A 1A3X:B 1E0T:A 1E0T:B 1E0T:C 1E0T:D 1PKY:A 1PKY:B 1PKY:C 1PKY:D7 1E0U:A 1E0U:B 1E0U:C 1E0U:D	66	2h 25m	2h 32m
Set 7	1SW3:A 1SW3:B 1WYI:A 1WYI:B 2JK2:A 2JK2:B 1R2T:A 1R2T:B 1R2R:A 1R2R:B 1M5W:A 1M5W:B 1M5W:C	13	1m 22s	1m 26s

shows how the new algorithm can superimpose proteins having the TIM barrel supermotifs. Fig. 13 is an LIPR of the same alignment, for convenience. The result indicates these proteins have structurally highly conserved regions since all 8 helices and 8 beta sheets have been aligned. Set 7 has been taken from the gold standard manually curated SCOP database. The proteins are taken from different superfamilies but, as Fig. 14 suggests, MASCOT is still able to align the barrel motifs on top of each other, with an *cRMSD* of 3.76Å.

4   Conclusions

MASCOT is a fast and elegant algorithm that succeeds in overcoming the major hurdles inherent in the multiple structure alignment problem by using a sum-of-pairs heuristic to choose a center protein and aligning all the other proteins with it. The excellent *cRMSD* scores obtained in a variety of experiments support our claim.

MASCOT can be extended to include the following functionalities in future:

1. Improve accuracy for aligning theoretical proteins.
2. Incorporate protein flexibility into the algorithm.
3. Derive a common core structure from the aligned input proteins for use as a template for protein threading.

A web application based on MASCOT has been developed by an University of Windsor undergraduate student Michael Salvadore. It will soon be made publicly available.

Acknowledgement. The authors wish to acknowledge the anonymous reviewers for their helpful comments.

References

1. Berman, H.M., Westbrook, J., Feng, Z., Gilliland, G., Bhat, T.N., Weissig, H., Shindyalov, I.N., Bourne, P.E.: The protein data bank. Nucleic Acids Research 28(1), 235–242 (2000)
2. Konagurthu, A.S., Whisstock, J.C., Stuckey, P.J., Lesk, A.M.: MUSTANG: a multiple structural alignment algorithm. Proteins: Structure, Function, and Bioinformatics 64(3), 559–574 (2006)
3. Shealy, P., Valafar, H.: Multiple structure alignment with msTALI. BMC Bioinformatics 13(1), 105 (2012)
4. Guda, C., Lu, S., Scheeff, E.D., Bourne, P.E., Shindyalov, I.N.: CE-MC: a multiple protein structure alignment server. Nucleic Acids Research 32(suppl. 2), W100–W103 (2004)
5. Menke, M., Berger, B., Cowen, L.: Matt: local flexibility aids protein multiple structure alignment. PLoS Computational Biology 4(1), e10 (2008)
6. Ye, Y., Godzik, A.: Multiple flexible structure alignment using partial order graphs. Bioinformatics 21(10), 2362–2369 (2005)
7. Shatsky, M., Nussinov, R., Wolfson, H.J.: MultiProt - A multiple protein structural alignment algorithm. In: Guigó, R., Gusfield, D. (eds.) WABI 2002. LNCS, vol. 2452, pp. 235–250. Springer, Heidelberg (2002)
8. Dror, O., Benyamini, H., Nussinov, R., Wolfson, H.: Mass: multiple structural alignment by secondary structures. Bioinformatics 19(suppl. 1), i95–i104 (2003)
9. Ye, J., Ilinkin, I., Janardan, R., Isom, A.: Multiple structure alignment and consensus identification for proteins. In: Bücher, P., Moret, B.M.E. (eds.) WABI 2006. LNCS (LNBI), vol. 4175, pp. 115–125. Springer, Heidelberg (2006)
10. Sun, H., Sacan, A., Ferhatosmanoglu, H., Wang, Y.: Smolign: a spatial motifs-based protein multiple structural alignment method. IEEE/ACM Transactions on Computational Biology and Bioinformatics 9(1), 249–261 (2012)
11. Wang, S., Zheng, W.M.: Fast multiple alignment of protein structures using conformational letter blocks. Open Bioinformatics Journal 3, 69–83 (2009)
12. Ye, J., Janardan, R.: Approximate multiple protein structure alignment using the sum-of-pairs distance. Journal of Computational Biology 11(5), 986–1000 (2004)
13. Micheletti, C., Orland, H.: MISTRAL: a tool for energy-based multiple structural alignment of proteins. Bioinformatics 25(20), 2663–2669 (2009)
14. Hedstrom, L.: Serine protease mechanism and specificity. Chemical Reviews 102(12), 4501–4524 (2002)
15. Koch, I., Lengauer, T., Wanke, E.: An algorithm for finding maximal common subtopologies in a set of protein structures. Journal of Computational Biology 3(2), 289–306 (1996)
16. Alesker, V., Nussinov, R., Wolfson, H.J.: Detection of non-topological motifs in protein structures. Protein Engineering 9(12), 1103–1119 (1996)
17. Alexandrov, N.N., Fischer, D.: Analysis of topological and nontopological structural similarities in the pdb: new examples with old structures. Proteins: Structure, Function, and Bioinformatics 25(3), 354–365 (1996)
18. Grindley, H.M., Artymiuk, P.J., Rice, D.W., Willett, P.: Identification of tertiary structure resemblance in proteins using a maximal common subgraph isomorphism algorithm. Journal of Molecular Biology 229(3), 707–721 (1993)
19. Lu, G.: Top: a new method for protein structure comparisons and similarity searches. Journal of Applied Crystallography 33(1), 176–183 (2000)

20. Panigrahi, S.C., Mukhopadhyay, A.: An eigendecomposition method for protein structure alignment. In: Basu, M., Pan, Y., Wang, J. (eds.) ISBRA 2014. LNCS, vol. 8492, pp. 24–37. Springer, Heidelberg (2014)
21. Kabsch, W., Sander, C.: Dictionary of protein secondary structure: pattern recognition of hydrogen-bonded and geometrical features. Biopolymers 22, 2577–2637 (1983)
22. Needleman, S.B., Wunsch, C.D.: A general method applicable to the search for similarities in the amino acid sequence of two proteins. Journal of Molecular Biology 48(3), 443–453 (1970)
23. Gusfield, D.: Efficient methods for multiple sequence alignment with guaranteed error bounds. Bulletin of Mathematical Biology 55(1), 141–154 (1993)
24. Kabsch, W.: A solution for the best rotation to relate two sets of vectors. Acta Crystallographica Section A: Crystal Physics, Diffraction, Theoretical and General Crystallography 32(5), 922–923 (1976)
25. Roy, K., Panigrahi, S.C., Mukhopadhyay, A.: Multiple alignment of structures using center of proteins. CoRR abs/1412.8093 (2014)
26. Lupyan, D., Leo-Macias, A., Ortiz, A.R.: A new progressive-iterative algorithm for multiple structure alignment. Bioinformatics 21(15), 3255–3263 (2005)
27. Bernardo, M.M., Meng, Y., Lockett, J., Dyson, G., Dombkowski, A., Kaplun, A., Li, X., Yin, S., Dzinic, S., Olive, M., et al.: Maspin reprograms the gene expression profile of prostate carcinoma cells for differentiation. Genes & Cancer 2(11), 1009–1022 (2011)

NRRC: A Non-referential Reads Compression Algorithm

Subrata Saha and Sanguthevar Rajasekaran[✉]

Department of Computer Science and Engineering,
University of Connecticut, Storrs, USA
{subrata.saha,rajasek}@engr.uconn.edu

Abstract. In the era of modern sequencing technology, we are collecting a vast amount of biological sequence data. The technology to store, process, and analyze the data is not as cheap as to generate the sequencing data. As a result, the need for devising efficient data compression and data reduction techniques is growing by the day. Although there exist a number of sophisticated general purpose compression algorithms, they are not efficient to compress biological data. As a result, we need specialized compression algorithms targeting biological data. Five different NGS data compression problems have been identified and studied. In this article we propose a novel algorithm for one of these problems. We have done extensive experiments using real sequencing reads of various lengths. The simulation results reveal that our proposed algorithm is indeed competitive and performs better than the best known algorithms existing in the current literature.

1 Introduction

Nowadays Next Generation Sequencing (NGS) technologies are producing millions to billions of short reads simultaneously in a single run. To process and analyze the sequencing reads, at first we need to efficiently store this vast amount of data. Specifically the increase in sequencing data generation rate is outpacing the rate of increase in disk storage capacity. Furthermore, when the size of the data transmitted through the internet increases, the transmission cost and congestion in the network also increase. Thus it is vital to devise efficient algorithms to compress biological data. General purpose algorithms do not consider some inherent properties of sequencing data, e.g., repetitive regions, identical reads, etc. Exploiting these properties one can devise better algorithms compared to general purpose data compression algorithms. In this paper we offer a novel algorithm to compress biological sequencing reads effectively and efficiently. Our algorithm achieves better compression ratios than the currently best performing algorithms in the domain of reads compression. By compression ratio we mean the ratio of the uncompressed data size to the compressed data size.

The following five versions of compression have been identified in the literature: 1) *Genome compression with a reference.* Here we are given many (hopefully very similar) genomic sequences. The goal is to compress all the sequences

© Springer International Publishing Switzerland 2015
R. Harrison et al. (Eds.): ISBRA 2015, LNBI 9096, pp. 297–308, 2015.
DOI: 10.1007/978-3-319-19048-8_25

using one of them as the reference. The idea is to utilize the fact that the sequences are very similar. For every sequence other than the reference, we only have to store the difference between the reference and the sequence itself; 2) *Reference-free Genome Compression*. This is the same as problem 1, except that there is no reference sequence. Each sequence has to be compressed independently; 3) *Reference-free Reads Compression*. Reference-free reads compression algorithms are needed in biological applications where there is no clear choice for a reference; 4) *Reference-based Reads Compression*. In this technique complete read data need not be stored but only the variations with respect to a reference genome are stored; and 5) *Metadata and Quality Scores Compression*. In this problem we are required to compress quality sequences associated with reads as well as metadata such as read name, platform, and project identifiers.

In this paper we focus on problem 3. We present an effective reference-free reads compression algorithm namely NRRC: *Non-Referential Reads Compression Algorithm*. This algorithm takes any FASTQ file as input and outputs the compressed reads in FASTA format. To begin with, reads are clustered based on a hashing scheme. Followed by this clustering, a representative string is chosen from each cluster of reads. Compression is independently done for each cluster. In particular, the representative string in any cluster is used as a reference to compress the other reads in this cluster. Simulation results show that our proposed algorithm performs better than the best known algorithms existing in the current literature.

The rest of this paper is organized as follows: Section 2 has a literature survey. Section 3 describes the proposed algorithm and analyses its time complexity. Our experimental platform is explained in Section 4. This section also contains the experimental results and discussions. Section 5 concludes the paper.

2   Related Works

We now briefly survey some of the algorithms that have been proposed in the literature to solve the problems of biological data compression.

In referential genome compression the goal is to compress a set S containing a large number of similar sequences. The core idea of reference-based compression can be described as follows. We first choose a reference sequence R from S. Then we compress every other sequence $s \in S$ by comparing it with R. Brandon et al. [3] have used various coders like Golomb, Elias, and Huffman to encode the mismatches. Christley, et al. [8] have proposed the DNAzip algorithm that exploits the human population variation database, where a variant can be a single-nucleotide polymorphism (SNP) or an indel (an insertion or a deletion of multiple bases). In contrast to DNAzip, Wang et al. [28] have presented a *de novo* compression program, GRS, which obtains variation information by using a modified UNIX diff program. The algorithm GReEn [24] employs a probabilistic copy model that calculates target base probabilities based on the reference. Given the base probabilities as input, an arithmetic coder was then used to encode the target.

Reference-free genome compression algorithms compress a single sequence at a time by identifying repeats in the given sequence and replacing these repeats with short codes. For example, BioCompress [12] and BioCompress-2 [13] methods are based on Lempel-Ziv (LZ) style substitutional algorithm to compress exact repeats and palindromes. BioCompress-2 also utilizes an order-2 context-based arithmetic encoding scheme to store the non-repetitive regions. Alternatively, GenCompress [6] and DNACompress [7] identify approximate repeats and palindromes so that a large fraction of the target sequence can be compressed to get a high compression ratio. Similarly GeNML [17] divides a sequence into blocks of fixed size. Then, blocks rich in approximate repeats are encoded with an efficient normalized maximum likelihood (NML) model; otherwise, plain or order-1 context based arithmetic encoding is used. Some other well-known reference-free genome compression algorithms are BIND [2], DELIMINATE [21], COMRAD [18], DNAEnc3 [22], and XM [5].

Reads compression methods can also be categorized into reference-based and non reference-based, similar to genome compression techniques. Next we survey some of the non-referential reads compression algorithms. Tembe et al. [27] proposed G-SQZ method. At first G-SQZ computes the frequency of each unique tuple formatted as <base, quality>. Huffman coding scheme is then used to encode each tuple. The more frequent a tuple is the less number of bits is needed to encode it. At the final step the encoded tuples along with metadata are written to a file in binary format. In DSRC algorithm [10] input is divided into blocks of 32 records. Every 512 blocks are then grouped to make a superblock. Each of the superblocks is then indexed and compressed independently using LZ77 coding scheme. An arithmetic encoding scheme based on high order Markov chains is used by Quip [15] to compress the input sequences. Bonfield and Mahoney [1] proposed Fqzcomp and Fastqz where the sequences are compressed by using an order-N context model and an arithmetic coder. BEETL proposed by Coax et al. [9] identifies repeats among the reads using the Burros-Wheeler Transform (BWT) data structure [4]. General purpose compression algorithms (like gzip, bzip2, or 7-zip) are used to compress the transformed data. A similar mechanism is followed by SCALCE [14]. In this method a consistent parsing algorithm initially proposed by [26] is used to find an identical longest 'core' substring from the clustered reads. The reads within a cluster are then compressed using other standard compression algorithms as stated above.

In a reference-based technique complete read data need not be stored but only the variations with respect to a reference genome are stored. Reference-based algorithms typically run in two steps. In the first step all the reads are aligned to a reference genome of interest by using one of the NGS aligners such as Bowtie [19], BWA [20], Novoalign (http://www.novo craft.com), etc. In the second step, the mapped positions and variations are encoded using methods such as arithmetic coding and Huffman coding. Some of the algorithms in this domain are GenCompress, SlimGene, CRAM [11], Quip, NGC [25], Samcomp, etc. In addition to the mapping and encoding procedure, CRAM uses a de Bruijn graph based assembly approach where assembled contigs are used to map all the unaligned reads. Quip assembles the genomic sequence from the given reads using a de Bruijn graph based de novo assembler.

3   Methods

In this section we present our novel algorithm for problem 3. Specifically, we present an algorithm for the following problem: Reference-free Reads Compression. **For this version of compression, our algorithm achieves better compression ratios than the currently best known algorithms for this problem.** We provide details of our Non-Referential Reads Compression (NRRC) algorithm next. There are 4 basic steps in our novel algorithm. At first NRRC clusters the given set of reads based on overlaps and similarity. For each cluster a consensus sequence is created. All the reads in the cluster are compressed using the consensus as the reference. I.e., for each read we only store its difference with the consensus. Two reads are said to be *neighbors* of each other if they have a large overlap (with a small Hamming distance in the overlapping region). We find the neighbors of each read in steps 1 and 2 and use this neighborhood information to cluster the reads and perform compression (in step 3 and 4). In step 1 we find the *potential neighbors* of each read and in step 2 we find the true neighbors of each read. More details follow.

3.1   Finding Potential Neighbors

Potential neighbors for any read are found using two hashings. In each hashing we generate all the k-mers (for some suitable value of k) of all the reads and hash them based on these k-mers. A read is a potential neighbor of another read if they are hashed into the same value in at least one of the two hashings. For every read we collect all the potential neighbors from the two hashings and merge them. Followed by this, in step 2 we find the neighbors of each read as explained in the next section. Having two different hashings enables us to maximize the chances of finding as many of the neighbors as possible for each of the reads.

In the first hashing, we generate the k_1-mers in each read and hash the reads based on these k_1-mers (for some suitable value of k_1). Let $h_1(.)$ be the hash function employed. By a hash bucket (or simply a bucket) we mean all the reads that have the same hash value with respect to at least one of the k_1-mers in them. If read R_1 has a k_1-mer x, read R_2 has a k_1-mer y, and if $h_1(x) = h_1(y)$ then we say that the reads R_1 and R_2 fall into the same bucket. Any read R will be hashed into at most $r - k_1 + 1$ buckets, where $r = |R|$. For every read R we collect *potential neighbors* from the buckets that R falls into. All the reads that fall into at least one of the buckets that R falls into will be called *potential neighbors* of R. We perform one more such hashing by generating k_2-mers of reads. Here k_1 and k_2 are appropriate integers chosen to optimize performance. In the second hashing also we collect potential neighbors for each read. The potential neighbor lists collected for each read from the two hashings are merged together.

Whenever we hash k-mers (where k is either k_1 or k_2) we record necessary information about each of the k-mers such as the read associated with it and its starting position in the read. We need this information to find and align overlapping reads. To reduce memory usage we only record a unique read id associated with any read and an integer corresponding to the starting position of the k-mer. As no hash function is perfect, two similar k-mers may be hashed into two different buckets. Also, two dissimilar k-mers might be hashed into the same bucket. In this case we will loose some potential neighbors which could play important roles in compression. Having two different hashings enables us to maximize the chances of finding as many of the neighbors as possible for each of the reads. After finding the potential neighbors of each of the reads by traversing the hash buckets, we merge the neighbors information.

3.2   Finding Neighbors

After having collected potential neighbors for each read, we do some pruning to eliminate those potential neighbors that are not likely to be neighbors. Let R be any read and let R' be another read that has a sufficient overlap with R. For instance, a suffix of R could overlap with a prefix of R'. In the pruning step we compute the Hamming distance between the two reads in the overlapping region. If this distance is less than a threshold, we will keep R' as a neighbor R. If not, we will prune R' from the neighbor list of R. Note that the same two reads might fall into more than one buckets together. In this case we will identify and use the largest overlap between the pair. This is how merging of neighbors' information between two hash buckets is also done.

3.3   Aligning and Building a Consensus String

If R is any read and $L(R)$ is the list of neighbors of R, we correct R using $L(R)$. If R' is any read in $L(R)$ we already have found the maximum overlap between R and R' in Step 2. In Step 2 we have also ensured that the Hamming distance between R and R' in the overlapping region is within a small threshold. We align every R' (from $L(R)$) with R in a greedy manner using the overlapping region. The **greedy alignment** is done as follows. Let R' be a potential neighbor of R as stated above. As R and R' are potential neighbors, they must share at least one identical $k-$mer. We align R' with R using this k-mer as the anchor. We then extend this match on both sides as much as possible. Specifically, let $R = x_1 x_2 \cdots x_r$ and $R' = y_1 y_2 \cdots y_r$. Let the common k-mer between R and R' be $x_i x_{i+1} \cdots x_{i+k-1}$ and $y_j y_{j+1} \cdots y_{j+k-1}$. We identify the least $i_1 \leq i$ and the largest $i_2 \geq (i + k - 1)$ such that the Hamming distance between $x_{i_1} x_{i_1+1} \cdots x_i \cdots x_{i+k-1} \cdots x_{i_2}$ and $y_{j-(i-i_1)} y_{j-(i-i_1)+1} \cdots y_{j+i_2-i}$ is $\leq d$. If $(i_2 - i_1 + 1) \geq \frac{r}{2}$, we call R' as a neighbor of R. Since the error rate in NGS technology is very small, we expect that if R and R' come from the same region of the genome, then they will share more than one $k-$mers. While processing the buckets of the hash table, we keep track of the $k-$mer for which the size of the overlapping region between R and R' (i.e., $i_2 - i_1 + 1$) is the biggest and at the same time the Hamming distance between them is within d. We align R and R' based on this k-mer. Note that the **greedy alignment** that we do does not take much time. After aligning the reads in $L(R)$ with R, we construct a consensus sequence by taking the calculated order of most frequent residues (here nucleotides), found at each position in the alignments.

Values of parameters such as k_1, k_2, etc. have been optimized to get the best results. To speed up the proposed algorithm we have used several techniques. After neighborhood calculation, if a read has less than two neighbors, we discard the read as the read is potentially too erroneous to be corrected. Also, if the size of any hash bucket is very large, we omit the entire bucket from further consideration as it potentially corresponds to repeated regions of the genomic sequence. Furthermore, if the size of a bucket is greater than a certain threshold we randomly pick some of the reads and discard the others from that bucket.

3.4   Compressing and Encoding the Reads

Each read in $L(R)$ and R are compressed using the consensus string. Specifically, for each read we only store its difference with the consensus string. Note that a read may appear in more than one cluster. In this case, the read is compressed only in that cluster where its compressed length will be the least. After compressing the reads using the consensus, they are encoded using Lempel-Ziv-Markov chain algorithm (LZMA) - a lossless data compression algorithm. It is basically a dictionary based compression algorithm having large dictionary sizes. The output of the dictionary is encoded using a range encoder (a variant of entropy encoding method). To predict the probability of each bit, it uses a complex probability model roughly similar to the arithmetic encoding technique. In brief LZMA performs the compression in two basic steps. At first it detects matches using sophisticated and highly efficient dictionary-based data structure and generates a stream of literal symbols and phrase references. These are then encoded one bit at a time using a range encoder in the second step. It searches the space of many encoding and chooses the best one using a dynamic programming algorithm. Steps of the algorithm are shown in Algorithm 1.

Algorithm 1: Non-Referential Reads Compressor (NRRC)
Input: A set S of reads
Output A set S' of compressed reads

begin
1. Generate k-mers of each read and hash the reads based on these k-mers. Equal k-mers fall into the same bucket. If R is any read, any other read that falls into at least one of the buckets that R falls into is treated as a potential neighbor of R. For every read $R \in S$ create a list $P(R)$ of potential neighbors.
   Perform the above task twice with two different values for k and for every read merge the lists of potential neighbors from the two hash tables. If the size of a bucket is larger than a threshold, only a subset of the bucket is included in the potential neighbors identification process.
2. Let R be any read. Align every read in $P(R)$ with R. Let R' be any read in $P(R)$. If R and R' overlap sufficiently and if in the overlapping

region the Hamming distance between R and R' is small, then we treat R' as a neighbor of R. For every read $R \in S$ construct a list $L(R)$ of neighbors of R in this fashion.

3. Let R be any read. The neighbors of R reside in $L(R)$. Greedily align R' with R for every $R' \in L(R)$.

4. Make the consensus string R_C and compress R along with reads in $L(R)$ using R_C as a reference. Perform this step for every read $R \in S$. In this step, a read may be present in more than one cluster. In this case, the read is compressed in that cluster where its compressed length will be the least.

5. Encode the already compressed reads using Lempel-ZivMarkov chain algorithm (LZMA).

   end

3.5   Time Complexity Analysis

In this section we analyze the time complexity of NRRC. Let n be the number of reads and r be the read length. In the first step of NRRC, we build hash tables and identify potential neighbors. The number of k-mers (k could either be k_1 or k_2) generated from each read is $r - k + 1$. Let $h(.)$ be the hash function employed. We think of the hash table as an array of buckets (or lists). Each bucket has an integer as its index. If the size of the array is N, then the index of any bucket is an integer in the range $[1, N]$. If k is small enough, one could employ direct hashing such that each k-mer is hashed into a bucket whose index is the k-mer itself (thought of as an integer). In this case, the hash array should be of size 4^k. If k is large, direct hashing may not be feasible. The expected size of each bucket is $\frac{(r-k+1)n}{N} = O\left(\frac{rn}{N}\right)$. The total time spent in hashing of Step 1 is $O(rn)$.

In the first step we also find potential neighbors of each read. A read falls into at most $r - k + 1 < r$ buckets and hence the expected number of potential neighbors for each read is $O\left(\frac{r^2 n}{N}\right)$. For every bucket we spend an expected $O\left(\left(\frac{rn}{N}\right)^2\right)$ time. Thus the total time spent in Step 1 has an expected value of $O\left(rn + \frac{r^2 n^2}{N}\right)$.

In Steps 2 and 3 we align reads. Specifically, if R is any read and $P(R)$ is the list of potential neighbors of R, then the expected size of $P(R)$ is $O\left(\frac{r^2 n}{N}\right)$. For every read $R' \in P(R)$, we align R' with R and compute the Hamming distance between R and R' in the overlapping region. Thus for every $R' \in P(R)$ we spend $O(r)$ time. As a result, the total time spent in Step 2 and 3 for each read is expected to be $O\left(\frac{r^3 n}{N}\right)$. Summing this over all the reads, the total expected time spent in Step 2 and 3 is $O\left(\frac{r^3 n^2}{N}\right)$.

In step 4, we form a consensus corresponding to the neighbors of each read. Let R be any read. Since the expected number of reads in $P(R)$ is $O\left(\frac{r^2 n}{N}\right)$, the expected time to build a consensus is $O\left(\frac{r^3 n}{N}\right)$. Subsequently, each read is compressed using the consensus. The expected time for this is also $O\left(\frac{r^3 n}{N}\right)$. Summing this over all the reads, the expected time spent in step 4 is $O\left(\frac{r^3 n^2}{N}\right)$.

In summary, the expected run time of NRRC (excluding the time for LZMA) is $O\left(rn + \frac{r^3 n^2}{N}\right)$.

4   Simulation Results and Discussion

4.1   Experimental Setup

We have compared our algorithm with the best known algorithms currently existing in the domain of reads compression. In this section we summarize the results. All the experiments were done on an Intel Westmere compute node with 12 Intel Xeon X5650 Westmere cores and 48 GB of RAM. The operating system running was Red Hat Enterprise Linux Server release 5.7 (Tikanga). NRRC compression and decompression algorithms are written in C++ and standard Java programming language, respectively. To compile the C++ source code we used g++ compiler (gcc version 4.6.1) with the -O3 option. Java source code was compiled and run by Java Virtual Machine (JVM) 1.6.0.

4.2   Datasets and Algorithms used for Comparisons

We have employed real datasets in our evaluation. Real datasets used are Illumina-generated short reads of various lengths. The nine experimental datasets listed in Table 1 have been taken from Sequence and Read Archive (SRA) at NCBI. To prove the effectiveness of our algorithm, we choose two different types of data. Datasets D1 to D4 consist of RNA-seq reads generated from transcriptomes of different species (i.e. human (D1-D2), mouse (D3), and bacterium (D4)). The rest

Table 1. Illumina generated reads from human, mouse, and various organisms

Dataset	Accession Number	# of Reads	Length	Description
D1	SRR037452	11,671,179	35	Human brain tissue
D2	SRR635193.1	26,065,855	54	Pooled amnion
D3	SRR689233.1	16,407,945	90	Mouse oocyte
D4	SRR519063.1	26,905,342	51	*Pseudomonas aeruginosa*
D5	SRR001665.1	10,408,224	36	*Escherichia coli*
D6	SRR361468	7,093,045	35	*Treponema pallidum*
D7	SRR353563	7,061,388	100	*Leptospira interrogans*
D8	SRR022866.1	12,775,858	76	*Staphylococcus aureus*
D9	SRR065202.1	11,954,555	42	*Haemophilus influenzae*

Table 2. Compression sizes of different reads compression algorithms in Bytes. Best results are shown in bold. Please note that although PathEnc uses a reference for compression, in this table we have not added the size of the compressed reference to the size of the compressed file. (If we do this then NRRC's compression ratios will be better than PathEnc's on all datasets)

Dataset	2-Bit	SCALCE	fastqz	PathEnc	NRRC
D1	102,122,816	66,558,377	85,493,834	**45,180,142**	67,522,391
D2	351,889,042	95,474,107	179,766,179	**47,592,829**	74,048,376
D3	369,178,762	87,419,349	–	59,497,698	**50,795,839**
D4	342,043,110	23,210,258	90,128,271	15,797,769	**10,634,801**
D5	93,674,016	33,882,183	59,210,245	20,918,160	**17,339,502**
D6	62,064,143	19,403,853	35,952,162	13,064,384	**10,648,375**
D7	176,534,700	36,065,779	48,856,429	22,728,610	**15,201,009**
D8	242,741,302	110,790,886	128,966,807	91,259,052	**90,442,321**
D9	125,522,827	25,523,344	56,519,690	14,962,205	**10,424,120**

Table 3. Compression ratios of different reads compression algorithms with respect to 8-bit encoding. Please note that although PathEnc uses a reference for compression, in this table we have not added the size of the compressed reference to the size of the compressed file. (If we do this then NRRC's compression ratios will be better than PathEnc's on all datasets)

Dataset	2-Bit	SCALCE	fastqz	PathEnc	NRRC
D1	4	6.14	4.79	**9.04**	6.05
D2	4	14.74	7.83	**29.57**	19.01
D3	4	16.89	N/A	24.82	**29.05**
D4	4	58.95	15.18	86.61	**128.65**
D5	4	11.06	6.33	17.91	**21.61**
D6	4	12.79	6.91	19.00	**23.31**
D7	4	19.58	14.45	31.07	**46.45**
D8	4	8.76	7.53	10.64	**10.74**
D9	4	19.67	8.88	33.56	**48.17**

(i.e. D5 to D9) are short read datasets generated from DNA molecules of different organisms.

We have compared our algorithm NRRC with three other well-known algorithms based on biological reads. Currently, MFCompress [23], PathEnc [16], SCALCE [14], and fastqz [1] are some of the most efficient reads compression algorithms available in the literature. Every algorithm we have compared against, except for PathEnc, is a *de novo* compression algorithm. PathEnc needs a reference genome to generate a statistical, generative model of reads. It is then employed in a fixed-order context, adaptive arithmetic coder. It does not align reads with the reference. In this context PathEnc falls in between reference based and non-reference based reads compression algorithms.

4.3   Discussion

Now we discuss how we have used other methods to compare with our algorithm. SCALCE version 2.7 executable was used with its default parameters. It encodes sequence data without considering the positions of Ns. As SCALCE is targeted for compressing FASTQ files, it generates three output files with extension .scalcen (for read names), .scalcer (reads), and .scalceq (qualities). We report only the size of .scalcer file it produced. Fastqz compression tool can compress FASTQ files using a refernce genome. We report the results of the *de novo* version of fastqz compression algorithm. It produces three files namely .fxh (header/metadata information), .fxb (reads), and .fxq (quality scores). The file size we report is the size of the .fxb file. Fastqz was not able to run on D3 dataset. PathEnc was used with its default parameter settings. The file sizes reported are the sizes of all its output files. Since PathEnc needs a reference sequence to build the model, we have provided a specific biological sequence of interest for each of the datasets. For datasets D1-D4 we used a set of human transcriptomes as the reference (as was done by PathEnc). We used Sanger-assembled genomic sequences of interest for the rest. MFCompress [23] was also run with specific parameter settings. But the results have been omitted due to its consistently poor performance.

We have done extensive experiments to realize that our algorithm NRRC is indeed an effective and competitive reads compression tool. Please, see Table 2 and Table 3 for detailed simulation results. Table 2 and Table 3 present the compressed sizes and compression ratios produced by different algorithms including NRRC, respectively. The algorithm that has the best compression ratio is shown in bold face. Clearly, our proposed algorithm is competitive and performs better than all the best known algorithms in a majority of the datasets. Specifically, NRRC produces poor results in D1 and D2 datasets. Although PathEnc uses a reference for compression, we did not add the compressed size of the reference in the end result. If we add the compressed size of the reference for each of the end results, NRRC will perform better than PathEnc on every dataset. NRRC does not record the identifiers of the reads in the compressed file (similar to other algorithms we have compared). So, the reads will not be in the same order as in the original file. It also discards duplicate reads (similar to PathEnc) if any. Clearly, these will not affect any downward analysis of reads. Since SCALCE and fastqz are FASTQ compression algorithms, these algorithms compress metadata, reads, and quality scores in a single run. So, for a fair comparison we have not shown the run times of our algorithm. NRRC is a single-core algorithm. On the contrary PathEnc is a multi-core algorithm. If we consider linear speed-up and CPU-hour, NRRC is generally faster than PathEnc.

5 Conclusions

Data compression is a vital problem in biology especially for NGS data. Five different NGS data compression problems have been identified and studied in the literature. In this paper we have presented a novel algorithm for one of these problems, namely, reference-free reads compression. From the simulation results it is evident that our algorithm indeed achieves compression ratios that are better than those of the currently best known algorithms. We plan to investigate the possibility of employing the techniques we have introduced in this paper for solving the other four compression problems.

Acknowledgments. This work has been supported in part by the following grants: NIH R01-LM010101 and NSF-1447711.

References

1. Bonfield, J.K., Mahoney, M.V.: Compression of FASTQ and SAM format sequencing data. PLoS One 8, e59190 (2013)
2. Bose, T., Mohammed, M.H., Dutta, A., Mande, S.S.: BIND - An algorithm for loss-less compression of nucleotide sequence data. J. Biosci. 37, 785–789 (2012)
3. Brandon, M.C., Wallace, D.C., Baldi, P.: Data structures and compression algorithms for genomic sequence data. Bioinformatics 25, 1731–1738 (2009)
4. Burrows, M., Wheeler, D.J.: A block-sorting lossless data compression algorithm. SRC Research Report (1994)
5. Cao, M.D., Dix, T.I., Allison, L., Mears, C.: A simple statistical algorithm for biological sequence compression. In: Proceedings of the 2007 IEEE Data Compression Conference (DCC 2007), pp. 43–52 (2007)
6. Chen, X., Kwong, S., Li, M.: A compression algorithm for DNA sequences and its applications in genome comparison. Genome Informat Ser. 10, 51–61 (1999)
7. Chen, X., Li, M., Ma, B., Tromp, J.: DNACompress: fast and effective DNA sequence compression. Bioinformatics 8, 1696–1698 (2002)
8. Christley, S., Lu, Y., Li, C., Xiaohui, X.: Human genomes as email attachments. Bioinformatics 25, 274–275 (2009)
9. Cox, A.J., Bauer, M.J., Jakobi, T., Rosone, G.: Large-scale compression of genomic sequence databases with the Burrows-Wheeler transform. Bioinformatics 28, 1415–1419 (2012)
10. Deorowicz, S., Grabowski, S.: Compression of DNA sequence reads in FASTQ format. Bioinformatics 27, 860–862 (2011)
11. Fritz, M.H.-Y., Leinonen, R., Cochrane, G., Birney, E.: Efficient storage of high throughput DNA sequencing data using reference-based compression. Genome Res. 21, 734–740 (2011)
12. Grumbach, S., Tahi, F.: Compression of DNA sequences. In: Proceedings of the 1993 IEEEData Compression Conference (DCC 1993), Snowbird, Utah, pp. 340–350 (1993)
13. Grumbach, S., Tahi, F.: A new challenge for compression algorithms. Genet. Seq. Inform. Process. Manag. 30, 875–886 (1994)
14. Hach, F., Numanagic, I., Alkan, C., Sahinalp, S.C.: SCALCE: Boosting sequence compression algorithms using locally consistent encoding. Bioinformatics 28, 3051–3057 (2012)

15. Jones, D.C., Ruzzo, W.L., Peng, X., Katze, M.G.: Compression of nextgeneration sequencing reads aided by highly efficient de novo assembly. Nucleic Acids Res. 40, e171 (2012)
16. Kingsford, C., Patro, R.: Compression of short-read sequences using path encoding. bioRxiv (2014)
17. Korodi, G., Tabus, I., Rissanen, J., Astola, J.D.: sequence compression - based on the normalized maximum likelihood model. IEEE Sign Process Mag. 24, 47–53 (2007)
18. Kuruppu, S., Beresford-Smith, B., Conway, T., Zobel, J.: Iterative dictionary construction for compression of large DNA data sets. IEEE-ACM Trans Computat Biol Bioinformatics 9, 137–149 (2012)
19. Langmead, B., Trapnell, C., Pop, M., Salzberg, S.L.: Ultrafast and memory-efficient alignment of short DNA sequences to the human genome. Genome Biol. 10, R25 (2009)
20. Li, H., Durbin, R.: Fast and accurate short read alignment with Burrows-Wheeler transform. Bioinformatics 25, 1754–1760 (2009)
21. Mohammed, M.H., Dutta, A., Bose, T., Chadaram, S., Mande, S.S.: DELIMINATE-a fast and efficient method for loss-less compression of genomic sequences. Bioinformatics 28, 2527–2529 (2012)
22. Pinho, A.J., Ferreira, P.J.S.G., Neves, A.J.R., Bastos, C.A.C.: On the representability of complete genomes by multiple competing finite-context (Markov) models. PLoS One 6, e21588 (2011)
23. Pinho, A.J., Pratas, D.: MFCompress: a compression tool for FASTA and multi-FASTA data. Bioinformatics 30, 117–118 (2014)
24. Pinho, A.J., Pratas, D., Garcia, S.P.: GReEn: a tool for efficient compression of genome resequencing data. Nucleic Acids Res. 40, e27 (2012)
25. Popitsch, N., Haeseler, A.V.N.: lossless and lossy compression of aligned high-throughput sequencing data. Nucleic Acids Res 41, e27 (2013)
26. Sahinalp, S.C., Vishkin, U.: Efficient approximate and dynamic matching of patterns using a labeling paradigm. In: Proceedings of the 37th Annual Symposium on Foundations of Computer Science, pp. 320–328 (1996)
27. Tembe, W., Lowey, J., Suh, E.: G-SQZ: compact encoding of genomic sequence and quality data. Bioinformatics 26, 2192–2194 (2010)
28. Wang, C., Zhang, D.: A novel compression tool for efficient storage of genome resequencing data. Nucleic Acids Res. 39, E45-U74 (2011)

New Heuristics for Clustering Large Biological Networks

Md. Kishwar Shafin[1]([✉]), Kazi Lutful Kabir[1], Iffatur Ridwan[1],
Tasmiah Tamzid Anannya[1], Rashid Saadman Karim[1],
Mohammad Mozammel Hoque[1], and M. Sohel Rahman[2]

Department of CSE, MIST, Mirpur Cantonment, Dhaka-1216, Bangladesh
kishwar.shafin@gmail.com
AℓEDA Group, Department of CSE, BUET, Dhaka-1205, Bangladesh

Abstract. In analysis of large biological networks traditional clustering
algorithms exhibit certain limitations. Specifically, these are either slow
in execution or unable to cluster. As a result, faster methodologies are
always in demand. In this context, some more efficient approaches have
been introduced most of which are based on greedy techniques. Clusters
produced as a result of implementation of any such approach are highly
dependent on the underlying heuristics. It is expected that better heuristics will yield improved results. As far we are concerned, SPICi can handle large protein-protein interaction (PPI) networks well. In this paper,
we have proposed two new heuristics and incorporate those in SPICi.
The experimental results exhibit improvements on the performance of
the new heuristics.

Keywords: Biological network · Clustering · Heuristics

1 Introduction

Clustering is considered to be an important tool in the context of biological network analysis. However, traditional clustering algorithms do not perform well in
the analysis of large biological networks being either extremely slow or even unable to cluster [1]. On the other hand, recent advancement of the state of the art
technologies along with computational predictions have resulted in large scale
biological networks for numerous organisms [2]. As a result, faster clustering
algorithms are of tremendous interest. There exist a number of clustering algorithms that work well on small to moderate biological networks. For instance, a
number of algorithms in the literature can guarantee that they generate clusters
with definite properties (e.g., Cfinder [3], [4], [5], [6]). They are however computationally very intensive and hence do not scale well as the size of the biological
network increases.

To this end, some more efficient approaches have been introduced most of
which are based on greedy techniques (e.g., SPICi [7], DPClus [8] etc.). However,

M. Sohel Rahman – Commonwealth Academic Fellow, supported by the UK government. Currently, on a sabbatical leave from BUET.

© Springer International Publishing Switzerland 2015
R. Harrison et al. (Eds.): ISBRA 2015, LNBI 9096, pp. 309–319, 2015.
DOI: 10.1007/978-3-319-19048-8_26

algorithms like MGclus [9] suit relatively well for clustering of large biological networks with dense neighborhood. In most cases, clusters produced by greedy approaches are highly dependent on the heuristic(s) employed. It is expected that a better heuristic will yield even more improved results. This motivates us to search for a better heuristic to devise an even better clustering algorithm that not only runs faster but also provides quality solutions.

SPICi [7] can be considered as new approach among the greedy techniques that can cluster large biological networks. After carefully studying the implementation of SPICi, we have discovered that some essential modification in the heuristics employed can bring drastic change in the clusters' quality. In this paper, we have proposed a couple of new heuristics with an aim to devise an even better clustering algorithm. The results obtained via analysis produce better performance for the heuristics proposed.

The residue of the paper will be unfolded in the following sequence : Section II covers Background Study, Proposed heuristics in Section III, Section IV represents Experiments and results, Section V with Conclusion followed by Supplementary materials.

2   Background

We start this section with some preliminaries of the algorithmic framework of SPICi. We will also briefly review the heuristics used in SPICi and subsequently discuss the new heuristics. A biological network is modeled as an undirected graph $G = (V, E)$ where each edge $(u, v) \in E$ has a *confidence score* $(0 < w_{u,v} \leq 1)$, also called the *weight* of the edge. We say that, $w_{u,v} = 0$, if the two vertices u, v have no edge between them. The *weighted degree* of each vertex u, denoted by $d_w(u)$, is the sum of the confidence scores of all of its incident edges, i.e., $d_w(u) = \sum_{(u,v) \in E} w_{u,v}$. Based on the confidence scores or weights of the edges, we can define the term *density* for each set of vertices $S \subseteq V$ as follows. The density $\mathcal{D}(S)$ of a set $S \subseteq V$ of vertices is defined as the sum of the weights of the edges that have both end vertices belonging to S divided by the total number of possible edges in S. In particular,

$$\mathcal{D}(S) = \frac{\sum_{u,v \in S} w_{u,v}}{|S| \times (|S|-1)/2}$$

For each vertex u and a set $S \subseteq V$, *support* of u by S is denoted by $\mathcal{S}(u, S)$ and is defined as the sum of the confidence scores of the edges of u that are incident to the vertices in S. To be particular,

$$\mathcal{S}(u, S) = \sum_{v \in S} w_{u,v}$$

Given a weighted network, the goal of SPICi is to output a set of disjoint dense sub-graphs. SPICi uses a greedy heuristic approach that builds one cluster at a time and expansion of each cluster is done from an original protein seed pair. SPICi depends on two parameters, namely, the *support threshold, T_s* and the *density threshold, T_d*. The use of these two parameters will be justified subsequently. Now, we briefly review how SPICi employs its heuristic strategies. In fact, SPICi first selects two seed nodes and then attempt to expand the clusters.

Seed Selection. While selecting the seed vertices, SPICi uses a heuristic. Very briefly, at first it chooses a vertex u in the network that has the highest weighted degree. Then it divides the neighboring vertices of u into five bins according to their edge weights, namely, $(0, 0.2], (0.2, 0.4], (0.4, 0.6], (0.6, 0.8]$ and $(0.8, 1.0]$. Then the vertex with the highest weighted degree belonging to the highest non-empty bin is chosen as the second seed, v. The edge (u, v) is referred to as the seed edge.

Cluster Expansion. For cluster expansion, SPICi follows a procedure similar to that of [10]. It works with a vertex set S for the cluster initially containing the two selected seed vertices. It uses a heuristic approach to build the clusters and it builds one cluster at a time. In the cluster expansion step, SPICi searches for the vertex u such that $\mathcal{S}(u, S)$ is maximum amongst all the unclustered vertices that are adjacent to a vertex in S. If $\mathcal{S}(u, S)$ is smaller than a threshold then u is not added to S and $\mathcal{D}(S)$ is updated accordingly. However, if the calculated $\mathcal{D}(S)$ turns out to be smaller than the density threshold T_d then SPICi does not include u in the cluster and output S.

3 Proposed Heuristics

The two heuristics SPICi employs are implemented in the form of two procedures, namely, **Search** and **Expand**. In the **Search** procedure, node with the highest outdegree is chosen as the seed and in **Expand**, node with the highest support is selected as the candidate to be added to the cluster. In this paper, we have proposed two heuristics and we combine our heuristics with the heuristics of SPICi to have two new versions of SPICi. We will refer to these two versions as **SPICi$_+^1$** and **SPICi$_+^2$**. To be specific, we employ a new heuristic and modify the **Expand** procedure of SPICi to get **Expand+**. Similarly, we employ another new heuristic and modify the **Search** procedure of SPICi to get **Search+**. In **SPICi$_+^1$**, we combine **Expand+** with **Search** and in **SPICi$_+^2$**, we combine **Expand** with **Search+**. In essence, our first heuristic is to choose the node with the highest weighted degree among the neighbors as the first seed and second one is to choose the node with the highest average weighted degree as the candidate to join the cluster.

The heuristics employed by SPICi are developed based on an observation that two vertices are more likely to be in the same cluster if the weight of the edge between them is higher [7]. Below, we illustrate an example to identify the shortcoming of this heuristic and to establish the necessity to introduce a new heuristic.

3.1 Average Edge Weight

Consider Figure 1. Here assume that, the current cluster set is $S = \{1, 2, 3\}$ and the set of candidate nodes is $\{4, 5\}$. The goal at this point is to expand the current cluster. Now SPICi calculates $\mathcal{S}(4, S) = 1.4$ and $\mathcal{S}(5, S) = 1.5$ and since

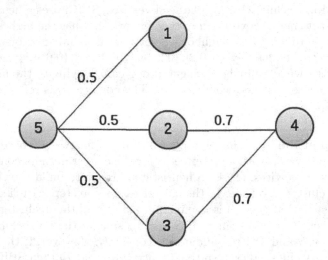

Fig. 1. An example to illustrate the necessity of a new measure

$S(5, S) > S(4, S)$, it will include node 5 in the cluster. However, two vertices are more likely to be in the same module if the weight on the edge between them is higher [7]. In Figure 1, we can see that the average weight with which node 4 is connected with nodes 2 and 3 is higher than the same with which node 5 is connected with nodes 1, 2 and 3. Although node 4 is not connected to node 1 it is more likely to find the set $\{4, 2, 3\}$ in the same module rather than the set $\{5, 1, 2, 3\}$. Hence, it seems more useful to include node 4 in the cluster as the average weight of node 4 is better than that of node 5. To make a better decision, we introduce a new heuristic measure which we refer to as the *Average Edge Weight* as follows. For each node u and a set $S \subseteq V$, let $Q \subseteq S$ be the set of vertices u is connected with. The average edge weight of u by S is defined as follows,

$$AverageEdgeWeight(u, S) = \frac{\sum_{v \in Q} w_{u,v}}{|Q|}$$

Now let us refer back to the scenario illustrated in the Figure 1. Using our new heuristic measure, we calculate *Average Edge Weight(4,S)*= 0.7 and *AverageEdgeWeight(5,S)*=0.5. Since *AverageEdgeWeight(4,S) > AverageEdgeWeight(5,S)*, in contrast to SPICi, we choose node 4 as desired. The modified **Expand** procedure, i.e., **Expand+** is provided at the end of this section.

3.2 Weighted Degree of Neighbors

For each vertex u, the weighted degree of its neighbors, denoted by $A_w(u)$ is simply the summation of the weighted degrees of all of its neighbors. So, we have $A_{w(u)} = \sum_{(u,v) \in E} d_w(v)$. To illustrate the usefulness of this heuristic measure,

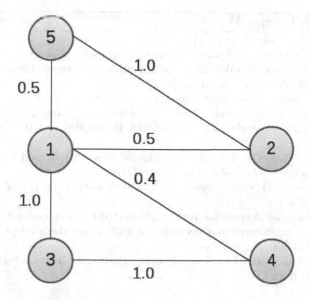

Fig. 2. Another illustration denoting the necessity of a new measure

let us consider Figure 2. While selecting the first seed SPICi groups the set of nodes with similar outdegrees. In Figure 2, SPICi would have two group of nodes {1, 2, 3, 5} and {4}. The grouping is as such, because node 1 has outdegree of 2.4 which is rounded off to 2 and similarly the outdegrees of 2, 3, 5 are 1.5, 2, 1.5 which are also rounded to 2. As these nodes have the same outdegree after rounding off, they are in the same group. And node 4 has outdegree of 1.4 so it is rounded off to 1 and it creates a new group. Now, SPICi will select a random node from the highest weight group. Suppose 5 is selected. But, clearly 5 is in a weak neighborhood, i.e., it does not have a dense group around it. But if we use the weighted degree of all the neighbors for a particular node then for node 1 we have the highest weighted degree $(A_w(1)=2+1.4+1.5+1.6=6.5)$. And, choosing the node with the highest weighted degree of neighbors is likely to enhance the probability to select the most promising node as the first seed. This ensures that we are selecting the node from a dense neighborhood, so in the expand process we will always start in a dense population. However, note that this heuristic is associated with an overhead that causes the complexity to rise by an $O(E)$ component. This is because we need to execute an extra loop to find the summation of the weighted degrees of the neighbors of a particular node. The modified **Search** procedure is provided at the end of this section.

Algorithm 1. Expand+(u, v)

---

initialize the cluster $S = \{u, v\}$
initialize *CandidateQ* to contain vertices neighboring u or v
initialize *AverageEdgeWeightHeap* to contain vertices neighboring u or v
while *CandidateQ* is not empty **do**
  find the largest non-empty bin of *Average Edge Weight Heap* nodes in *CandidateQ* among the bins with *AverageEdgeWeight(t,S)* range of $(0.8, 1], (0.6, 0.8], (0.4, 0.6], (0.2, 0.4], (0.0, 0.2]$.

  extract t from *CandidateQ* with highest $support(t, S)$ and t belongs to largest non-empty bin of *AverageEdgeWeightHeap*
  if $support(t, S) \geq T_s*|S|*density(S)$ and $density(S \cup \{t\}) > T_d$ **then**
    $S = S + \{t\}$
    increase the *support* for vertices connected to t in *CandidateQ*
    for all unclustered vertices adjacent to t, insert them into *CandidateQ* if not present
    for all unclustered vertices adjacent to t, update *AverageEdgeWeightHeap*
    break from loop
  end if
end while
return S

Algorithm 2. Search+

Initialize **DegreeQ** to be V
while DegreeQ is not empty **do**
  Extract u from **DegreeQ** with largest weighted degree of neighbors
  if u has adjacent vertices in **DegreeQ then**
    Find from u's adjacent vertices the second seed protein v
    S=Expand(u, v)
  else
    $S=\{u\}$
  end if
  $V=V - S$
  Delete all vertices in S from **DegreeQ**
  For each vertex in t in **DegreeQ** that is adjacent to a vertex in S, decrement its weighted degree by support(t,S)
end while

---

4 Experiments and Results

We have conducted our experiments on a PC having Intel 2.40 GHz core i5 processor with 4GB memory. The coding has been done in C++ and in Codeblocks 10.05 IDE. All the analysis was done in Linux (Ubuntu 12.04) environment. For all the experiments we set both T_s and T_d to 0.5, the same value used in SPICi. For the analysis we had to convert the gene names and we used various sources to convert and extract the gene names from [11], [12] and [13]. The overall conversion of the gene names affects the analysis. But as all the experiments were

donc using the same procedure, the overall penalty borne by the clusters are the
same. The reported runtimes of SPICi, SPICI$^1_+$ and SPICI$^2_+$ are wall clock times
as the same was done in [7].

4.1   Network Datasets

Same as SPICi our analyis is concentrated on four networks. Two networks
for yeast and two networks for human. These are the same networks used by
SPICi. The properties of these networks are reported in Table 1. Experimentally
determined physical and genetic interactions are found in the two Biogrid networks [14]. On the other hand, functional association between proteins that are
derived from data integration are found in the two STRING networks [15]. These
datasets are available at [16]. SPICi did another analysis on the Human bayesian
network which we could not get from the authors. For the Biogrid networks all
non-redundant interaction pairs are extracted which include the protein genetic
and physical interactions and all weighted interactions for the STRING networks
were used as reported in [7] .

Table 1. Test set of biological networks

	Biogrid yeast	STRING YEAST	Biogrid Human	STRING Human
Vertices	5361	6371	7498	18670
Edges	85866	311765	23730	1432538

4.2   GO Analysis

The GO analysis conducted here is based on the analysis done in [7]. They used
a framework described in [1] to evaluate the obtained clusters. We have used the
same framework to compare among the clusters we have obtained using different
heuristics. To construct the functional modules' reference set we have used Gene
Ontology (GO) in the same way as it was done in [7]. The Gene Ontology [17] is an
external measure to derive functional modules. For a GO biological process (BP)
or cellular component (CC) functional term, a module contains all the proteins
that are annotated with that term. Evaluation of clustering algorithms is done
by judging how well the clusters correspond to the functional modules as derived
from either GO BP or GO CC annotations. Following the work of [7], we have
considered the GO terms that annotate at most 1000 proteins for each organism.
For a particular GO annotation A, G_A is the functional module set having all
genes that are annotated with A. To measure the similarity between GO functional
modules and derived clusters, [1] uses the following three measures.

– Jaccard: Consider a cluster C. With each GO derived functional module
  group G_A the Jaccard value of C is computed as $\frac{|C \cap G_A|}{|C \cup G_A|}$. The maximum
  Jaccard value we get for cluster C over all GO term A is considered the
  Jaccard value of C.

- PR (Precision Recall): Consider a cluster C. With each GO derived functional module G_A, its PR value is computed as $\frac{|C \cap G_A|}{|G_A|} \frac{|C \cap G_A|}{|C|}$. The maximum PR value we get for cluster C over all GO term A is considered the PR value of C. terms A.
- Semantic density: Average semantic similarity between each pair of annotated proteins it is computed for each cluster. For two proteins $p1$ with annotations $A(p1)$ and $p2$ with annotations $A(p2)$. The semantic similarity of their GO annotations is defined as:

$$\frac{2 * \min_{a \in A(p1) \cap A(p2)} \log(p(a))}{\min_{a \in A(p1)} \log(p(a)) + \min_{a \in A(p1)} \log(p(a))}$$

where $p(a)$ is the fraction of annotated proteins with annotation a in the organism [18], [1]. For semantic density calculations, all GO terms that are annotating even more than 1000 proteins are also considered [7].

Table 2. GO analysis of clusters output by SPICi, SPICi$^1_+$ and SPICi$^2_+$

Network	Algorithm	BP sDensity	BP Jaccard	BP PR	CC sDensity	CC Jaccard	CC PR
Biogrid Yeast	SPICi	0.351	0.189	0.158	0.291	0.156	0.129
	SPICi$^2_+$	0.358	0.205	0.171	0.271	0.176	0.145
	SPICi$^1_+$	0.365	0.191	0.158	0.302	0.155	0.126
	MGclus	0.126	0.176	0.136	0.117	0.171	0.131
Biogrid Human	SPICi	0.191	0.117	0.093	0.097	0.059	0.038
	SPICi$^2_+$	0.182	0.135	0.107	0.086	0.067	0.042
	SPICi$^1_+$	0.203	0.128	0.099	0.107	0.062	0.039
	MGclus	0.061	0.126	0.092	0.026	0.079	0.046
STRING Yeast	SPICi	0.431	0.225	0.196	0.333	0.169	0.149
	SPICi$^2_+$	0.455	0.246	0.213	0.334	0.183	0.158
	SPICi$^1_+$	0.531	0.224	0.195	0.429	0.175	0.153
	MGclus	0.076	0.138	0.098	0.079	0.118	0.081
STRING Human	SPICi	0.384	0.099	0.187	0.518	0.049	0.059
	SPICi$^2_+$	0.384	0.113	0.198	0.512	0.054	0.065
	SPICi$^1_+$	0.409	0.214	0.187	0.533	0.078	0.058
	MGclus	-	-	-	-	-	-

The measures above have a range between 0 and 1. The higher the values, the better is the result of uncovering clusters satisfying the functional modules corresponding to GO. We calculate jaccard, PR and semantic density for each cluster and for both BP and CC ontology. These measures are attributed to all proteins in the cluster. Singleton cluster genes are penalized by assigning 0 to each of the three measure. Lastly, we compute average value of the six measures (three CC and three BP) over all proteins of the network.

From the Table 2, we observe that the changes in the heuristic of the algorithm affect the quality of the clusters. We know that the higher the values of measurement, the better the result of uncovering clusters in correspondence with functional modules. We can see that SPICi$^2_+$ has higher values than of SPICi in most of the cases. We can observe that in STRING networks which are quite denser SPICi$^1_+$ performs well. It is mentionable that, in all of the cases either SPICi$^1_+$ or SPICi$^2_+$ has the higher value than that of SPICi which clearly indicates that improvement in heuristics will result in better clusters. On the other hand, the MGclus [9] algorithm is not capable to uncover quality clusters though it works on strongly interconnected networks. Though the results are moderate in small sized network, the algorithm is unable to produce quality clusters in large network and it fails to cluster STRING Human networks in particular.

4.3   Run Time Analysis

A clustering algorithm should cluster large biological networks fast while maintaining the quality of the cluster. The time represented in Table 3 are the times required to cluster networks. As can be seen from the table, Average Weighted Degree heuristic was applied in SPICi$^1_+$ uses the same time complexity but only the number of calculations increases for which the change in runtime is negligible. On the other hand while using the other heuristic in seed selection in SPICi$^2_+$ it takes an extra pre-calculation of $O(E)$ which affects the run time heavily. However, in many cases it provides us higher quality clusters (Table 2). Run time of MGclus [9] algorithm is quite high, mainly because it uses Java Library and it can not cluster STRING Human network which is the largest network in the considered data-set. Finally, it can be noted that SPICi uses a faster way to read the data from the file.

Table 3. Run time analysis

	Biogrid Yeast	STRING Yeast	Biogrid Human	String Human
SPICi	1s	1s	1s	9s
SPICi$^2_+$	1s	3s	1s	11s
SPICi$^1_+$	1s	2s	2s	9s
MGclus	32s	5s	85s	Unable to Cluster

5   Conclusion

In bioinformatics clustering algorithms are considered to be one of the most important tools. Although there are a number of clustering algorithms that can cluster biological network, most of them fail to handle large biological networks. In this paper we have proposed two new heuristics for SPICi [7]. Our experimental results and observations indicate better performance in comparison to SPICi. This particular attempt is an ongoing research work. We are working on robustness analysis, more efficient qualitative and quantitative analysis against some new clustering methods, cluster size analysis to validate this effort as a more effective one.

Supplementary Materials

The source codes of our proposed heuristics are freely available at http://goo.gl/e9du1Y.

References

1. Song, J., Singh, M.: How and when should interactome-derived clusters be used to predict functional modules and protein function? Bioinformatics 25, 3143–3150 (2009)
2. Brun, M.-C., Herrmann, C., Guénoche, A.: Clustering proteins from interaction networks for the prediction of cellular functions. BMC Bioinformatics 5, 95 (2004)
3. Adamcsek, B., Palla, G., Farkas, I.J., Derényi, I., Vicsek, T.: Cfinder: locating cliques and overlapping modules in biological networks. Bioinformatics 22, 1021–1023 (2006)
4. Palla, G., Derényi, I., Farkas, I., Vicsek, T.: Uncovering the overlapping community structure of complex networks in nature and society. Nature 435, 814–818 (2005)
5. Colak, R., Hormozdiari, F., Moser, F., Schönhuth, A., Holman, J., Ester, M., Sahinalp, S.C.: Dense Graphlet Statistics of Protein Interaction and Random Networks. In: Pacific Symposium on Biocomputing, vol. 14, pp. 178–189 (2009)
6. Georgii, E., Dietmann, S., Uno, T., Pagel, P., Tsuda, K.: Enumeration of condition-dependent dense modules in protein interaction networks. Bioinformatics 25, 933–940 (2009)
7. Jiang, P., Singh, M.: SPICi: a fast clustering algorithm for large biological networks. Bioinformatics 26, 1105–1111 (2010)
8. Altaf-Ul-Amin, M., Tsuji, H., Kurokawa, K., Asahi, H., Shinbo, Y., Kanaya, S.: DPClus: A density-periphery based graph clustering software mainly focused on detection of protein complexes in interaction networks. BMC Bioinformatics 7, 150–156 (2006)
9. Frings, O., Alexeyenko, A., Sonnhammer, E.L.: Mgclus: network clustering employing shared neighbors. Molecular BioSystems 9(7), 1670–1675 (2013)
10. Altaf-Ul-Amin, M., Shinbo, Y., Mihara, K., Kurokawa, K., Kanaya, S.: Development and implementation of an algorithm for detection of protein complexes in large interaction networks. BMC Bioinformatics 7, 207 (2006)
11. STRING-Known and Predicted Protein-Protein Interactions, http://www.string-db.orgl (last accessed on March 16, 2014)
12. THE SYNERGIZER, http://llama.mshri.on.ca/synergizer/translate/ (last accessed on May 11, 2014)
13. BioGRID 3.2 Help and Support Resources, http://thebiogrid.org (last accessed on April 26, 2014)
14. Chatr-aryamontri, A., Breitkreutz, B.-J., Heinicke, S., Boucher, L., Winter, A.G., Stark, C., Nixon, J., Ramage, L., Kolas, N., O'Donnell, L., Reguly, T., Breitkreutz, A., Sellam, A., Chen, D., Chang, C., Rust, J.M., Livstone, M.S., Oughtred, R., Dolinski, K., Tyers, M.: The BioGRID interaction database: 2013 update. Nucleic Acids Research 41, 816–823 (2013)
15. Jensen, L.J., Kuhn, M., Stark, M., Chaffron, S., Creevey, C.J., Muller, J., Doerks, T., Julien, P., Roth, A., Simonovic, M., Bork, P., von Mering, C.: STRING 8 - a global view on proteins and their functional interactions in 630 organisms. Nucleic Acids Research 37, 412–416 (2009)

16. SPICi: Speed and Performance In Clustering,
    http://compbio.cs.princeton.edu/spici/ (last accessed on May 21, 2014)
17. Ashburner, M., Ball, C.A., Blake, J.A., Botstein, D., Butler, H., Cherry, J.M.,
    Davis, A.P., Dolinski, K., Dwight, S.S., Eppig, J.T., Harris, M.A., Hill, D.P., Issel-
    Tarver, L., Kasarskis, A., Lewis, S., Matese, J.C., Richardson, J.E., Ringwald, M.,
    Sherlock, G.M.R.G.: Gene Ontology: tool for the unification of biology. Nature
    Genetics 25, 25–29 (2000)
18. Lord, P.W., Stevens, R.D., Brass, A., Goble, C.A.: Semantic Similarity Measures as
    Tools for Exploring the Gene Ontology. In: Pacific Symposium on Biocomputing,
    pp. 601–612 (2003)

A Novel Algorithm for Glycan *de novo* Sequencing Using Tandem Mass Spectrometry

Weiping Sun[(✉)1], Gilles A. Lajoie[2], Bin Ma[3], and Kaizhong Zhang[1]

[1] Department of Computer Science, The University of Western Ontario, London, Ontario, N6A 5B7, Canada
{wsun63,kzhang}@csd.uwo.ca
[2] Department of Biochemistry, The University of Western Ontario, London, Ontario, N6A 5C1, Canada
glajoie@uwo.ca
[3] David R. Cheriton School of Computer Science, University of Waterloo, Waterloo, Ontario, N2L 3G1, Canada
binma@uwaterloo.ca

Abstract. Glycosylation is one of the most important and prevalent post-translational modifications of proteins. Identifying the structures of such protein-linked glycans has become necessary in biochemistry analysis. In the past decade, tandem mass spectrometry (MS/MS) has gradually served as an effective technique in glycoproteomics analysis because of its high throughput and sensitivity. Different approaches have emerged to address the challenges in computational analysis of mass spectrometry based glycoproteomics data. However, there are only a few available software tools characterizing glycans using the spectra produced from intact glycopeptides, which can conserve glycosylation site information. Furthermore, with the development of advanced mass spectrometry techniques, more accurate and complete spectra like HCD spectra can be applied to identify glycopeptides. In this paper, we proposed a heuristic algorithm for glycan *de novo* sequencing from HCD MS/MS spectra of N-linked glycopeptides. Experiments conducted on a dataset comprising of 46 MS/MS spectra showed that our results were comparable with those identified by GlycoMaster DB, which is designed based on database searching method.

Keywords: Mass spectrometry · Glycan *de novo* sequencing · Computational proteomics

1 Introduction

Glycosylation is one of the most common and important post-translational modifications (PTMs) of proteins, and over half of eukaryotic proteins in nature are estimated to be glycoproteins [1]. The carbohydrate chains covalently attached to glycoproteins or glycolipids, or as free forms in the cell plasma are usually referred as oligosaccharides or glycans. Unlike other simple PTMs which have a fixed mass change, glycans constitute a significant amount of mass and structural

© Springer International Publishing Switzerland 2015
R. Harrison et al. (Eds.): ISBRA 2015, LNBI 9096, pp. 320–330, 2015.
DOI: 10.1007/978-3-319-19048-8_27

variation in biological systems [2]. Glycoproteins serve in a variety of processes such as recognition between cell types or cellular structures. Research has suggested that the decrease in the activity of certain glycosylation-related enzymes can lead to serious physiological disorders [3]. Therefore, the characterization of glycoproteins is becoming an increasingly challenging aspect in the emerging proteomics research [4,5].

There are mainly two types of protein glycosylation, N-linked and O-linked glycosylation. In N-linked glycoproteins, carbohydrate groups are linked to polypeptide chains via the amide nitrogen of an asparagine residue; while in O-linked glycoproteins, glycans are linked via the hydroxyl groups of serine, threonine, or hydroxylysine residues [6]. Analysis of protein sequence databases have revealed that in most cases N-linked oligosaccharides are attached to a sequence motif Asn-X-Ser/Thr, where X denotes any amino acid except proline [7]. It is also occasionally observed that the consensus tripeptide Asn-X-Cys can also act as an acceptable sequon in N-glycosylation [8]. All mammalian N-linked glycans share a common core structure composed of two N-acetylglucosamine residues linked to a branched mannose triad [9]. Many other sugar units may be attached to each of the mannose residues of this branched core and the resulting structures fall into three main categories of N-linked glycoforms: high mannose, complex and hybrid [6,8]. The work of this paper is based on the analysis of N-linked glycopeptides.

During the past decade, tandem mass spectrometry (MS/MS) has gradually served as a major technique to determine glycan and glycopeptide primary structures because of its high sensitivity and throughput [10]. Depending on whether glycans and peptides are separated or not before the mass spectrometry analysis, there are generally two different strategies to analyze glycoproteins in biological samples. In one approach, deglycosylation is firstly applied to glycoproteins in order to obtain glycans and deglycosylated glycopeptides separately. And then after MS/MS analysis, their tandem mass spectra are collected to do database search or peptide/glycan identification [11]. In another method, glycoproteins are digested into glycopeptides by trypsin first, and then the resulting intact glycopeptides are characterized by tandem mass spectrometry [12,13]. In MS/MS, different fragmentation methods result in ion dissociation occurring at different sites and generate different types of dominant fragment ions. Collision-induced dissociation (CID) and higher-energy collision dissociation (HCD) usually break glycosidic bonds and yield b-ions and y-ions. HCD spectra is featured with a predominance of y-ions; b-ions and a-ions or other smaller species obtained from further fragmentations could be less frequently observed [14]. Electron-capture dissociation (ECD) and electron-transfer dissociation (ETD) often lead to cleavages at peptide backbone and produce c-ions and z-ions. Ideally, ETD/ECD spectra can provide information of both the peptide sequence and the glycosylation site because it keeps the attached glycan intact on the peptide backbone [15].

Currently, numerous approaches have emerged to automatically interpret mass spectrometry data of glycans or glycopeptides. One extensively studied method is to find the best matching glycans by searching the glycan database and comparing theoretical mass spectra with the experimental ones. Several software packages have been designed based on this strategy, such as GlycoSearchMS [16], GlycoPep DB[17], GlycoPeptideSearch (GPS) [18], GlyDB [19], and Glyco-Master DB [20]. Another approach, called *de novo* sequencing, is essential to determine novel or unknown glycans. The computation of *de novo* sequencing does not rely on database knowledge, instead the algorithms directly construct glycan structures from MS/MS spectra. There have been several attempts to characterize glycan structures using this method. Tang *et al.* proposed a dynamic programming algorithm GLYCH [21] to determine oligosaccharide structures from tandem mass spectra. This algorithm was designed for MS/MS spectra of released glycans only and cannot handle glycopeptide data. Shan *et al.* developed a software program called GlycoMaster [22] for glycan *de novo* sequencing from CID MS/MS. In [22], it proved that glycan *de novo* sequencing is a NP-hard problem under the condition that each mass value of the spectrum could be used only once and then provided a heuristic algorithm to solve the problem. In [23], Böcker *et al.* presented an exact algorithm based on fixed-parameter algorithmics to solve this NP-hard problem.

Compared with CID spectrum, HCD spectrum has a larger detecting range which could provide relatively complete and intensive peaks in the larger mass side of the spectrum. In this paper, we present a new strategy for N-linked glycan structure determination. By utilizing the quality peaks in the larger mass side of HCD spectrum, the glycan *de novo* sequencing problem was modeled as a top-down tree constructing process in a heuristic manner, in which the process started at the glycosylation site and the peptide was regarded as the root of the tree.

2   Mathematical Model

In MS/MS, six types of fragmented ions are commonly observed. From the reducing end, there are x-, y-, and z-ions; while in the non-reducing end, fragments are labelled with a-, b-, and c-ions. The a- and x-ions are generated by cross-ring cleavage [24]. Practically, y-ions dominate the fragment ions in HCD spectra. Therefore, we only consider y- and b-ions when generating possible glycan candidates. In the evaluation, we also take the internal fragment ions into consideration. When modeling glycopeptide structures, studies showed that a glycan structure was usually abstracted by a labelled rooted tree with node labels representing monosaccharide types [21,22,23]. In such representation, b-ions correspond to subtrees and y-ions correspond to the remaining glycopeptide with the subtrees removed. In this section, we will first formulate the glycopeptide structure and then describe the mathematical model for the glycan *de novo* sequencing problem.

For an N-linked glycopeptide, there is only one glycan attached to the peptide at the glycosylation site. The total mass of a glycopeptide consists of the residue mass of peptide, glycan and an extra water. Let Σ_a be the alphabet of different types of amino acids and $P = a_1 a_2 \ldots a_m$ be the string of peptide. For an amino acid $a \in \Sigma_a$, we define $\|a\|$ as its residue mass, then the residue mass of the peptide is $\|P\| = \sum_{1 \leq i \leq m} \|a_i\|$.

The sugar units that constitute a glycan through glycosidic bonds are called monosaccharides. It has been observed that some of the monosaccharides are epimers, which means that they differ only in their configurations rather than their mass. By *de novo* sequencing method, we cannot distinguish two epimers because they can hardly be separated from MS/MS spectra. Therefore, we only consider six types of common monosaccharides in this study based on the datasets we have, as shown in Table 1. We use Σ_g to denote the alphabet of different types (different mass) of monosaccharides. For a monosaccharide $g \in \Sigma_g$, $\|g\|$ is used to symbolize its residue mass value.

Table 1. Monosaccharide types used in the experiments

Monosaccharide	Composition	Monoisotopic mass		Symbol
		Intact	Residue	
Xylose (Xyl)	$C_5H_{10}O_5$	150.0528	132.0423	★
Fucose (Fuc)	$C_6H_{12}O_5$	164.0685	146.0579	▲
Hexose (Hex)	$C_6H_{12}O_6$	180.0634	162.0528	○
N-Acetyl hexosamine (HexNAc)	$C_8H_{15}NO_6$	221.0899	203.0794	□
N-Acetyl neuraminic acid (NeuAc)	$C_{11}H_{19}NO_9$	309.1060	291.0954	◆
N-Glycolyl neuraminic acid (NeuGc)	$C_{11}H_{19}NO_{10}$	325.1009	307.0903	◇

An N-linked glycan tree T is an unordered tree with its root linked to a peptide P. Each node of T represents a monosaccharide, labelled by an element from Σ_g. The degree of a glycan tree is bounded by four because there are at most five linkages for one monosaccharide. Given a glycan tree T that include n monosaccharides, its mass can be represented as $\|T\| = \sum_{1 \leq i \leq n} \|g_i\|$. The actual mass of a glycopeptide G which consists of a glycan T and a peptide P is $\|G\| = \|P\| + \|T\| + \|H_2O\|$.

Assume that t_i is the subtree of T rooted at ith node and its mass is $\|t_i\|$, then the mass of the b-ion associated with t_i is $b_i = \|t_i\| + 1$.[1] Let $y_{\{i_1,i_2,\dots,i_k\}}$ denote the mass value of the y-ion corresponding to subtrees of T rooted at peptide P with $t_{i_1}, t_{i_2}, \dots, t_{i_k}$ removed, where $t_{i_1}, t_{i_2}, \dots, t_{i_k}$ are nonoverlapping subtrees of t_i respectively. And y_0 denotes the mass of the y-ion generated at the cleavage of glycosylation site, without glycan included. In HCD spectrum, peptide is maintained intact, thus the mass value of y-ion only depends on the composition of glycan tree T.

Theoretically, the mass value set of b-ions and y-ions generated from a glycopeptide with n monosaccharides is,

$$I(G) = \bigcup_{1 \leq i \leq n} \{b_i\} \bigcup_{\{i_1,i_2,\dots,i_k\}} y_{\{i_1,i_2,\dots,i_k\}} \tag{1}$$

---

[1] There is a proton added to the ion in the ionization process.

Assume that $\mathcal{M}$ is used to denote the peak list of a glycopeptide spectrum, then we have $\mathcal{M} = \{(m_i, h_i)|i = 1, 2, \ldots, n\}$, where m_i and h_i represent the mass and the intensity of a peak respectively. Intuitively, the evidence that a spectrum $\mathcal{M}$ is generated from a glycopeptide G is that the more and higher peaks in $\mathcal{M}$ match with ion fragments of G. We use M to denote a set of peaks from the spectrum $\mathcal{M}$ that match with the theoretical ion mass values of G within the error tolerance δ,

$$M(G) = \{(m_i, h_i) \in \mathcal{M}|\exists m \in I(G), |m - m_i| \leq \delta\} \tag{2}$$

For each peak in the set $M(G)$, a scoring function can be defined according to its mass value m and intensity h. Here we use $f(m, h)$ to denote the function and one simple function can be defined as $f(m, h) = \log(h)$.

Therefore, the GLYCAN DE NOVO SEQUENCING problem can be defined as follows: Given a spectrum $\mathcal{M}$, a precursor mass value M_p, a predefined error bound δ, and a peptide mass $\|P\|$, construct a glycan tree T such that $|\|T\| + \|P\| + \|H_2O\| + 1 - M_p| \leq \delta$, and the score $S(T)$ is maximized,

$$S(T) = \sum_{(m_i, h_i) \in M(G)} f(m_i, h_i) \tag{3}$$

The equation above is used to evaluate how likely a tree structure matches with a spectrum. Several factors can be considered in the scoring function $f(m, h)$, such as mass value, intensity and ion types. Researchers can choose different factors to formulate the scoring function according to the fragmentation techniques used to generate MS/MS. It is worthy to notice that the set $M(G)$ is used to denote the peak list, which indicates that different ion fragments with the same mass value refer to the same peak in spectrum and are only counted once in computing the scoring function.

3 Algorithm

It has been proved that the complexity of the glycan *de novo* sequencing problem is NP-hard, under the condition that each mass value in spectrum cannot be repeatedly used [22]. Previous methods in [21] and [22] both constructed good solutions for smaller size trees and then assemble the reported trees into larger ones. Such strategy is suitable for glycan sequencing from CID spectra. While in HCD spectra, numerous quality peaks in the lager mass side can be used to explore new tree construction strategies. Therefore, in this paper, we provide a heuristic algorithm which construct the glycan tree from root to leaves based on HCD spectra.

A glycan tree with n vertices can be represented as $T = \langle v_1, v_2, \ldots, v_n \rangle$, where v_i denotes a node of the tree T. If the whole peptide is treated as a node, the notation v_0 can represent such node attached with an empty tree. $d(v_i)$ is used to denote the degree of the subtree rooted at v_i, and $m(T)$ represents the summation of monosaccharide residue mass values of the glycan tree T. We use $F(n)$ to denote a set of glycan trees with n nodes,

$$F(n) = \{ \, T \mid T \text{ is a glycan tree}, |V_T| = n \}$$

Given a glycan tree T, and a monosaccharide $g \in \Sigma_g$, v is a node of T and it has less than four children, i.e., $d(v) < 4$. We use $T_{v \otimes g}$ to represent a new tree generated from T, where g is a new node added to the tree through node v. Thus, v is the parent node of g, and g becomes a leaf node of the newly constructed tree. In addition, we use $T \otimes g$ to represent the set containing all the possible glycan trees generated by g and $T = \langle v_1, v_2, \ldots, v_n \rangle$,

$$T \otimes g = \{ T_{v_i \otimes g} \mid v_i \in V_T, d(v_i) < 4 \} \tag{4}$$

As we mentioned before, using HCD fragmentation method, the peptide can be kept intact during fragmentation. Thus, each y-ion fragment corresponds to a subtree of the glycan tree T that rooted at v_0. During the construction of the glycan tree, we need to find out all those subtrees to calculate the theoretical mass values of y-ions. We use r to denote a root-preserving subtree of T and the mass value of its corresponding y-ion is $\|r\| + \|P\| + \|H_2O\| + 1$.

Let $RPST$ denote the set of all the root-preserving subtrees of a glycan tree T, then we have,

$$RPST(T) = \{ r \mid r \text{ is a root-preserving subtree of } T \}$$

Therefore, the set of root-preserving subtrees of $T \otimes g$ can be represented as follows,

$$RPST(T \otimes g) = \bigcup_{v_i \in V_T} RPST(T_{v_i \otimes g})$$

Lemma 1. *Given a glycan tree T, and a monosaccharide $g \in \Sigma_g$, the set of root-preserving subtrees of the newly generated trees in $T \otimes g$ can be calculated in the following way,*

$$RPST(T \otimes g) = RPST(T) \cup \{ r_{v_i \otimes g} | \forall v_i \in V_T, s.t. \ d(v_i) < 4, v_i \in V_r, r \in RPST(T) \}$$

Proof. The set of the root-preserving subtrees for a new tree $T_{v \otimes g}$ generated from a tree T and a node g via node v should contain all the subtrees from $RPST(T)$. Besides, those newly generated trees derived from $RPST(T)$ by adding the node g to node v are also included in the set. Thus we have $RPST(T_{v \otimes g}) = RPST(T) \cup \{ r_{v \otimes g} \mid d(v) < 4, v \in V_r, r \in RPST(T) \}$. Furthermore, the computation of $RPST(T \otimes g)$ corresponds to the calculation of $RPST(T_{v \otimes g})$ over all the nodes in V_T. $\square$

The glycan structures are abstracted as unordered rooted trees in our method. The direct computation of the set $T \otimes g$ and $RPST(T \otimes g)$ from Equation (4) and Lemma 1 will generate duplicate trees. The removal of identical trees should be taken into consideration. Otherwise, the size limit of the candidate set would be reached quickly, yet the correct result is not included. To solve this problem, a string is assigned to each newly generated tree to represent its structure and nodes information. The strings are computed based on the tree isomorphism determination algorithm described in [26], which can be used to determine the isomorphism of two n-vertex labeled trees in $O(n)$ time. Therefore, we can eliminate the duplications of trees in a set in $O(nN \log N)$ time, where N is the size of the set.

Based on the mathematical model described above, we now introduce our heuristic algorithm for the GLYCAN DE NOVO SEQUENCING problem. The whole glycan tree structure is gradually constructed by adding one node during each round in the computation. For each round with n nodes, a fixed number of glycan trees with highest scores are maintained in $F(n)$. $F(n)$ is computed in two steps. Firstly, compute a set of candidate glycan structures $F_c(n) = \cup_{g \in \Sigma_g} \cup_{T \in F(n-1)} T \otimes g$; Secondly, compute the score of each structure in $F_c(n)$ by evaluating how its theoretical ion masses match with peaks in mass spectrum $\mathcal{M}$, then put the top $|F|$ glycans in $F(n)$ and remove those glycans with low scores. During the construction process, if the mass of one generated glycan satisfies the desired glycan mass value M', then this glycan will be put into the candidate results set R. When the program finished, a fix number of glycans sorted by scores from high to low can be obtained in R.

Algorithm 1. Glycan *de novo* Sequencing

INPUT: Given a spectrum $\mathcal{M}$, and glycopeptide precursor mass value M_p, and peptide mass value m_p, and a predefined error bound δ.

OUTPUT: A set R consists of candidate glycan structures with their scores, and each glycan tree T in R satisfies $|\|T\| + \|P\| + \|H_2O\| + 1 - M_p| \leq \delta$.

1:   $M' = M_p - m_p - 19$, $M_{min} = m_p$
2:   **while** $M_{min} + \min\|g\| \leq M'$ **do**
3:      $F_c(n) = \emptyset$
4:      **for** $T \in F(n-1)$ **do**
5:         **for** $v_i \in V_T$ **do**
6:            **for** $g \in \Sigma_g$ **do**
7:               $T' = T_{v_i \otimes g}$
8:               $RPST(T') = RPST(T)$
9:               **for** $r \in RPST(T)$ **do**
10:                  **if** $v_i \in V_r$ and $d(v_i) < 4$ **then**
11:                     $RPST(T') = RPST(T') \cup r_{v_i \otimes g}$
12:               $F_c(n) = F_c(n) \cup T'$
13:      Score each glycan tree in $F_c(n)$ according to $RPST(T)$, put top $|F|$ in $F(n)$
14:      Find the minimum tree mass M_{min} in $F(n)$
15:      Select the trees from $F(n)$ that satisfies mass requirement and put them in R

Theoretically, according to Lemma 1, the complexity to compute all the root-preserving subtrees for a tree is not polynomial. Because the number of all possible rs in $RPST$ of a rooted tree with n vertices and maximal out degree d is exponential to n. However, in practice during our computation for glycan structures, the maximal size of $RPST$ was less than 50. The main reason may be that the degree of most nodes in a glycan tree is less than four and the identical trees in each tree set had been removed. For simplicity, the number of elements in set $RPST(T)$ can be regarded as a constant C. Therefore, the time complexity of the algorithm proposed above is $O(n^3 \times |\Sigma_g| \times |F| \times C \times \log(n \times |F|))$, where n is the total number of the vertices for a glycan tree, which is less than 20 practically.

4 Experiments and Discussion

The algorithm proposed above was implemented in the experiments to test its performance and the top 1000 candidates were selected during each round computation, *i.e.*, $|F| = 1000$. The error bound $\delta = 0.2$Da were used in the experiment.

4.1 Datasets

The glycopeptide samples used in the experiments were derived from three kinds of protein samples: Alpha-1-acid glycoprotein of Bos taurus (Bovine), Ovomucoid of Gallus gallus (Chicken), and Ig gamma-3 chain C region of Homo sapiens (Human). Experiments were carried on a Thermo Scientific Orbitrap Elite hybrid mass spectrometer and HCD fragmentation technique was used.

The newly developed software tool GlycoMaster DB [20] was used for comparison. GlycoMaster DB can analyze mass spectra produced with HCD fragmentation and identify N-linked glycans by searching against the glycan structure database GlycomeDB [25]. The main reason we choose a database searching method for comparison is that the algorithms mentioned in [21,22,23] which designed based on *de novo* sequencing method cannot handle glycopeptide data or can only analyze CID spectra. Besides, the results identified by database searching method are relatively reliable.

Our experimental dataset contained 46 HCD spectra of glycopeptides that were identified from the collected MS/MS spectra by GlycoMaster DB. The reported glycan structures were used to benchmark the performance of our proposed method.

4.2 Experimental Results

For each MS/MS spectrum in the dataset, top 10 candidates of glycan structures were reported by GlycoMaster DB. Among those results, the highest ranked glycan structure was treated as the reference structure, and this structure was compared with all the results constructed by our algorithm. Table 2 shows the ranking status of the reference structures observed in our reported results for those 46 MS/MS spectra.

Table 2. Performance of our algorithm compared with GlycoMasterDB

Rank	No.1	No.2	No.3-10	No.>10	Can't find
Number of glycans	35	6	1	2	2
Ratio(%)	76.09	13.04	2.17	4.35	4.35

As one can see from Table 2, there are 35 glycans with highest scores generated by our proposed method have the same structures as those top-ranked glycans interpreted by GlycoMaster DB. In addition, if the case that the corresponding reference structure ranking top two in our reported results is deemed correct, then the accuracy rate of our proposed method can reach to 89.13%. Among the results that the reference structures ranked greater than 10, the lowest rank observed in our results is 32. However, for the associated spectra, the top ranked glycan structure reported by our method and the second ranked structure identified by GlycoMaster DB were identical. And the scores of those top two results reported by GlycoMaster DB were indeed very close.

There are two entries that the reference glycan structures cannot be observed in the results provided by our proposed algorithm. However, our reported glycans with highest score were only partially different from the related reference structures. Figure 1 shows the difference of these two pairs of results.

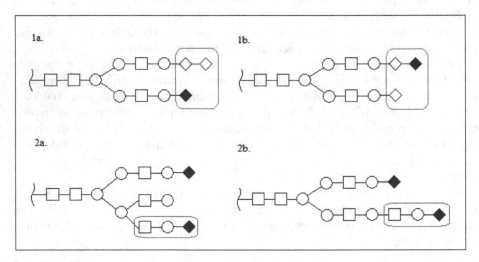

Fig. 1. Comparison of two pairs of glycan structures identified by GlycoMaster DB and our method respectively.

Each pair of glycans in the same row shown in Fig.1. were interpreted by the same HCD spectra. Glycans 1a. and 2a. were identified by GlycoMaster DB with highest scores, while 1b and 2b were ranked first in the results of our method. As one can see from the figure, in each row, the two glycans have high resemblance and one can be converted to the other one through a few steps of operations. This indicated that although our algorithm did not find out the same glycan structures as GlycoMaster DB did, our results were only partially different and can be used to assist the filtration of tentative candidates generated by database searching method. Our reported results were reasonable because they were also supported well by each corresponding spectrum. In future, we will combine our *de novo* sequencing method with database searching method to improve the result accuracy of the glycan characterization from MS/MS spectra.

Acknowledgments. The authors would like to thank Dr. Lin He for providing his software GlycoMaster DB. This work was supported in part by the NSERC Discovery Grant and a Discovery Accelerator Supplements Grant. WS was partially supported by the CSC (China Scholarship Council) Scholarship.

Reference

1. Apweiler, R., Hermjakob, H., Sharon, N.: On the frequency of protein glycosylation, as deduced from analysis of the SWISS-PROT database. Biochimica et Biophysica Acta (BBA)-General Subjects 1473(1), 4–8 (1999)
2. Ohtsubo, K., Marth, J.D.: Glycosylation in cellular mechanisms of health and disease. Cell 126(5), 855–867 (2006)
3. Dwek, R.A., Butters, T.D., Platt, F.M., Zitzmann, N.: Targeting glycosylation as a therapeutic approach. Nature Reviews Drug Discovery 1(1), 65–75 (2002)
4. Wuhrer, M., Catalina, M.I., Deelder, A.M., Hokke, C.H.: Glycoproteomics based on tandem mass spectrometry of glycopeptides. Journal of Chromatography B 849(1), 115–128 (2007)
5. Von Der Lieth, C.W., Bohne-Lang, A., Lohmann, K.K., Frank, M.: Bioinformatics for glycomics: status, methods, requirements and perspectives. Briefings in Bioinformatics 5(2), 164–178 (2004)
6. Garrett, R., Grisham, C.M.: Biochemistry. Saunders College, Philadelphia (1995)
7. Gavel, Y., von Heijne, G.: Sequence differences between glycosylated and non-glycosylated Asn-X-Thr/Ser acceptor sites: implications for protein engineering. Protein Engineering 3(5), 433–442 (1990)
8. Blom, N., Sicheritz-Pontén, T., Gupta, R., Gammeltoft, S., Brunak, S.: Prediction of post translational glycosylation and phosphorylation of proteins from the amino acid sequence. Proteomics 4(6), 1633–1649 (2004)
9. Dell, A., Morris, H.R.: Glycoprotein structure determination by mass spectrometry. Science 291(5512), 2351–2356 (2001)
10. Ma, B.: Challenges in computational analysis of mass spectrometry data for proteomics. Journal of Computer Science and Technology 25(1), 107–123 (2010)
11. Pan, S., Chen, R., Aebersold, R., Brentnall, T.A.: Mass spectrometry based glycoproteomics from a proteomics perspective. Molecular & Cellular Proteomics 10(1), R110-003251 (2011)
12. Dalpathado, D.S., Desaire, H.: Glycopeptide analysis by mass spectrometry. Analyst 133(6), 731–738 (2008)
13. Woodin, C.L., Maxon, M., Desaire, H.: Software for automated interpretation of mass spectrometry data from glycans and glycopeptides. Analyst 138(10), 2793–2803 (2013)

14. Frese, C.K., Altelaar, A.M., Hennrich, M.L., Nolting, D., Zeller, M., Griep-Raming, J., Mohammed, S., et al.: Improved peptide identification by targeted fragmentation using CID, HCD and ETD on an LTQ-Orbitrap Velos. Journal of Proteome Research 10(5), 2377–2388 (2011)

15. Alley, W.R., Mechref, Y., Novotny, M.V.: Characterization of glycopeptides by combining collision induced dissociation and electron transfer dissociation mass spectrometry data. Rapid Communications in Mass Spectrometry 23(1), 161–170 (2009)

16. Lohmann, K.K., von der Lieth, C.W.: GlycoFragment and GlycoSearchMS: web tools to support the interpretation of mass spectra of complex carbohydrates. Nucleic Acids Research 32(suppl. 2), W261–W266 (2004)

17. Go, E.P., Rebecchi, K.R., Dalpathado, D.S., Bandu, M.L., Zhang, Y., Desaire, H.: GlycoPep DB: a tool for glycopeptide analysis using a smart search. Analytical Chemistry 79(4), 1708–1713 (2007)

18. Pompach, P., Chandler, K.B., Lan, R., Edwards, N., Goldman, R.: Semi-automated identification of N-Glycopeptides by hydrophilic interaction chromatography, nano-reverse-phase LC MS/MS, and glycan database search. Journal of Proteome Research 11(3), 1728–1740 (2012)

19. Ren, J.M., Rejtar, T., Li, L., Karger, B.L.: N-Glycan structure annotation of glycopeptides using a linearized glycan structure database (GlyDB). Journal of Proteome Research 6(8), 3162–3173 (2007)

20. He, L., Xin, L., Shan, B., Lajoie, G.A., Ma, B.: GlycoMaster DB: Software To Assist the Automated Identification of N-Linked Glycopeptides by Tandem Mass Spectrometry. Journal of Proteome Research 13(9), 3881–3895 (2014)

21. Tang, H., Mechref, Y., Novotny, M.V.: Automated interpretation of MS/MS spectra of oligosaccharides. Bioinformatics 21(suppl. 1), i431–i439 (2005)

22. Shan, B., Ma, B., Zhang, K., Lajoie, G.: Complexities and algorithms for glycan sequencing using tandem mass spectrometry. Journal of Bioinformatics and Computational Biology 6(01), 77–91 (2008)

23. Böcker, S., Kehr, B., Rasche, F.: Determination of glycan structure from tandem mass spectra. IEEE/ACM Transactions on Computational Biology and Bioinformatics (TCBB) 8(4), 976–986 (2011)

24. Zaia, J.: Mass spectrometry of oligosaccharides. Mass Spectrometry Reviews 23(3), 161–227 (2004)

25. Ranzinger, R., Herget, S., Wetter, T., Von Der Lieth, C.W.: GlycomeDB-integration of open-access carbohydrate structure databases. BMC Bioinformatics 9(1), 384 (2008)

26. Aho, A.V., Hopcroft, J.E., Ullman, J.D.: The Design and Analysis of Computer Algorithms. Addison-Wesley, Reading (1974)

Community Detection-Based Feature Construction for Protein Sequence Classification

Karthik Tangirala[✉], Nic Herndon, and Doina Caragea

Department of Computing and Information Sciences
Kansas State University, Manhattan, KS 66502, USA
{karthikt,nherndon,dcaragea}@ksu.edu

Abstract. Machine learning algorithms are widely used to annotate biological sequences. Low-dimensional informative feature vectors can be crucial for the performance of the algorithms. In prior work, we have proposed the use of a community detection approach to construct low dimensional feature sets for nucleotide sequence classification. Our approach uses the Hamming distance between short nucleotide subsequences, called k-mers, to construct a network, and subsequently uses community detection to identify groups of k-mers that appear frequently in a set of sequences. While this approach worked well for nucleotide sequence classification, it could not be directly used for protein sequences, as the Hamming distance is not a good measure for comparing short protein k-mers. To address this limitation, we extend our prior approach by replacing the Hamming distance with substitution scores. Experimental results in different learning scenarios show that the features generated with the new approach are more informative than k-mers.

Keywords: Community detection · Feature construction · Feature selection · Dimensionality reduction · Protein sequence classification · Supervised learning · Semi-supervised learning · Domain adaptation

1 Introduction

Machine learning has been extensively used to address prediction and classification problems in the field of bioinformatics. Advancements in sequencing technologies have led to the availability of large amounts of sequential data (mostly unlabeled), which can benefit learning algorithms. In general, most learning algorithms require a vectorial representation of the data in terms of features. Representing the data through low-dimensional informative feature sets is critical for the performance of the algorithms, in terms of both accuracy and complexity.

However, for many biological problems it is not yet understood which features are informative. In the absence of known informative features, it is common to represent the sequences as the count of k-mers generated using a sliding window-based approach. To do this, a window of a particular size, k, is traversed across the sequence, and at each step in the traversal, the fragment of the sequence within the window is captured. All such possible unique subsequences/fragments

© Springer International Publishing Switzerland 2015
R. Harrison et al. (Eds.): ISBRA 2015, LNBI 9096, pp. 331–342, 2015.
DOI: 10.1007/978-3-319-19048-8_28

(referred to as k-mers) are used as features to represent sequences. As informative features can have variable length, the size, k, of the window is varied. However, variable length k-mers result in high-dimensional feature sets, increased computational complexity and sometimes decreased classification accuracy.

Feature selection is one of the techniques widely used to reduce the dimensionality of the input feature space, while retaining most of the informative features. Most of the feature selection techniques use the available labeled data to estimate feature-class dependency scores for all features. The features are then filtered based on the corresponding feature-class dependency scores. In theory, feature selection can be applied not only in supervised learning (large amounts of labeled data is used in the learning process), but also in semi-supervised learning (small amounts of labeled and large amounts of unlabeled data are used) and domain adaptation (large amounts of labeled data from a source domain, along with small amounts of labeled data and large amounts of unlabeled data from a target domain are used to learn classifiers for the target data). However, in the semi-supervised and domain adaptation, as the amount of available (target) labeled data is small, feature selection may not capture the feature-class dependencies accurately. Furthermore, when the number of features is very large, feature selection techniques might be computationally expensive. Therefore, alternative methods to generate a reduced set of informative features can presumably benefit supervised, semi-supervised and domain adaptation algorithms.

Towards this goal, in [16], we have introduced the idea of using a community detection algorithm to generate a low-dimensional informative sequential feature set for classifying nucleotide sequences (specifically, for the problem of classifying exons as either alternatively spliced or constitutive). Our approach extended TFBSGroup [15], an unsupervised approach to identify transcription factor binding sites in a small number of nucleotide sequences, based, in turn, on the community detection algorithm proposed in [23]. The worst case running time of TFBSGroup is quartic in the total number of sequences and the length of each sequence in the dataset. As a result, running TFBSGroup on large sets of sequences has high computational cost. We proposed a fast and novel extension to TFBSGroup [16], which makes it possible to generate features for large sets of nucleotide sequences. Our approach is based on randomly sampling small subsets of sequences (as opposed to using all the sequences at once) and finding informative features in each set separately. The final set of informative features is obtained by taking the union of the individual sets found using TFBSGroup. Although our prior approach [16] was successfully used to identify low-dimensional informative features (referred to as c-mers) for nucleotide sequences, it cannot directly be applied for protein sequences given the large size of the protein alphabet and the short length of the informative protein k-mers.

To address this limitation, in this paper, we further extend the approach in [16] to protein sequences by making use of amino acid substitution scores in place of the Hamming distance, under the assumption that the substitution scores are better than the Hamming distance when comparing short protein subsequences. We have applied the proposed approach to the problem of classifying protein sequences based on their localization. To evaluate the predictive power of c-mers in classifying protein sequences, we have conducted experiments in three different learning scenarios: supervised, semi-supervised and domain adaptation. Experimental results in all three learning scenarios suggest that the features generated with the community detection approach are more informative than k-mers in classifying protein sequences.

The rest of the paper is organized as follows: The related work on applications that have used k-mers, feature selection and community detection approaches is described in Section 2. The proposed approach of using a community detection algorithm to generate features for biological sequences is discussed in Section 3. Section 4 lists the research questions that we are addressing through this work, along with details about the set of experiments conducted, and the datasets used. The results of the experiments conducted are presented in Section 5, followed by conclusions in Section 6.

2   Related Work

In bioinformatics, and especially biological sequence classification, the sliding window approach is frequently used, sometimes together with feature selection or a different dimensionality reduction method, to generate k-mers and represent biological sequences as vectors of k-mers [1,2,3]. As an alternative to feature selection, we propose to use community detection to select a small set of informative features (specifically, k-mers that appear frequently in a set of sequences).

To find communities, Grivan and Newman [8,22] proposed a hierarchical divisive algorithm, that iteratively removes edges between nodes based on their "betweenness", until the modularity of a partition reaches the maximum. The "betweenness" measure defines the total number of shortest paths between any two nodes that pass through an edge. The authors estimated the modularity of a partition, referred to as the Newman-Girvan modularity, by comparison with a null model (random graph). Their algorithm is believed to be the first of modern day community detection approaches. Clauset et al. [4] proposed a fast community detection approach that uses the Newman-Girvan modularity gain. Their approach starts with a set of isolated nodes, and the nodes are iteratively grouped based on the modularity gain. While some techniques use exhaustive optimization to better estimate the final maximum modularity, at the expense of computational cost [9,10,11,12], more efficient techniques have also been proposed to identify communities from large complex networks [23,24,25,26,27].

In bioinformatics, community detection has been mainly used in the context of protein-protein interaction networks and prediction of functional families [18,19,20,21]. Jia et al. [15] used community detection to identify transcription factor binding sites in a small set of nucleotide sequences (approach referred to as TFBSGroup). In [16], we have extended TFBSGroup to construct sequential features for classifying large sets of nucleotide sequences. To the best of our knowledge, community detection algorithms have not been used to construct sequential features for classifying protein sequences in a machine learning framework.

3   Feature Construction Using Community Detection

3.1   Community Detection Algorithm

Complex network analysis has gained a lot of attention among researchers interested in identifying hidden structural and relational properties within a large system. A network, similar to a graph, comprises of a set of V nodes, $\{n_1, n_2, \cdots, n_V\}$, along with a set of E edges, $\{(n_i, n_j) \mid 1 \leq i \neq j \leq V\}$. Many complex systems can be represented using a network, with nodes being the elementary components of the system and the relationship between the components being the links.

A community is a sub-network whose nodes are highly connected with each other, as compared to other nodes outside the community. Thus, a community reflects a group of closely related nodes. Identifying communities can uncover structural properties of a network. From the methods available to identify communities, we use a technique based on modularity, proposed by Blondel et al. [23].

The modularity of a network (denoted by Q) measures the structure of a network by defining the strength of the network when divided into modules (sub-networks or communities). High modularity suggests that the nodes within each community are densely connected when compared to other nodes. The algorithm proposed in [23] identifies communities by optimizing the modularity gain. It is a fast and efficient approach to identify high modularity partitions in a large network, which can be seen as a two-phase iterative process.

In the first phase, each node is assigned to a different community. Then, for each node, n_i, the algorithm computes the gain in modularity, ΔQ, achieved by removing n_i from its community and placing it in the community of n_j, where n_j is a neighboring node of n_i. It then assigns n_i to the community of that n_j, for which the maximum modularity gain is obtained. In the second phase, a new network is constructed, with the nodes being the communities identified in the first phase. The weights of the edges between the new nodes are computed as the sum of the weights of the edges between nodes of the corresponding two communities. Edges among nodes of the same community form self-loops in the new network. These two phases are iterated until there is no further improvement in the modularity gain, and, then, the final set of communities is returned.

3.2   Identifying Motifs Using Community Detection

Jia et al. [15] introduced the idea of using community detection to identify transcription factor binding sites (a.k.a., motifs) in a set of nucleotide sequences. A motif is a pattern that is widespread across different sequences, and potentially has biological significance. Consequently, a motif can be obtained by aligning a set of subsequences that occur across different sequences (called motif instances), which are highly correlated to each other. The motif is also referred to as the consensus of its motif instances. The approach proposed by Jia et al. [15], called TFBSGroup, aims at identifying motifs under the ZOMOPS constraint (Zero, One or Multiple Occurrences of a motif Per Sequence). The motifs identified have length k, and there are at most d mismatches between motif instances and the motif consensus. For a set of N sequences of maximum length L, the TFBSGroup approach also works in three phases/steps.

(Step 1) The first step deals with the construction of an N-partite network and detection of communities in that network. The nodes of the network represent all possible k-mers (subsequences of length k) of the input sequences. Therefore, for a set of N sequences, each of length L, there are $(N * (L - k + 1))$ nodes. Two nodes are connected by an edge only if the Hamming distance between the k-mers corresponding to the two nodes is no more than x (a parameter that the TFBSGroup algorithms takes). Given that the maximum Hamming distance allowed between a motif instance and the motif consensus is d (another TFBSGroup parameter), it follows that the maximum Hamming distance between any two motif instances is $2d$. Therefore, while constructing the network, the maximum value that x can be given is $2d$. We should note that there is no edge between nodes (k-mers) belonging to the same sequence, which means that a set of N sequences results in an N-partite network.

(Step 2) After constructing the network, all possible communities of size at least q (another parameter) are identified. Then, from each community, a motif consensus is generated by aligning all k-mers from that particular community.

(Step 3) Finally, each motif consensus is greedily refined towards a final motif, and a significance score is calculated for it. The top t motifs (default $t=10$) are then selected based on the significance score (see [15] for more details).

According to [15], the worst-case time complexity of the TFBSGroup algorithm is quartic in terms of the total number of input sequences, N, and the length of the sequences, L: $O(p(k, x)^2 \times N^4 \times L^4)$, where $p(k, x)$ is the probability of two random k-mers having Hamming distance at most x. Although TFBSGroup can successfully identify transcription factor binding sites in a small set of sequences, it cannot be applied for generating features for classification problems, due to the large number of sequences involved. To address this problem, in [16], we proposed an approach for scaling up TFBSGroup, as described below.

3.3   Feature Construction for Large Nucleotide Sequence Datasets

To extend TFBSGroup to generate features for sequence classification problems, in [16], we proposed to run TFBSGroup on a set of randomly selected R samples, each of S sequences, from the available data consisting of N sequences, where $S \ll N$. The time complexity of running TFBSGroup on R *samples* reduced to:

$$O(p(k, x)^2 \times S^4 \times L^4 \times R) \ll O(p(k, x)^2 \times N^4 \times L^4),$$ when $(R \times S^4) \ll N^4$.

We choose R and S that satisfy the condition above to achieve scalability. Furthermore, when generating the R samples, we allow overlap between samples, but there is no overlap between sequences within a sample. The reason for this is that we are interested in finding patterns/motifs that are frequent across sequences, but not necessarily within a sequence. By allowing samples to overlap, we can essentially link subsequences in different samples, and get higher coverage.

We run TFBSGroup on each individual sample and select the top t motifs from each sample. All the resulting motifs are merged together to form the final set of motifs. As a result, the final set of motifs contains a total of $t \times R$ motifs (for a particular length of the motif, k). The frequency count representation of all unique motif instances present in the final set of motifs is then used to represent sequences for classification. We refer to the set of motif instances as the set of c-mers given that they are identified based on community detection. This approach has been successfully used to construct informative features for learning classifiers from large sets of nucleotide sequences [16]. However, it cannot be directly used to construct features for protein sequences, as the Hamming distance does not capture well differences between short protein k-mers.

3.4   Feature Construction for Large Protein Sequence Datasets

For protein sequences, motifs of shorter length carry better information than motifs of longer length [2]. When the length of the motif is small (*e.g.*, $k = 1$, 2 or 3), the probability of two protein k-mers having Hamming distance less than a particular threshold, x, is high, as $x \approx k$, and thereby the resulting network in very dense. For longer k-mers (*e.g.*, $k = 6$, 7 or 8), usually the desired threshold x is smaller than k. In such cases, given the large alphabet size of protein sequences, the probability of having an edge between two nodes is very low, thereby resulting in a very sparse network. Therefore, when Hamming distance is used to construct a network of protein subsequences, the resulting network is either too sparse or too dense. To address this issue, we propose to use substitution scores when constructing sequential features for protein sequences. Substitution scores are computed using substitution matrices for amino acids, which take into account the divergence time as well as the substitution rate for each possible alignment of amino acids. Based on the default parameters for BLAST, in this work, we used PAM30 matrix [5,13] to compute the substitution scores for pairs of k-mers.

Similar to the Hamming distance, the substitution score for a pair of k-mers is computed based on the alignment of the amino acids at the respective positions of the k-mers. However, when using substitution matrices, as opposed to the Hamming distance, the score of an alignment at a particular position is affected not only by the match/mismatch of the respective amino acids, but also by the degree of match/mismatch as captured by the substitution matrix. For example, consider two pairs of *3*-mers: $\{JQK, LZK\}$ and $\{PGD, RGD\}$. For pair 1, the Hamming distance is 2 and the substitution score is 19, and for pair 2, the Hamming distance is 1 and the substitution score is 10 (where substitution scores are computed using PAM30 matrix [5]). We should note that the substitution scores represent similarity scores, as opposed to distances. The higher the substitution score values, the more similar the sequences. Thus, based on the Hamming distance, the k-mers of pair 2 are more similar than the k-mers of pair 1. Contrarily, based on substitution scores, the k-mers of pair 1 are more similar when compared to pair 2. Given the fact that substitution scores capture the degree of match/mismatch, they are preferable to the Hamming distance when interested in identifying similar protein sequences.

In this work, we find protein motifs with the property that the substitution score between any motif instance and the corresponding motif consensus is at least s (a parameter of the algorithm). When constructing the subsequence network, a pair of nodes (protein subsequences of length k) are connected by an edge only if the substitution score of the two k-mers is greater than a particular threshold p (to avoid spurious edges, in our experiments, we chose $p = s/2$). After constructing the network, all possible communities of size at least q are identified in the network using the community detection algorithm, and the k-mers corresponding to each community are aligned to form the motif consensus. Subsequently, each motif consensus is greedily refined towards the final motif for that community, using the substitution scores, and, for each community, the refined motif along with the motif instances are returned, together with a normalized substitution score. The above process is repeated for all communities identified by the algorithm. The top t motifs (default t=10) from all the resulting motifs are then selected based on the normalized substitution scores. The unique motif instances belonging to the final motifs are used as classification features and will be denoted by c-mers, similar to how the community detection-based features for nucleotide sequences are denoted in our previous work.

4 Experimental Setup

In Section 4.1, we present the research questions addressed through our work. The datasets used are described in Section 4.2, and the details about the experimental setup are presented in Section 4.3.

4.1 Research Questions

We addressed the following research questions:

1. *How does the number of c-mers compare to the number of all possible k-mers?* The set of c-mers generated using the community detection algorithm satisfy the ZOMOPS constraint (Zero, One or Multiple Occurrences of the motifs Per Sequence). Therefore, we expect the dimensionality of the set of c-mers to be very small when compared to the dimensionality of the set of all k-mers.

2. *How does the predictive power of c-mers compare to that of k-mers?* To investigate the predictive power of c-mers, we compare the performance of the classifiers learned from sequences represented using c-mers with that of the classifiers learned from sequences represented using an equal number of k-mers (obtained via feature selection from the total number of k-mers). Given that the community detection approach used to generate c-mers is not supervised (i.e., does not make use of sequence labels), we have conducted experiments in three different learning scenarios: supervised, semi-supervised and domain adaptation. While larger amounts of data are possibly available in the supervised learning scenario, the assumption in the semi-supervised and domain adaptation scenarios is that only small amounts of labeled data are available for the domain of interest. Therefore, we expect the features obtained using community detection to be more informative than the k-mers, at least in these scenarios, as feature-class dependencies may not be well captured by feature selection when the amount of labeled data is limited.

4.2   Datasets

In this work, we targeted the problem of classifying protein sequences based on their respective localization. We used four different protein sequence datasets:

- **PSORTb datasets [7]:** The Gram-negative (GN) dataset consists of 1444 sequences belonging to one of the five classes: cytoplasm (278), cytoplasmic membrane (309), periplasm (276), outer membrane (391) and extracellular (190). The Gram-positive (GP) dataset consists of 541 sequences belonging to one of the four classes: cytoplasm (194), cytoplasmic membrane (103), cellwall (61) and extracellular (183).
- **TargetP datasets [6]:** The plant (P) dataset consists of 940 sequences belonging to one of the four classes: chloroplast (141), mitochondrial (368), secretory pathway/signal peptide (269) and other (consisting of 54 proteins labeled nuclear and 108 proteins labeled cytosolic). The non-plant (NP) dataset consists of 2738 sequences belonging to one of the three classes: mitochondrial (361), secretory pathway/signal peptide (715) and other (consisting of 1224 proteins labeled nuclear and 438 proteins labeled cytosolic).

4.3   Experiments

As mentioned above, we conducted experiments in three different scenarios: supervised, semi-supervised, and domain adaptation. We used the naïve Bayes multinomial (NBM) classifier for the supervised scenario; the co-training iterative-based algorithm, with NBM as the base classifier, for the semi-supervised scenario; and the algorithm proposed in [14], derived from the NBM classifier, for the domain adaptation scenario. All experiments are conducted using 5-fold cross-validation, with four folds used for training and the remaining fold for testing. In the supervised scenario, all the training data was assumed to be labeled and was used to learn the classifiers. In the semi-supervised scenario, we assumed 20% of the training data to be labeled and up to 80% to be unlabeled (specifically, we experimented with 20%, 40%, 60%, and 80% of the training data as unlabeled). Finally, for the domain adaptation scenario, we assumed a source domain with labeled data to be available, in addition to the target domain labeled and unlabeled data. We conducted experiments with the following pairs of source→target domains: $GP \rightarrow GN$, $GN \rightarrow GP$, $P \rightarrow NP$ and $NP \rightarrow P$, respectively. Only the overlapping classes within a pair of domains were used in the domain adaptation scenario (i.e., cytoplasm, cytoplasmic membrane and extracellular for the GP/GN pairs, and mitochondrial, secretory pathway/signal peptide and others for P/NP pairs). For each run, we used all the data from the source domain, and we split the training target data into 20% as labeled and up to 80% as unlabeled (i.e., 20%, 40%, 60%, and 80%).

We evaluated the performance of the classifiers using the area under the receiver operating characteristic curve (AUC), as the class distribution is relatively balanced for all data sets. We report the average AUC over the five runs.

Our goal is to compare two feature representations, c-mers and k-mers. The details of how these sets of features were generated are provided below:

- For c-mers, we invoke the proposed approach (Section 3.4) with the following parameter values: length of the motif $k \in \{2, 3, 4\}$, minimum community size $q = 5$, minimum substitution score $s = 15$, number of samples $R = 50$, sample size $S = 10$, and number of motifs selected $t = 10$. The algorithm returns the set of c-mers. We denote the number of c-mers by N_{c-mers}. We should note that the total number of unique sequences from all the R samples, N_{RS}, can be smaller than the total number of all the sequences in the dataset N, as the samples generated can have overlapping sequences.
- For k-mers, we use the sliding window approach. To make a fair comparison between c-mers and k-mers, we generate k-mers of the same length $k \in \{2, 3, 4\}$, on the same set of training sequences, N_{RS}. In addition, when comparing the performance, we apply feature selection on k-mers, using the labeled data only, to select top k-mers, such that the number of k-mers used, N_{k-mers}, is the same as the number of, c-mers, N_{c-mers}. For feature selection, we use *Entropy based Category Coverage Difference* (ECCD), proposed in [17], as this measure makes use of both the distribution of the sequences containing the features and the frequency of the occurrence of a feature value within the sequence, to compute the feature-class dependency scores.

5   Results

Following are the total number of features (#c-mers, #k-mers) generated, averaged over five folds for all four datasets: GP (1976, 74203), GN (1823, 83060), P (1684, 77091) and NP (1751, 102815). As can be seen, the total number of c-mers is much smaller than the total number of k-mers. As the set of c-mers represents a reduced set of k-mers (c-mers $\subset k$-mers), this means that our proposed approach can be seen as a dimensionality reduction technique for k-mers.

Table 1 shows the AUC values for the supervised, semi-supervised, and domain adaptation scenarios, respectively. As can be seen, the AUC values are higher when using c-mers as compared to k-mers in all scenarios, for most experiments, specifically, in 3 out of 4 cases for the supervised scenario, and 15 out of 16 cases for the semi-supervised scenario as well as for the domain adaptation scenario. In semi-supervised/domain adaptation learning scenarios, given that the amount of available labeled data is small, feature selection may not estimate the feature-class dependencies accurately, thereby selecting a set of possibly uninformative features, while filtering out informative features. Thus, in cases when the amount of available labeled data is small, c-mers are expected to outperform k-mers selected using feature selection. Surprisingly, we observed a similar behavior also in the supervised learning scenario, the scenario where we presumably have sufficient labeled data to estimate the feature-class dependencies accurately. The reason for this might be that feature selection can still leave out informative features, as the size of the final set is limited by the number of c-mers, and it possibly includes some uninformative features. On the other hand, the set of c-mers, having the same size as the set of selected k-mers, capture much better features that carry information about classes.

Table 1. AUC values on the four datasets: Gram-positive (GP), Gram-negative (GN), plant (P), and non-plant (NP). For the semi-supervised and domain adaptation scenarios, the amount of labeled data is fixed to 20%, while the amount of unlabeled data is varied from 20% to 80%. For domain adaptation, the source and target domains are indicated as source→target. Note that for each scenario, in most cases, the classifiers had higher AUCs when using c-mers.

	Supervised learning scenario							
Unlabeled	GP		GN		P		NP	
	c-mers	k-mers	c-mers	k-mers	c-mers	k-mers	c-mers	k-mers
0	**0.925**	0.869	**0.929**	0.915	**0.837**	0.754	0.834	**0.874**

	Semi-supervised learning scenario							
Unlabeled	GP		GN		P		NP	
	c-mers	k-mers	c-mers	k-mers	c-mers	k-mers	c-mers	k-mers
20	**0.847**	0.746	**0.877**	0.793	**0.72**	0.663	**0.825**	0.787
40	**0.831**	0.748	**0.882**	0.793	**0.705**	0.656	**0.793**	0.788
60	**0.852**	0.742	**0.87**	0.783	**0.698**	0.655	0.773	**0.774**
80	**0.822**	0.749	**0.851**	0.788	**0.705**	0.659	**0.773**	0.77

	Domain adaptation scenario							
Unlabeled	GN→GP		GP→GN		NP→P		P→NP	
	c-mers	k-mers	c-mers	k-mers	c-mers	k-mers	c-mers	k-mers
20	**0.877**	0.785	**0.911**	0.899	**0.802**	0.78	**0.829**	0.763
40	**0.856**	0.755	**0.91**	0.893	**0.748**	0.728	**0.821**	0.777
60	**0.852**	0.731	**0.902**	0.896	**0.739**	0.728	**0.777**	0.73
80	**0.839**	0.741	0.892	**0.895**	**0.734**	0.727	**0.803**	0.744

Together, the small dimensionality of the set of c-mers and the performance results in Table 1 suggest that our approach is successful in retaining informative/predictive features, while reducing the dimensionality by a large extent in all learning scenarios considered.

6   Conclusion

We have investigated the predictive power of the features generated using a community detection approach, for classifying proteins based on their respective localizations. As the original approach of using Hamming distance [15,16] to generate nucleotide features does not work for sequences of a large alphabet size (such as proteins), we proposed a novel idea of using substitution scores as a similarity metric between two protein k-mers in the process of constructing the protein subsequence network. The resulting c-mers are associated with a set of motifs which represent groups of similar subsequences that occur frequently in the set of sequences. As opposed to that, the set of k-mers generated with a sliding window take into account all possible subsequences of a certain length occurring in the sequences. Both approaches are unsupervised, as they do not make use of class labels. To evaluate the predictive power of the features generated using our proposed approach (specifically, the predictive power of c-mers), we have conducted experiments in supervised, semi-supervised and domain adaptation learning scenarios. The results of the experiments show that our proposed approach generated low-dimensional informative features in supervised, semi-supervised and domain adaptation scenarios. Furthermore, those features have resulted in improved performance as opposed to k-mers selected based on feature-class dependency scores, even in the supervised scenario, where presumably there is enough labeled data to accurately estimate the scores.

Acknowledgments. This work was supported by an Institutional Development Award (IDeA) from the National Institute of General Medical Sciences of the National Institutes of Health under grant number P20GM103418. The content is solely the responsibility of the authors and does not necessarily represent the official views of the National Institute of General Medical Sciences or the National Institutes of Health. The computing for this project was performed on the Beocat Research Cluster at Kansas State University, which is funded in part by grants MRI-1126709, CC-NIE-1341026, MRI-1429316, CC-IIE-1440548.

References

1. Leslie, C.S., Eskin, E., Cohen, A., Weston, J., Noble, W.S.: Mismatch string kernels for discriminative protein classification. Bioinformatics 20(4), 467–476 (2004)
2. Caragea, C., Silvescu, A., Mitra, P.: Protein sequence classification using feature hashing. Proteome Science 10(1), 1–8 (2012)
3. Sun, L., Luo, H., Bu, D., Zhao, G., Yu, K., Zhang, C., Liu, Y., Chen, R., Zhao, Y.: Utilizing sequence intrinsic composition to classify protein-coding and long non-coding transcripts. Nucleic Acids Research (2013)
4. Clauset, A., Newman, M.E.J., Moore, C.: Finding community structure in very large networks. Physical Review E, 1–6 (2004)
5. Dayhoff, M.O., Schwartz, R.M., Orcutt, B.C.: A model of evolutionary change in proteins. Atlas of Protein Sequence and Structure 5(suppl. 3), 345–351 (1978)
6. Emanuelsson, O., Nielsen, H., Brunak, S., Heijne, G.: Predicting subcellular localization of proteins based on their n-terminal amino acid sequence. Journal of Molecular Biology 300(4), 1005–1016 (2000)
7. Gardy, J.L., Laird, M.R., Chen, F., Rey, S., Walsh, C.J., Ester, M., Brinkman, F.S.L.: Psortb v.2.0: Expanded prediction of bacterial protein subcellular localization and insights gained from comparative proteome analysis. Bioinformatics 21(5), 617–623 (2005)
8. Girvan, M., Newman, M.E.J.: Community structure in social and biological networks. Proceedings of the National Academy of Sciences 99(12), 7821–7826 (2002)
9. Guimera, R., Sales-Pardo, M., Amaral Modularity, L.A.N.: from fluctuations in random graphs and complex networks. Phys. Rev. E 70(025101) (2004)
10. Massen, C.P., Doye, J.P.K.: Identifying communities within energy landscapes. Phys. Rev. E 71(046101) (2005)
11. Medus, A., Acuna, G., Dorso, C.: Detection of community structures in networks via global optimization. Physica A: Statistical Mechanics and its Applications 358(2), 593–604 (2005)

12. Guimera, R., Amaral, L.A.N.: Functional cartography of complex metabolic networks. Nature 433(7028), 895–900 (2005)
13. Henikoff, S., Henikoff, J.G.: Amino acid substitution matrices from protein blocks. Proceedings of the National Academy of Sciences 89(22), 10915–10919 (1992)
14. Herndon, N., Caragea, D.: Naïve Bayes Domain Adaptation for Biological Sequences. In: Proceedings of the 4th International Conference on Bioinformatics Models, Methods and Algorithms, BIOINFORMATICS 2013, pp. 62–70 (2013)
15. Jia, C., Carson, M., Yu, J.: A fast weak motif-finding algorithm based on community detection in graphs. BMC Bioinformatics 14(1), 1–14 (2013)
16. Tangirala, K., Caragea, D.: Community detection-based features for sequence classification. In: Proceedings of the 5th ACM Conference on Bioinformatics, Computational Biology, and Health Informatics (BCB 2014). ACM (2014)
17. Largeron, C., Moulin, C., Géry, M.: Entropy based feature selection for text categorization. In: Proc. of the 2011 ACM Symp. on Applied Computing, SAC 2011, pp. 924–928 (2011)
18. Dongfang, N., Xiaolong, Z.: Prediction of hot regions in protein-protein interactions based on complex network and community detection. In: IEEE International Conference on Bioinformatics and Biomedicine (BIBM), pp. 17–23 (December 2013)
19. Mahmoud, H., Masulli, F., Rovetta, S., Russo, G.: Community detection in protein-protein interaction networks using spectral and graph approaches. In: Formenti, E., Tagliaferri, R., Wit, E. (eds.) CIBB 2013. LNCS, vol. 8452, pp. 62–75. Springer, Heidelberg (2014)
20. Mallek, S., Boukhris, I., Elouedi, Z.: Predicting proteins functional family: A graph-based similarity derived from community detection. In: Filev, D., Jabłkowski, J., Kacprzyk, J., Krawczak, M., Popchev, I., Rutkowski, L. (eds.) Intelligent Systems'2014. AISC, vol. 323, pp. 629–639. Springer, Heidelberg (2015)
21. van Laarhoven, T., Marchiori, E.: Robust community detection methods with resolution parameter for complex detection in protein protein interaction networks. In: Shibuya, T., Kashima, H., Sese, J., Ahmad, S. (eds.) PRIB 2012. LNCS, vol. 7632, pp. 1–13. Springer, Heidelberg (2012)
22. Newman, M.E.J., Girvan, M.: Finding and evaluating community structure in networks. Phys. Rev. E 69(026113) (2004)
23. Blondel, V., Guillaume, J., Lambiotte, R., Mech, E.: Fast unfolding of communities in large networks. J. Stat. Mech, P10008 (2008)
24. Donetti, L., Muñoz, M.A.: Improved spectral algorithm for the detection of network communities. In: Proceedings of the 8th Granada Seminar - Computational and Statistical Physics, pp. 1–2 (2005)
25. Raghavan, U.N., Albert, R., Kumara, S.: Near linear time algorithm to detect community structures in large-scale networks. Physical Review E 76(3) (September 2007)
26. Rosvall, M., Bergstrom, C.T.: Maps of random walks on complex networks reveal community structure. Proceedings of the National Academy of Sciences 105(4), 1118–1123 (2008)
27. Radicchi, F., Castellano, C., Cecconi, F., Loreto, V., Parisi, D.: Defining and identifying communities in networks. Proceedings of the National Academy of Sciences of the United States of America 101(9), 2658–2663 (2004)

Curvilinear Triangular Discretization of Biomedical Images

Jing Xu[✉] and Andrey N. Chernikov

Department of Computer Science, Old Dominion University, Norfolk, VA, USA
{jxu,achernik}@cs.odu.edu

Abstract. Mesh generation is a useful tool for obtaining discrete descriptors of biological objects represented by images. The generation of meshes with straight sided elements has been fairly well understood. However, in order to match curved shapes that are ubiquitous in nature, meshes with high-order elements are required. Moreover, for the processing of large data sets, automatic meshing procedures are needed. In this work we present a new technique that allows for the automatic construction of high-order curvilinear meshes. This technique allows for a transformation of straight-sided meshes to curvilinear meshes with C^2 smooth boundaries while keeping all elements valid as measured by their Jacobians. The technique is demonstrated with examples.

Keywords: Biomedical image processing · High-order mesh generation

1 Introduction

Discretization of complex shapes into simple elements are widely used in various computing areas that require a quantitative analysis of spatially dependent attributes. One, traditional, area is the finite element analysis [12] which is used to numerically solve partial differential equations derived using solid mechanics and computational fluid dynamics approaches. With this approach one starts with the knowledge of the constitutive physical laws and initial (boundary) conditions and obtains a prediction of the properties of objects of interest. Another, emerging, area is the use of discretization for delineating homogeneous spatial zones within objects that can be represented as units for an overall object description. With this approach one starts with the knowledge of object properties and uses statistical methods to infer the processes that govern the formation of the object. Therefore, the second approach can be viewed as a reversal of the first approach, that still relies on a similar discretization technique. This second approach is a useful tool for bioinformatics applications, for example gene expression pattern analysis [10,11,4].

Said discretizations of objects are usually called meshes, and the simple elements that they consist of are either triangles and tetrahedra (in two and three dimensions, respectively), or quadrilaterals and hexahedra. Furthermore, elements can have either straight or curved sides. In our previous work [10,11] we

© Springer International Publishing Switzerland 2015
R. Harrison et al. (Eds.): ISBRA 2015, LNBI 9096, pp. 343–354, 2015.
DOI: 10.1007/978-3-319-19048-8_29

used triangular meshes with straight sides to discretize images of fruit fly embryos. However, the embryos, like most biological objects, have curved shapes, and their discretizations with straight-sided elements have limited accuracy. To obtain much higher accuracy one needs to use curved-sided elements that match the curves of object boundaries.

In this paper we build the methodology for automatically generating valid high-order meshes to represent curvilinear domains with smooth global mesh boundaries. Cubic Bézier polynomial basis is selected for the geometric representation of the elements because it provides a convenient framework supporting the smooth operation and mesh validity verification. We highlight the three contributions of this paper:

1. Curved mesh boundary is globally smooth. It satisfies the C^2 smoothness requirement, i.e., the first and second derivatives are continuous.
2. A new procedure was developed to efficiently verify the validity. It is formulated to work in an arbitrary polynomial order.
3. Our proposed approach is robust in the sense that all the invalid elements are guaranteed to be eliminated.

The procedure starts with the automatic construction of a linear mesh that simultaneously satisfies the quality (elements do not have arbitrarily small angles) and the fidelity (a reasonably close representation) requirements. The edges of those linear elements which are classified on the boundary are then curved using cubic Bézier polynomials such that these boundary edges constitute a cubic spline curve. Once our validity verification procedure detects invalid elements, the meshing procedure next curves the interior elements by iteratively solving for the equilibrium configuration of an elasticity problem until all the invalid elements are eliminated.

Various procedures have been developed and implemented to accomplish the generation of a curvilinear mesh. Sherwin and Peiro [8] adopted three strategies to alleviate the problem of invalidity: generating boundary conforming surface meshes that account for curvature; the use of a hybrid mesh with prismatic and tetrahedral elements near the domain boundaries; refining the surface meshes according to the curvature. However, these strategies are intuitive solutions that are not guaranteed to generate valid high-order meshes. The mesh spacing is decided by a user defined tolerance ϵ related to the curvature and a threshold to stop excessive refinement. Persson and Peraire [6] proposed a node relocation strategy for constructing well-shaped curved meshes. Compared to our method which iteratively solves for the equilibrium configuration of a linear elasticity problem, they use a nonlinear elasticity analogy, and by solving for the equilibrium configuration, vertices located in the interior are relocated as a result of a prescribed boundary displacement. Luo et al. [5] isolate singular reentrant model entities, then generate linear elements around those features, and curve them while maintaining the gradation. Local mesh modifications such as minimizing the deformation, edge or facet deletion, splitting, collapsing, swapping as well as shape manipulation are applied to eliminate invalid elements whenever they are introduced instead of our global node relocation strategy. George and Borouchaki [7] proposed a method for constructing tetrahedral meshes of degree two from a polynomial surface mesh of degree two. *Jacobian* is introduced for guiding the correction of the invalid curved elements. When the polynomial degree is higher, it is complicated to calculate the *Jacobian*, so we develop a procedure suitable for a polynomial of any degree. Furthermore, none of the above algorithms generates C^1 and C^2 smooth mesh boundaries.

The rest of the paper is organized as follows. in Section 2, we review some basic definitions. Section 3 gives a description of the automatic construction of a graded linear mesh and the transformation of the linear mesh into a valid high-order mesh. We present meshing results in Section 4 and conclude in Section 5.

2 Bézier Curves and Bézier Triangles

2.1 Bézier Curves

We express Bézier curves in terms of Bernstein polynomials. A nth order Bernstein polynomial is defined explicitly by

$$B_i^n(t) = \binom{n}{i} t^i (1 - t)^{n-i}, \quad i = 0, ..., n, \quad t \in [0, 1],$$

where the binomial coefficients are given by

$$\binom{n}{i} = \begin{cases} \frac{n!}{i!(n-i)!} & \text{if } 0 \le i \le n \\ 0 & \text{else.} \end{cases}$$

One of the important properties of the Bernstein polynomials is that they satisfy the following recurrence:

$$B_i^n(t) = (1 - t)B_i^{n-1}(t) + tB_{i-1}^{n-1}(t),$$

with

$$B_0^0(t) \equiv 1, \quad B_j^n(t) \equiv 0 \quad for \quad j \in 0, ..., n.$$

Then the Bézier curve of degree n in terms of Bernstein polynomial can be defined recursively as a point-to-point linear combination (linear interpolation) of a pair of corresponding points in two Bézier curves of degree $n - 1$. Given a set of points $P_0, P_1, ..., P_n \in E^2$, where E^2 is a two-dimensional Euclidean space, and $t \in [0, 1]$, set

$$b_i^r(t) = (1 - t)b_i^{r-1}(t) + tb_{i+1}^{r-1}(t) \quad \begin{cases} r = 1, ..., n \\ i = 0, ..., n - r \end{cases}$$

and $b_i^0(t) = P_i$. Then $b_0^n(t)$ is the point with parameter value t on the Bézier curve b^n. The set of points $P_0, P_1, ..., P_n$ are called *control points*, and the polygon P formed by points $P_0, P_1, ..., P_n$ is called *control polygon* of the curve b^n.

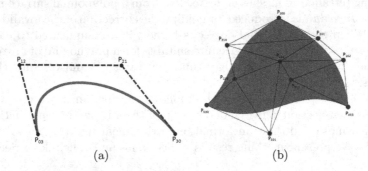

Fig. 1. (a) An example of the cubic Bézier curve with its control polygon formed by four control points. (b) An example of the cubic Bézier triangle with its control net formed by ten control points.

An explicit form of a n-th order Bézier curve can be defined as

$$b^n(t) = \sum_{i=0}^{n} B_i^n(t) P_i.$$

The barycentric form of Bézier curves demonstrates its symmetry property nicely. Let u and v be the barycentric coordinates, $u \in [0,1]$ and $v \in [0,1]$, $u + v = 1$, then

$$b^n(u,v) = \sum_{i+j=n} B_{ij}^n(u,v) P_{ij},$$

where $B_{ij}^n(u,v) = \frac{n!}{i!j!} u^i v^j$, $P_{ij} \in E^2$ are the control points, and $i + j = n$.

Specifically, the cubic Bézier curve can be written in terms of the barycentric coordinates,

$$b^3(u,v) = \sum_{i+j=3} B_{ij}^3(u,v) P_{ij} = u^3 P_{03} + 3u^2 v P_{12} + 3uv^2 P_{21} + v^3 P_{30},$$

Fig. 1a gives an example of the cubic Bézier curve with its control polygon.

2.2   Bézier Triangles

Univariate Bernstein polynomials are the terms of the binomial expansion of $[t + (1-t)]^n$. In the bivariate case, a n-th order Bernstein polynomial is defined by

$$B_{\boldsymbol{i}}^n(\boldsymbol{u}) = \binom{n}{\boldsymbol{i}} u^i v^j w^k,$$

where

$$\boldsymbol{i} = \{i,j,k\}, \quad |\boldsymbol{i}| = n, \quad \boldsymbol{u} = \{u,v,w\},$$

$u \in [0,1], v \in [0,1]$ and $w \in [0,1]$ are the barycentric coordinates and $u+v+w = 1$. It follows the standard convention for the *trinomial coefficients* $\binom{n}{i} = \frac{n!}{i!j!k!}$.

This leads to a simple definition of a Bézier triangle of degree n

$$\mathcal{T}^n(u,v,w) = \sum_{i+j+k=n} B_{ijk}^n(u,v,w)P_{ijk},$$

where P_{ijk} is a control point. Specifically, the Bézier triangle of degree three can be written as

$$\mathcal{T}^3(u,v,w) = \sum_{i+j+k=3} B_{ijk}^3(u,v,w)P_{ijk}$$
$$= P_{300}u^3 + P_{030}v^3 + P_{003}w^3 + 3P_{201}u^2w + 3P_{210}u^2v$$
$$+ 3P_{120}uv^2 + 3P_{102}uw^2 + 3P_{021}v^2w + 3P_{012}vw^2 + 6P_{111}uvw.$$
$$(1)$$

Fig. 1b gives an example of the cubic triangular patch with its control net formed by its ten control points.

3   Mesh Generation for Curvilinear Domains

Given a bounded curved domain $\Omega \subset \mathcal{R}^2$, the algorithm outputs a curvilinear mesh of the *interior* of Ω with global smooth boundary. Fig. 2 illustrates the main steps performed by our algorithm. The details are elaborated below.

The mesh has to provide a close approximation of the object shape, and we measure the closeness by the fidelity tolerance, the two-sided Hausdorff distance from the mesh to the image and the image to the mesh. For image boundary I and mesh boundary M, the one-sided distance from I to M is given by

$$H(I \to M) = \max_{i \in I} \min_{m \in M} d(i,m),$$

where $d(\cdot, \cdot)$ is the regular Euclidean distance. The one-sided distance from M to I is given similarly by

$$H(M \to I) = \max_{m \in M} \min_{i \in I} d(m,i).$$

The two-sided distance is:

$$H(I \leftrightarrow M) = \max\{H(I \to M), H(M \to I)\}.$$

To generate the initial linear mesh, we adopt the image-to-mesh conversion algorithm [3], for four reasons: (1) it allows for a guaranteed angle bound (quality), (2) it allows for a guaranteed bound on the distance between the boundaries of the mesh and the boundaries of the object (fidelity), (3) it coarsens the mesh to a much lower number of elements with gradation in the interior, (4) it is formulated to work in both two and three dimensions.

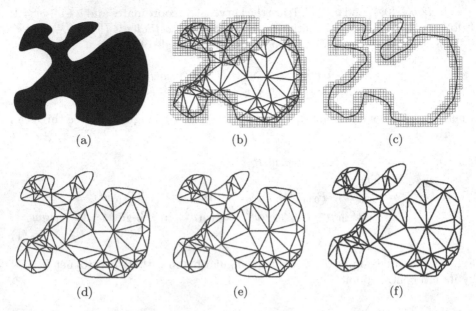

(a)   (b)   (c)

(d)   (e)   (f)

Fig. 2. An illustration of the main steps performed by our algorithm. (a) The input two-dimensional image. It shows a curvilinear domain to be meshed. (b) A linear mesh which satisfies the user specified quality and fidelity tolerances. The shaded region represents the fidelity tolerance. (c) Those edges that are classified on the mesh boundary are curved such that the C^2 smoothness requirement is satisfied. (d) When there are curved edges on the boundary and linear edges in the interior, the mesh validity need to be verified. (e) The red triangles are invalid elements detected by our verifying procedure. (f) To fix these invalid triangles, the interior edges are curved by iteratively solving for the equilibrium configuration of an elasticity problem, and a valid mesh is obtained.

We transform the linear boundary edges followed by curving the interior edges to eliminate the invalid elements. Bézier curve basis is selected because its mathematical descriptions are compact, intuitive, and elegant. It is easy to compute, easy to use in higher dimensions (3D and up), and can be stitched together to represent any shape.

Smoothness of the resulting curve is assured by imposing one of the continuity requirements. A curve or surface can be described as having C^n continuity, n being the measure of smoothness. Consider the segments on either side of a point on a curve:

C^0: The curves touch at the joint point;
C^1: First derivatives are continuous;
C^2: First and second derivatives are continuous.

A cubic polynomial is the lowest degree polynomial that can guarantee a C^1 or a C^2 curve. Biomedical objects usually have naturally smooth boundaries, and can be approximated by either a C^1 or a C^2 curve. In our previous work [9],

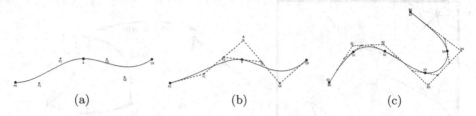

(a)   (b)   (c)

Fig. 3. An illustration of the construction of the cubic spline curve. (a) An example of a cubic spline curve, formed by two Bézier curves with control points P_0, P_1, P_2, S and S, Q_1, Q_2, Q_3. (b) An A-frame is a structure in which P_2 is the midpoint of $\overline{AP_1}$, Q_1 is the midpoint of $\overline{AQ_2}$ and S is the midpoint of $\overline{P_2Q_1}$. (c) The cubic spline curve constructed with the help of the B-spline points shown in green. The black points are S points, which are the endpoints of the boundary edges of the linear mesh. The red points are the control points need to be calculated to form the cubic spline curve.

we present how to construct a C^1 Bézier curve; in the following we will address how to generate a C^2 Bézier curve. By counting incorrectly classified pixels (i.e., inside vs. outside the shape) in the final mesh, the most suitable curve can be determined.

We aim to find a cubic spline curve passing through all the mesh boundary points given in order. It is a piecewise cubic curve that is composed of pieces of different cubic curves glued together, and it is so smooth that it has a second derivative everywhere and the derivative is continuous. Fig. 3a gives an example of a cubic spline curve.

If two Bézier curves with control points P_0, P_1, P_2, S and S, Q_1, Q_2, Q_3 are touched at point S, both their first and second derivatives match at S if and only if their control polygons fit an A-frame, which is a structure in which P_2 is the midpoint of $\overline{AP_1}$, Q_1 is the midpoint of $\overline{AQ_2}$ and S is the midpoint of $\overline{P_2Q_1}$ as Fig. 3b shows.

To fit the A-frame in the set of cubic curves, one easy approach is to use B-spline as an intermediate step. In Fig. 3c, the S points (shown in black) are known, they are the endpoints of the boundary edges of the linear mesh. What still needs to be calculated are the red control points. If the B-spline points (the apexes of the A-frames, shown in green) are known, the control points (shown in red) can be easily calculated by computing the one third and two thirds positions between the connection of every two adjacent B-spline points. The B-spline points can be computed by the relationship between S points:

$$6S_i = B_{i-1} + 4B_i + B_{i+1}.$$

By solving a linear system of equations, the coordinates of B-spline points can be obtained.

The naive high-order mesh generation does not ensure that all the elements of the final curved mesh are valid. Fig. 4a gives an example of this critical issue: some of the curvilinear triangular patches have tangled edges. Thus, it is

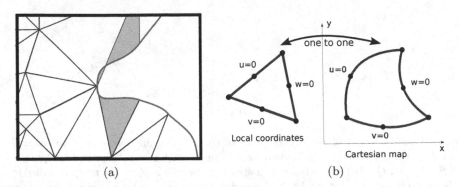

Fig. 4. (a) An example of invalid mesh. The red line is the curved mesh boundary, and the blue lines are straight mesh edges in the interior. The curved triangles that are tangled are highlighted in gray. (b) An illustration of the local u, v, w coordinates are distorted into a new, curvilinear set when plotted in global Cartesian x, y space. A general principle for the transformation: an one-to-one correspondence between Cartesian and curvilinear coordinates.

necessary to verify the validity and eliminate all the invalid elements as a post-processing step once the curved mesh has been constructed. When elements of the basic types will be mapped into distorted forms, a general principle is that an one-to-one correspondence between Cartesian and curvilinear coordinates can be established (illustrated in Fig. 4b).

The Jacobian matrix carries important information about the local behavior of the transformation from linear elements to curved elements. A violation of the condition that the determinant of the Jacobian Matrix is strictly positive every-where means the violation of the bijection general principle. One way to detect the element validity is evaluating the sign of the determinant of the Jacobian matrix throughout the element. In our previous work [9], we took advantage of the properties of Bézier triangle, formulated the *Jacobian* expression and calculated the tight lower bound of the *Jacobian* value at the element level. However, although we used the third order polynomial, it is computationally and geometrically complicated. To get the tight bound, we recursively refined the convex hull of the *Jacobian* (it is a forth order Bézier triangle) using the Bézier subdivision algorithm. In this paper, an efficient element validity verification procedure is developed for polynomials of arbitrary order. An element is invalid if and only if the control net of the Bézier triangle is twisted, meaning that at least one of the control triangles (shadowed triangles in Fig. 5) of the control net is inverted.

It is usually not enough to curve only the mesh boundary because some control points may be located such that element distortions occur in the interior of the mesh. In such case, interior mesh edges should also be curved to eliminate the invalidity or improve the curved element quality.

We move the control points of the interior mesh edges using a finite element method [12]. The geometry of the domain to be meshed is represented as an

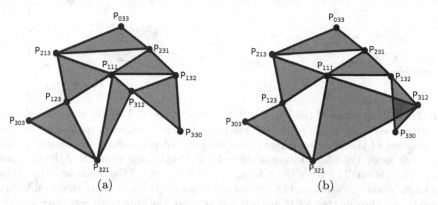

Fig. 5. (a) A valid control net with the uninverted control triangles. (b) A twisted control net with one inverted control triangle $\triangle P_{012}P_{102}P_{003}$.

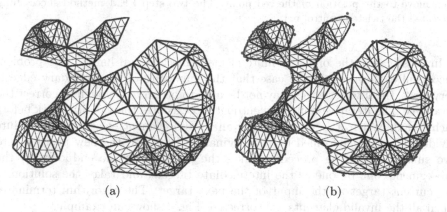

Fig. 6. (a) The control nets of the linear mesh elements is the undeformed geometry. (b) The red control points of the smooth curved boundary edges are the external loadings.

elastic solid. For each linear mesh edge, the positions of the two points which are located in the one third and two thirds ratio of each edge are computed. These positions are the original positions of the control points of the interior edges before deformation. These points form the control nets of the linear mesh elements. The control nets sticking together as a whole is the undeformed geometry (shown in Fig. 6a). The external loadings are the control points (red points in Fig. 6b) of the smooth curved boundary edges. The control nets are deformed such that the control points of the boundary edges of the linear mesh move to the corresponding control points of the curved boundary edge. By solving for the equilibrium configuration of an elasticity problem, the finial configuration is determined and the new positions of the control points of the interior mesh edges after deformation are obtained. Fig. 6 illustrates these steps.

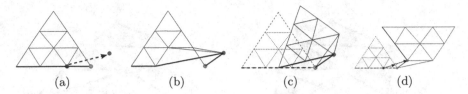

Fig. 7. An illustration of the iterative finite element method. (a) In this example there is only one element in the mesh, the black border line represents the mesh boundary, the position of the blue point represents the original position of one control point of the linear mesh. (b) The position of the red point is the new position of the blue point after deformation. The green point is one of the endpoints of the boundary edge, thus it has to maintain its position. The control net is invalid because there exists an inverted triangle. (c) The one step FEM method was applied, the blue point was directly moved to the position of the red point. After solving for the equilibrium configuration, the control net is still twisted. (d) The iterative way: make the position of the yellow point as the intermediate target, first move the blue point to the position of the yellow point, then move to the position of the red point. The two step FEM method successfully corrected the twisted control net.

In most cases, the one step finite element method can handle this problem successfully. However, in the case that the curvature of the boundary edge is very large, the interior edges may not be able to be curved enough to correct the invalidity. The iterative finite element method executes the validity check before each round. When it is reported that an invalid element exists, the procedure divides the segment formed by the original position and the new position into two subsegments. The procedure takes the positions of the endpoints of the subsegments one by one as the intermediate targets, and takes the solution of the current target as the input of the next target. The algorithm terminates when all the invalid elements are corrected. Fig. 6 shows an example.

4   Mesh Examples

The input data to our algorithm is a two-dimensional image. The procedure for mesh untangling and quality improvement was implemented in MATLAB. All the other steps were implemented in C++ for efficiency.

We meshed a region of a slice of mouse brain atlas [1] and a region of a fruit fly embryo [2]. The size of the first input is $2550 * 2050$ pixels, the size of the second input is $1900 * 950$ pixels. Each pixel has side lengths of 1 unit in x and y dimensions, respectively. In each example, we show the linear mesh result and the high-order mesh result (see Fig. 8 and Fig. 9). For both examples, the fidelity tolerance for the linear mesh was specified by two pixels and the angle quality bound was specified by $20°$. In the first example, after boundary edges were curved, there were two invalid elements in the mesh interior. In the second example, the number of invalid elements is one. After the mesh untangling procedure, all the invalid elements were corrected in both examples. The incorrectly

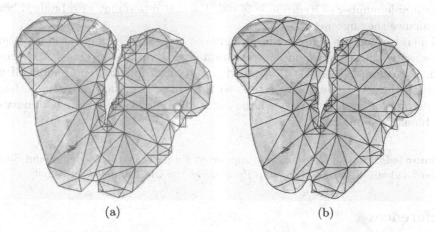

(a) (b)

Fig. 8. Meshing results for a slice of mouse atlas [1]. (a) Linear mesh with two pixels fidelity tolerance and 19° angle bound. (b) Final curvilinear mesh in which all the elements are valid.

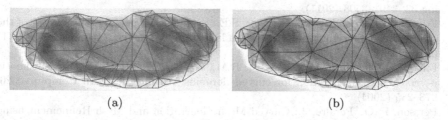

(a) (b)

Fig. 9. Meshing results for a fruit fly embryo [2]. (a) Linear mesh with two pixels fidelity tolerance and 19° angle bound. (b) Final curvilinear mesh in which all the elements are valid.

classified pixels (include both background pixels in the mesh elements and tissue pixels outside the mesh) of the high-order mesh in both examples were improved about 10 percent compared to that of the linear mesh.

5 Conclusion

We presented a new approach for automatically constructing a guaranteed quality curvilinear mesh to represent geometry with smooth boundaries.

Our future work will include the run time improvement. Compared to the linear mesh generation we present in Section 3, the construction of the high-order mesh is slow. One reason of the inefficiency is the hybrid code we implemented both in MATLAB and C++. The most critical reason is that, in the procedure of invalid mesh correction, we can not anticipate how many iterations we need to eliminate all the invalid elements. As a result, we repeat this procedure until the suitable number of iteration is found. Parallelization may also be used later to enhance the efficiency.

The other concern is further improving the accuracy. Our next step is designing a more suitable linear mesh generation algorithm for curvilinear discretization. For example, besides the two-sided Hausdorff fidelity requirement, if we warp all the mesh boundary vertices to the image boundary, after smoothing, the high-order mesh boundary will be naturally approaching to the boundary of the biomedical objects.

Acknowledgments. This work was supported (in part) by the Modeling and Simulation Graduate Research Fellowship Program at the Old Dominion University.

References

1. Allen brain atlas (2014)
2. Berkeley drosophila genome project (2014)
3. Chernikov, A., Chrisochoides, N.: Multitissue tetrahedral image-to-mesh conversion with guaranteed quality and fidelity. SIAM Journal on Scientific Computing 33, 3491–3508 (2011)
4. Frise, E., Hammonds, A.S., Celniker, S.E.: Systematic image-driven analysis of the spatial drosophila embryonic expression landscape. Molecular Systems Biology 6(1) (2010)
5. Luo, X.J., Shephard, M.S., O'Bara, R.M., Nastasia, R., Beall, M.W.: Automatic p-version mesh generation for curved domains. Engineering with Computers 20, 273–285 (2004)
6. Persson, P.-O., Peraire, J.: Curved Mesh Generation and Mesh Refinement using Lagrangian Solid Mechanics. In: Proceedings of the 47th AIAA Aerospace Sciences Meeting and Exhibit, Orlando, FL (January 2009)
7. George, P.L., Borouchaki, H.: Construction of tetrahedral meshes of degree two. Int. J. Numer. Mesh. Engng. 90, 1156–1182 (2012)
8. Sherwin, S.J., Peiro, J.: Mesh generation in curvilinear domains using high-order elements. Int. J. Numer., 1–6 (2000)
9. Xu, J., Chernikov, A.: Automatic curvilinear mesh generation with smooth boundary driven by guaranteed validity and fidelity. In: International Meshing Roundtable, London, UK, pp. 200–212. Elsevier (October 2014)
10. Zhang, W., Feng, D., Li, R., Chernikov, A., Chrisochoides, N., Osgood, C., Konikoff, C., Newfeld, S., Kumar, S., Ji, S.: A mesh generation and machine learning framework for Drosophila gene expression pattern image analysis. BMC Bioinformatics 14, 372 (2013)
11. Zhang, W., Li, R., Feng, D., Chernikov, A., Chrisochoides, N., Osgood, C., Ji, S.: Evolutionary soft co-clustering: formulations, algorithms, and applications. In: Data Mining and Knowledge Discovery, pp. 1–27 (2014)
12. Zienkiewicz, O.C., Taylor, R.L., Zhu, J.Z.: The Finite Element Method: Its Basis and Fundamentals, 6th edn. Butterworth-Heinemann, Oxford (2005)

The Role of miRNAs in Cisplatin-Resistant HeLa Cells

Yubo Yang[1], Cuihong Dai[1], Zhipeng Cai[2], Aiju Hou[1], Dayou Cheng[1],
and Dechang Xu[1(✉)]

[1] School of Food Science and Engineering,
Harbin Institute of Technology, Harbin, China
[2] Department of Computer Science, Georgia State University, Atlanta, USA

Abstract. Chemotherapy is the main strategy in the treatment of cancer, however, the development of drug-resistance is the obstacle in long-term treatment of cervical cancer. Cisplatin is one of the most common drugs used in cancer therapy. Recently, accumulating evidence suggests that miRNAs are involved in various bioactivities in oncogenesis. It is not unexpected that miRNAs play a key role in acquiring of drug-resistance in the progression of tumor. In this study, we induced and maintained four levels of cisplatin-resistant HeLa cell lines (HeLa/CR1, HeLa/CR2, HeLa/CR3 and HeLa/CR4). According to the previous studies and exiting evidence, we selected five miRNAs (miR-183, miR-182, miR-30a, miR-15b and miR-16) and their potential target mRNAs as our research targets. The real-time RT-PCR was used to detect the relative expression of miRNAs and their mRNAs. The results show that miR-182 and miR-15b were up-regulated in resistant cell lines while miR-30a was significantly down-regulated. At the same time, the targets they regulated are related to the drug-resistance. The expression alteration of selected miRNAs in resistant cell lines compared to their parent HeLa cell line suggests that HeLa cell drug resistance is associated with distinct miRNAs, which indicates that miRNAs may be one of the therapy targets in the treatment of cervical cancer by sensitizing cell to chemotherapy.

Keywords: HeLa cell · Cisplatin-resistance · miRNAs · Chemotherapy

1   Introduction

Cisplatin (CDDP) is one of the most commonly used drugs in human malignancies therapy with the capability to damage DNA in normal and cancer cells. The main mechanism of cisplatin activation is through the combination to double-stranded DNA, and then form a DNA adducts to interfere in DNA replication and RNA transcription, resulting in apoptosis. The bioactivities of cisplatin determined its high efficient in the treatment of cervical cancer, and it also widely   used in other types of cancer such as bladder, testicular, head and neck, small and non-small cee lung, *etc*[1].

Although chemotherapy with cisplatin is a basic strategy for the treatment of various cancers, the acquiring of drug-resistance becomes the main obstacle in cancer. The formation and aggravation of drug-resistance involves different molecular mechanism, including the increase of DNA repair activity and anti-apoptotic adjustment

© Springer International Publishing Switzerland 2015
R. Harrison et al. (Eds.): ISBRA 2015, LNBI 9096, pp. 355–365, 2015.
DOI: 10.1007/978-3-319-19048-8_30

factor activity, the reduce of drug accumulation, the increase of cisplatin adducts tolerance, the changes of post-translational modification and microtubule protein, inactivation by thiol containing species and the failure of cell death [2]. Although the aforementioned mechanisms are cell-line dependent, according to the studies over the past decade, miRNAs play a key role in the formation of drug resistance in tumor cells [3].

MicroRNAs (miRNAs) are large class single-stranded RNA molecules approximately 22 nucleotide (nt) length. MiRNAs play a key role in many cellular processes. MiRNAs are linked to tumor firstly due to its location on the chromosome very close to tumor-associated genomic regions and chromosomal breakpoints [4], and its abnormal expression in many malignant tumors. At current study, about 3% of human genome encodes miRNAs; however, there are more than 1/3 of protein-coding genes regulated by miRNAs [5]. Current studies confirmed many tumor-associated miRNAs, which targets are verified by experiments, thus further understanding of tumor-associated miRNAs is important to tumor prediction, diagnosis and treatment.

Most recently, there is an increasing number of miRNAs that have been shown to link to drug-resistance in different tumor cells. It is noteworthy that some miRNAs can mediate the formation of tumor stem cell and epithelial- mesenchymal transition, which is important for the development of drug-resistance [6]. At the same time, some miRNAs target chemosensitivity related genes and then change the sensitivity of tumor cells to anticancer drugs [7].

To further understand of the potential role of miRNAs in drug resistance, and gain a primary perspective on the effective of miRNAs in the development of cisplatin resistance in HeLa cells, we studied a limit set of miRNAs which may play a role in the formation of cisplatin resistance in HeLa by target some mRNAs. At the same time, tumor suppressor gene p53 have been proven to play a role in the cellular resistance to cisplatin. It can enhance cisplatin-induced apoptosis and sensitize HeLa cell to cisplatin [8]. Therefore, to confirm our cisplatin-resistant cell lines indeed acquired the resistance ability, we also analyzed the mRNA expression level of *p53*. In our study, we found the specific miRNAs that were differential expressed in different level of cisplatin-resistant HeLa cells.

2   Material and Methods

2.1   Cell Lines and Cultures

Human cervical cancer cell line HeLa (obtained from Cell Resource Center of Shanghai Institute of Life Science, Chinese Academy of Science, Shanghai, China) and its different level of cisplatin-resistant variants HeLa/CR1, HeLa/CR2, HeLa/CR 3, HeLa/CR4 (induced and maintained in our laboratory) were cultured in RPMI-1640 medium supplemented with 10% heat-inactivated fetal bovin serum (FBS) and 1% penicillin and streptomycin mixture. All the cells were cultured in a humid atmosphere with 5% CO_2 at 37°C.

2.2   Induce of Cisplatin-Resistant HeLa Cells

To mimic the process of cell resistance development and obtain different levels of resistant cells , the HeLa cells were induced using a concentration gradient method. When the parental cells can growth stably at low concentration of cisplatin solution, then stepwise raise the concentration of cisplatin from 0.2μg/mL to 1.0μg/mL. Eventually four types of HeLa cells with different resistance levels were established.

2.3   Detection of Drug-Resistant Ability

The yellow dye 3(4, 5-dimethylthiazol2-yl)-2, 5-diphenyl tetrazolium bromide (MTT) is a common method to assess the viability of cells because the absorbance is proportional to the number of viable cells [9, 10]. The single cell suspension $(5.0*10^4/\text{mL})$ of different variant HeLa cells and normal cell in logarithmic phase were seeded in 96-well plates Each well added 180μL and 5 repeat wells were used for each experimental condition and adherent culture 24h. Ten microliters of the MTT（5 mg/ml）solution was then added to each culture well, and the plates were incubated for 4h at 37°C. Carefully aspirated the medium in well, 150μl DMSO added to each well, set the plate into the shaker with low oscillation speed 10min, and make crystals fully dissolved. Absorbance was measured with an automatic enzyme-linked immunosorbent assay reader at a wavelength of 490 nm. The abilities of each cisplatin-resistant cell lines were calculated by the formula (1) according to the IC_{50} of each cell lines.

Multiple of cisplatin resistance $=IC_{50}$ of cisplatin-resistant cells/IC_{50} of HeLa cell   (1)

2.4   Selection of miRNAs and miRNA Target Prediction

Combined with the existing literature, we initially picked out five candidate miRNAs associated with drug resistance and predicted the targets of the candidate miRNAs (Table 1) according to preliminary test results , existing theoretical knowledge , targetscan (http://genes.mit.edu/targetscan/) [11] and Diana-microT-CDS [12, 13] database.

Table 1. Candidate microRNAs and their predicted targets

Candidate microRNAs	Predicted targets
miR-183	KIAA1199, PDCD4
miR-182	PDCD4
miR-30a	Beclin1
miR-15b	Bcl-2
miR-16	Bcl-2

2.5   Total RNA Extraction

Total RNA of different type cell lines was isolated with Trizol regent (Invitrogen) according to the manufacturer's instruction and treated with 10 U DNase (RNase free) for

30 min at 37°C. DNase was removed with phenol-chloroform. The RNA concentrations were calculated according to the absorbance determined using a spectrophotometer at 260 nm.

2.6   Real-Time PCR

For target mRNA level, reverse transcription (RT) was performed by using High Capacity cDNA Reverse Transcription Kit (Applied Biosystems) with 1microgramme total RNA. The mRNA level was detected by Real-time PCR performed in Applied Biosystems Prism 7300 Real-time PCR system with SYBR green PCR Premix (Takara) according to the manufacturer's proposal. As for the level of miRNA, the transcription of miRNA was performed using TaqMan MicroRNA Reverse Transcription Kit under the instruction of manufacturer. The level of miRNA was assessed using TaqMan Universal Master Mix II. For normalization purpose, U6 snRNA and GADPH were determined as the internal references in miRNAs and target mRNA quantitative PCR respective. All reactions were performed in triplicate. The relative expression level of each mRNA and miRNA was calculated using the $2^{-\Delta\Delta Ct}$ method.

2.7   Statistical Analysis

The software SPSS 22 was utilized for statistical analysis. The means of the relative expression level of miRNAs and mRNAs were compared to the control group (normal HeLa cell) with paired Student's t-test. A 5% level of probability can be recognized to be significant.

3   Results

3.1   Cisplatin Resistance Ability of Four Cisplatin-Resistant HeLa Cell Lines

The cisplatin-resistant cells were induced and maintained in different concentration of cisplatin liquid. After the cells in different cisplatin concentration medium grew stably, using MTT method, we detected the cisplatin resistance ability of four different level resistant variants by comparing with the normal HeLa cell. As shown in Fig.1, the cell induced and cultured in higher cisplatin concentration acquired the higher resistance ability. It also shows that the resistant abilities of four variants are gradually increase, which can simply mimic the development of the drug resistance of HeLa cells.

3.2   Real-Time RT-PCR Analysis of Candidate miRNAs in Cisplatin-Resistant HeLa Cell Lines

To detect whether the candidate miRNAs are involved in the development of drug resistance in HeLa cell, we induced four types cisplatin-resistant HeLa cell, HeLa/CR1, HeLa/CR2, HeLa/CR3 and HeLa/CR4, and performed real time qPCR of the candidate miRNAs in four resistant cell lines and their parental cell line HeLa. It was shown that the relative expression levels of five miRNAs altered in resistant cell

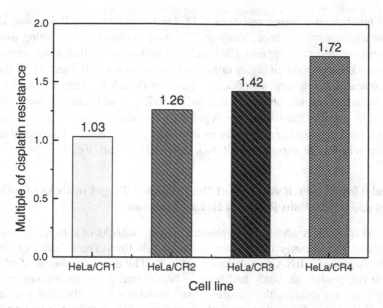

Fig. 1. Multiple of cisplatin resistance of different cisplatin-resistant cells compared to their parent HeLa cell

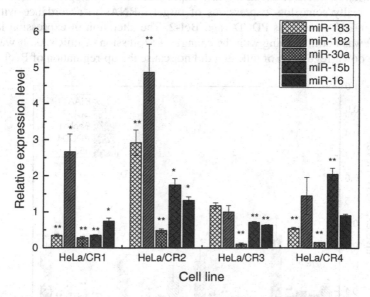

Fig. 2. Real-time RT-PCR analysis of miR-183, miR-182, miR-30a, miR-15b and miR-16, data are shown as fold changes of miRNA levels in cisplatin-resistant HeLa cell lines relative to HeLa cell line, which is set as 1 (data are represent as: mean ± SD). *p<0.05; **p<0.01.

lines compared to parental cell line HeLa (Fig.2). In the five candidate miRNAs, miR-182 was upregulated in all resistant cell lines compared to HeLa cell, this is in accordance with the research that miR-183~miR-96~miR-182 cluster was upregulated in multidrug resistant   Ehrlich ascites tumor cell lines [21]. On the contrary, miR-30a was down-regulated in all resistant cell lines. Study in [22] gave evidence that targeting miR-30a enhances imatinib resistance against CML cells, which suggests that lower expression level of miR-30a may results in higher drug-resistance in tumor cell lines. All of the five miRNAs expression levels were significantly altered in HeLa/CR2 The cell initially demonstrated cisplatin resistance ability. In HeLa/CR4 miR-182 and miR-15b were upregulated. MiR-183 and mi-30a were down-regulated. The expression profiles of the five miRNAs demonstrated that they may play an important role in the development of cisplatin resistance in HeLa cell, especially miR-183, miR-182 and miR-30a.

3.3   Real-Time RT-PCR Analysis of the Predicted Target mRNAs of miRNAs and p53 in Cisplatin-Resistant HeLa Cell Lines

The real-time RT-PCR analysis results showed that all mRNAs of selected genes were down-regulated significantly in the four resistant cell lines The relative expression level of all researched mRNA were below 0.5(Fig.3). The down-regulation of p53 was in line with our prediction. Study had been indicated that in cisplatin-resistant HeLa cell line p53 did not enhance the sensitivity and could not show the ability to induce apoptosis. This may indicate that in cisplatin-resistant HeLa cell lines the apoptosis induced by p53 may be avoided [15]. According to our prediction, the high expression level of miRNAs would result in the low expression level of target mRNAs. In fact, there were really some low expressions of target mRNAs in accordance with their moderator miRNAs, such as PDCD4 and Bcl-2. The alteration of expression level of KIAA1199 was not in keeping with the change of expression of miRNAs. It was notable that the down-regulation of miR-30a did not cause the up-regulation of Beclin1.

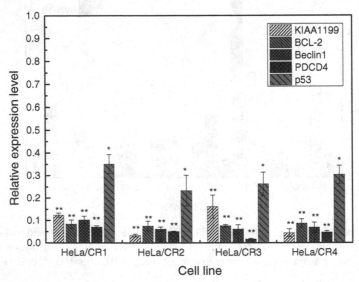

Fig. 3. Real-time RT-PCR analysis of p53 and the mRNAs of predicted targets KIAA1199, BCL-2 Beclin1 and PDCD4 of candidate miRNAs in cisplatin-resistant HeLa cell lines relative to HeLa cell line, which is set as 1 (data are represent as: mean ± SD). *p<0.05; **p<0.01.

4   Discussion

Our previous study shows that KIAA1199 is the target of has-miR-183. Existing researches show that KIAA1199 is related to cancer pathways. Study in [23] confirmed that the expression level of KIAA1199 in breast cancer is higher than normal breast tissue. Researches proved thatKIAA1199 is involved in various signal pathways, including Wnt pathway, cell apoptosis, and DNA repair, and cell cycle, cancer cell migration [23, 24, 25, and 26]. Knockout KIAA1199 can cause the up-regulation of BAX protein, and BAX can active rapid programmed cell apoptosis by promoting Caspase-3 [23]. High level of BAX is related to the enhancing of chemotherapy response. Therefore, the expression of miR-183 should be down-regulated theoretically, and the expression level of KIAA1199 should be up-regulated corresponding in resistant cell. However, there is a high level of miR-183 in HeLa/CR2 and HeLa/CR3, especially in HeLa/CR2. This may due to the drug response in cell induced by cisplatin. Moreover, as the target of miR-183, KIAA1199 should be up-regulated when miR-183 is down-regulated. In fact, the level of KIAA1199 in cisplatin-resistant cell lines still is down-regulated. The explanation for this contradictory phenomenon may be the dynamic nature of tumor cells and the variability expression of cellular molecules, and there may be other regulators that can mediate the expression of KIAA1199. From the evidence, we can forecast that miR-183 plays a suppressive role in the development of drug resistance.

Programmed Cell Death Protein 4 (PDCD4) is a pro-apoptotic protein and it is involved in proliferation and cell cycle progression. Research has shown that over-expression of PDCD4 can sensitize cell to cisplatin in prostate cancer [27].PDCD4 is one of the targets of carcinogenic miRNAs, studies have shown that PDCD4 is involved in TGF- TGF-β1-induced apoptosis in hepatocellular carcinoma cells, and is also the target of miR-183 in HCC cells. Study in [28] found that miR-183 was significantly up-regulated, and PDCD4 was down-regulated by miR-183 in HCC cells. MiR-182 is another PDCD4-targeting miRNA. Studies in [29] reported that overexpression of miR-182 can down-regulate PDCD4 and make non-small cell lung cancer (NSCLC) cell more resistant to cisplatin. Similarly, increasing miR-182 results in resistance of Ovarian cancer cells to cisplatin and Taxol and decreases PDCD4 [30]. All these indicate that miR-183 and miR-182 play an oncogenic role via suppressing the PDCD4 expression. For this reason, we can predict the up-regulation of miR-183 and miR-182 in cisplatin-resistant HeLa cell lines. From the results of our study, the over-expression of miR-182 and down-expression of PDCD4 were detected in resistant cells, however, there was no significantly over-expression of miR-183 to be detected. Study also proved that inhibiting the miRNA cluster miR-183/96/182 can sensitizes gliomas cells to chemotherapy [31], which suggests that miR-183 and miR-182 may play a synergistic role. Overall, the expression of miR-183 and miR-182 is up-regulated, this is accordant with our prediction. As for the lower expression level of miR-183, the explanation may be that miR-183 plays both suppressor and promoter in the development of drug-resistance by targeting KIAA1199 and PDCD4.

Study in [32] demonstrated that PDCD4 can sensitize gastric cancer cells to tumor necrosis factor-related apoptosis-inducing ligand (TRAIL), and inhibiting Akt by phosphoinositide 3-kiase (PI3K) inhibitor can active PDCD4 so that increase TRAIL sensitivity in TRAIL-resistant gastric   cancer cell line. The study suggests that PDD4 can enhance the drug sensitivity of gastric cancer cell and lead to apoptosis by inter-fere PI3K/Akt signaling pathway. Moreover, activation of FOXO3A, functioning downstream of PI3K/PTEN/AKT signaling cascade, can mediate the cytostatic and cytotoxic effect of cancer therapy drug, such as cisplatin, lapatinib and doxorubicin [33]. In study [34], the expression level of miR-182 was higher in glucocorticoid (GC) resistant cell lines than GC-sensitive cell lines in lymphoblastic malignancies, and miR-182 can enhance GC resistance by targeting FOXO3A. From the evidence, we can speculate that miR-182 makes contribution to the development of cisplatin resistance through inhibiting PI3K/PTEN/AKT signal pathway by targeting PDCD4.

Autophagy is an important homeostatic catabolic progress that plays a dual role in cancer cells. Studies have shown that autophagy is upregulated by some tumor inhibi-tor genes in breast cancer [35]. This indicates that autophagy may suppress the tumor-genesis. Beclin1 as a tumor suppressor gene can mediate the autophagy. Paradoxically, autophagy could promote cell proliferation and drug resistance in some cancer cells [36]. Evidence suggests that it is activated in cancer cells, under various stresses, such as chemotherapy. During the chemotherapy of some cancer, autophagy contributes the chemotherapy resistance. Study has shown that miR-30a is targeting of beclin1 can decrease the autophagic activity by down-regulating the expression of beclin1, which will sensitize cell to cisplatin [9].

It is notable that expression of miR-30a changes with cell types, in breast cancer, the expression is up, but in colorectal cancer, the expression is down [32].In this study, miR-30a was down-regulated in cisplatin-resistant cell lines. This is in accordance with the existing evidence. However, the expression of beclin1 did not keep with our pre-diction. Explanations for this contradictory phenomenon are because the expression of beclin1 is regulated by several factors. Accumulating evidence has shown that nuclear factor (NF)-κB, E2F transcription factors (E2F) are involved in the beclin1-dependent autophagy by regulating the expression of beclin1 [38], thus miR-30a is not the unique regulator of beclin1 so that the effect of miR-30a is not significant.   As a tumor sup-pressor gene, the expression of beclin1 is likely low, which suggests that in the devel-opment of cisplatin-resistance in HeLa cell, beclin1 may play a role of suppressor but not promoter. The last speculation is that it may not the beclin1 that contributes in the acquisition of cisplatin-resistance in HeLa cell, for the reason that miR-30a also can mediate the autophagy gene *ATG5* and enhance imatinib-induced cytotoxicity [33]. Therefore, we can speculate that miR-30a is related in drug sensitivity by targeting the genes of autophagy pathway, but not just targeting *Beclin1*.

B cell lymphoma 2 (Bcl-2) is a crucial inhibitor in apoptosis in eukaryotic cells, therefore the overexpression of Bcl-2 is involved in oncogenesis [34]. Study has shown that Bcl-2 is one of the targets of miR-15b and miR-16, which are reported down-regulation in about 80% in prostate cancer compared to normal tissue [35]. This suggests that miR-15b and miR-16 function as tumor suppressors. Research reported that overexpression of miR-16 and miR-15b can sensitize multidrugresistant gastric cancer cell line SGC7901/VCR to VCR-induced apoptosis by targeting the posttran-scriptional expression of Bcl-2 [36].

In our results, miR-15b and miR-16 show slight down-regulation in HeLa/CR1 and HeLa/CR2 while the expression of Bcl-2 did not been up-regulated. However, the level of Bcl-2 is stable in four cell lines. This is inconsistent with our speculation, therefore further researches are needed to confirm the results.

In conclusion, from the results we acquired at present, miR-182 may play a role in the development of cisplatin-resistance in HeLa cell by targeting the expression of PDCD4 and involved in the PI3K/AKT pathway, while miR-183 may main target KIAA1199 compared with PDCD4 in resistant cell lines. Moreover, the significant down-regulation of miR-30a can suggest that it may play a central role in the acquisition of drug-resistance in HeLa cell indirectly by involving in autophagy pathway. Therefore, altering the expression level of miR-182 and miR-30a may be the therapy strategy to overcome drug resistance. Though the expression profiles of targets mRNAs are not in according to the miRNAs expression profiles and our prediction perfectly, it may due to the regulation lag between miRNAs and mRNAs. Furthermore, due to the dualism of miRNAs and regulators as well as the difference between different types of cancer, to investigate the exact mechanism and pathway of the development of drug-resistance in HeLa cell, further studies are desired.

Acknowledgements. This work is partially supported by the China Natural Science Foundation (Grant Number: 30771371, 31271781, 6110030), the National High-tech R&D Program of China (863 Program) (Grant Number: 2001AA231091, 2004AA231071), Heilongjiang Province Science Foundation (Grant Number: 2004C0314), Heilongjiang Province key scientific and technological project (Grant Number: WB07C02), HIT Science Foundation (Grant Number: HIT. 2003. 38), National MOST special fund (Grant Number: KCSTE- 2000-JKZX- 021, NCSTE- 2007- JKZX- 022, 2012EG111228) "China Postdoctoral Science Foundation" (AUGA4100207411) and "the Fundamental Research Funds for the Central Universities" (Grant No. HIT. NSRIF.2014092).

Reference

1. Florea, A.M., Busselberg, D.: Cisplatin as an anti-tumor drug: cellular mechanisms of activity, drug resistance and induced side effects. Cancers 3, 1351–1371 (2011)
2. Cepeda, V., Fuertes, M.A., Castilla, J., Alonso, C., Quevedo, C., Pérez, J.M.: Biochemical mechanisms of cisplatin cytotoxicity. Anti-Cancer Agents in Medicinal Chemistry (Formerly Current Medicinal Chemistry-Anti-Cancer Agents) 7(1), 3–18 (2007)
3. Calin, G.A., Sevignani, C., Dumitru, C.D., Hyslop, T., Noch, E., Yendamuri, S., Shimizu, M., Rattan, S., Bullrich, F., Negrini, M., Croce, C.M.: Human microRNA genes are frequently located at fragile sites and genomic regions involved in cancers. Proceedings of the National Academy of Sciences of the United States of America 101, 2999–3004 (2004)
4. Saito, Y., Jones, P.M.: Epigenetic Activation of Tumor Suppressor MicroRNAs in Human Cancer Cells. Cell Cycle 5, 2220–2222 (2006)
5. Sarkar, F.H., Li, Y., Wang, Z., Kong, D., Ali, S.: Implication of microRNAs in drug resistance for designing novel cancer therapy. Drug Resistance Updates: Reviews and Commentaries in Antimicrobial and Anticancer Chemotherapy 13, 57–66 (2010)

6. Blower, P.E., Chung, J.H., Verducci, J.S., Lin, S., Park, J.K., Dai, Z., Liu, C.G., Schmittgen, T.D., Reinhold, W.C., Croce, C.M., Weinstein, J.N., Sadee, W.: MicroRNAs modulate the chemosensitivity of tumor cells. Molecular Cancer Therapeutics 7, 1–9 (2008)

7. Zou, Z., Wu, L., Ding, H., Wang, Y., Zhang, Y., Chen, X., Chen, X., Zhang, C.Y., Zhang, Q., Zen, K.: MicroRNA-30a sensitizes tumor cells to cis-platinum via suppressing beclin 1-mediated autophagy. The Journal of Biological Chemistry 287, 4148–4156 (2012)

8. Weller, M.: Predicting response to cancer chemotherapy: the role of p53. Cell and Tissue Research 292, 435–445 (1998)

9. Minagawa, Y., Kigawa, J., Itamochi, H., Kanamori, Y., Shimada, M., Takahashi, M., Terakawa, N.: Cisplatin-resistant HeLa Cells Are Resistant to Apoptosis via p53-dependent and-independent Pathways. Cancer Science 90, 1373–1379 (1999)

10. Abate, G., Mshana, R.N., Miörner, H.: Evaluation of a colorimetric assay based on 3-(4, 5-dimethylthiazol-2-yl)-2, 5-diphenyl tetrazolium bromide (MTT) for rapid detection of rifampicin resistance in Mycobacterium tuberculosis. The International Journal of Tuberculosis and Lung Disease 2, 1011–1016 (1998)

11. Cai, Z., Zhang, T., Wan, X.-F.: A Computational Framework for Influenza Antigenic Cartography. PLoS Comput Biol 6(10), e1000949 (2010)

12. Sorrentino, A., Liu, C.G., Addario, A., Peschle, C., Scambia, G., Ferlini, C.: Role of microRNAs in drug-resistant ovarian cancer cells. Gynecologic Oncology 111, 478–486 (2008)

13. Reczko, M., Maragkakis, M., Alexiou, P., Grosse, I., Hatzigeorgiou, A.G.: Functional microRNA targets in protein coding sequences. Bioinformatics 28, 771–776 (2012)

14. Paraskevopoulou, M.D., Georgakilas, G., Kostoulas, N., Vlachos, I.S., Vergoulis, T., Reczko, M., Filippidis, C., Dalamagas, T., Hatzigeorgiou, A.G.: DIANA-microT web server v5.0: service integration into miRNA functional analysis workflows. Nucleic Acids Research 41, W169–W173 (2013)

15. Husted, S., Sokilde, R., Rask, L., Cirera, S., Busk, P.K., Eriksen, J., Litman, T.: Micro-RNA expression profiles associated with development of drug resistance in Ehrlich ascites tumor cells. Molecular Pharmaceutics 8, 2055–2062 (2011)

16. Sui, X., Chen, R., Wang, Z., Huang, Z., Kong, N., Zhang, M., Han, W., Lou, F., Yang, J., Zhang, Q., Wang, X., He, C., Pan, H.: Autophagy and chemotherapy resistance: a promising therapeutic target for cancer treatment. Cell Death & Disease 4, e838 (2013)

17. Jami, M.S., Hou, J., Liu, M., Varney, M.L., Hassan, H., Dong, J., Ding, S.J.: Functional proteomic analysis reveals the involvement of KIAA1199 in breast cancer growth, motility and invasiveness. BMC Cancer 14, 194 (2014)

18. Birkenkamp-Demtroder, K., Maghnouj, A., Mansilla, F., Thorsen, K., Andersen, C.L., Oster, B., Hahn, S., Orntoft, T.F.: Repression of KIAA1199 attenuates Wnt-signalling and decreases the proliferation of colon cancer cells. British Journal of Cancer 105, 552–561 (2011)

19. Evensen, N.A., Kuscu, C., Nguyen, H.L., Zarrabi, K., Dufour, A., Kadam, P., Hu, Y.J., Pulkoski-Gross, A., Bahou, W.F., Zucker, S., Cao, J.: Unraveling the role of KIAA1199, a novel endoplasmic reticulum protein, in cancer cell migration. Journal of the National Cancer Institute 105, 1402–1416 (2013)

20. Kuscu, C.: Epigenetic Regulations and Promoter Characterization of CERIG (Cancer Endoplasmic Reticulum Gene-KIAA1199) (2012)

21. Shiota, M., Izumi, H., Tanimoto, A., Takahashi, M., Miyamoto, N., Kashiwagi, E., Kidani, A., Hirano, G., Masubuchi, D., Fukunaka, Y., Yasuniwa, Y., Naito, S., Nishizawa, S., Sasaguri, Y., Kohno, K.: Programmed cell death protein 4 down-regulates Y-box binding protein-1 expression via a direct interaction with Twist1 to suppress cancer cell growth. Cancer Research 69, 3148–3156 (2009)

22. Li, J., Fu, H., Xu, C., Tie, Y., Xing, R., Zhu, J., Qin, Y., Sun, Z., Zheng, X.: miR-183 inhibits TGF-beta1-induced apoptosis by downregulation of PDCD4 expression in human hepatocellular carcinoma cells. BMC Cancer 10, 354 (2010)
23. Ning, F.L., Wang, F., Li, M.L., Yu, Z.S., Hao, Y.Z., Chen, S.S.: MicroRNA-182 modulates chemosensitivity of human non-small cell lung cancer to cisplatin by targeting PDCD4. Diagn. Pathol. 9, 143 (2014)
24. Wang, Y.Q., Guo, R.D., Guo, R.M., Sheng, W., Yin, L.R.: MicroRNA-182 promotes cell growth, invasion, and chemoresistance by targeting programmed cell death 4 (PDCD4) in human ovarian carcinomas. Journal of Cellular Biochemistry 114, 1464–1473 (2013)
25. Tang, H., Bian, Y., Tu, C., Wang, Z., Yu, Z., Liu, Q., Xu, G., Wu, M., Li, G.: The miR-183/96/182 cluster regulates oxidative apoptosis and sensitizes cells to chemotherapy in gliomas. Current Cancer Drug Targets 13, 221–231 (2013)
26. Wang, W.Q., Zhang, H., Wang, H.B., Sun, Y.G., Peng, Z.H., Zhou, G., Shi, M.Y., Wang, R.Q., Fang, D.C.: Programmed cell death 4 (PDCD4) enhances the sensitivity of gastric cancer cells to TRAIL-induced apoptosis by inhibiting the PI3K/Akt signaling pathway. Molecular Diagnosis & Therapy 14, 155–161 (2010)
27. Wilson, M.S., Brosens, J.J., Schwenen, H.D., Lam, E.: FOXO and FOXM1 in cancer: the FOXO-FOXM1 axis shapes the outcome of cancer chemotherapy. Current Drug Targets 12, 1256–1266 (2011)
28. Yang, A., Ma, J., Wu, M., Qin, W., Zhao, B., Shi, Y., Jin, Y., Xie, Y.: Aberrant microRNA-182 expression is associated with glucocorticoid resistance in lymphoblastic malignancies. Leukemia & Lymphoma 53, 2465–2473 (2012)
29. Gong, C., Bauvy, C., Tonelli, G., Yue, W., Delomenie, C., Nicolas, V., Zhu, Y., Domergue, V., Marin-Esteban, V., Tharinger, H., Delbos, L., Gary-Gouy, H., Morel, A.P., Ghavami, S., Song, E., Codogno, P., Mehrpour, M.: Beclin 1 and autophagy are required for the tumorigenicity of breast cancer stem-like/progenitor cells. Oncogene 32, 2261–2272, 2272e 2261–2211 (2013)
30. Aredia, F., Scovassi, A.I.: Manipulation of autophagy in cancer cells: an innovative strategy to fight drug resistance. Future Medicinal Chemistry 5, 1009–1021 (2013)
31. Kong, Y.W., Ferland-McCollough, D., Jackson, T.J., Bushell, M.: microRNAs in cancer management. The Lancet Oncology 13, e249–e258 (2012)
32. Kang, R., Zeh, H.J., Lotze, M.T., Tang, D.: The Beclin 1 network regulates autophagy and apoptosis. Cell Death and Differentiation 18, 571–580 (2011)
33. Yu, Y., Yang, L., Zhao, M., Zhu, S., Kang, R., Vernon, P., Tang, D., Cao, L.: Targeting microRNA-30a-mediated autophagy enhances imatinib activity against human chronic myeloid leukemia cells. Leukemia 26, 1752–1760 (2012)
34. Yang, J., Cao, Y., Sun, J., Zhang, Y.: Curcumin reduces the expression of Bcl-2 by upregulating miR-15a and miR-16 in MCF-7 cells. Medical Oncology 27, 1114–1118 (2010)
35. Aqeilan, R.I., Calin, G.A., Croce, C.M.: miR-15a and miR-16-1 in cancer: discovery, function and future perspectives. Cell Death and Differentiation 17, 215–220 (2010)
36. Xia, L., Zhang, D., Du, R., Pan, Y., Zhao, L., Sun, S., Hong, L., Liu, J., Fan, D.: miR-15b and miR-16 modulate multidrug resistance by targeting BCL2 in human gastric cancer cells. International Journal of Cancer. Journal International du Cancer 123, 372–379 (2008)

DNA AS X: An Information-Coding-Based Model to Improve the Sensitivity in Comparative Gene Analysis

Ning Yu[1]([✉]), Xuan Guo[1], Feng Gu[2], and Yi Pan[1]([✉])

[1] Department of Computer Science, Georgia State University
25 Park Place, Atlanta, GA 30319, USA
{nyu4,xguo9}@student.gsu.edu,
yipan@cs.gsu.edu
http://www.cs.gsu.edu
[2] Department of Computer Science, College of Staten Island
2800 Victory Blvd., Staten Island, NY 10314, USA
Feng.Gu@csi.cuny.edu
http://www.csi.cuny.edu

Abstract. In recent studies [1–3], lots of hidden homology in DNA genome are not found by current comparative tools despite decades of research. Many scholars modeled the genome as a monotonous string, which limits and probably obstructs the discovery of some significant patterns. We propose an information-coding-based model called DNA As X (DAX) to improve the sensitivity in comparative genomic studies by integrating the principles and concepts of other disciplines including information coding theory and signal processing into genome analysis. The proposed DNA As X model uses character-analysis-free (CAF) techniques, where X is the intermediate for analysis that can be digit, code, signal, vector, tree, graph network and so on. It provides novel and comprehensive perspectives to further analyze and recognize the critical patterns hidden in DNA genomes. Comparing with traditional character-analysis-based (CAB) methods, DAX not only enriches the tools and the knowledge library of computational biology but also extends the domain from 1-D character string analysis to 2-D spatial/temporal domain. Furthermore, by applying the DAX model to the issue of exon prediction as an evaluation, we illustrate the insights behind this model. The experimental results show that the DAX methodology can improve the sensitivity in genome analysis by using the novel information-coding techniques.

Keywords: DNA AS X (DAX) · Character-analysis-free (CAF) technique · Numeric-based method · Numerical representation · Transformation · Signal processing · Information coding theory · Error-tolerance · Vectors · Two-dimensional spatial/temporal domain

© Springer International Publishing Switzerland 2015
R. Harrison et al. (Eds.): ISBRA 2015, LNBI 9096, pp. 366–377, 2015.
DOI: 10.1007/978-3-319-19048-8_31

1   Introduction

The purpose of DNA As X (DAX) design is to improve the performance in comparative genomics where conserved sequence searching is a preliminary way to recognize and analyze the hidden conservation in nucleotide sequences. In recent studies [1–3], lots of conservation in genome sequences cannot be detected by current comparative methods even though the research in this field have developed for long time. Combinations of nucleotide mutation, insertion/deletion, gap, shift, repeat and error in DNA genome sequence lead to numerous aberrant patterns where traditional character-query tools cannot exert their efforts. Such deficiencies in fundamental methods inevitably result in that some significant patterns can not be recognized, which has been ceaselessly reported by latest findings on coding genes and conservation [4–7].

Since the first DNA-based genome was sequenced in the 1970s, by tacit agreement people have gotten used to characters for sequence analysis. Indeed the character-annotation-based or character-analysis-based (CAB) methods have played a significant role in analyzing and annotating genome data. Because of the nature of characters, four human-readable characters, A, T/U, C and G, respectively annotating the nucleotides of Adenine, Thymine/Uracil, Cytosine and Guanine, are conveniently combined together to denote enormous variations of genes, proteins, RNAs and DNAs. A large percent of practitioners in bioinformatics study these combinations and arrangements of four characters to conduct their research on genomes. However, with the advent of high-throughput sequencing technologies, CAB methods manifest their gaps/limitations on dealing with exponentially increasing demands on genome-data mining. The main reason is that CAB methods are limited to the linear analysis, in which the only object is the monotonic and inflexible characters and existing numerous numeric-based methods in engineering can hardly be exerted. Classical alignment methods based on conventional character-analysis have shown their drawbacks on handling increasing issues such as sequence assembly [8], whole genome analysis [9, 10], regulatory ncRNA prediction [11], genome-wide association studies [12, 13], evolutionary pathway [14] and so forth. Unfortunately, the work on developing novel numerical representations have not drawn enough attentions from scientists although the related work have emerged for a while. Targeting at the above notch, the generic framework of DAX transforms the character sequences into a series of signal waves, vectors and graphs by adopting the knowledge and methods in information coding theory and signal processing primarily. As an illustrative example, we demonstrate an implementation of the Character-Analysis-Free (CAF) technique for the improvement of the sensitivity in comparative gene studies and conduct the related experiments used to evaluate our proposed model. The result shows that the information-coding-based model has achieved desirable performances in improving the sensitivity of comparative gene studies.

The rest of the paper is organized as follows. Section 2 introduces the related work, especially those on character-analysis-free methods. Section 3 presents the details of DAX model and framework as well as fundamental concepts. Section 4

provides the implementation and Section 5 shows the experiments with evaluations. At last, Section 6 concludes the work and discusses the future work that DAX
can expand.

2   Related Work

The methods of Digital-signal-processing (DSP) applied in genomics adopt the
numerical representations as the intermediates that convert the strings into various numerical values. For example, the Voss binary representation [15, 16] is
currently the most used scheme that projects the nucleotide sequence into four
binary sequences denoted as $x_A[n]$, $x_C[n]$, $x_G[n]$ and $x_T[n]$ with the presence
(denoted as 1) or absence (denoted as 0) for the corresponding nucleotides.
This representation aims to facilitate the power spectra computation and spectral transformation. Complex-based methods [17] are employed for each of four
bases and each corresponding sequence so as to reflect the complementary features of nucleotides from mathematical perspectives. A typical assignment is
defined: $A = 1 + j$, $T = 1 - j$, $G = -1 + j$ and $C = -1 - j$. The geometric
interpretation imposes a structure that the Euclidean distance between A and
C is greater than the distance between A and T [18]. Alternatively, pulse amplitude modulation (PAM) scheme [19] uses non-integer representations to preserve
DNA's reverse complementary properties. For example, (A, G, C, T) is assigned
to (-1.5, -0.5, 0.5, 1.5) so that G-C and A-T base pairs are complementary in
arithmetic. However, in this scheme the bio-chemical association is ambiguous.

Encoding-based numerical representation can transform the nucleotide sequence into a series of encoding digits. As the simplest example, Kent et al
[20] use 2-bit format to compress and store the DNA sequences in a compact
randomly-accessible format, which gives a 16-byte header to contain the encoding information and pack each DNA nucleotide to two bits per base, T:00, C:01,
A:10 and G:11. However, this type of arbitrary assignment is criticized [21] for
that it cannot provide real signals to understand biological research. Instead, a
weight-based assignment in [21] is employed to the spectral transformation where
the weight coefficients derive from the enthalpy analysis for each nucleotide pair.

Another emerging method in information coding theory is Galois field (GF)
mapping [18], which maps nucleotides to Galois Field $GF(4)$ and transform the
sequence into orthogonal (n, k) codes. This method manifests the advantage of
information coding in genome analysis that error-correcting coding structure
reflects the nature of genome coding and it also shows the efficient effects on
detecting genome redundancy and gene mutations [22].

Discrete Fourier transformation (DFT) is the most commonly used method in
genomic signal processing that takes advantage of DNA numerical series for the
genome spectral analysis. The numerical data are mathematically transferred
into the sums of sine and cosine functions through Fourier transform. After the
DFT calculations, the spectral information of genome sequence appear on a 2 dimensional plot. It benefits the periodicity detection, especially the identification
of amino acid 3-periodicity, since the periodical signals can be resolved from the

3-periodicity amino acide codens. Thus, setting an appropriate spectral threshold is one of keys to accurately find the exon boundaries. DFT usually uses a window to calculate the spectral value for each position, resulting in either the high computing load if the size is large or the signal loss if the size is small. Digital signal filters, such as infinite impulse response (IIR) or finite impulse response (FIR), can be applied to cope with these problems [23].

However, the applicable areas of DFT and other transforms are relatively narrow only limited to the periodicity-related detection. Our information-coding-based model is not limited to detect the periodicity but aim to fill the gaps in the preceding literature and establish a generic framework for genome analysis.

3   Methodology

3.1   Bio-chemical Model

Assuming four nucletotides are distributed equiprobably, the entropy brings the maximum information capacity to the sequence. In terms of the definition of information entropy, we have the entropy value of 2 bit. Compatible to modern computing system, we consider the Galois Field GF(2) and the extension of GF(4) for any GF(2) pair [18]. DNA sequences can be encoded to binary codes [24] based on the principles in information coding theory. Obviously, two important rules are needed to consider, nucleotide biochemical properties and the features of binary codes. According to the chemical and biological enthalpy values of the nearest nucleotide combinations [25], four nucleotide acids can be placed in the order of (C,T,A,G) so that the bio-chemical dynamics manifest the symmetric properties as shown in Figure 1. Also, in the ascendant order of molecular physical size and weight, C, T/U, A, G are the best placement corresponding to symmetric codes. As mapping these properties to features of binary numeric coding, we encode C, T, A, and G to 00, 01, 10, and 11 respectively in two bits of binary codes.

On the other hand, the mutations/changes between four nucleotides differentiate the transition and the transversion. Relatively, transition (A-G, C-T) takes place more frequently than transversion (C-G, T-A) as the different colors (light-dark) respectively shown in Figure 1. Corresponding to the enthalpy values in Figure 1, we can see the weak bonds between pairs of transitions comparing with the strong bonds between nucleotide pairs of transversions. Mapping to the binary codes, the Hamming distance and the Euclidean distance between two codes precisely reflect the differences between transition and transversion. The Hamming distance (d_h) and the Euclidean distance (d_e) are defined for any two codes c and c' as Equation 1 ($c, c' \in GF(4)$).

$$d_h = c \oplus c', d_e = c - c'. \tag{1}$$

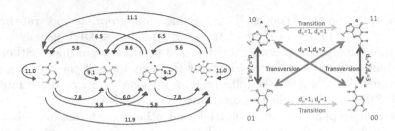

Fig. 1. Left: Symmetric thermodynamics pattern [25] in terms of the enthalpy values of thermodynamic interactions between two molecules. Unit of measurement is kcal/mol. Right: Molecule coding to reflect the bio-chemical relations [26].

3.2   Encoding and Signalizing Model

In terms of algebraic coding theory, a single nucleotide can be encoded into c that $c = \psi_1 x + \psi_0$, where ψ_1 and $\psi_0 \in GF(2)$. Any nucleotide in a sequence is encoded into that $c = \psi_{2i+1} x^{2i+1} + \psi_{2i} x^{2i}$ according to the extended $GF(4)$ where $i \in \{0, 1, ..., n-1\}$ indicates the location information of any nucleotide in this sequence.

Since ψ_i and $\psi_i' \in GF(2)$, $\psi_{2i+1} x^{2i+1} + \psi_{2i} x^{2i}$ and $\psi_{2i+1}' x^{2i+1} + \psi_{2i}' x^{2i} \in GF(4)$, $i \in \{0, 1, ..., n-1\}$ and n is the length of this DNA sequence, assuming that two DNA reads/tuples with the length of k are denoted as u and u' respectively and that $2k - 1$ $(k > 0)$ is the degree of the expressed polynomial with coefficients from the extension of GF(2) [27], u and u' can be expressed as follows.

$$u = \psi_{2k-1} x^{2k-1} + \psi_{2k-2} x^{2k-2} + ... + \psi_1 x^1 + \psi_0 x^0 = \sum_{i=0}^{2k-1} \psi_i x^i \qquad (2)$$

$$u' = \psi_{2k-1}' x^{2k-1} + \psi_{2k-2}' x^{2k-2} + ... + \psi_1' x^1 + \psi_0' x^0 = \sum_{i=0}^{2k-1} \psi_i' x^i \qquad (3)$$

The hamming distance and the Euclidean distance denoted as D_h and D_e between u and u' are

$$D_h(u, u') = \|u \oplus u'\| = \sum_{i=0}^{2k-1} (\psi_i \oplus \psi_i') \qquad (4)$$

and

$$D_e(u, u') = \|u - u'\| = \sum_{i=0}^{2k-1} (\psi_i - \psi_i') x^i. \qquad (5)$$

$D_h(u, u')$ is useful to identify the errors between two nucleotide reads/messages with the length of k. Assuming that m indicates the hamming distance/error tolerated by k-tuple messages/reads, by assigning the different (k, m), $0 \leq m \leq 2k$, we can achieve the goal of tolerating the distance in various patterns between reads.

For example, $D_h(u, u') \leq m$ means that the difference between u and u' is less or equal to m.

Assuming that two DNA string sequences s and s' have the nucleotide lengths of n and n' respectively, w and w' are the numerical representations of string sequences s and s'. We denote the transformation and the inverse transformation as follows:

$$w = T(s), s = T^{-1}(w). \tag{6}$$

Assuming that u and u' are polynomials with the same degree of $2k - 1$ ($k > 0$) and coefficients from the extension of $GF(2)$, w and w' can be represented as two series, $w = \{u_0, u_1, ..., u_{n-k}\}$ and $w' = \{u'_0, u'_1, ..., u'_{n'-k}\}$. Therefore, sequential codes u_{i-1}, u_i and u_{i+1} in sequence w can be expressed as the Equations 7 and 8.

$$\psi_{2(k+i)-1} x^{2k-1} + \psi_{2(k+i)-2} x^{2k-2} + \frac{u_{i-1}}{x^2} = u_i \tag{7}$$

$$u_{i+1} x^2 + \psi_{2i+1} x + \psi_{2i} = u_i \tag{8}$$

For sequence $w = \{u_0, u_1, ..., u_{n-k}\}$ and sequence $w' = \{u'_0, u'_1, ..., u'_{n'-k}\}$ with the same polynomial degree for u and u', we denote $v_l = \langle u_{il,jl}, u'_{i'l,j'l} \rangle$ as a vector that satisfies (k, m) for sequential pairs of u and u' ($u \in [u_{il}, u_{jl}]$, $u' \in [u'_{i'l}, u'_{j'l}]$). We further denote the set of possible vectors as $F = \{v_0, v_1, ..., v_q\}$ ($q \geq 0$). Therefore, the alignment between w and w' can be represented by a set f that contains a series of vectors, $f \subseteq F$.

3.3   Framework

The generic framework of the DAX model consists of four main parts that represent four processing phases respectively, as shown in the dashed rectangles of Figure 2, including transformation, feature extraction, signal processing, and inverse transformation.

(1) Transformation. This phase is responsible for transforming the DNA sequences into signals, where encoding and signalizing may vary dependent on the selected encoding model.

(2) Feature extraction. Common features can be extracted by matching two series of signal waves and further form the coding vectors that can be processed on the stage of signal processing. These extracted features may be raw and unpruned, which will be further refined in the next phase.

(3) Signal processing. Object X in this phase represents quantitative vectors. Because of the nature of vectors, they can form undirected trees/graphs where a vector may be contained in multiple paths. Chaining these vector into a larger path is the goal of global comparative methods [28]. The one with the maximum coverage will be selected as the best path.

(4) Inverse transformation. As the counterpart of transformation, inverse transformation is responsible for converting the intermediate results, graphs, trees, paths, vectors, signals, codes and digits, back into human-readable character sequences, denoted as Equation 6.

4   Implementation

Following the DAX framework mentioned in last section, main methods adopted in our application are illustrated in Figure 2. DAX framework can be implemented through various methods not limited to the described methods here.

The encoding and signalizing spends the computing time of $O(n)$ time. The information tree is numeric-based hash table to improve the searching efficiency. Each amplitude needs to take time of $O(logn)$ to find the corresponding location and for all signals it takes time of $O(nlogn)$. The construction of information tree takes the computing time of $O(n)$ only. Thus, the time complexity of the entire section is $O(nlogn)$.

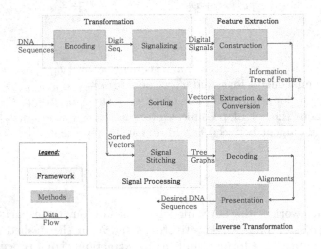

Fig. 2. Methods in DAX framework

The efficiency of signal stitching depends on the number of discovered vectors. Signal stitching problem is equivalent to exon chaining problem. Using dynamic programming and vector graphs, stitching the candidate vectors takes $O(n)$ time.

5   Experiment and Evaluation

In order to evaluate the performance of the DAX algorithm, we use the ROSETTA dataset which contains 140 orthologous gene pairs and 1,160 cross-species exons from human and mouse [29]. The purpose of the evaluation is to provide comparative experimental results to show the quality of the DAX algorithm and further validate the effectiveness of the DAX framework.

The DAX program is written in C/C++ and these experiments are conducted on the system of Intel i7 1.8GHz, 8G RAM, 500G HD and Ubuntu 12.4.

5.1   Sensitivity

The sensitivity of exon prediction is a significant indicator to evaluate an algorithm, which can be measured by several parameters: the number of conservation, the length of exon, the coverage of exon and sequence similarity [30]. Among them, the coverage percentage of exon is the most important measurement to evaluate the sensitivity. In [29], the coverage percentage of exon is calculated by a strict rule: only if the two ends of the exon are simultaneously predicted, that is, only if one hundred percent of the length of exon is predicted, the exon is counted as being predicted; if only one end is predicted, we calculate how much percentage of the length of the exon is predicted and classify the percentage to the corresponding classes, such as ninety percent or seventy percent. Adopting partial data in [30], as shown in Figure 3, we can see that the DAX algorithm based on character-analysis-free techniques has a desirable overall performance although the 100 percent sensitivity is a little inferior to others.

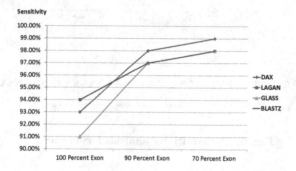

Fig. 3. Comparative results of exon coverage

In order to show the comprehensive evaluation on the conservation, the length of exon, the coverage of exon and sequence similarity, we scrutinize nine genes from [29] to have a breakdown analysis on the evaluation of our algorithm in the sensitivity. In Figure 4, DAX algorithm shows good results except Ribosom6al S24 gene, which contains two exon regions of only 3-bp that all algorithms fail to predict. In four genes, Casein kinase, Skeletal alpha-actin, Hsc70 and Int-1 oncogene, one end of each gene is missed by GLASS [29]. For more details of DAX performance, in Figure 5, the curves of quantitative results for 39 exons are plotted, including original exon's similarity, predicted exon's coverage percentage and predicted exon's similarity as well as the exon's lengths. We can see that the DAX can detect almost various lengths of exons with the range from 20+ nt to 1k+ nt except extremely short ones (3 nt and 6 nt).

5.2   Coding Length, Execution Time and Sensitivity

Despite of the less importance of computing speed comparing with that of sensitivity, we evaluate the execution time of the DAX program by adjusting the

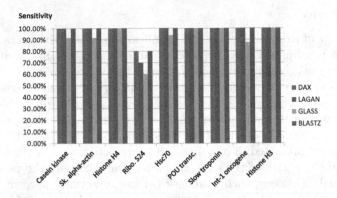

Fig. 4. Percentage of predicted exons for nine genes

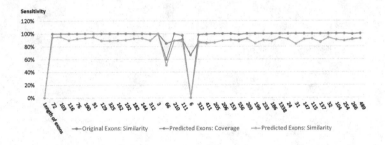

Fig. 5. The sensitivity analysis for 39 exons

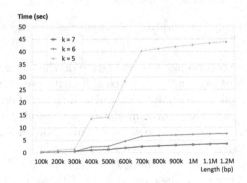

Fig. 6. Relation between coding length and running time

magnitude of dataset and the coding length k in u to see the mutual relationship between these impact factors. In Figure 6, the dataset of 1.2M can be computed for exon prediction in 3.5 seconds if the coding length is set to 7 while the running time is double if coding length is set to 6. From the trend of the chart, we can conclude that the larger magnitude of coding length may result in the

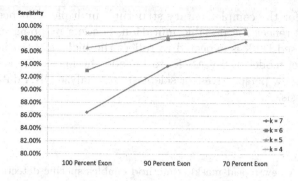

Fig. 7. Relation between coding length and sensitivity

faster speed. However, the large magnitude of coding length does not contribute to the sensitivity of predictive results. As illustrated in Figure 7, a larger coding length leads to lower sensitivities, which means the lower granularity of common features. On the other hand, although a small coding length often brings a better resolution of prediction, it probably results in devastating computing loads and the problem of over-representation. Thus, properly choosing the coding length can determine the success of the experimental result. By observing the experimental data, we choose coding length equal to 6 as a compromise between the execution time and the sensitivity.

From the above experiments, we see that the DAX method can meet our expectation of improving the sensitivity in exon prediction and its partial performance outperform other existing methods. Since finding the conserved sequence areas is one of the fundamental issues in bioinformatics, we believe that more productive outcomes can be obtained if we apply the DAX method as an effective computational tool to other issues.

6   Conclusion

DNA AS X (DAX) is a systematic framework that adopts information-coding-based techniques in the analysis and the pattern recognition in genome data. By constructing a generic coding framework for the character-to-X transformation, the character-analysis-free techniques process and analyze the intermediates, such as signal waves, vectors, graph network and so forth, through existing numeric methods. We implement the DAX framework on detecting exons between two homologous sequences of human and mouse and the experimental results show that our DAX model based on character-analysis-free techniques can improve the sensitivity of exon prediction. A few questions in this project are remaining for the future work, such as integrating the splicing sites to improve the specificity, applying to more dataset, etc. Besides, we plan to apply the DAX methodology to other issues in the future, such as RNA secondary and tertiary structure prediction where the complementary codes can model the folding process for the complementary structure, multiple-sequence alignments where the codes reflect the possible mutation, genome mapping and assembly where coding-based DAX is expected to reflect the biological truth. In sum, we introduce a novel model, DNA AS X, to encode the nucleotide sequences into codes and bring new perspectives for scientists to analyze DNA data in the area of genome analysis.

References

1. Frith, M.C.: A new repeat-masking method enables specific detection of homologous sequences. Nucleic Acids Research 39(4), e23 (2011)
2. Frith, M.C., Noé, L.: Improved search heuristics find 20 000 new alignments between human and mouse genomes. Nucleic Acids Research 42(7), e59 (2014)
3. Trimble, W., Keegan, K., D'Souza, M., Wilke, A., Wilkening, J., Gilbert, J., Meyer, F.: Short-read reading-frame predictors are not created equal: sequence error causes loss of signal. BMC Bioinformatics 13(1), 183 (2012)
4. Djebali, S., Davis, C.A., Merkel, A., Dobin, A., Lassmann, T., Mortazavi, A.M., Schlesinger, F.: Landscape of transcription in human cells. Nature 489(7414), 101–108 (2012)
5. ENCODE. An integrated encyclopedia of dna elements in the human genome. Nature 489(7414), 57–74 (September 2012)
6. Hiller, M., Schaar, B.T., Bejerano, G.: Hundreds of conserved non-coding genomic regions are independently lost in mammals. Nucleic Acids Research (2012)
7. Klimke, W., O'Donovan, C., White, O., Brister, J.R., Clark, K., Fedorov, B., Tatusova, T.: Solving the problem: Genome annotation standards before the data deluge. Standards in Genomic Sciences 5(1), 168–193 (2011)
8. Li, H., Homer, N.: A survey of sequence alignment algorithms for next-generation sequencing. Briefings in Bioinformatics 11(5), 473–483 (2010)
9. Wu, X., Cai, Z., Wan, X.-F., Hoang, T., Goebel, R., Lin, G.: Nucleotide composition string selection in HIV-1 subtyping using whole genomes. Bioinformatics 23(14), 1744–1752 (2007)
10. Cai, Z., Goebel, R., Salavatipour, M., Lin, G.: Selecting dissimilar genes for multi-class classification, an application in cancer subtyping. BMC Bioinformatics 8(1), 206 (2007)
11. Tesorero, R.A., Yu, N., Wright, J.O., Svencionis, J.P., Cheng, Q., Kim, J.-H., Cho, K.H.: Novel regulatory small rnas in streptococcus pyogenes. PLoS One 8(6), e64021(2013)
12. Guo, X., Meng, Y., Yu, N., Pan, Y.: Cloud computing for detecting high-order genome-wide epistatic interaction via dynamic clustering. BMC Bioinformatics 15(1), 102 (2014)
13. Yang, K., Cai, Z., Li, J., Lin, G.: A stable gene selection in microarray data analysis. BMC Bioinformatics 7(1), 228 (2006)
14. Cai, Z., Duan, Y., Li, Y., Lin, G., Ozden, M., Wan, X.F.: Ipminer: a progenitor gene identifier for influenza a virus. Influenza Other Respi. Viruses 5(suppl. 1), 413–415 (2011)
15. Silverman, B.D., Linsker, R.: A measure of DNA periodicity. Journal of Theoretical Biology 118(3), 295–300 (1986)
16. Voss, R.F.: Evolution of long-range fractal correlations and $1/f$ noise in dna base sequences. Phys. Rev. Lett. 68, 3805–3808 (1992)

17. Cristea, P.D.: Genetic signal representation and analysis. In: Proc. SPIE, vol. 4623, pp. 77–84 (2002)
18. Rosen, G.L.: Signal Processing for BiBiological-inspired Gradient Source Localization and DNA Sequence Analysis. PhD thesis, Georgia Institute of Technology, School of Electrical and Computer Engineering (August 2006)
19. Chakravarthy, N., Spanias, A., Iasemidis, L.D., Tsakalis, K.: Autoregressive modeling and feature analysis of DNA sequences. EURASIP Journal on Advances in Signal Processing 2004(1), 952689 (2004)
20. Kent, W.J., Sugnet, C.W., Furey, T.S., Roskin, K.M., Pringle, T.H., Zahler, A.M., Haussler, D.: UCSC genome browser. Genome Res 12(6), 996–1006 (2002)
21. Kauer, G., Blöcker, H.: Applying signal theory to the analysis of biomolecules. Bioinformatics 19(16), 2016–2021 (2003)
22. Rosen, G.L.: Examining coding structure and redundancy in DNA. IEEE Engineerin. In: Medicine and Biology Magazine, Special Issue on Communication Theory, Coding Theory, and Molecular Biology 62–68 (January/February 2006)
23. Yoon, B.J.: Hidden markov models and their applications in biological sequence analysis. Current Genomic 10, 402–415 (2009)
24. Blahut, R.E.: Algebraic Codes for Data Transmission, 2nd edn. Cambridge University Press, Cambridge (2003)
25. Breslauer, K.J., Frank, R.: Predicting DNA duplex stability from the base sequence. Proceedings of the National Academy of Sciences 83(11), 3746–3750 (1986)
26. Crick, F.: Codon and anticodon pairing: the wobble hypothesis. Journal of Molecular Biology 19, 548–555 (1966)
27. Lin, S., Costello, D.J.: Error control coding: fundamentals and applications, vol. 114. Pearson-Prentice Hall, Upper Saddle River (2004)
28. Dubchak, I., Poliakov, A., Kislyuk, A., Brudno, M.: Multiple whole-genome alignments without a reference organism. Genome Res. 19, 682–689 (2009)
29. Batzoglou, S., Pachter, L., Mesirov, J.P., Berger, B., Lander, E.S.: Human and mouse gene structure: Comparative analysis and application to exon prediction. Genome Res. 10, 950–958 (2000)
30. Brudno, M., Do, C.B., Cooper, G.M., Kim, M.F., Davydov, E., Green, E.D., Sidow, A., Batzoglou, S.: LAGAN and Multi-LAGAN: efficient tools for large-scale multiple alignment of genomic DNA. Genome Res., 13 (April 2003)

A Distance-Based Method for Inferring Phylogenetic Networks in the Presence of Incomplete Lineage Sorting

Yun Yu[✉] and Luay Nakhleh

Department of Computer Science, Rice University, 6100 Main Street, Houston, TX 77005, USA
{yy9,nakhleh}@rice.edu

Abstract. Hybridization and incomplete lineage sorting (ILS) are two evolutionary processes that result in incongruence among gene trees and complicate the identification of the species evolutionary history. Although a wide array of methods have been developed for inference of species phylogeny in the presence of each of these two processes individually, methods that can account for both of them simultaneously have been introduced recently. However, these new methods are based on the optimization of certain criteria, such as parsimony and likelihood, and are thus computationally intensive. In this paper, we present a novel distance-based method for inferring phylogenetic networks in the presence of ILS that makes use of pairwise distances computed from multiple sampled loci across the genome. We show in simulation studies that the method infers accurate networks when the estimated pairwise distances have good accuracy. Furthermore, we devised a heuristic for post-processing the inferred network to remove potential false positive reticulation events. The method is computationally very efficient and is applicable to very large data sets.

1 Introduction

Understanding the evolutionary history of a set of species and the intricate relationships between the evolution of genes and genomes are two central questions in biology. It has long been acknowledged that the evolutionary history of a genomic region from a set of species is not necessarily congruent with that of the species [16], which is the classic gene tree/species tree problem. The incongruence among gene trees and species tree may be caused by various evolutionary processes. Incomplete lineage sorting (ILS), which is a result of random genetic drift in populations, is one common process, especially in evolutionary scenarios that involve rapid speciation and/or large population sizes. The occurrence of ILS and its extent have been reported in various data studies of very diverse sets of organisms; e.g., [29,24,14,32,4,33,8,30]. A large variety of methods have been developed to deal with it; see [26,5,15,23] for recent surveys of such methods.

A second evolutionary process that results in gene tree incongruence is reticulation, which includes horizontal gene transfer in asexual species and hybridization in sexual species. Hybridization is believed to play an important role in several groups of eukaryotic species [1,2,17,18,27]. Not only does hybridization result in gene tree incongruence, but it also results in non-treelike phylogenetic relationships among species, that are best represented by *phylogenetic networks*. The structure of a phylogenetic network

© Springer International Publishing Switzerland 2015
R. Harrison et al. (Eds.): ISBRA 2015, LNBI 9096, pp. 378–389, 2015.
DOI: 10.1007/978-3-319-19048-8_32

is a rooted, directed acyclic graph, which allows for nodes with more than one parents. Many methods have been devised to infer these phylogenetic networks by making use of gene tree incongruence; see [22,11,23] for recent surveys of such methods.

With increasingly available genomic data, patterns of cooccurrence of hybridization and incomplete lineage sorting are being observed, or suspected, in the data [7,6,28,3,20]. This has called for developing methods that can take both hybridization and incomplete lineage sorting into account. Methods that assume only ILS as the cause of incongruence would completely miss the possibility of hybridization, whereas methods that infer phylogenetic networks without accounting for ILS would end up grossly overestimating the amount of hybridization when ILS is also at play. To address this issue, several methods were proposed recently. However, given the the complexity of modeling such scenarios in general, most of these methods focused on special cases of the problem (typically with limited complexity); e.g., [31,9,19,13,12,38]. More recently, methods for inferring general networks based on parsimony and likelihood criteria were developed [35,34,37,36]. The applicability of these inference methods is currently limited to small data sets, given the hardness of the inference problems under these two criteria.

Distance-based methods have long been some of the fastest methods in phylogenetics, producing very good estimates on phylogenies with thousands of taxa in minutes. Even when the accuracy of inferences made by these methods is not very high, trees produced by distance-based methods are still used as initial trees for the most computationally intensive and detailed methods, such as maximum likelihood. Thus, distance-based method provide a very good tool in phylogenetics. In this paper, we introduce a novel distance-based method that infers a phylogenetic network from pairwise distance data in the presence of both hybridization and ILS. Our method builds on the GLASS method [21] that was recently introduced to infer species trees from pairwise distances obtained from multiple loci under the assumption that all incongruence is due to ILS. We studied the performance of our method on simulated data and found that it produces very good results, even when we perturbed the pairwise distances so as to simulate error in distance estimates. We also devised a heuristic for potentially eliminating false positive reticulations in order to minimize the overestimation of the number of reticulations.

It is important to note that accurate estimates of pairwise distances based on multiple loci is a requirement for a good performance of our method (just like they are a requirement for a good performance of GLASS). We view this as a major obstacle facing the application of this method to real data. Nevertheless, as we pointed out above, this method can still be used to quickly generate a good phylogenetic network to initialize the search employed by computationally intensive methods such as [37,36].

2 Methods

2.1 Phylogenetic Networks

In order to account for both hybridization and incomplete lineage sorting in the evolutionary history of a set of species (or, genomes), we use an evolutionary (rather than "data-display") *phylogenetic network* model [22]. For a node v in a digraph, we denote by $d^-(v)$ and $d^+(v)$ the in- and out-degree of v. A (binary) phylogenetic $\mathscr{X}$-network N is a rooted, directed, acyclic graph whose node-set $V(N)$ is partitioned into four sets:

- $\{r\}$, the root of N, with $d^-(r) = 0$ and $d^+(r) = 2$;
- The leaf-set $V_L = \{v \in V(N) : d^-(r) = 1, d^+(r) = 0\}$, which are bijectively labeled by $\mathscr{X}$;
- The internal tree nodes $V_T = \{v \in V : d^-(r) = 1, d^+(r) = 2\}$; and
- The reticulation nodes $V_N = \{v \in V : d^-(r) = 2, d^+(r) = 1\}$.

Every structure inferred by our algorithm (described below) is a phylogenetic network. However, it is important to point out that there are phylogenetic networks that cannot be inferred by our algorithm. This is not a limitation of the algorithm, but rather has to do with the reconstructibility of certain reticulation scenarios (e.g., a reticulation edge involving two nodes one of which falls on the path from the root to the other node). More generally, let us denote by $L(v)$ the set of taxa that label leaves that are descendants of node v. Given a phylogenetic network N, for each node v in $\{r\} \cup V_T$, we define the set $dp(v) = \{L(v_1) - L(v_2), L(v_2) - L(v_1)\}$ where v_1 and v_2 are the two children of v. Then if a phylogenetic network contains two nodes $u, v \in \{r\} \cup V_T$ where $dp(u) = dp(v)$, one of these two nodes cannot be inferred by our method.

2.2 Inferring a Network from a Distance Matrix

We denote by D_L a distance matrix over a set of taxa L where $D_L(i, j)$ is the distance between taxa i and j in L. With respect to the nodes of a phylogenetic network, we define two functions $DMax(u, v, S)$ and $DMin(u, v, S)$, where u and v are two nodes and S is a set of nodes, to be $DMax(u, v, S) = \max\{D_L(a, b) : a \in L(u) - L(v), b \in L(v) - L(u), \nexists w \in S \text{ s.t.} \{a, b\} \subseteq L(w)\}$ and $DMin(u, v, S) = \min\{D_L(a, b) : a \in L(u) - L(v), b \in L(v) - L(u), \nexists w \in S \text{ s.t.} \{a, b\} \subseteq L(w)\}$. See Figure 1 for an illustration.

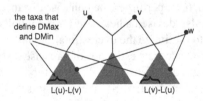

Fig. 1. An illustration of $DMax(u, v, S)$ and $DMin(u, v, S)$ computation on $S = \{u, v, w\}$

Assuming the pairwise distances are realizable by a phylogenetic network, the basic idea of our method is we start with a set of nodes S, each labeled by a taxon in L and then we do the following until S has only one node:

1. Let X and Y be two nodes in S that have the minimum $DMin(X, Y, S)$.
2. If $DMin(X, Y, S) = DMax(X, Y, S)$, a speciation event is considered. We remove X and Y in S and add node XY.
3. If $DMin(X, Y, S) \neq DMax(X, Y, S)$, a hybridization event is considered. We find the most parsimonious way to make a reticulation node(s), which can be one of the following:

- A reticulation node, say u, is added onto an edge whose tail is a descendant of X. We remove Y from S and add a new node whose children are u and Y.
- A reticulation node, say u, is added onto an edge whose tail is a descendant of Y. We remove X from S and add a new node whose children are u and X.
- Two reticulation nodes, say u_x and u_y, are added onto an edge whose tail is a descendant of X and an edge whose tail is a descendant of Y, respectively. We add a new node whose children are u_x and u_y to S.

More details can be found in Alg.1. See Fig. 2 for an example.

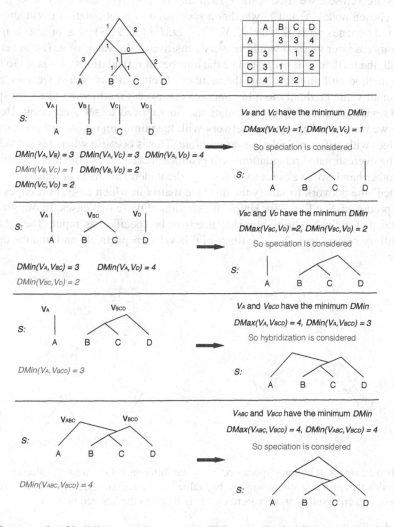

Fig. 2. An example of building a species network given true pairwise distances. The true species network and distance matrix are given on the top. For simplicity, the third parameter S is omitted in $DMin$ and $DMax$ since the context is clear.

In practice, the pairwise distances are estimated from gene data, and it is important to account for the error inherent in these estimates. The GLASS method of [21] uses pairwise distances that are computed as the minimum interspecific coalescence times across all loci and then builds a species tree using simple clustering. The rationale behind this method is that when the number of loci goes to infinity, the minimum interspecies coalescence times across all loci should converge to the speciation times. Here, given multiple loci data, we computed pairwise distances exactly like what GLASS does which is using minimum interspecific coalescence times across all loci. Now suppose we have a node (u, v) with time t in a species tree. Then if the number of loci is large enough, we should see $D(a, b)$ very close to t for all $a \in L(u), b \in L(v)$. To quantify if two numbers are "close", we used some ε such that if $|x - y| \le \varepsilon$ we say $x \approx y$. Then for two chosen nodes X and Y whether a speciation or a reticulation event should be considered depends on if $DMax(X, Y, S) - DMin(X, Y, S) \le \varepsilon$ or not. It is clear that in this case our method would be very sensitive to the value of ε. If ε is set to be too small, the method will overestimate the number of reticulations; if ε is set to be too big, the method will underestimate the number of reticulations. When we vary ε from a very small value gradually to a big one, we can expect the method to return species networks with fewer and fewer reticulations. So we need to set a criterion. Here we say that we want to infer a species network with the minimum number of unreasonably short edges with as few reticulations as possible. This is because when ε is set to be too small, the overestimated reticulations will produce short edges in the inferred network. On the other hand, when ε is set to be too big, the underestimation of reticulations will "squashed" the network to satisfy the distance matrix in which case short edges might also be produced. See Fig. 3 for simple illustration of these two cases. In our program, a value σ that defines "short" branches needs to be specified as input. Then the program will try ε equal to $1, 2, \ldots, k$ times of this value respectively and find the optimal network.

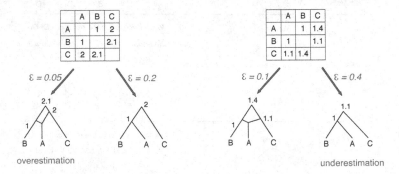

Fig. 3. Two examples of building species networks on different ε. Left: a small value of ε caused an overestimation of reticulations. Right: a big value of ε caused an underestimation of reticulations. Both of them result in short branches (of length 0.1) in the inferred network.

The details of our method are shown in Alg. 1. It takes a distance matrix D_L, a value σ that defines short branches, and a value k that sets the values of ε as we discussed above as input, and returns an inferred phylogenetic network. It reflects the basic idea of our method. In fact, we found that when we tried to find two nodes X and Y in S that have the minimum $DMin(X, Y, S)$, there might be multiple pairs of nodes that share the same minimum value. To address this issue, we kept a stack in the program so that every time there were more than one optimal pair we added a flag in the stack. After a network was built from choosing one of the optimal pairs, the program read the flag on the top of the stack and rolled back to the point where that flag was added and tried another optimal pair, until all optimal pairs were tried. All equally optimal species networks would be returned.

Input: A distance matrix D_L, σ, k.
Output: A phylogenetic network N.
$numCloseNodes \leftarrow 0$;
$numReticulations \leftarrow 0$;
$N \leftarrow NIL$;
for $i = 1$ **to** k **do**
    $\varepsilon \leftarrow i \times \sigma$;
    Let S be a set of nodes each labeled by a taxon in L and each node has time 0;
    while $|S| > 1$ **do**
        Let X and Y be two nodes in S that has the minimum $DMin(X, Y, S)$;
        $t_{min} \leftarrow DMin(X, Y, S)$;
        if $DMax(X, Y, S) - t_{min} \leq \varepsilon$ **then**
            Remove X and Y from S;
            Add a new node (X, Y) with time t_{min} to S;
        end
        else
            $(w_x, v_x) \leftarrow \arg\max_{(w,v)}\{|L(v) - L(Y)| : DMin(v, Y, S) = t_{min}, DMax(v, Y, S) - t_{min} \leq \varepsilon, w$ is a descendant of $X\}$;
            if $edge$ (w_x, v_x) $does\ not\ exist$ **then**
                $(w_y, v_y) \leftarrow \arg\max_{(w,v)}\{|L(v) - L(X)| : DMin(X, v, S) = t_{min}, DMax(X, v, S) - t_{min} \leq \varepsilon, w$ is a descendant of $Y\}$;
                if $edge$ (w_y, v_y) $does\ not\ exist$ **then**
                    $(w_x, v_x), (w_y, v_y) \leftarrow \arg\max_{(w_1,v_1),(w_2,v_2)}\{|L(v_1) - L(v_2)| + |L(v_2) - L(v_1)| : DMin(v_1, v_2, S) = t_{min}, DMax(v_1, v_2, S) - t_{min} \leq \varepsilon, w_1$ is a descendant of X and w_2 is a descendant of $Y\}$;
                    Add a new node whose children are u_x and u_y with time t_{min} to S where u_x and u_y are newly added nodes on (w_x, v_x) and (w_y, v_y) respectively;
                end
                else
                  Remove X from S;
                  Add a new node whose children are X and u with time t_{min} to S where u is a newly added node on (w_y, v_y);
                end
            end
            else
                Remove Y from S;
                Add a new node whose children are u and Y with time t_{min} to S where u is a newly added node on (w_x, v_x);
            end
        end
    end
    Let N' be the network that rooted at the only node in S;
    Let c be the number of branches of N' whose branch length is less than σ;
    Let r be the number of reticulations of N';
    if $N = NIL$, or $c < numCloseNodes$, or $c = numCloseNodes$ and $r < numReticulations$
    then
        $N = N'$; $c = numCloseNodes$; $r = numReticulations$;
    end
end
return N;

Algorithm 1. inferNetworkFromDistanceMatrix

2.3 Removing Reticulations with Low Support

In our simulation study (see Results section), we found that our method tended to over-estimate the number of reticulations, especially when the number loci is small. To address this issue, we employed a heuristics to remove reticulations with low bootstrap support. More specifically, assuming the original data contained n loci, we randomly sampled n loci with replacement and used them as the input of our method to infer a species network. This process was repeated 100 times. Then we removed reticulations of the species network inferred from the original dataset that were not well supported by the 100 species networks obtained from bootstrap. To do so, we first defined a function called **computeBootstrapSupport**, which takes a target species network and a set of bootstrap species networks and returns the target species network N with bootstrap support for every edge. The support of an edge in the species network is calculated as the percentage of that edge present in the bootstrap networks. To see whether one edge in network N_1 exists in network N_2, we simply computed the hardwired cluster [11] induced from that edge and then check if there is any edge in N_2 inducing the same hardwired cluster. The detailed algorithm for removing reticulations of a species network with low support given a set of bootstrap networks and a bootstrap threshold is shown in Alg. 2, where $Support(u, v)$ means the support of edge (u, v).

Input: A species network N, a set of bootstrap networks BN, $threshold$.
Output: a species network N'
$N' \leftarrow$ **computeBootstrapSupport**(N, BN) ;
Let $numLowSupport$ be the number of edges in N' that has low support;
foreach *edge (u, v) visited when post-traversing N'* **do**
    if $Support(u, v) < threshold$ **then**
        foreach *child node w of v that is also a reticulation node* **do**
            $N'' \leftarrow N'$;
            Remove reticulation edge (v, w) of N'';
            $N'' \leftarrow$ **computeBootstrapSupport**(N'', BN) ;
            Let $tnls$ be the number of edges in N'' that has low support;
            if $numLowSupport > tnls$ **then**
                return removeLowSuportEdges$(N'', BN, threshold)$;
            end
        end
    end
end
return N';

Algorithm 2. removeLowSuportEdges

---

3   Results

We used synthetic datasets to test the performance of our method. We first generated 2 datasets, each consisting of 100 random species trees with 10 taxa of height 8 and 20 taxa of height 16 respectively using PhyloGen [25]. The height of tree is the total branch lengths from the root of the tree to any of its leaf. Then for each species tree, we randomly added 1, 2, 3, 4 and 5 reticulations respectively. To add a reticulation to a species network, we randomly chose two edges in the network and add an edge between their midpoints from the higher one to the lower one. Then the lower one became a new reticulation node and we randomly assigned an inheritance probability from 0.1 to 0.9. Within the branches of each species network we simulated 25, 50, 100, 200, 500, 1000, 2000 and 5000 gene trees respectively using program ms [10].

We run our method on these gene trees and compared the inferred species networks with true ones using hardwired cluster distance [11]. Note that in all simulations, we set parameters as $\sigma = 0.1$ and $k = 5$ (see Alg. 1). σ was set to be 0.1 because it is a good threshold of "short" branch when branch lengths are in coalescent units. We also tried different values and found that varying it slightly did not have much affect on results. For the setting of k, we found that in our simulations most optimal species networks were found at $\varepsilon = 2\sigma$ or $\varepsilon = 3\sigma$, and setting k to be more than 5 would only change the results very slightly. The result is shown in Fig. 4. Note that when multiple equally optimal networks were returned, the average distance of those tie networks was calculated. We can see that overall our method made very accurate inferences. As expected, for both datasets, the error of the inferred networks increased slightly with the number of reticulations, because for the same number of taxa increasing the number of reticulations made the inference problem harder. Also, for both datasets, as the number of gene trees increased, the accuracy of the inferred networks increased. When comparing the results from the two datasets, we can see that the 20-taxon dataset actually produced slightly better result. This is because for the same number of reticulations the reticulations are expected to be more independent from each other on a network with more taxa, which makes the inference problem easier.

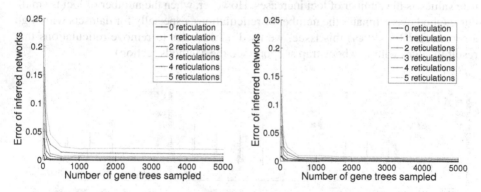

Fig. 4. Accuracy of the method using true gene trees. Results of the 10-taxon dataset and the 20-taxon dataset are shown in the left and right panels, respectively. The errors of the inferred networks were computed using hardwired cluster distance [11]. The results were averaged over 100 repetitions.

In order to test the robustness of our method to error in pairwise distance estimates, we synthetically perturbed the true distances. More specifically, the pairwise distances obtained above underwent 5 different perturbation experiments $i = 1, 2, 3, 4, 5$: In experiment i, each pairwise distance was multiplied by a (uniformly distributed) random number in the range $[1, 1 + i\epsilon]$ for $\epsilon = 0.1$. For example, in the results, the "30% error" data sets were obtained by multiplying each pairwise distance by a random number in $[1, 1.3]$ (each pairwise distance was multiplied by a potentially different number). The inference method was then applied to the perturbed data sets. The results of using these perturbed pairwise distances are shown in Fig. 5. We can see that overall our method still produced accurate results. As expected, on the same dataset, the accuracy of the inferred network decreased as the value of $i\epsilon$ increased. Further, the effect of the distance error on the network accuracy decreased with increasing the number of gene trees. It is important to note that the error has more impact on the "harder" datasets, that is, the ones with more reticulation nodes.

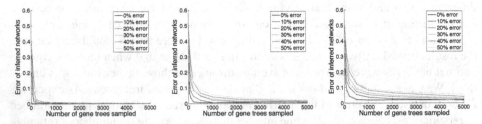

Fig. 5. Accuracy of the method using perturbed pairwise distances on 10-taxon dataset. Results of datasets containing true species networks with 0, 3 and 5 reticulations are shown from left to right columns, respectively. The errors of the inferred networks were computed using hardwired cluster distance [11]. The results were averaged over 100 repetitions.

We also examined the number of reticulations in the inferred species networks; see Fig. 6. As the results show, the estimates of the numbers of reticulations tend to the true values as the number of loci increases. However, when the number of loci is small, our method overestimates the number of reticulations, especially for datasets with high error values. To address this issue, we used a heuristics to remove reticulations that result in edges with low bootstrap support (see the Methods section).

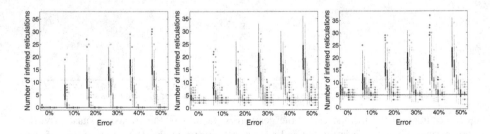

Fig. 6. The number of reticulations in inferred species networks of 10-taxon dataset. Results of datasets containing true species networks with 0, 3 and 5 reticulations are shown from left to right columns, respectively. In each subfigure, boxes from left to right (from black to cyan) in each group corresponds to datasets consisting of 25, 50, 100, 200, 500, 1000, 2000 and 5000 gene trees respectively. The solid horizontal black line in each subfigure indicates the true number of reticulations.

In Fig. 7, we show results based on the "hardest" dataset where the true species networks contain 10 taxa and 5 reticulations and the pairwise distances of taxa from gene trees were randomly perturbed by at most 50%. When multiple equally optimal species networks were returned, we chose a random one to which to apply the heuristic.

We varied the bootstrap threshold by using values 70, 80 and 90. As the results show, the heuristics successfully reduced the number of reticulations in the inferred species networks, especially for datasets with a small number of loci. For datasets with 25 gene trees, the mean number of reticulations was reduced from 15 to 7 when a bootstrap threshold of 70 was used. As expected, when a larger bootstrap threshold was used, the inferred species networks had fewer reticulations. On the other hand, the accuracy of the inferred species networks increased after reducing the number of reticulations.

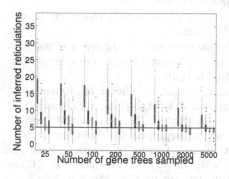 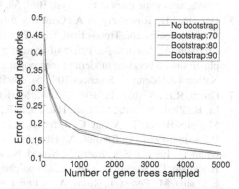

Fig. 7. Results of using heuristics to remove reticulations that result in edges with low bootstrap support based on 10-taxon and 5-reticulation datasets where distance matrices of gene trees were randomly perturbed by at most 50%. Left: the number of reticulations in the inferred species networks. The solid horizontal black line indicates the true number of reticulations. Right: the error of the inferred species networks.

In terms of the running time, for the largest and most complex dataset (20 taxa, 5 reticulations and 5000 gene trees), the program took, on average, around 3 minutes to complete the inference. For most of the datasets with 10 taxa, 5 reticulations and 5000 gene trees, the program finished in 10 seconds or less.

4 Conclusions

In this paper, we proposed a simple, yet effective distance-based method for inferring phylogenetic networks from pairwise distances in the presence of incomplete lineage sorting. Our method is a simple extension of the GLASS method [21]. It is important to note, though, that while GLASS has theoretical guarantees (the authors proved its statistical consistency), our method makes heuristic decisions and currently lack any theoretical guarantees. However, our simulation study demonstrate the method can obtain very good results, even when noise is added to the distance estimates. In practice, distance-based methods in general suffer from the lack of accurate methods for estimating pairwise distances. As the amount of molecular sequence data increases and more sophisticated methods are developed for more accurate estimates of pairwise distances, the application of distance-based methods would become more common, particularly for large data sets. Nevertheless, the speed of these methods make them appealing for rapid generation of a relatively accurate network to initialize the search of a more accurate, and computationally intensive method, such as maximum likelihood or Bayesian inference.

References

1. Arnold, M.L.: Natural Hybridization and Evolution. Oxford University Press, Oxford (1997)
2. Barton, N.H.: The role of hybridization in evolution. Molecular Ecology 10(3), 551–568 (2001)
3. The Heliconius Genome Consortium: Butterfly genome reveals promiscuous exchange of mimicry adaptations among species. Nature 487(7405), 94–98 (2012)
4. Cranston, K.A., Hurwitz, B., Ware, D., Stein, L., Wing, R.A.: Species trees from highly incongruent gene trees in rice. Syst. Biol. 58, 489–500 (2009)
5. Degnan, J.H., Rosenberg, N.A.: Gene tree discordance, phylogenetic inference and the multispecies coalescent. Trends Ecol. Evol. 24(6), 332–340 (2009)
6. Eriksson, A., Manica, A.: Effect of ancient population structure on the degree of polymorphism shared between modern human populations and ancient hominins. Proceedings of the National Academy of Sciences 109(35), 13956–13960 (2012)
7. Green, R.E., Krause, J., Briggs, A.W., Maricic, T., Stenzel, U., Kircher, M., Patterson, N., Li, H., Zhai, W., Fritz, M.H.-Y., Hansen, N.F., Durand, E.Y., Malaspinas, A.-S., Jensen, J.D., Marques-Bonet, T., Alkan, C., Prfer, K., Meyer, M., Burbano, H.A., Good, J.M., Schultz, R., Aximu-Petri, A., Butthof, A., Hber, B., Hffner, B., Siegemund, M., Weihmann, A., Nusbaum, C., Lander, E.S., Russ, C., Novod, N., Affourtit, J., Egholm, M., Verna, C., Rudan, P., Brajkovic, D., Kucan, E., Guic, I., Doronichev, V.B., Golovanova, L.V., Lalueza-Fox, C., de la Rasilla, M., Fortea, J., Rosas, A., Schmitz, R.W., Johnson, P.L.F., Eichler, E.E., Falush, D., Birney, E., Mullikin, J.C., Slatkin, M., Nielsen, R., Kelso, J., Lachmann, M., Reich, D., Pbo, S.: A draft sequence of the Neandertal genome. Science 328(5979), 710–722 (2010)
8. Hobolth, A., Dutheil, J., Hawks, J., Schierup, M., Mailund, T.: Incomplete lineage sorting patterns among human, chimpanzee, and orangutan suggest recent orangutan speciation and widespread selection. Genome Research 21, 349–356 (2011)
9. Holland, B.R., Benthin, S., Lockhart, P.J., Moulton, V., Huber, K.T.: Using supernetworks to distinguish hybridization from lineage-sorting. BMC Evol. Biol. 8, 202 (2008)
10. Hudson, R.R.: Generating samples under a Wright-Fisher neutral model of genetic variation. Bioinformatics 18, 337–338 (2002)
11. Huson, D.H., Rupp, R., Scornavacca, C.: Phylogenetic Networks: Concepts, Algorithms and Applications. Cambridge University Press, New York (2010)
12. Joly, S., McLenachan, P.A., Lockhart, P.J.: A statistical approach for distinguishing hybridization and incomplete lineage sorting. Am. Nat. 174(2), E54–E70 (2009)
13. Kubatko, L.S.: Identifying hybridization events in the presence of coalescence via model selection. Syst. Biol. 58(5), 478–488 (2009)
14. Kuo, C.-H., Wares, J.P., Kissinger, J.C.: The Apicomplexan whole-genome phylogeny: An analysis of incongruence among gene trees. Mol. Biol. Evol. 25(12), 2689–2698 (2008)
15. Liu, L., Yu, L.L., Kubatko, L., Pearl, D.K., Edwards, S.V.: Coalescent methods for estimating phylogenetic trees. Mol. Phylogenet. Evol. 53, 320–328 (2009)
16. Maddison, W.P.: Gene trees in species trees. Syst. Biol. 46(3), 523–536 (1997)
17. Mallet, J.: Hybridization as an invasion of the genome. Trends Ecol. Evol. 20(5), 229–237 (2005)
18. Mallet, J.: Hybrid speciation. Nature 446, 279–283 (2007)
19. Meng, C., Kubatko, L.S.: Detecting hybrid speciation in the presence of incomplete lineage sorting using gene tree incongruence: A model. Theor. Popul. Biol. 75(1), 35–45 (2009)
20. Moody, M.L., Rieseberg, L.H.: Sorting through the chaff, nDNA gene trees for phylogenetic inference and hybrid identification of annual sunflowers (Helianthus sect Helianthus). Molecular Phylogenetics And Evolution 64, 145–155 (2012)

21. Mossel, E., Roch, S.: Incomplete lineage sorting: consistent phylogeny estimation from multiple loci. IEEE/ACM Transactions on Computational Biology and Bioinformatics (TCBB) 7(1), 166–171 (2010)
22. Nakhleh, L.: Evolutionary phylogenetic networks: models and issues. In: Heath, L., Ramakrishnan, N. (eds.) The Problem Solving Handbook for Computational Biology and Bioinformatics, pp. 125–158. Springer, New York (2010)
23. Nakhleh, L.: Computational approaches to species phylogeny inference and gene tree reconciliation. Trends in Ecology & Evolution 28(12), 719–728 (2013)
24. Pollard, D.A., Iyer, V.N., Moses, A.M., Eisen, M.B.: Widespread discordance of gene trees with species tree in Drosophila: evidence for incomplete lineage sorting. PLoS Genet. 2(10), e173 (2006)
25. Rambaut, A.: Phylogen v1.1 (2012), http://tree.bio.ed.ac.uk/software/phylogen/
26. Rannala, B., Yang, Z.: Phylogenetic inference using whole genomes. Annu. Rev. Genomics Hum. Genet. 9, 217–231 (2008)
27. Rieseberg, L.H.: Hybrid origins of plant species. Annu. Rev. Ecol. Syst. 28, 359–389 (1997)
28. Staubach, F., Lorenc, A., Messer, P.W., Tang, K., Petrov, D.A., Tautz, D.: Genome patterns of selection and introgression of haplotypes in natural populations of the house mouse (mus musculus). PLoS Genet. 8(8), e1002891 (2012)
29. Syring, J., Willyard, A., Cronn, R., Liston, A.: Evolutionary relationships among Pinus (Pinaceae) subsections inferred from multiple low-copy nuclear loci. Am. J. Bot. 92, 2086–2100 (2005)
30. Takuno, S., Kado, T., Sugino, R.P., Nakhleh, L., Innan, H.: Population genomics in bacteria: A case study of staphylococcus aureus. Molecular Biology and Evolution 29(2), 797–809 (2012)
31. Than, C., Ruths, D., Innan, H., Nakhleh, L.: Confounding factors in HGT detection: statistical error, coalescent effects, and multiple solutions. J. Comput. Biol. 14, 517–535 (2007)
32. Than, C., Sugino, R., Innan, H., Nakhleh, L.: Efficient inference of bacterial strain trees from genome-scale multi-locus data. Bioinformatics 24, i123–i131 (2008)
33. White, M.A., Ane, C., Dewey, C.N., Larget, B.R., Payseur, B.A.: Fine-scale phylogenetic discordance across the house mouse genome. PLoS Genetics 5, e1000729 (2009)
34. Yu, Y., Barnett, R.M., Nakhleh, L.: Parsimonious inference of hybridization in the presence of incomplete lineage sorting. Systematic Biology 62, 738–751 (2013)
35. Yu, Y., Degnan, J.H., Nakhleh, L.: The probability of a gene tree topology within a phylogenetic network with applications to hybridization detection. PLoS Genetics 8, e1002660 (2012)
36. Yu, Y., Dong, J., Liu, K., Nakhleh, L.: Maximum likelihood inference of reticulate evolutionary histories. Proceedings of the National Academy of Sciences 111, 16448–16453 (2014)
37. Yu, Y., Ristic, N., Nakhleh, L.: Fast algorithms and heuristics for phylogenomics under ils and hybridization. BMC Bioinformatics 14, S6 (2013)
38. Yu, Y., Than, C., Degnan, J.H., Nakhleh, L.: Coalescent histories on phylogenetic networks and detection of hybridization despite incomplete lineage sorting. Systematic Biology 60, 138–149 (2011)

Predicting Protein Functions Based on Dynamic Protein Interaction Networks

Bihai Zhao[1,2], Jianxin Wang[1(✉)], Fang-Xiang Wu [1,3], and Yi Pan[4]

[1]School of Information Science and Engineering, Central South University,
Changsha 410083, China
jxwang@mail.csu.edu.cn
[2]Department of Information and Computing Science, Changsha University,
Changsha 410003, China
[3]Department of Mechanical Engineering and Division of Biomedical Engineering,
University of Saskatchewan, Saskatoon, SK S7N 5A9, Canada
[4]Department of Computer Science, Georgia State University, Atlanta, GA 30302-4110, USA

Abstract. Accurate annotation of protein functions plays a significant role in understanding life at the molecular level. With accumulation of sequenced genomes, the gap between available sequence data and their functional annotations has been widening. Many computational methods have been proposed to predict protein function from protein-protein interaction (PPI) networks. However, the precision of function prediction still needs to be improved. Taking into account the dynamic nature of PPIs, we construct a dynamic protein interactome network by integrating PPI network and gene expression data. To reduce the negative effect of false positive and false negative on the protein function prediction, we predict and generate some new protein interactions combing with proteins' domain information and protein complex information and weight all interactions. Based on the weighted dynamic network, we propose a method for predicting protein functions, named PDN. After traversing all the different dynamic networks, a set of candidate neighbors is formed. Then functions derived from the set of candidates are scored and sorted, according to the weighted degree of candidate proteins. Experimental results on four different yeast PPI networks indicate that the accuracy of PDN is 18% higher than other competing methods.

Keywords: Protein-protein interaction · Functions prediction · Dynamic networks · PDN

1 Introduction

Proteins are biological macromolecules responsible for a wide range of activities in living cells, tissues, organs, and bodies. Proteins carry out their functions in the context of environment they are in. This environment includes other macromolecules such as proteins, DNA, or RNA. The function annotation of a protein is an important issue in post-genomics due to the critical roles of proteins in various biological processes. General methods for protein function prediction are based on experimental or computational approaches. Although experimental techniques have been developed

© Springer International Publishing Switzerland 2015 2015
R. Harrison et al. (Eds.): ISBRA 2015, LNBI 9096, pp. 390–401, 2015.
DOI: 10.1007/978-3-319-19048-8_33

to predict protein functions, these methods cannot scale up to accommodate the vast amount of sequence data because of their inherent difficulties such as the requirement of well trained technicians, labors, and time consumptions. As a result, a wide range of computational approaches have been proposed in the last decades. Traditional computational methods such as identifying domains or finding Basic Local Alignment Search Tool (BLAST) [1] hit among proteins with experimentally determined functions. The recent availability of protein-protein interaction (PPI) data for many species has created new opportunities for function prediction. Many methods have been proposed to predict function from PPI) networks. Function annotation methods based on PPI networks can be classified into five categories: neighborhood counting methods [2-4], graph theoretic methods [5, 6], Markov random field based methods [7, 8], module-assisted methods [9, 10] and integrating multiple information sources methods [11, 12].

The neighborhood counting method determines functions of a protein based on known functions of its immediate neighborhood [2-4]. A PPI network can be modeled as a simple graph, in which a vertex represents a protein and an edge represents an interaction between two distinct proteins, so it is natural to apply graph algorithms for its functions analysis. Two main approaches have been suggested in this context: cut-based approaches [5] and a flow-based algorithm [6]. A lot of probabilistic methods to the annotation problem have been proposed [7, 8]. All of them rely on a Markovian assumption: the function of a protein is independent of all other proteins given the functions of its immediate neighbors. Module-assisted approaches [9, 10] predict functional modules of related proteins firstly and then annotate each module based on known functions of its members. A significant proportion of PPI networks obtained from high-throughput experiments have been found to contain false positives and false negatives, which have negative effects on the prediction of functions. To overcome these limitations, some researchers have integrated data from multiple sources for the annotation task. These approaches differ in the way the sources are combined. Network topological properties are combined with sequence, gene expression, domain information, protein complexes and others to predict protein function from PPI networks [11, 12]. Although a great progress has been made on the computational methods, the prediction of function based on PPI network is still very challenging.

While under new conditions or stimuli, not only the number and location of proteins would be changed, but also their interactions. A protein might interact with others in different phases or time points. Interactions are vital in every biological process in a living cell. Many significant molecular processes are performed by large and sophisticated multi-molecular machines such as anaphase-promoting complexes [14], RNA splicing [15], polyadenylation [16], protein exports and transport complexes [17]. Therefore, stable modules might be developed among proteins, while temporary and dynamic modules would be formed by proteins with changing interactions. Proteins would join in different modules in various phases of molecular process to perform some common functions with other ones. In other words, proteins would have diverse functions under new conditions or in different time points. Therefore, it is ineffective to predict protein functions based on static networks.

Interactions observed in experiments might not take place within organisms, or might occur in cells at certain moment or period. Interactions among proteins within biological networks would change with time, external conditions, stimuli or cell stages, which is the dynamic character of PPI networks. Therefore, an effective way to improve prediction performance is to construct dynamic protein interaction networks and predict dynamic biological modules under different conditions or in different time points to conduct protein function annotations. The experiment results in our previous research prove that predicting functions is imperfect only based on PPI networks or certain protein features, because functions between proteins might be overlapped and differentiated.

Yook et al. [18] have concluded that the most functional classes appear in the form of isolating subnets within PPI networks. Current function prediction methods are based on the analysis of static networks, without exploring meta-organizations as well as interaction subnets or considering dynamic features of modules and functions. In this paper, we construct a dynamic interactome network by combining PPI network and gene expression data. Based on the constructed dynamic network, a new method is proposed to predict protein functions, called PDN (Predicting protein function based on Dynamic Network). Since PPI data obtained from high-throughput experiments might contain false negative to some extent, the PDN method integrates multiple biological data such as protein domain information and protein complexes, in order to increase accuracy of prediction. The results show that PDN outperforms current algorithms for protein function prediction.

2   Method

2.1   Construction of Dynamic Networks

Dynamic networks can be viewed as ordered graphic sequences of a complex network, i.e., snapshots [19] at different time points. The formal definition of dynamic networks follows as:

Definition 1: Dynamic Networks. A dynamic network G is defined as a series of networks $\{G_1, G_2,..., G_i,...,G_k\}$. $G_i= (V_i, E_i)$ is a sample network at the ith condition, where $V_i= \{v_{i1}, v_{i2},..., v_{in}\}$ represents a set of proteins, $E_i = \{e_{i1}, e_{i2},... , e_{im}\}$ represents a set of interactions. We assume that $e^+=(u, v)\in (E_i\backslash E_{i-1})$ represents a new interaction emerged at the ith condition relative to the (i-1) the condition, $e^-=(u, v)\in (E_{i-1}\backslash E_i)$ is an interaction existing at the $(i-1)$th condition while disappearing at the ith condition.

Dynamic networks could be classified into two categories: spatial dynamic networks and time dynamic networks, according to the sampling conditions. Spatial dynamic networks mean different interactions among proteins under diversely spatial or other conditions, like various cell locations. Time dynamic networks refer to different interactions among proteins in different sampling time points, for instance, different protein expression levels of genes or proteins at various moments lead to dynamic changes of interactions among proteins. This study mainly focuses on the construction of time dynamic networks and their applications to protein function prediction. Benjamin et al. have found that diploid yeast strain spontaneously begins respiratory cycles as measured by oxygen consumption, after growth to high density followed by a brief starvation period [20]. They have performed micro-array analysis of gene expression further and confirmed that genes are expressed periodically.

Gene expression data could reflect dynamic features of proteins under various conditions or at different phrases in a biological process. Thus gene expression data at varying time points or under different conditions could provide another way to explore dynamic changes of protein interactions. Tang [21] et al. have constructed time course protein-protein interaction (TC-PPI) networks by incorporating time course gene expression into PPI networks, and applied it successfully to the identification of function modules. As to two interactive proteins in PPI networks, if their gene expression data at one moment exceed a certain fixed threshold, it is believed that they co-express at the moment and there would be an edge in TC-PPI networks at that moment.

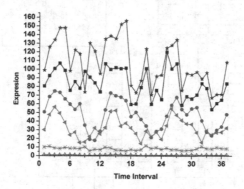

Fig. 1. Interval analysis of gene expression data

As shown in Figure 1, proteins expression would vary dramatically, with the minimum approaching to 0 and the maximum over 150. The expression patterns would also change with genes. For example, some proteins play an important role but have relatively low expression levels during the whole cell cycle. Hence, it is inappropriate for us to filter out all proteins using a uniform threshold when constructing a dynamic network. An active threshold for each protein based on the characteristics of its expression curve is more valid for investigating whether a protein is active at certain moment. Wang [22] et al. have proposed the three-sigma rule to construct dynamic networks, and obtained the thresholds based on the average values and standard deviations of gene expressions of specific proteins. In this paper, we design the threshold via expression levels of individual genes, which is different from the three-sigma rule. Gene expressions in 36 time points [20] could be divided into three cycles, i.e., every 12 time points per cycle. However, researchers used to construct a PPI network within 36 time points, greatly distinct from the network constructed per 12 time points in this paper. The value of gene expression at certain time point is the average value in three cycles, which is calculated as follows:

$$T'(i) = \frac{T(i) + T(i+12) + T(i+24)}{3} \quad (i \in [1,12]) \tag{1}$$

As for the threshold, it is simply set as the average expression value of each gene in this study. Figure 2 describes the process of a dynamic network construction. The first step, combine three cycles of gene expressions into one cycle according to equation (1) and filter out proteins with gene expression values below their average value. The second step, construct a dynamic network, if two genes co-express at the ith ($i \in [1,12]$) time point and interact with each other in PPI networks, an interaction would be added in the dynamic network at this time point.

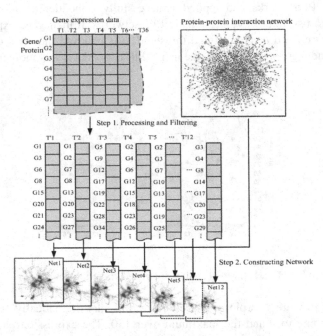

Fig. 2. Schematic of construction of dynamic networks

2.2   PDN Method

Researches [23] show that PPI data obtained through high-throughput biological experiments contains relatively high rates of false positives and false negatives. The false positives would be an obstacle to the precision of prediction algorithm, because the number of mismatched functions would rise. The false negatives would lead to the loss of interaction data, and continue to inhibit the increase of the number of functions correctly matched, so the recall is hard to be lifted. Thus reducing false positives and false negatives is the key to improve the performance of protein function prediction. Currently, a few function prediction algorithms have incorporated topological characteristics and multiple biological data, which aim at reducing the impact of false negatives by adding new protein interactions based on other biological data. However, those methods would inevitably result in the rise of false positives in the meanwhile. That is to say, the PPI data predicted on the basis of multiple biological data might be false positives.

This study aims at achieving the balance between false positives and false negatives by incorporating multiple biological data and constructing dynamic protein interaction networks. The multiple biological data includes proteins' domain, protein complexes and gene expression data. In our previous work [24], we have successfully predicted protein functions by combining proteins' domain and topological characteristics of PPI networks and improved the performance of prediction compared to other methods. Besides, domain is widely applied in prevailing methods to predict functions, such as Zhang [11], DCS [12], and PON [13], which demonstrate the close relationship between domains and functions of proteins.

To reduce the influence of false positives and false negatives, we utilize the information regarding protein domain as well as protein complexes, topological characteristics of PPI networks and the constructed dynamic networks. As to false negatives, we add interactions to the original PPI networks on the basis of domain and complex information; while as to false positives, we delete protein interactions having no co-expressions based on the newly constructed dynamic protein networks. The PDN algorithm is shown in Figure 3 and consists of three stages.

Input: A PPI network $G = (V, E)$

Output: The set of predicted functions PF

1. Generate a weighted network $G = (V, E, W)$ by Equation (2);
2. Construct a new weighted network $G' = (V, E', W')$ by Equation (5);
3. Get a set of dynamic networks $GS = \{G_1, G_2, ... , G_{12}\}$
4. FOR each un- annotated protein u DO
5.   Get a set of candidate proteins $P = \{p_1, p_2, ..., p_n\}$ Score_Protein(p_i)
6.   Get a set of candidate functions $F = \{f_1, f_2, ..., f_m\}$
7.   FOR each function f_i in F, calculate
8. $Score\_Function(f_j) = \sum_{i=1}^{n} Score\_Proten(p_i) \times t_{ij}, \quad (i \in [1, n])$
9.   END FOR
10.   Order all functions in F in descending order by their score
11. $PF = \{f_1, f_2, ..., f_k\}$;
12.   Output PF
13. END FOR

Fig. 3. PDN algorithm

The first stage of PDN, interaction weighting, weights all interactions using domain information, protein complex information and topological characteristics of PPI networks, removes and adds interactions. PDN weights interactions based on topology of networks firstly. Generally speaking, for a pair of interacting proteins, the probability of the interaction can be reflected by the number of common neighbors of them. PDN uses AdjustECC to calculate the weight of protein pairs, which is defined as:

$$AdjustECC(v_i,v_j)=\begin{cases}\dfrac{|N_i\cap N_j|^2}{(|N_i|-1)*(|N_j|-1)} & ,\ |N_i|>1\ and\ |N_j|>1 \\ 0 & ,\ |N_i|=1\ or\ |N_j|=1\end{cases}\tag{2}$$

Actually, AdjustECC is a variant of ECC, where N_i and N_j are the neighborhood sets of v_i and v_j, respectively. If there is not an interaction between v_i and v_j, AdjustECC $(v_i, v_j)=0$.

Next, PDN supplements interactions using domains information and protein complexes to the original PPI network. For a pair of proteins v_i and v_j, D_i and D_j are sets of domains of v_i and v_j, respectively. PD (v_i, v_j) represents the probability of sharing domains, which is calculated as follows:

$$PD(v_i,v_j)=\begin{cases}\dfrac{|D_i\cap D_j|^2}{|D_i|*|D_j|} & ,\ |D_i|>0\ and\ |D_j|>0 \\ 0 & ,\ |D_i|=0\ or\ |D_j|=0\end{cases}\tag{3}$$

In equation (3), $D_i\cap D_j$ denotes the set of common domains between v_i and v_j. In a similar way, PC (v_i, v_j), the probability of sharing complexes between v_i and v_j can be calculated as follow:

$$PC(v_i,v_j)=\begin{cases}\dfrac{|C_i\cap C_j|^2}{|C_i|*|C_j|} & ,\ |C_i|>0\ and\ |C_j|>0 \\ 0 & ,\ |C_i|=0\ or\ |C_j|=0\end{cases}\tag{4}$$

where C_i and C_j is the set of protein complexes that contained v_i and v_j, respectively and $C_i\cap C_j$ denotes the set of common protein complexes of them.

Given a pair of proteins v_i and v_j in the PPI network $G=(V, E)$, the weight of interaction between them is calculated as:

$$Weight(v_i,v_j)=\begin{cases}AdjustECC(v_i,v_j)+PD(v_i,v_j)+PC(v_i,v_j) & ,\ (v_i,v_j)\in E \\ PD(v_i,v_j)+PC(v_i,v_j) & ,\ (v_i,v_j)\notin E\end{cases}\tag{5}$$

$E'=\{(v_i,v_j)\,|\,v_i,v_j\in V,(v_i,v_j)\notin E, PD(v_i,v_j)+PC(v_i,v_j)>0\}$ is the set of new interactions generated according to domain information and protein complex information. $G'=(V, E\cup E', W)$ is the constructed network, where $W=\{w(e_1), w(e_2),\dots, w(e_m)\}$, $w(e_i)$ represents the weight of e_i.

The second stage, dynamic networks construction, takes as input the weighted PPI network, generates 12 dynamic weighted networks using the gene expression profiles according to equation (1). At the ith time point ($i\in[1,12]$), if a pair of proteins interact in the PPI network and co-express, an interaction would be added between them in the dynamic weighted network.

The third stage is predicting and scoring. The set of candidate proteins is formed according to neighbors of tested proteins from these 12 sub- networks. The algorithm scores for every candidate proteins. Scores of candidate proteins are the sum of weight of interaction between the candidate proteins and the testing proteins in all 12 sub- networks. Given a set of dynamic networks set $GS = \{G_1, G_2,..., G_{12}\}$, $G_i= (V, E_i, W_i)$ ($i\in [1,12]$), u is a testing protein with unknown functions, v is a protein with known functions in the PPI network. The score of v can be calculated by the following formula:

$$Score\_\mathrm{Pr}otein(v) = \sum_{i=1}^{12}Weight(v,u)\times t_i, \quad \text{where } t_i = \begin{cases} 1 & , \quad if\ (v,u)\in E_i \\ 0 & , \quad otherwise \end{cases} \qquad (6)$$

After the set of candidate proteins is formed, PDN algorithm generates a set of candidate functions and computes the ranking scores for candidate functions. It is assumed that $P= \{p_1, p_2,..., p_n\}$ is a set of candidate proteins , $F= \{f_1, f_2,..., f_m\}$ is a list of functions of all proteins in P. The score of a candidate function f_j in F can be calculated as follow:

$$Score\_Function(f_j) = \sum_{i=1}^{n}Score\_\mathrm{Pr}oten(p_i)\times t_{ij}, \quad (i\in [1,n]) \qquad (7)$$

In equation (7), if a protein p_i has the function f_j then t_{ij} equals to 1, otherwise t_{ij} equals to 0. PDN sorts all candidate functions in descending order by their ranking scores and selects top N of the ranked functions for the testing protein with unknown functions. $N= min\ (l, m)$, where l is the number of predicted functions and m is the maximum number of functions of candidate proteins.

3   Results and Discussion

3.1   Experimental Data

The *Saccharomyces cerevisiae* (yeast) PPI networks are widely used in the network-based function prediction methods, because the species of yeast has been well characterized by knockout experiments and is the most complete and convincible. Here, we also adopt the yeast PPI network to test our method. The DIP [25] dataset, updated to Oct.1, 2014, consists of 5,017 proteins and 23,115 interactions among proteins. The self-interactions and the repeated interactions are filtered out in DIP data. The annotation data of proteins used for method validation is the latest version (2012.3.3) downloaded from GO official website [26]. To avoid too special or too general, only those GO terms that annotate at least 10 and at most 200 proteins will be kept in the experiments. After processing by this step, the number of GO terms is 267.

The domain data is derived from Pfam database [27], including 1107 different types of domains among 3,042 proteins. As for the protein complex information, we used the dataset CYC2008 [28] which consists of 408 protein complexes involving 1,492 proteins in the yeast PPI network. The gene expression data of yeast [20] for the construction of dynamic networks contains 6,776 genes and 36 samples in total, with 4,898 genes involved in the DIP network. For proteins which have no corresponding gene expression data, we simply set zero values. In addition, the GO data and Pfam domain data are transformed to use the ensemble genome protein entries because the original PPI network uses such a labeling system.

In order to evaluate the performance of our proposed new protein function prediction method, PDN, We have applied our method and other five competing algorithms, including Zhang [11], DCS [12], DSCP [12], NC [2] and PON [13], to the datasets mentioned above. DSCP is a variant of DSC, which combines protein complex information. NC is a classic functions prediction algorithm while Zhang, DCS, DSCP, NC and PON integrate domain information.

3.2   Cross Validation

The proteins in the PPI network are partitioned into two subsets, the training set and the testing set. Functions are hidden from the part of proteins in the PPI network artificially. These proteins consist of the testing set and the rest proteins form the training set. Functions of proteins in the testing set are predicted, using functional information of proteins in the training set. Then predicted functions are compared with actual functions to evaluate the performance of protein function prediction algorithms. In this study, we put one protein into the testing set and the remaining proteins into the training set. Each function of proteins in the testing set is assigned with a probability, according to the functions of proteins in the training set. Then a number of top-ranking functions are selected to annotate the protein with unknown functions. The quality of prediction depends on the matching results of predicted functions with actual ones. There are two widely used criteria to measure the predicted results. The one is the precision which measures the percentage of predicted functions that match the known functions. The other is the recall which measures the fraction of known functions that are matched by the predicted ones. They can be calculated as follows:

$$\mathrm{Precison} = \frac{TP}{TP+FP}, \qquad \mathrm{Recall} = \frac{TP}{TP+FN} \qquad (8)$$

where TP is the number of predicted functions matched by known functions. FP is the number of predicted functions that are not matched by known functions. FN is the number of known functions that are not matched by predicted functions. F-measure, as the harmonic mean of precision and recall, is another measure to evaluate the performance of a method synthetically.

In the DIP PPI network, 2870 proteins among all 5017 proteins have been annotated by known functions. We analyze the overall prediction performance of PDN and other five methods for these 2870 proteins, firstly. Figure 4 shows the average precision, recall and F-measure of various algorithms.

From Figure 4 we can see that PDN archives the highest precision and F-measure, the second-highest recall after NC. F-measure of PDN is 86.04%, 31.95%, 109.64%, 184.83% and 15.4% higher than Zhang, DCS, NC, PON and DSCP, respectively. PDN is the only one method that F-measure over 0.4 among all methods. At the same time, we consider the number of proteins annotated at least one function correctly. The number of proteins which have been matched at least one function by

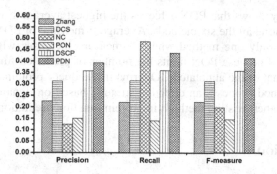

Fig. 4. Overall performance comparisons of various algorithms

PDN, Zhang, DCS, NC, PON and DSCP is 1491, 810, 1148, 1619, 572 and 1298, respectively. PDN archives the second largest number of matched proteins after NC. The number of matched proteins of PDN is 84.07%, 29.88%, 160.66% and 14.87% more than Zhang, DCS, PON and DSCP, respectively. Thus it can be seen that PDN has the highest accuracy than other algorithms.

All these methods take the different strategies for selecting the number of functions for proteins. It may not be sufficient to evaluate these methods by comparing their precision and recall directly. To have a more objective comparison of the performances of these methods, we adopt the F-measure curve to evaluate the global performance of each method in terms of the different strategies of function selection adopted by these six prediction methods. We try to choose the same number of functions for each method, i.e. to choose for each protein respectively the top K functions of each prediction method. As for Zhang, DCS and DSCP methods, we choose the top M ($M<=K$) proteins with the highest similarity value and then select the top K functions from the function list as predictive functions which are in descending order according to the maximum value of protein similarity. As for PDN, NC and PON methods, we choose the top K GO terms to assign functional annotations for those unknown proteins (K ranges from 1 to 50). Average F-measure will be calculated respectively under different K values and shown in Figure 5.

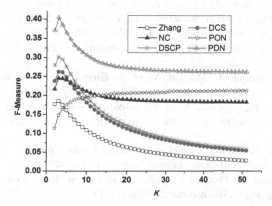

Fig. 5. Comparison of average F-measure of various methods

Figure 5 clearly shows that PDN achieves the highest average F-measure from beginning to end among all the six methods. Average F-measure of PDN falls into (0.26, 0.4]. PON is the only one method whose F-measure increases with increasing the value of K. This is because PON limits the number of predicted functions to be less than or equal to that of the annotated GO terms in the query protein. All of facts confirm that our method proposes an effective strategy based on dynamic PPI networks and outperforms other existing methods in protein function prediction.

4   Conclusion

At present, plenty of recognizable genes and proteins have not gained characteristics in experiments, so their functions have been unknown yet. The algorithm of functional prediction is generally based on static PPI network. However, interactions change with time, external conditions, stimulations and different stages of cells. Combining PPI network with multiple sources of biological data, we have constructed a dynamic weighting PPI network by using the time course gene expression profiles, and put forward a function prediction method named PDN aiming at reducing the negative effects of false positives and false negatives in PPI networks for protein function prediction. The experimental results have verified the effectiveness of PDN method.

Acknowledgements. This work is supported in part by the National Natural Science Foundation of China under Grant No. 61232001, No. 61128006, No. 61073036, No. 61003124, No. 61428209 and the Pro-gram for New Century Excellent Talents in University under Grant NCET-10-0798, NCET-12-0547.

References

1. Jones, D.T., Swindells, M.B.: Getting the most from PSI–BLAST. Trends in Biochemical Sciences 27(3), 161–164 (2002)
2. Schwikowski, B., Uetz, P., Fields, S.: A network of protein–protein interactions in yeast. Nature Biotechnology 18, 1257–1261 (2000)
3. Hishigaki, H., Nakai, K., Ono, T., et al.: Assessment of prediction accuracy of protein function from protein–protein interaction data. Yeast 18(6), 523–531 (2001)
4. Chua, H.N., Sung, W.K., Wong, L.: Exploiting indirect neighbours and topological weight to predict protein function from protein-protein interactions. Bioinformatics 22(13), 1623–1630 (2006)
5. Vazquez, A., Flammini, A., Maritan, A., et al.: Global protein function prediction from protein-protein interaction networks. Nature Biotechnology 21(6), 697–700 (2003)
6. Nabieva, E., Jim, K., Agarwal, A., et al.: Whole-proteome prediction of protein function via graph-theoretic analysis of interaction maps. Bioinformatics 21(suppl. 1), i302–i310 (2005)
7. Deng, M., Zhang, K., Mehta, S., et al.: Prediction of protein function using protein-protein interaction data. Journal of Computational Biology 10(6), 947–960 (2003)
8. Letovsky, S., Kasif, S.: Predicting protein function from protein/protein interaction data: a probabilistic approach. Bioinformatics 19(suppl. 1), i197–i204 (2003)

9. Yeang, C.H., Mak, H.C., McCuine, S., et al.: Validation and refinement of gene-regulatory pathways on a network of physical interactions. Genome Biol. 6, R62 (2005)
10. Scott, J., Ideker, T., Karp, R.M., et al.: Efficient algorithms for detecting signaling pathways in protein interaction networks. J. Comput. Biol. 13, 133–144 (2006)
11. Zhang, S., Chen, H., Liu, K., Sun, Z.: Inferring protein function by domain context similarities in protein-protein interaction networks. BMC Bioinformatics 10, 395 (2009)
12. Peng, W., Wang, J., Cai, J., et al.: Improving protein function prediction using domain and protein complexes in PPI networks. BMC Systems Biology 8(1), 35 (2014)
13. Liang, S., Zheng, D., Standley, D.M., et al.: A novel function prediction approach using protein overlap networks. BMC Systems Biology 7(1), 61 (2013)
14. Peters, J.M.: The anaphase-promoting complex: proteolysis in mitosis and beyond. Molecular Cell 9(5), 931–943 (2002)
15. Black, D.L.: Mechanisms of alternative pre-messenger RNA splicing. Annual Review of Biochemistry 72(1), 291–336 (2003)
16. LaCava, J., Houseley, J., Saveanu, C., et al.: RNA degradation by the exosome is promoted by a nuclear polyadenylation complex. Cell 121(5), 713–724 (2005)
17. Spirin, V., Mirny, L.A.: Protein complexes and functional modules in molecular networks. Proceedings of the National Academy of Sciences 100(21), 12123–12128 (2003)
18. Yook, S.H., Oltvai, Z.N., Barabási, A.L.: Functional and topological characterization of protein interaction networks. Proteomics 4(4), 928–942 (2004)
19. Mantzaris, A.V., Bassett, D.S., Wymbs, N.F., et al.: Dynamic network centrality summarizes learning in the human brain. Journal of Complex Networks 1(1), 83–92 (2013)
20. Tu, B.P., Kudlicki, A., Rowicka, M., et al.: Logic of the yeast metabolic cycle: temporal compartmentalization of cellular processes. Science 310(5751), 1152–1158 (2005)
21. Tang, X., Wang, J., Liu, B., et al.: A comparison of the functional modules identified from time course and static PPI network data. BMC Bioinformatics 12(1), 339 (2011)
22. Wang, J., Peng, X., Peng, W., et al.: Dynamic protein interaction network construction and applications. Proteomics 14(4-5), 338–352 (2014)
23. Zhao, B.H., Wang, J.X., Li, M., et al.: Detecting Protein Complexes Based on Uncertain Graph Model. IEEE/ACM Transactions on Computational Biology and Bioinformatics 11(3), 486–497 (2014)
24. Zhao, B.H., Wang, J.X., Li, M., et al.: Prediction of essential proteins based on overlapping essential modules. IEEE Transactions on NanoBioscience 13(4), 415–424 (2014)
25. Xenarios, X., et al.: DIP: the database of interacting proteins. Nucleic Acids Research 28, 289–291 (2000)
26. Ashburner, M., Ball, C.A., Blake, J.A., et al.: Gene Ontology: tool for the unification of biology. Nature Genetics 25(1), 25–29 (2000)
27. Finn, R.D., Mistry, J., Tate, J., et al.: The Pfam protein families database. Nucleic Acids Research, gkp985 (2009)
28. Pu, S., Wong, J., Turner, B., et al.: Up-to-date catalogues of yeast protein complexes. Nucleic Acids Research 37(3), 825–831 (2009)

An Iterative Approach for Phylogenetic Analysis of Tumor Progression Using FISH Copy Number

Jun Zhou[1,3], Yu Lin[2], William Hoskins[1], and Jijun Tang[1(✉)]

[1] University of South Carolina, Columbia, SC 29205, USA
jtang@cse.sc.edu
[2] University of California, San Diago, La Jolla, CA 92093, USA
[3] Tianjin Key Laboratory of Cognitive Computing and Application, Tianjin University, Tianjin 300072, China

Abstract. Copy number variants are an underlying factor in human evolution and in many diseases, especially in cancer. Tumors generally contain cells with a varying number of gene copies, and the variance in the number of gene copies follows a pattern formed by an evolutionary process. The Fluorescence in situ hybridization (FISH) provides researchers a reliable technique to measure the copy numbers of preselected genes in a group of cells. Recently, Chowdhury et al. successfully modeled the progression of tumor progression using FISH copy number to the Rectilinear Steiner Minimum Tree (RSMT) problem, and proposed both exact and heuristic algorithms to reconstruct phylogenetic trees modeling the development of cancer cell patterns [1]. We proposed new heuristics to attack the RSMT problem, which is inspired by iterative approaches to approximate solutions to the Steiner tree in the "small phylogeny" problem [2,3]. Experimental results from both simulated and real tumor data show that our approach outperforms the previous heuristic algorithm in approximating better solutions for the RSMT problem.

Keywords: FISH · Tumor phylogentic inference · Median problem · Small phylogeny problem

Introduction

It is well known that the occurrence of cancer is driven by somatic genetic changes, including single-nucleotide variants, insertions and deletions, copy number aberrations, structural variants, and gene fusions [4]. Unique cancer clones are created as a consequence of accumulating changes in the progeny of a single most recent common ancestor. Ongoing linear and branching evolution results in the creation of multiple varying subclones, mapping the evolutionary history of a tumor to a tree [5]. The clonal evolution tree is shaped by genetic changes, as well as competition between subclones under the pressures of environmental selection. Inferring the evolutionary history of a particular type of tumor can help pinpoint important changes that lead to the recurrence of some genome aberrations [6]. There has already been a lot of research on identifying important

© Springer International Publishing Switzerland 2015
R. Harrison et al. (Eds.): ISBRA 2015, LNBI 9096, pp. 402–412, 2015.
DOI: 10.1007/978-3-319-19048-8_34

genes that are related to the evolution of cancer and may contribute developing better cancer treatment, among which phylogenetic inference plays a significant role [7,8,9,10].

During tumor development, the gene copy number can increase or decrease, due to various mechanisms [11,12,13]. FISH (Fluorescent In Situ Hybridization) developed by bio-medical researchers in the early 1980s has been used to detect and localize the presence or absence of specific DNA sequences and to visualize the genomic diversity of chromosome aberrations [14]. While the single cell sequencing (SCS) technique has the potential to count the number of specific genes or specific regions for a group of cells, the highly non-uniform coverage, the admixture signal, and relatively high cost make the current SCS technique not very suitable to resolve accurate gene copy numbers.

Phylogenetic inferences have facilitated the study of cancer initiation, progression, treatment, and resistance by regarding cancer as the product of evolutionary processes [15] and are usually limited to providing an average signal from a population of cells based on sequencing data [16,17]. Although FISH yields single-cell resolution profiles, previous studies are limited to a small number of preselected gene probes, e.g., two [18] and three [8]. More recently, Chowdhury et al. successfully modeled the progression of tumor progression using FISH copy number to the Rectilinear Steiner Minimum Tree (RSMT) problem, and proposed both exact and heuristic algorithms to reconstruct phylogenetic trees modeling the development of cancer cell patterns [1]. However, both algorithms do not scale well with the number of gene probes, making them impractical to handle dozens of gene probes — a typical number of genes in one complicated signal pathway. A similar model based on the Steiner Minimum Tree has also been introduced to study the "small phylogeny" problem at both the sequence level [2] and the gene order level [3]. A special case of the "small phylogeny" problem is called the median problem — given three sequences (or permutations), find the configuration of a median genome to minimize the sum of the pairwise distances between the median and three input ones [19]. Sankoff et al. proposed iterative approaches to approximate solutions to the Steiner tree, which iteratively solve the median problem for one internal vertex at a time, and to make improvement until a local optimum is found [2,3]. We propose similar heuristics to approximate solutions to the RSMT problem through iteratively optimizing the median version of RSMT problem. Moreover, our iterative approach contains new procedures for the generation of median instances and the order of iterative optimizations of median instances, which generalizes the previous approaches of using median solvers to approximate solutions to the Steiner tree [2,3] and takes into consideration specific characterization and challenging in the RSMT problem. Experimental results from both simulated and real tumor data show that our approach outperforms the previous heuristic algorithm in approximating better solutions for the RSMT problem.

Methods

Below we described our approach for building a phylogenetic tree by using copy number change information from FISH data. In FISH data, such as cancer data, each cell has some non-negative integer count of each gene probe. Given two cell count patterns $(x_1, x_2, \ldots, x_d)$ and $(y_1, y_2, \ldots, y_d)$, the pairwise distance under the rectilinear metric (or L_1 distance) is defined as $|x_1 - y_1| + |x_2 - y_2| + \ldots + |x_d - y_d|$, where $x_i, y_i \in \mathbb{N}$. The weight of a tree with nodes labeled by cell count patterns is defined as the sum of all branch lengths under the rectilinear metric. Since the distance between two cell count patterns under the rectilinear metric represents the number of single gene duplication and loss events between them, a minimum weight tree, including Steiner nodes if needed, explains the k observed cell count patterns d probes with minimum total number of single gene duplication and loss events, from a single ancestor, e.g., cell count pattern with a copy number count of 2 for each gene probe (i.e., a healthy diploid cell) [1]. The RSMT problem is NP-complete [20] and Chowdhury *et al.* proposed an inefficient exact algorithm and a heuristic algorithm based on the median-joining algorithm for maximum parsimony phylogenetics [1].

Rectilinear Steiner Minimum Tree (RSMT) Problem

The RSMT problem for phylogenetic inference from FISH data is defined as follows.

Definition: RSMT(k, d)

Input: FISH data of k cell count patterns on d probes for a given patient

Output: A minimum weight tree with the rectilinear metric (or L_1 distance) including all the observed k cell count patterns and, as needed, unobserved Steiner nodes along with their cell count patterns for d probes

Median Version and Iterative Optimization for the RSMT Problem

The median version of RSMT problem can be solved in linear time.

Theorem 1. *RSMT(3,d) can be solved in time $O(d)$.*

Proof. Given three original cell count patterns $(x_1^1, x_2^2, \ldots, x_d^3)$, $(x_1^2, x_2^2, \ldots, x_d^2)$ and $(x_1^3, x_2^3, \ldots, x_d^3)$, RSMT$(3, d)$ returns a cell count pattern $(m_1, m_2, \ldots, m_d)$ such that $\sum_{i=1}^{3} \sum_{j=1}^{d} |x_j^i - m_j|$ is minimized, where $x_j^i, m_j \in \mathbb{N}$. Since the count for each gene probe is independent, we can optimize m_j independently which minimizes $\sum_{i=1}^{3} |x_j^i - m_j|$, respectively, and m_j simply equals to the median of x_j^1, x_j^2 and x_j^3. Thus $(m_1, m_2, \ldots, m_d)$ can be constructed in time $O(d)$ and if it differs from all three input cell count patterns then a Steiner node with cell count pattern $(m_1, m_2, \ldots, m_d)$ has to be introduced. On the other hand,

$\sum_{j=1}^{d} min_{y \in \mathbb{N}} \sum_{i=1}^{3} |x_j^i - y|$ is a lower bound for the minimum weight of any Steiner

tree on three input cell count patterns, and $\arg\min_{y \in \mathbb{N}} \sum_{i=1}^{3} |x_j^i - y| = m_j$, thus

the above construction is optimum under the rectilinear metric. $\square$

Two instances of RSMT(3,d) are shown in Figure 1. Given three cell count patterns in Figure 1(a), a Steiner node is introduced in Figure 1(b) with reduced the weight of the tree (i.e., the number of single gene duplication and loss events) from 7 to 4. Figure 1(c) shows an instance that no Steiner node is introduced.

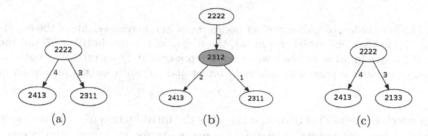

(a)   (b)   (c)

Fig. 1. Instances of RSMT(3,4) and the introduction of the Steiner node as the median. Each white node represents an input cell count pattern, and each green node represent an inferred Steiner node. Branch lengths are shown in Blue.

In the study of "small phylogeny" problem, Sankoff et al. studied iterative approaches to approximate solutions to the Steiner tree, which solve the median problem for one internal vertex at a time, and iteratively make improvement until a local optimum is found [2,3]. For each internal node in the current (binary) tree, the input for a median instance consists its three immediate neighbors [3]. While the above generation of median instances works for iterative approaches in attacking the "small phylogeny" problem in which the tree topology is fixed, it is not directly applicable to the study of the RSMT problem since the solution for the RSMT problem may change the tree topology by adding Steiner nodes to the input tree. Our approach thus checks all potential triplets in the tree, instead of only the triplets introduced by immediate neighbors of internal nodes, which is more suitable to deal with varying tree topologies and may also help to escape from the local optimum.

The order how the Steiner nodes are added to the tree may also affect the weight of the resulting tree. Figure 2(a) shows the original tree before iterative optimization, and Figure 2 (b) and (c) show the introduction of Steiner nodes through two different orders. Compared to Figure 2 (c), Figure 2 (b) first introduces a Steiner node 21422282 which prevents adding new potential Steiner nodes in the later stage. We define an interference score for each potential Steiner node to model the interference between potential Steiner nodes. The *Steiner count* of any node in the current tree is defined as the number of

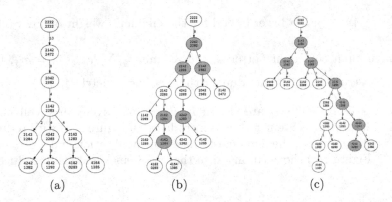

(a)   (b)   (c)

Fig. 2. Different orders of adding Steiner nodes result in different weights of the resulting trees. From the same initial tree in (a), the weight of the tree in (b) is 37 and the weight of the tree in (c) is 35. Each white node represents an input cell count pattern, and each green node represent an inferred Steiner node. Branch lengths are shown in Blue.

triplets which contains this node and requires the introduction of a Steiner node to optimize the tree weight. The *interference score* for each potential Steiner node with respect to a triplet is thus defined as the sum of *Steiner counts* of three nodes in that triplet. At each iteration, the potential Steiner node with minimum *interference score* is added to minimize the inference upon other potential Steiner nodes with respect to the current tree. An example is shown in Figure 3.

Our iterative algorithm starts from a Minimum Spanning tree built from the set of input cell count patterns, select a median instance at a time, and iteratively

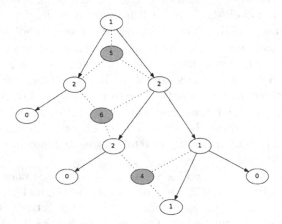

Fig. 3. The definition of Steiner count of the node in the current tree and the interference score of potential Steiner nodes to be added. Each input cell count pattern is represented by a white node labeled by its Steiner counts, and each potential green node is represented by a green node labeled by its interference score.

make improvement until a local optimum is found. The detailed description is shown in Algorithm 1.

Algorithm 1: An iterative algorithm to approximate solutions for $\text{RSMT}(k, d)$

Input: a set of k cell count patterns on d gene probes

Output: a tree with additional Steiner nodes if needed and k nodes that correspond to k input cell count patterns respectively

Initialization: the initial tree T_0 = a Minimum Spanning tree on k cell count patterns under the rectilinear metric

Iteration: from tree $T_i(V_i)$ on node set V_i to $T_{i+1}(V_{i+1})$ on node set V_{i+1}

  Identify the set S of the potential Steiner nodes from all possible triplets in T_i

  While S is not empty

    Select the potential Steiner node p with minimum interference score in S

    Build a Minimum Spanning tree on $\{V_i \cup p\}$ as $T(V_i \cup p)$

    If the weight of $T(V_i \cup p)$ is lower than the weight of $T_i(V_i)$, restart the iteration

      $T_{i+1}(V_{i+1}) = T(V_i \cup p)$

    Else

      $S = S \setminus \{p\}$

Exit condition: S is empty

Datasets

We used both real cervical cancer and breast cancer data samples and simulation samples generated through the same process described in the the supplemental material of the previous study by Chowdhury *et al.* [1]. The cervical cancer data contain four gene probes LAMP3, PROX1, PRKAA1 and CCND1, and the breast cancer data contain eight gene probes COX-2, MYC, CCND1, HER-2, ZNF217 ,DBC2, CDH1 and p53. All those genes are chosen because they are considered as important factors for cancer growth inhibition or promotion. The cervical cancer data is from 16 lymph positive patients (both primary and metastasis tumors) and 15 lymph negative patients, making 47 samples in total. The breast cancer data is from 12 patients with both IDC and DCIS and 1 patient with only DCIS, making 25 samples in total. More details of this FISH data set can be found in Chowdhury *et al.* [1].

Chowdhury *et al.* proposed an inefficient exact algorithm and an efficient heuristic algorithm to reconstruct phylogenetic trees modeling the development of cancer cell patterns [1]. Since the inefficient exact algorithm can not finish most of the test samples with a reasonable amount of time, we compare our iterative approach to the efficient heuristic algorithm [1]. In the following text, we refer to the efficient heuristic algorithm as FISHtrees [1], and refer to our iterative approach as iFISHtrees.

Real Cancer Dataset

There are 25 data samples for the breast cancer dataset, and our iterative approach iFISHtrees performs better than FISHtrees in 14 sample, ties in 10 samples, and performs worse in 1 sample. Table 1 summarizes the comparison between FISHtrees and iFISHtrees (ties are not included due to the space limit and the better tree weights are shown in bold). Figure 4 shows two trees constructed by FISHtrees and our iFISHtrees reconstructed from the DCIS cancer sample from patient 13, respectively. For example, the Steiner node 44423334 is introduced by iteratively checking all potential triplets in iFISHtrees, which allows iFISHtrees to escape from the local optimum that has trapped FISHtrees.

Table 1. Comparison on the real dataset for real breast cancer samples

Case #	Initial		FISHtrees		iFISHtrees	
	Node #	Tree weight	Node #	Tree weight	Node #	Tree weight
B1_IDC	119	230	135	213	132	**212**
B1_DCIS	143	259	158	**241**	159	242
B2_IDC	104	238	124	217	123	**216**
B3_DCIS	106	72	80	100	80	**98**
B4_IDC	110	232	129	214	129	**213**
B6_IDC	85	116	90	112	90	**111**
B7_IDC	59	128	73	116	71	**113**
B7_DCIS	76	202	84	186	83	**184**
B9_IDC	94	251	121	222	119	**217**
B9_DCIS	76	177	89	164	89	**162**
B10_DCIS	95	154	89	146	89	**145**
B11_DCIS	80	144	87	136	84	**135**
B12_IDC	112	212	124	201	123	**200**
B13_IDC	84	140	92	133	92	**131**
B13_DCIS	43	66	47	63	47	**62**

Similarly, our iterative approach iFISHtrees performs better than FISHtrees in 15 sample, ties in 30 samples, and performs worse in 2 samples, out of 47 samples in the cervical cancers datasets. Table 2 summarizes the comparison between FISHtrees and iFISHtrees (ties are not included due to the space limit and the better tree weights are shown in bold).

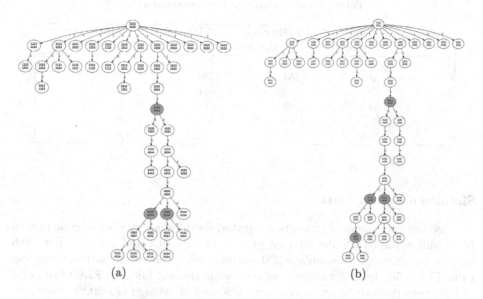

(a)   (b)

Fig. 4. Phylogenetic trees constructed by `FISHtrees` (a) and `iFISHtrees` (b) from the DCIS breast cancer sample of patient 13, respectively. Each node in the tree is labeled by a cell count pattern of eight gene probes COX-2, DBC2, MYC, CCND1, CDH1, p53, HER-2 and ZNF217. Each white node represents an input cell count pattern, and each green node represent an inferred Steiner node. Branch lengths are shown in Blue.

Table 2. Comparison on the real dataset for real cervical cancer samples

Case #	Initial		FISHtrees		iFISHtrees	
	Node #	Tree weight	Node #	Tree weight	Node #	Tree weight
C5	140	208	153	**195**	151	196
C9	130	144	131	143	132	**142**
C10	72	87	72	87	73	**86**
C12	63	72	63	72	64	**71**
C15	66	75	67	74	68	**73**
C21	63	77	67	**73**	65	74
C27	49	60	50	59	52	**57**
C29	76	85	78	83	78	**82**
C32	160	216	167	209	169	**207**
C34	67	88	72	83	73	**82**
C37	71	74	72	73	73	**72**
C42	157	207	164	199	166	**198**
C45	126	183	136	172	140	**169**
C46	87	116	92	110	93	**109**
C49	128	166	132	162	133	**161**
C53	64	82	66	80	66	**79**
C54	123	152	129	146	130	**145**

Table 3. Comparison on simulated datasets

Probe #	Growth factor	FISHtrees =iFISHtrees	FISHtrees >iFISHtrees	FISHtrees <iFISHtrees
4	0.4	176	23	1
6	0.4	161	30	9
8	0.4	162	31	7
4	0.5	182	18	0
6	0.5	160	31	9
8	0.5	162	32	6

Simulated Cancer Data

We also test on simulated datasets generated for different number of gene probes (4, 6 and 8) and for different tree growth factors (0.4 and 0.5) [1]. For each pair of parameters, we simulate 200 samples with cell count patterns varying from 75 to 150. Table 3 summarizes the comparison of between FISHtrees and iFISHtrees from these simulation datasets, and in average iFISHtrees outperforms FISHtrees on all of them. Moreover, we also generate simulated datasets for relatively larger number of gene probes (e.g., 12 and above), FISHtrees started taking too much time to produce solutions while our iterative approach iFISHtrees still scales well with dozens of gene probes, and thus we did not include the comparison results for larger number of gene probes.

Conclusions

Chowdhury et al. successfully modeled the progression of tumor progression using FISH copy number to the Rectilinear Steiner Minimum Tree (RSMT) problem, and proposed both exact and heuristic algorithms to reconstruct phylogenetic trees modeling the development of cancer cell patterns [1]. We show that the RSMT problem can be solved in linear time when there are only three input cell count patterns. Inspired by the iterative approaches to approximate solutions to the Steiner tree in the "small phylogeny" problem [2,3], we propose a new iterative algorithm to approximate solutions of the RSMT problem. Moreover, our new iterative approach extends the generation of median instances, and also takes into account the order of iterative optimizations of median instances. problem. Experimental results from both simulated and real tumor data show that our approach outperforms the previous heuristic algorithm in approximating better solutions for the RSMT problem and may provide insights into more likely tumor progression pathways. Chowdhury et al. recently constructed a phylogenetic model of tumor progression that include copy number changes not only at the scale of single genes, but also at the scale of entire chromosomes and the whole genome [21]. Extensions of our iterative approach for this more general model are possible, but remain to be thoroughly tested.

Acknowledgements. We thank Lingxi Zhou, Bin Feng,and Yan Zhang for helpful comments. JZ, WH and, JT were funded by NSF IIS 1161586 and an internal grant from Tianjin University, China. YL was supported by a fellowship of the Swiss National Science Foundation (grant no. 154563). The funders had no role in study design, data collection and analysis, decision to publish, or preparation of the manuscript.

References

1. Chowdhury, S.A., Shackney, S.E., Heselmeyer-Haddad, K., Ried, T., Schäffer, A.A., Schwartz, R.: Phylogenetic analysis of multiprobe fluorescence in situ hybridization data from tumor cell populations. Bioinformatics 29(13), 189–198 (2013)
2. Sankoff, D., Cedergren, R.J., Lapalme, G.: Frequency of insertion-deletion, transversion, and transition in the evolution of 5s ribosomal rna. Journal of Molecular Evolution 7(2), 133–149 (1976)
3. Blanchette, M., Bourque, G., Sankoff, D.: Breakpoint phylogenies. Genome Informatics 8, 25–34 (1997)
4. Futreal, P.A., Coin, L., Marshall, M., Down, T., Hubbard, T., Wooster, R., Rahman, N., Stratton, M.R.: A census of human cancer genes. Nature Reviews Cancer 4(3), 177–183 (2004)
5. Yates, L.R., Campbell, P.J.: Evolution of the cancer genome. Nature Reviews Genetics 13(11), 795–806 (2012)
6. Baudis, M.: Genomic imbalances in 5918 malignant epithelial tumors: an explorative meta-analysis of chromosomal CGH data. BMC Cancer 7(1), 226 (2007)
7. Pleasance, E.D., Cheetham, R.K., Stephens, P.J., McBride, D.J., Humphray, S.J., Greenman, C.D., Varela, I., Lin, M.-L., Ordóñez, G.R., Bignell, G.R., et al.: A comprehensive catalogue of somatic mutations from a human cancer genome. Nature 463(7278), 191–196 (2009)
8. Martins, F.C., De, S., Almendro, V., Gönen, M., Park, S.Y., Blum, J.L., Herlihy, W., Ethington, G., Schnitt, S.J., Tung, N., et al.: Evolutionary pathways in BRCA1-associated breast tumors. Cancer Discovery 2(6), 503–511 (2012)
9. Navin, N., Krasnitz, A., Rodgers, L., Cook, K., Meth, J., Kendall, J., Riggs, M., Eberling, Y., Troge, J., Grubor, V., et al.: Inferring tumor progression from genomic heterogeneity. Genome Research 20(1), 68–80 (2010)
10. Cheng, Y.-K., Beroukhim, R., Levine, R.L., Mellinghoff, I.K., Holland, E.C., Michor, F.: A mathematical methodology for determining the temporal order of pathway alterations arising during gliomagenesis. PLOS Computational Biology 8(1), 1002337
11. Caldecott, K.W.: Single-strand break repair and genetic disease. Nature Reviews Genetics 9(8), 619–631 (2008)
12. Hastings, P., Lupski, J.R., Rosenberg, S.M., Ira, G.: Mechanisms of change in gene copy number. Nature Reviews Genetics 10(8), 551–564 (2009)
13. Cleaver, J.E., Lam, E.T., Revet, I.: Disorders of nucleotide excision repair: the genetic and molecular basis of heterogeneity. Nature Reviews Genetics 10(11), 756–768 (2009)
14. Langer-Safer, P.R., Levine, M., Ward, D.C.: Immunological method for mapping genes on drosophila polytene chromosomes. Proceedings of the National Academy of Sciences 79(14), 4381–4385 (1982)
15. Attolini, C.S.-O., Michor, F.: Evolutionary theory of cancer. Annals of the New York Academy of Sciences 1168(1), 23–51 (2009)

16. Greenman, C.D., Pleasance, E.D., Newman, S., Yang, F., Fu, B., Nik-Zainal, S., Jones, D., Lau, K.W., Carter, N., Edwards, P.A., et al.: Estimation of rearrangement phylogeny for cancer genomes. Genome Research 22(2), 346–361 (2012)
17. Shlush, L.I., Chapal-Ilani, N., Adar, R., Pery, N., Maruvka, Y., Spiro, A., Shouval, R., Rowe, J.M., Tzukerman, M., Bercovich, D., et al.: Cell lineage analysis of acute leukemia relapse uncovers the role of replication-rate heterogeneity and microsatellite instability. Blood 120(3), 603–612 (2012)
18. Pennington, G., Smith, C.A., Shackney, S., Schwartz, R.: Reconstructing tumor phylogenies from heterogeneous single-cell data. Journal of Bioinformatics and Computational Biology 5(02a), 407–427 (2007)
19. Fertin, G.: Combinatorics of Genome Rearrangements, pp. 133–149 (2009)
20. Garey, M.R., Johnson, D.S.: The rectilinear steiner tree problem is NP-complete. SIAM Journal on Applied Mathematics 32(4), 826–834 (1977)
21. Chowdhury, S.A., Shackney, S.E., Heselmeyer-Haddad, K., Ried, T., Schäffer, A.A., Schwartz, R.: Algorithms to model single gene, single chromosome, and whole genome copy number changes jointly in tumor phylogenetics. PLOS Computational Biology 10(7), 1003740 (2014)

Phenome-Based Gene Discovery Provides Information about Parkinson's Disease Drug Targets

Yang Chen[1,2] and Rong Xu[1(✉)]

[1] Center for Clinical Investigation, School of Medicine, Taipei, Taiwan
[2] Department of Electrical Engineering and Computer Science, School of Engineering
Case Western Reserve University, Cleveland OH, 44106, USA

Abstract. We propose a phenome-based strategy to identify novel disease associated genes for PD, and investigated the translational potential of the predicted genes in drug discovery. Different from previous studies, we incorporated multiple disease phenotypic similarity networks, and integrated them with a genetic network to infer novel candidate genes. We validated the approach in two experiments: the 15 known PD genes from OMIM were averagely ranked within the top 0.8%, and the top-ranked genes were enriched for 669 PD genes from GWAS. In addition, our approach prioritized the target genes for both FDA-approved PD drugs and candidate PD drugs in clinical trials. The result provides empirical evidence that our computational gene prediction approach has the translational potential in PD drug discovery.

1 Introduction

Parkinson's disease (PD) is a common neurodegenerative disorder with unclear disease mechanisms and limited effective drug treatments [1]. Detecting novel disease genes is useful in understanding PD and identifying new drug targets. Systematically studying disease phenotypes have the potential to uncover the underlying genetic factors for PD. However, disease phenotype data used in current gene prediction approaches remain largely incomplete. Recently, we explored a new phenotype data source in biomedical ontologies and constructed the disease manifestation network (DMN). We showed that DMN contains new knowledge [2] and is useful in disease gene prediction [3]. In this study, we integrated DMN with a widely-used phenotype network provided by human phenotype ontology (HPO), and identify candidate PD genes using the combined network. We demonstrated that the PD genes predicted by our phenome-based approach can provide information for PD drug targets.

2 Method

Our approach consists of two parts: (1) predict genes for PD and (2) investigate the translational potential of the predicted genes. We downloaded the phenotype network of HPO (7395 nodes) and DMN (2312 nodes), and connected them using maps between disease nodes. We also built a gene network (17,831 nodes) using

© Springer International Publishing Switzerland 2015
R. Harrison et al. (Eds.): ISBRA 2015, LNBI 9096, pp. 413–414, 2015.
DOI: 10.1007/978-3-319-19048-8

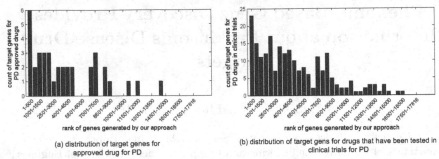

(a) distribution of target genes for approved drug for PD

(b) distribution of target gens for drugs that have been tested in clinical trials for PD

Fig. 1. Distribution of target genes for approved PD drugs and candidate PD drugs that are tested in clinical trials

1,971,371 gene functional relationships from STRING. We extracted 4021 and 1872 disease-gene associations from OMIM to connect HPO and DMN to the gene network, respectively.

We used PD and its 15 associated genes (in OMIM) as the seeds, and iteratively updated the score for every gene node at step k: $p_{k+1} = (1-\gamma)M^T p_k + \gamma p_0$, where p_0 is a vector of initial scores for each node, γ is the probability of restarting from the seeds, and M is the transition matrix of the entire combined network, which was normalized following the method in [3].

We validated our method in two ways: testing the ranks of seed genes in a "leave-one-out" cross validation, and examining the ranks of 669 PD associated genes obtained from genome-wide association studies (GWAS). Then we predict genes for PD and evaluated if the top-ranked genes are enriched for the genes targeted by the drugs that have been approved for PD and have been tested for PD in clinical trials.

3   Results

The top-ranked genes are relevant to PD. In the cross validation, the average rank for the retained seed genes is 147 (within the top 0.8%). The top 500 genes contain 99 PD genes from GWAS, which is a 4.95-fold enrichment compared with the average of 1000 random rankings ($p < e^{-8}$).

The top-ranked predicted genes provide information of PD drug targets. Our approach prioritized the drug targets for approved PD drugs (fig. 1(a)) and candidate PD drugs in clinical trials fig. 1(b). In addition, the top 500 genes are enriched for novel drug target genes, which offer unique opportunities for identifying candidate PD drugs through drug repositioning.

References

1. Olanow, C.W., Stern, M.B., Sethi, K.: The scientific and clinical basis for the treatment of Parkinson disease. Neurology 72(21 suppl.4), S1–S136 (2009)
2. Chen, Y., Zhang, X., Zhang, G.Q., Xu, R.: Comparative analysis of a novel disease phenotype network based on clinical manifestations. Journal of Biomedical Informatics 53, 113–120 (2014)
3. Chen, Y., Li, L., Zhang, G.Q., Xu, R.: Phenome-driven Disease Genetics Prediction Towards Drug Discovery. ISMB/ECCB (accepted, 2015)

Estimating Features with Missing Values and Outliers: A Bregman-proximal Point Algorithm for Robust Non-negative Matrix Factorization with Application to Gene Expression Analysis

Stéphane Chrétien[1]([✉]), Christophe Guyeux[2], Bastien Conesa[3],
Régis Delage-Mouroux[4], Michèle Jouvenot[4],
Philippe Huetz[5], and Françoise Descôtes[6]

[1] Laboratoire de Mathématiques de Besançon, Université de Franche-Comté,
16, route de Gray, 25000 Besançon, France
stephane.chretien@univ-fcomte.fr
[2] FEMTO-ST Institute, UMR 6174 CNRS, DISC Computer Science Department
Université de Franche-Comté, 16, route de Gray, 25000 Besançon, France
[3] ISIFC, 23 Rue Alain Savary, 25000 Besançon, France
[4] Université de Franche-Comté, EA 3922/IFR133 Université de Franche-Comté -
UFR Sciences et Techniques EA 3922/IFR133 - 25030 Besançon, France
[5] ABC&T, 33 rue Charles Nodier, 25000 Besançon, France
[6] Service de Biochimie et Biologie Moléculaire Sud, Pavillon 3D, Centre Hospitalier
Lyon Sud, 69495, Pierre Bénite Cedex, France

Finding a relevant dictionary for extracting the relevant features in a dataset is a very important task for many applications. The Non-negative Matrix Factorization (NMF) is a recent and very efficient method for achieving this goal in the case of non-negative data. Given a dataset consisting of n vectors $x_1, \ldots, x_n$ in $\mathbb{R}^d$, the NMF approach consists of building a matrix M whose columns are $x_1, \ldots, x_n$ and then factorize this matrix as

$$M = UV^t + E,$$

where E is an error term, U and V are componentwise non-negative, and U has a small number of columns. The columns of U represent the "features" present in the dataset and the interpretation of this decomposition is that each data consists of a mixture of the discovered features.

Since its study by Lee and Seung [7] in the late 90's the method, first explored in the chemometrics community, enjoyed a significant gain of interest from many application fields and especially in machine learning. It has been successfully used for document clustering [11], email surveillance [1], hyperspectral image analysis [5], face recognition [4], blind source separation [3], etc. It has recently also been applied to microarray data analysis [6] and biomedicine [8].

The approach proposed in the present article relies on Bregman-proximal iterations. Our goal is to extend the method to the case where data may be missing and/or corrupted by the occurrence of outliers. Our approach borrows ideas from robust PCA [2], where the matrix to approximate is decomposed into a low rank part and a sparse part:

© Springer International Publishing Switzerland 2015
R. Harrison et al. (Eds.): ISBRA 2015, LNBI 9096, pp. 415–416, 2015.
DOI: 10.1007/978-3-319-19048-8

$$M = L + S.$$

The low rank part L is intended to approximate the data set which is supposed to be of low rank, and the sparse part S represents the outliers. In the seminal article [2], the noise is not taken into account. However, in datasets such as gene expression data, the noise may be very large and one has to search for a low rank solution that removes the noise at the same time. In the present work, we propose an efficient method that denoises the data, guesses the missing values, and detects the outliers in the matrix M while performing a low-rank non-negative matrix factorization of the recovered matrix. For this purpose, we use a mixture of Bregman proximal methods and of the Augmented Lagrangian scheme as it is used in the Alternating Direction of Multipliers Method (ADMM). This mixture is also justified by [10], which presents a clear interpretation of the ADMM in terms of proximal method-type iterations.

References

1. Berry, M.W., Browne, M.: Email Surveillance Using Non-negative Matrix Factorization. Computational and Mathematical Organization Theory 11(3), 249–264 (2005)
2. Candès, E.J., Li, X., Ma, Y., Wright, J.: Robust Principal Component Analysis? Journal of ACM 58(1), 1–37 (2010)
3. Chan, T.H., Ma, W.K., Chi, C.Y., Wang, Y.: A convex analysis framework for blind separation of non-negative sources. IEEE Trans. on Signal Processing 56(10), 5120–5134 (2008)
4. Guillamet, D., Vitrià, J.: Non-negative matrix factorization for face recognition. In: Escrig, M.T., Toledo, F.J., Golobardes, E. (eds.) CCIA 2002. LNCS (LNAI), vol. 2504, pp. 336–344. Springer, Heidelberg (2002)
5. Jia, S., Qian, Y.: Constrained non-negative matrix factorization for hyperspectral unmixing. IEEE Trans. on Geoscience and Remote Sensing 47(1), 161–173 (2009)
6. Kim, H., Park, H.: Sparse non-negative matrix factorizations via alternating non-negativity-constrained least squares for microarray data analysis. Bioinformatics 23(12), 1495–1502 (2007)
7. Lee, D.D., Sebastian Seung, H.: Learning the parts of objects by non-negative matrix factorization. Nature 401(6755), 788–791 (1999)
8. Li, Y., Sima, D., Van Cauter, S., Croitor Sava, A., Himmelreich, U., Pi, Y., Van Huffel, S.: Hierarchical non-negative matrix factorization (hNMF): a tissue pattern differentiation method for glioblastoma multiforme diagnosis using MRSI. NMR in Biomedicine 26(3), 307–319 (2013)
9. Li, L., Lebanon, G., Park, H.: Fast Bregman divergence NMF using Taylor expansion and coordinate descent. In: Proc. of the 18th ACM SIGKDD Int. Conf. on Knowledge Discovery and Data Mining, pp. 307–315 (2012)
10. Parikh, N., Boyd, S.: Proximal Algorithms. Foundations and Trends in Optimization 1(3), 123–231 (2014)
11. Xu, W., Liu, X., Gong, Y.: Document clustering based on non-negative matrix factorization. In: Proceedings of the 26th Annual International ACM SIGIR Conference on Research and Development in Information Retrieval, pp. 267–273. Association for Computing Machinery, New York (2003)

Conservation and Network Analysis of the (4β+α) Fold of the Immunoglobulin-Binding B1 Domain of Protein G to Elucidate the Key Determinants of Structure, Folding and Stability

Jason C. Collins, John Bedford, and Lesley H. Greene[✉]

Old Dominion University, Department of Chemistry and Biochemistry Norfolk,
VA 23529, Norfolk, USA
{jccollin,jbedford,lgreene}@odu.edu

Keywords: GB1 · Protein Folding · Networks · Evolutionary Conservation

Introduction

Proteins fold from an ensemble of denatured states by a restriction of conformational space to form the initial native-like topology followed by further stabilizing secondary and tertiary interactions. It is believed that the formation of the initial native-like topology is guided by an evolutionarily conserved set of amino acids. Residues are typically conserved in a superfamily of proteins because they make critical interactions that are more important in maintaining the common fold. This could lead to residues clustering together in a hydrophobic core to stabilize the initial native-like topology [1, 2]. This network of conserved amino acids has been the target of computational and experimental research which seeks to investigate the link between conserved amino acids and how they facilitate rapid and correct folding of a protein into its native state [3, 4, 5]. Using bioinformatics approaches we can determine and assess which amino acids are conserved for the fold of a protein. This type of analysis is highly useful and important in understanding the tertiary structure of proteins and becomes significantly more powerful when supported with experimental data. The application of network science has also become important in the study of protein structure and folding [2, 3, 6, 7, 8, 9].

Results and Discussion

Our model system is the *Streptococcal* B1 immunoglobulin-binding domain of protein G (GB1) with a T2Q mutation to prevent methionine excision [10]. GB1 is a small, 56 amino acid bacterial immunoglobulin-binding protein with a 4β+α fold. This fold is composed of a two-layer sandwich consisting of a four-stranded β-sheet that packs against an α-helix. Using several bioinformatics approaches we investi-

R. Harrison et al. (Eds.): ISBRA 2015, LNBI 9096, pp. 417–419, 2015.
DOI: 10.1007/978-3-319-19048-8

gated which residues are key determinants in forming this fold. We identified nine structurally conserved amino acids using a conservation analysis and propose that they are critical to forming and stabilizing the 4β+α fold. The nine conserved residues form a predominantly hydrophobic nucleus within the core of GB1 based on the average hydrophobicity analysis. A network analysis of all the long-range interactions in the structure of GB1 in concert with a Betweenness-centrality (BC) analysis was conducted. It revealed the relative significance of each conserved amino acid residue based on the number and location of the interactions. The BC analysis identified four nodes (Tyr3, Leu5, Phe52 and Val54) with high betweenness. Interestingly, the 4 amino acids are found on the N- and C-termini β-strands and appear to be important in bringing the two hairpins together. Additionally, three of the four amino acids in the GB1 network are in conserved residue positions in the superfamily. This result indicates that these positions appear to be more centrally important to the network and may be of higher importance, forming first in the folding process. However, experimental studies are necessary to truly determine if their hypothesized role in the formation of the 4β+α fold is supported.

In summary, figure 1 highlights the results our conservation and network studies. This computational analysis provides an important foundation for the design of experimental work which is critical to solving the protein folding problem.

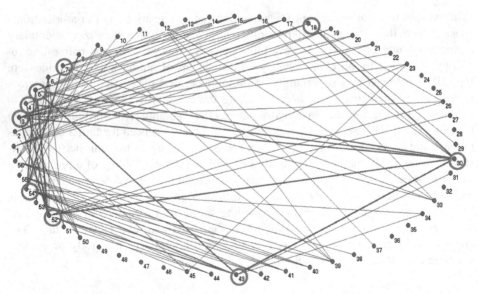

Fig. 1. Network of Long-Range Interactions in the Structure of GB1. Individual amino acids are the red colored nodes connected by long-range interactions shown as blue lines. Residue numbers are in black. The nine conserved amino acids based on a statistical conservation analysis are shown by orange circles. Red links indicate long-range interactions between the nine conserved amino acids. The methodology and equation can be found in references [2] and [11].

References

1. Schueler-Furman, O., Baker, D.: Conserved Residue Clustering and Protein Structure Prediction. Proteins 52, 225–235 (2003)
2. Greene, L.H., Hamada, D., Eyles, S.J., Brew, K.: Conserved Signature Proposed for Folding in the Lipocalin Superfamily. FEBS Lett. 553, 39–44 (2003)
3. Greene, L.H., Grant, T.M.: Protein Folding by 'Levels of Separation': A Hypothesis. FEBS Lett. 586, 962–966 (2012)
4. Higman, V.A., Greene, L.H.: Elucidation of Conserved Long-Range Interaction Networks in Proteins and Their Significance in Determining Protein Topology. Physica A 368, 595–606 (2006)
5. Li, H., Wojtaszek, J.L., Greene, L.H.: Analysis of Conservation in the Fas-Associated Death Domain Protein and the Importance of Conserved Tryptophans in Structure, Stability and Folding. BBA-Proteins Proteom. 1794, 583–593 (2009)
6. Greene, L.H.: Protein Structure Networks. Brief Funct Genomics 11, 469–478 (2012)
7. Rao, F., Caflisch, A.: The Protein Folding Network. J. Mol. Biol. 342, 299–306 (2004)
8. Lei, H., Su, Y., Jin, L., Duan, Y.: Folding Network of Villin Headpiece Subdomain. Biophys J. 99, 3374–3384 (2010)
9. Estrada, E.: Universality in Protein Residue Networks. Biophys J. 98, 890–900 (2010)
10. Bauer, M.C., Xue, W.F., Linse, S.: Protein Gb1 Folding and Assembly from Structural Elements. Int. J. Mol. Sci. 10, 1552–1566 (2009)
11. Collins, J.C.: Investigation into Protein Folding and Misfolding. Ph.D. Dissertation, Department of Chemistry and Biochemistry, Old Dominion University (2015)

Assessment of Transcription Factor Binding Motif and Regulon Transfer Methods

Sefa Kilic and Ivan Erill[✉]

University of Maryland Baltimore County
Department of Biological Sciences
1000 Hilltop Circle, Baltimore, Maryland 21250, USA
{sefa1,erill}@umbc.edu

Introduction. Comparative genomics has been leveraged in many studies to characterize transcriptional regulatory networks [1,2]. However, despite its fundamental importance in such studies, the effect of motif and regulon transfer methods remains largely unstudied. Thanks in large part to high-throughput experimental techniques, available experimental data has increased dramatically over the last few years and it has become possible for the first time to reliably assess methods used for regulatory network reconstruction. In this study, we describe three different transfer methods that define transcription factor (TF) binding motif in a target species given some regulatory activity information in a reference species. Motif-based transfer is performed using the reference binding motif to search for putative binding sites in the target genome on the assumption that, for a given TF, the binding motif is relatively well conserved across closely related species. This method has been shown to perform well at inferring existing regulatory networks in previously uncharacterized genomes [3,4]. The alternative source of prior information is the regulatory network itself. The putative regulon is then constructed based on orthologous transfer of the reference regulon and *de novo* motif discovery is performed on the promoter regions of putatively regulated target genes.

Methods. We compiled binding site data from several publicly available databases including CollecTF [5], a database of experimentally-validated sites. The first method that we tested is the direct transfer using the collection of known binding sites from a model species to build a position-specific scoring matrix (PSSM) which is used then to scan the promoter regions of the target genome to identify putative sites. The second method defines the motif by performing motif discovery on pre-searched candidate sequences. After the PSSM search, promoters with high scoring sites are given as input to the motif discovery algorithm with the motivation of capturing motifs slightly different from the reference one and mitigating the effects of inaccurate PSSM score threshold. The final method that we tested, called network transfer, does not assume motif conservation. The underlying hypothesis is that the regulon across two genomes might be functionally conserved to some degree even if the binding motif is not. To define the motif in target species through network transfer, the first step is to identify target regulon, the collection of genes that are orthologous to the ones in the reference regulon. In the next step, the promoters of operons in the target regulon are used for motif discovery. To assess the performance of different transfer methods quantitatively, we measured both

© Springer International Publishing Switzerland 2015
R. Harrison et al. (Eds.): ISBRA 2015, LNBI 9096, pp. 420–421, 2015.
DOI: 10.1007/978-3-319-19048-8

(a) Euclidean distance between the true motif and the inferred motif and (b) the area under ROC curve for the inferred motif. To assess the significance of performances, we computed the distance and area under ROC curve using a column-permuted version of the target motif as the inferred motif.

Results. We measured the performance of the transfer methods by applying them to all pairs of species with at least 10 binding sites for a particular TF, yielding 411 pairs of species belonging mostly to either Fur or LexA. Our results show that direct transfer and motif discovery on pre-searched promoters perform very similarly. Since these two methods are based on motif conservation, they perform well when the TF proteins in the reference and target species are highly similar. As the TF protein distance increases, their performances decrease dramatically. Although network transfer is capable of inferring conserved and non-conserved motifs for large protein distances in many cases, our permutation analysis showed that, overall, the network transfer method does not perform significantly well for any level of reference-target TF distance. This finding is consistent with previous studies reporting high plasticity in transcriptional regulatory networks [6]. Another reason for poor network transfer is the strictness of the orthology-based regulon transfer method. In the future, we intend to relax the network transfer method by using functional similarity (e.g., cluster of orthologous groups) for regulon transfer rather than direct orthology. Also, we plan to investigate whether combining the information from the extended network transfer with relaxed PSSM searches can enhance the performance of direct transfer as the similarity between reference and target motifs decays.

References

1. Ravcheev, D.A., Best, A.A., Sernova, N.V., Kazanov, M.D., Novichkov, P.S., Rodionov, D.A.: Genomic reconstruction of transcriptional regulatory networks in lactic acid bacteria. BMC Genomics 14, 94 (2013)
2. Meireles-Filho, A.C., Stark, A.: Comparative genomics of gene regulation conservation and divergence of cis-regulatory information. Current Opinion in Genetics & Development 19(6), 565–570 (2009)
3. Leyn, S.A., Kazanov, M.D., Sernova, N.V., Ermakova, E.O., Novichkov, P.S., Rodionov, D.A.: Genomic reconstruction of the transcriptional regulatory network in Bacillus subtilis. J. Bacteriol. 195(11), 2463–2473 (2013)
4. Leyn, S.A., Suvorova, I.A., Kholina, T.D., Sherstneva, S.S., Novichkov, P.S., Gelfand, M.S., Rodionov, D.A.: Comparative genomics of transcriptional regulation of methionine metabolism in Proteobacteria. PLoS ONE 9(11), e113714 (2014)
5. Kılıç, S., White, E.R., Sagitova, D.M., Cornish, J.P., Erill, I.: CollecTF: a database of experimentally validated transcription factor-binding sites in Bacteria. Nucleic Acids Research (2013)
6. Price, M.N., Dehal, P.S., Arkin, A.P.: Orthologous transcription factors in bacteria have different functions and regulate different genes. PLoS Comput Biol 3(9), e175 (2007)

Short Tandem Repeat Number Estimation from Paired-end Sequence Reads by Considering Unobserved Genealogy of Multiple Individuals

Kaname Kojima[✉], Yosuke Kawai, Naoki Nariai, Takahiro Mimori,
Takanori Hasegawa, and Masao Nagasaki[✉]

Department of Integrative Genomics, Tohoku Medical Megabank Organization,
Tohoku University, 2-1 Seiryo-machi, Aoba-ku, Sendai-shi, Miyagi 980-8573, Japan
{kojima,kawai,nariai,mimori,
t-hasegw,nagasaki}@megabank.tohoku.ac.jp
http://nagasakilab.csml.org/en/

The progress of next-generation sequencing (NGS) technologies enables whole genome sequencing for each individual in practical time and with reasonable cost. From NGS data, single nucleotide variants for more than a thousand of individuals were accurately detected in genome wide scale [6]. However, we still have difficulty in accurately detecting structural variations such as genome insertion and deletion, copy number variations, and short tandem repeat (STR) number polymorphisms, especially from low coverage NGS data. Repeat number polymorphisms are known to associate with various disease phenotypes such as the association of CAG repeat stretch in the Huntingtin gene with Huntington's disease. From NGS data, several approaches such as lobSTR [2] and RepeatSeq [3] have been proposed to estimate repeat numbers in STR regions by counting repeat patterns included in aligned sequence reads. Although these approaches can accurately detect STR variants and estimate their repeat numbers, STR regions longer than the length of sequence reads cannot be handled. Another strategy is to use paired-end reads aligned flanking regions of the target repeat in the reference genome [1]. Insert size inferred from the aligned paired-end reads is longer than its actual size if the repeat number is smaller than that in the reference genome and shorter if the repeat number is larger. Thus, the insert size inferred from the paired-end reads can be used to estimate repeat numbers. Since insert size is longer than sequence reads, this strategy can be used for estimating repeat numbers for relatively long STR regions that cannot be handled by the strategy counting repeat patterns in sequence reads. However, the accuracy of estimated repeat numbers from insert size is not high especially for low coverage NGS data, compared to the strategy counting repeat patterns.

We propose a new statistical model named coalescentSTR, which considers the unobserved genealogy of multiple individuals to estimate repeat numbers for these individuals using insert size information from paired-end reads. In the model, the genealogy is handled as coalescent trees, which describe the ancestral history of genomes for multiple individuals backward in time [4]. Because the change in repeat numbers in genealogy is naturally considered in our model, more accurate estimation of repeat numbers is expected. For the estimation of

© Springer International Publishing Switzerland 2015
R. Harrison et al. (Eds.): ISBRA 2015, LNBI 9096, pp. 422–423, 2015.
DOI: 10.1007/978-3-319-19048-8

the model, coalescent trees are sampled from phased genotypes around a target STR region with the Markov chain Monte Carlo (MCMC) method. We also propose a new belief propagation method that can calculate the loopy belief propagation and the mixed-product belief propagation [5] while handling sampled coalescent trees as hidden variables. By using the proposed belief propagation, we estimate repeat numbers by searching approximated maximum configuration of the posterior. In a simulation study with synthetically generated NGS data for STR regions mostly longer than read length, we compare the performance of coalescentSTR, coalescentSTR with genealogies from randomly shuffled haplotypes, coalescentSTR without genealogy information, lobSTR, RepeatSeq, and STRViper on datasets with various numbers of individuals. In the performance evaluation, a root mean squared error (RMSE) between true and estimated repeat numbers is considered. The effectiveness of our approach is also verified with real whole exome datasets of 33 HapMap JPT individuals in the 1000 Genomes Project (1KGP).

CoalescentSTR provides the least RMSE values in both simulation and real datasets. In addition, coalescentSTR with shuffled individuals provides worse results than coalescentSTR on average, and hence the effectiveness of considering genealogy is verified. From the comparison of computational time, coalescentSTR requires the most computational time, and it is mainly taken by sampling coalescent trees with MCMC. One idea for resolving this issue is to use MCMC with approximate Bayesian computation in order to avoid the calculation of likelihood for each sampled tree, which mainly takes the computational time in sampling. Although the recombination of genomes is usually not considered in coalescent theory, the use of ancestral recombination graph is one of the solutions for handling larger size of genome structural variations such as large size copy number variations. We are planning to extend our approach to handle larger size of genome events in future work.

References

1. Cao, M.D., et al.: Inferring short tandem repeat variation from paired-end short reads. Nucleic Acids Research 42(3) (2014)
2. Gymrek, M., et al.: lobSTR: A short tandem repeat profiler for personal genomes. Genome Research 6, 1154–1162 (2012)
3. Highnam, G., et al.: Accurate human microsatellite genotypes from high-throughput resequencing data using informed error profiles. Nucleic Acids Research 4(1) (2013)
4. Kingman, J.F.C.: On the genealogy of large populations. Journal of Applied Probability 19(A), 27–43 (1982)
5. Liu, Q., Ihler, A.: Variational algorithms for marginal MAP. Journal of Machine Learning Research 14, 3165–3200 (2013)
6. 1000 Genomes Project Consortium et al.: An integrated map of genetic variation from 1,092 human genomes. Nature 491(7422), 56–65 (2012)

PnpProbs: Better Multiple Sequence Alignment by Better Handling of Guide Trees

Yongtao Ye, Tak-Wah Lam, and Hing-Fung Ting

HKU-BGI Bioinformatics Algorithms and Core Technology Research Lab,
Department of Computer Science, University of Hong Kong, Hong Kong, China
{ytye,twlam,hfting}@cs.hku.hk

Due to the huge computational requirement for constructing optimal multiple sequence alignments, much research has been done on designing heuristics for generating good, but not necessarily optimal, alignments. One of the most widely used heuristics is *progressive alignments* [1], which (i) constructs optimal pairwise alignment for every pair of the input sequences, (ii) using these pairwise alignments constructs a guidetree, and (iii) based on the "order" suggested by the guidetree constructs a multiple alignment progressively as follows: the most related sequences are aligned first, and the more distant ones are aligned later. This paper proposes a general method for improving MSA tools that adopt this progressive alignment heuristic, or more specifically, proposes an adaptive approach for improving step (ii) for guidetree construction.

We first classify input sequences into two types: they are *normally related* if their similarity is above some threshold (default value is 18%); otherwise they are *distantly related*. For normally related sequences, we try to generate better guide trees for them. Note that many progressive alignment tools use the UPGMA method to construct guide trees, which merges clusters of sequences (represented as subtrees) into larger clusters (subtrees) iteratively, and in each iteration, the pair of clusters with the smallest distance will be chosen and merged. The distance $d_{\ell k}$ between clusters C_ℓ and C_k, where C_k is resulted from the merging of clusters C_i and C_j, is defined to be

$$d_{\ell k} = \frac{d_{\ell i}|C_i| + d_{\ell j}|C_j|}{|C_i| + |C_j|}, \qquad (*)$$

where $|C|$ denotes the size, i.e., the number of sequences, in cluster C. Intuitively, $d_{\ell k}$ estimates the average cost (e.g., the average number of gapped columns needed to be introduced) for aligning C_ℓ to a single sequence in C_k. This definition works fine in general, but it may not be the best for inputs with low *discrepancy*, i.e., for those sequences that differ mostly in some common regions, and are more or less identical in all the other positions. For these low discrepancy sequences, we propose to use the following distance definition: $d_{\ell k} = (d_{\ell i} + d_{\ell j})/2$. We think that (*) may overestimate the influence of the sizes, especially when one cluster is much larger than the other. To see why, suppose that C_i has low discrepancy. Then, its sequences look like one "meta-sequence" and the same set of gapped columns would be enough to align most of them; the average cost is not very sensitive to C_i's size. Note that (*) is reduced to our definition when C_i and C_j contain only one (meta-)sequence.

© Springer International Publishing Switzerland 2015
R. Harrison et al. (Eds.): ISBRA 2015, LNBI 9096, pp. 424–426, 2015.
DOI: 10.1007/978-3-319-19048-8

Table 1. Summary of results. Columns show the average sum of pairs scores (SP) and total column scores (TC) multiplied by 100. The best results in each column are shown in bold. The second best results in each column are marked with *. Table (a) groups the results according to the similarity of the families in OXBench. For (b), the Twilight Zone contain families with no more than 25% similarity, and the Superfamily contains those with similarities mostly between 20%-50%. For (c), RV11 contains families with less than 20%, and RV11 those with similarity between 20%-40%. The last column of table (a) shows the running time on OXBench (395 families) using a single CPU thread. All tools ran with default parameters.

(a) Mean SP and TC scores on OXBench

	ALL(0-100%)		0%-20%		20%-50%		50%-100%		Time
	SP	TC	SP	TC	SP	TC	SP	TC	mm:ss
PNPProbs	**90.41**	**82.23**	**48.98**	**24.88**	83.47*	**68.79**	**98.05**	**95.18**	2:58
GLProbs	90.38*	82.14*	47.29*	22.95*	**83.48**	68.65*	**98.05**	**95.18**	3:15
MSAProbs	90.07	81.75	44.83	22.08	82.77	67.74	98.01	95.08	4:04
Probalign	89.97	81.68	43.58	20.51	82.53	67.46	**98.05**	**95.18**	2:10
CONTRAlign	89.34	79.87	44.76	17.83	81.56	64.75	97.55	94.10	10:19
ProbCons	89.68	80.86	44.15	20.30	82.06	66.33	97.84	94.61	1:48
MUSCLE	89.50	80.67	45.64	21.90	81.75	66.15	97.63	94.28	0:19
MAFFT	88.00	77.96	37.82	13.27	78.99	60.86	97.41	93.68	0:19
T-Coffee	89.52	80.50	43.99	19.11	81.82	65.85	97.75	94.38	15:05
ClustalΩ	88.91	79.99	39.09	16.38	80.71	64.49	97.76	94.58	0:12
ClustalW	89.43	80.16	42.94	18.23	81.67	65.01	97.76	94.40	0:22
PicXAA	89.64	80.74	45.11	22.04	81.86	65.91	97.84	94.55	4:26
DIALIGN	83.97	72.41	26.03	8.07	72.67	52.57	95.21	89.54	3:17
Align-m	86.95	76.06	28.36	12.74	76.35	57.54	96.95	92.60	21:14

(b) Mean SP and TC scores on SABmark

	ALL		Twilight Zone		Superfamily	
	SP	TC	SP	TC	SP	TC
PnpProbs	61.37*	**41.70**	**44.40**	**24.80**	67.19*	**47.49**
GLProbs	**61.42**	41.36*	44.35*	24.30*	**67.27**	47.21*
MSAProbs	60.27	40.02	42.97	22.88	66.20	45.90
Probalign	59.53	38.63	42.42	22.64	65.39	44.11
CONTRAlign	57.45	35.59	39.01	17.69	63.77	41.73
ProbCons	59.69	39.17	42.81	22.78	65.47	44.79
MUSCLE	54.51	33.47	34.69	16.96	61.29	39.13
MAFFT	52.63	32.57	31.72	15.17	59.79	38.53
T-Coffee	59.14	39.53	41.66	23.29	65.13	45.10
ClustalΩ	55.02	35.47	35.55	18.10	61.69	41.42
ClustalW	51.92	31.37	31.45	15.09	58.93	36.95
PicXAA	59.37	39.11	41.05	21.51	65.65	45.14
DIALIGN	47.09	27.11	27.85	12.73	53.69	32.05
Align-m	46.19	31.07	25.72	16.28	53.21	36.14

(c) Mean SP and TC scores on BALiBASE

	ALL		RV11		RV12	
	SP	TC	SP	TC	SP	TC
PnpProbs	82.80*	**68.00**	68.91	**45.73**	94.79*	87.23*
GLProbs	**83.20**	67.59*	**69.72**	44.68	**94.84**	**87.38**
MSAProbs	82.35	66.83	68.13	44.02	94.63	86.52
Probalign	82.53	67.27	69.50*	45.34*	94.63	86.20
CONTRAlign	77.59	58.10	61.78	35.60	91.23	77.52
ProbCons	81.55	65.22	66.99	41.68	94.12	85.54
MUSCLE	75.60	58.27	57.15	32.06	91.53	80.89
MAFFT	72.46	52.58	52.96	26.19	89.30	75.38
T-Coffee	80.82	64.93	65.63	41.36	93.94	85.29
ClustalΩ	75.96	59.38	59.01	36.21	90.60	79.38
ClustalW	69.63	49.21	50.06	22.99	86.52	71.84
PicXAA	81.33	66.08	66.56	44.06	93.47	84.19
DIALIGN	68.63	48.22	49.72	26.81	84.18	65.81
Align-m	71.45	56.04	51.88	33.06	88.36	75.88

For distantly related sequences, they are not similar at all, except for some small local domains or motifs embedded in many long divergent regions. We find that for these sequences progressive alignment heuristic always introduces many mis-aligned columns, and early mistakes cannot be corrected and would be propagated and cause more mistakes for later alignments. Thus, for distantly related sequences, we propose to abandon the progressive heuristic and use instead non-progressive methods, which do not depend on global pairwise alignments and guidetrees. In particular, we propose to use the sequencence annealing technique [2] to this kind of sequences.

We have implemented a MSA tool called PnpProbs by modifying our previous progressive alignment tool GLProbs [3] according to our aforementioned ideas. We have tested PnpProbs extensively on three popular benchmark alignment databases, namely BAliBASE, OXBench and SABmark, comparing its performance with that of a dozen other leading multiple sequence alignment tools. As shown in Table 1, the quality of PnpProbs's alignments is in general signficantly better than those generated by the other tools, specially for distantly related sequences.

References

1. Feng, D.-F., Doolittle, R.F.: Progressive sequence alignment as a prerequisitetto correct phylogenetic trees. Journal of Molecular Evolution 25(4), 351–360 (1987)
2. Schwartz, A.S., Pachter, L.: Multiple alignment by sequence annealing. Bioinformatics 23(2), e24–e29 (2007)
3. Ye, Y., Cheung, D.W., Wang, Y., Yiu, S.-M., Zhan, Q., Lam, T.-W., Ting, H.-F.: GLProbs: Aligning multiple sequences adaptively. In: Proceedings of the International Conference on Bioinformatics, Computational Biology and Biomedical Informatics, p. 152. ACM (2013)

A Novel Method for Predicting Essential Proteins Based on Subcellular Localization, Orthology and PPI Networks

Gaoshi Li[1,2], Min Li[1], Jianxin Wang[1(✉)], and Yi Pan[1,3]

[1]School of Information Science and Engineering,
Central South University, Changsha 410083, China
[2]Guangxi Key Lab of Multi-source Information Mining and Security,
Guangxi Normal University, Guilin 541004, China
[3]Department of Computer Science, Georgia State University, Atlanta, GA 30302-4110, USA

Currently, a lot of computational methods to identify essential proteins have been presented. The typical network-based essential protein discovery methods include Degree Centrality (DC) [1], Betweenness Centrality (BC) [2], Closeness Centrality (CC) [3], Subgragh Centrality (SC) [4], Eigenvector Centrality (EC) [5], Information Centrality (IC) [6] and Edge Clustering Coefficient Centrality (NC) [7], etc. At present, in order to achieve higher accuracy, researchers try to combine the topological characters with different biological information, such as subcellular localization [8], evolutionary conservation [9], expression level [10]. For example, PeC[11] predicts essential proteins by using PPI network topology information and gene expression profiles. ION [12] combines the topological characters of PPI networks with orthologous information.

In this paper, by integrating subcellular localization, orthology with PPI network, a novel method, named SON, is proposed to predict essential proteins. Firstly, the relationship between subcellular localization, orthology and essentiality of proteins are analyzed. Based on the relation between subcellular localization and topology of PPI networks, the subcellular localization score is calculated. Orthologous score is the same as that introduced in ION [12]. The definition of edge clustering coefficients in [7] is used. . Calculation model of sorting score is an expansion of random walk model that is linear combination of three values. The essentiality of each protein is calculated by a linear combination of the subcellular localization score, orthologous score and NC.

To validate the effectiveness of the proposed method SON, we test SON by using the PPI network of S.cerevisiae. There are total of 5093 proteins and 24743 interactions in PPI network data set of S.cerevisiae. Essential proteins data set is integrated from MIPS, SGD, DEG and SGDP. There are 1167 essential proteins in PPI network in total. Subcellular localization data set includes 5095 yeast proteins and 206831 subcellular localization records. After preprocessing, there are still 3923 proteins in PPI network that have subcellular localization data. Orthologous proteins data set is taken from Version 7 of InParanoid that contains a set of pairwise comparisons between 100 whole genomes. The experimental results of SON compared with other nine essential protein discovery methods (DC, BC,CC, SC, EC, IC, NC, PeC, ION) are shown in Fig. 1. From Fig.1 we can see that SON can get higher prediction accuracy than other methods.

© Springer International Publishing Switzerland 2015
R. Harrison et al. (Eds.): ISBRA 2015, LNBI 9096, pp. 427–428, 2015.
DOI: 10.1007/978-3-319-19048-8

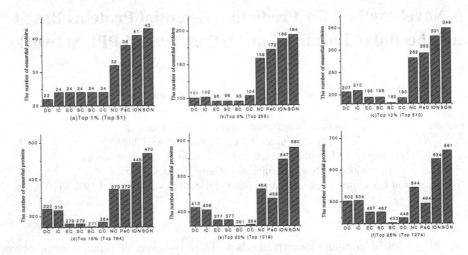

Fig. 1. SON compared with ION and eight centrality methods

Acknowledgment. This work is supported in part by the National Natural Science Foundation of China under Grant No. 61232001, No.61370024, No. 61420106009.

References

1. Hahn, M.W., Kern, A.D.: Comparative genomics of centrality and essentiality in three eukaryotic protein-interaction networks. Mol. Biol. Evol. 22, 803–806 (2005)
2. Joy, M.P., Brock, A., Ingber, D.E., Huang, S.: High-betweenness proteins in the yeast protein interaction network. Journal of Biomedicine and Biotechnology 2005, 96–103 (2005)
3. Wuchty, S., Stadler, P.F.: Centers of complex networks. J. Theor. Biol. 223, 45–53 (2003)
4. Estrada, E., Rodriguez-Velazquez, J.A.: Subgraph centrality in complex networks. Phys. Rev. E. 71, 056103 (2005)
5. Bonacich, P.: Power and centrality: A family of measures. Am. J. Sociol. 92, 12 (1987)
6. Karen, S., Zelen, M.: Rethinking centrality: Methods and examples. Social Networks 11, 37 (2002)
7. Wang, J.X., Li, M., Wang, H., Pan, Y.: Identification of Essential Proteins Based on Edge Clustering Coefficient. IEEE/ACM Transactions on Computational Biology and Bioinformatics 9, 1070–1080 (2012)
8. Acencio, M.L., Lemke, N.: Towards the prediction of essential genes by integration of network topology, cellular localization and biological process information. BMC Bioinformatics 10, 290 (2009)
9. Fraser, H.B., Hirsh, A.E., Steinmetz, L.M., Scharfe, C., et al.: Evolutionary rate in the protein interaction network. Science 296, 750–752 (2002)
10. Rocha, E.P.C., Danchin, A.: An Analysis of Determinants of Amino Acids Substitution Rates in Bacterial Proteins. Mol. Biol. Evol. 21, 108–116 (2004)
11. Li, M., Zhang, H., Wang, J.X., et al.: A new essential protein discovery method based on the integration of protein-protein interaction and gene expression data. BMC Syst. Biol. 6, 15 (2012)
12. Peng, W., Wang, J., Wang, W., et al.: Iteration method for predicting essential proteins based on orthology and protein-protein interaction networks. BMC Systems Biology 6, 87 (2012)

BASE: A Practical *de novo* Assembler for Large Genomes Using Longer NGS Reads

Binghang Liu[1,2], Ruibang Luo[1,2], Chi-Man Liu[1], Dinghua Li[1], Yingrui Li[2], Hing-Fung Ting[1], Siu-Ming Yiu[1], and Tak-Wah Lam[1(✉)]

[1] HKU-BGI Bioinformatics Algorithms and Core Technology Research Laboratory and Department of Computer Science, University of Hong Kong, Hong Kong, China
{bhliu,rbluo,cmliu,dhli,hfting,smyiu,twlam}@cs.hku.hk
[2] BGI-Shenzhen, Shenzhen, China
liyr@genomics.cn

De novo genome assembly is a fundamental problem in genomic research. When assembling relatively large genomes, time is often a very important concern, and one might have no choice but to use a more efficient assembler like SOAPdenovo2 [1] instead of a high-quality but prohibitively slow assembler (SPAdes[2] is a typical example). As the read length of high-throughput sequencers increases beyond 100bp, it has been expected that the quality issue of SOAPdenov2 can gradually improve. Yet SOAPdenov2, a typical *de Bruijn* graph based assembler, has inherent difficulty to fully utilize the advantage of longer reads (say, 150 bp and 250 bp from Illumina HiSeq and MiSeq, respectively). Other assemblers, such as the string graph assembler SGA [3] and the *multisized de Bruijn* graph assembler SPAdes, though more favorable for longer reads, are very slow and less popular. It is still up to the challenge how to better utilize longer reads to develop a fast-and-accurate assembler.

This paper presents a new assembler called BASE, whose assembly quality is approaching SPAdes for relatively longer reads. BASE is way faster than SPAdes and SGA, but slower than SOAPdenovo2. BASE is based on a simple seed-extension approach. Exploiting an efficient indexing of short reads (bi-directional BWT [4]), BASE firstly generates adaptive seeds with high probability of unique appearance in the genome and high sequencing qualities. Then rooted at such seeds, BASE constructs extension trees and gradually removes the branches with a novel method called reverse validation. In this method, by extensively utilizing information including read coverage and paired-end relationship, structures like *"Tips"*, *"butterfly"* and *"bubbles"*, as defined by DBG based assemblers (see, for example [1]), are recognized and simplified to obtain consensus sequences from the reads sharing the seeds. And these consensus sequences are further extended to form high-quality contigs.

Benchmark on several bacteria and human data sets clearly reflects our expected performance of BASE on speed and assembly quality when reads are getting longer. Our first benchmark was based on two data sets of deeply sequenced bacteria genomes (~240X), with read length 100 and 250 respectively. Among others, the correct N50 of

---

B. Liu and R. Luo – Joint first authors.

© Springer International Publishing Switzerland 2015
R. Harrison et al. (Eds.): ISBRA 2015, LNBI 9096, pp. 429–430, 2015.
DOI: 10.1007/978-3-319-19048-8

assembly was evaluated; for 100 bp reads, BASE is slightly better than SOAPdenov2 and SGA, but way below SPAdes (the correct N50 for BASE, SOAPdenov2, SGA and SPAdes are respectively 92,706, 82,495, 74,584 and 299,305). For 250 bp reads, BASE performs much better than SOAPdenov2 and SGA and is approaching SPAdes (precisely, 159,715, 88,858 and 95,711 and 169,978, respectively). Regarding speed, BASE is consistently a few times faster than SPAdes and SGA, but still slower than SOAPdenovo2. We have further compared BASE and SOAPdenov2 using human genome data sets with read length 100, 150 and 250. BASE consistently achieves a higher N50 for each data set, and the improvement becomes obvious when the read length increases to 250. SOAPdenovo2 uses relatively more memory when sequencing error is high.

In conclusion, BASE is an efficient assembler for constructing good-quality contigs, especially when reads are relatively longer. BASE could be extended easily to support scaffolding in the future.

References

1. Luo, R., Liu, B., Xie, Y., Li, Z., et al.: SOAPdenovo2: an empirically improved memory-efficient short-read de novo assembler. GigaScience 1, 18 (2012)
2. Bankevich, A., Nurk, S., Antipov, D., et al.: SPAdes: a new genome assembly algorithm and its applications to single-cell sequencing. Journal of Computational Biology: A Journal of Computational Molecular Cell Biology 19, 455–477 (2012)
3. Simpson, J.T., Durbin, R.: Efficient de novo assembly of large genomes using compressed data structures. Genome Research 22, 549–556 (2012)
4. Lam, T.W., Li, R.Q., Tam, A., Wong, S., Wu, E., Yiu, S.M.: High Throughput Short Read Alignment via Bi-directional BWT. In: 2009 IEEE International Conference on Bioinformatics and Biomedicine, pp. 31–36 (2009)
5. Peng, Y., Leung, H.C., Yiu, S.M., Chin, F.Y.: IDBA-UD: a de novo assembler for single-cell and metagenomic sequencing data with highly uneven depth. Bioinformatics 28, 1420–1428 (2012)
6. Bradnam, K.R., Fass, J.N., Alexandrov, A., et al.: Assemblathon 2: evaluating de novo methods of genome assembly in three vertebrate species. GigaScience 2, 10 (2013)
7. Gonnella, G., Kurtz, S.: Readjoiner: a fast and memory efficient string graph-based sequence assembler. BMC Bioinformatics 13, 82 (2012)
8. Magoc, T., Pabinger, S., Canzar, S., Liu, X., Su, Q., Puiu, D., Tallon, L.J., Salzberg, S.L.: GAGE-B: an evaluation of genome assemblers for bacterial organisms. Bioinformatics 29, 1718–1725 (2013)
9. Miyamoto, M., Motooka, D., Gotoh, K., Imai, T., Yoshitake, K., Goto, N., Iida, T., Yasunaga, T., Horii, T., Arakawa, K., Kasahara, M., Nakamura, S.: Performance comparison of second- and third-generation sequencers using a bacterial genome with two chromosomes. BMC Genomics 15, 699 (2014)

InteGO2: A Web Tool for Measuring and Visualizing Gene Semantic Similarities Using Gene Ontology

Jiajie Peng[1], Hongxiang Li[1*], Yongzhuang Liu[1], Liran Juan[2], Qinghua Jiang[2], Yadong Wang[1(✉)], and Jin Chen[3,4(✉)]

[1] School of Computer Science and Technology, Harbin Institute of Technology, Harbin, China
[2] School of Life Science and Technology, Harbin Institute of Technology, Harbin, China
[3] MSU-DOE Plant Research Laboratory, Michigan State University, East Lansing 48824, USA
[4] Department of Computer Science & Engineering, Michigan State University, East Lansing 48824, USA

The Gene Ontology (GO) is a widely used bioinformatics resource using ontologies to represent biological knowledge and describe function information for genes and gene products. GO has three categories shared by all organisms: molecular function, biological process and cellular component. As an integrated resource, GO provides rich information and a convenient way to study gene functional similarity [9].

Various methods have been proposed to measure gene functional similarities by comparing GO terms with which the genes are annotated. Based on the types of information in GO they use, these methods can be grouped into three categories [5]: 1) edge-based measurements are fully dependent on the structure of GO, and simply equalize the terms at the same topological level [11]; 2) node-based measurements consider the annotations and common ancestors, neglecting the complex topology of GO [7,6]; and 3) hybrid measurements [10] fully utilize the topological information in GO structure, but neglect gene annotations.

Since none of these measurements can take into account all the information in GO (structure, annotation, all common parents, most informative common parent, *etc.*), we have recently presented two integrative measurements successively to unite the strength of the existing measures [3,4]. Our model automatically selects and integrates seed measurements with a meta-heuristic search process in three steps. First, all the ranked similarity values of a background gene set are calculated with all the GO-based semantic similarity measurements that we would integrate. Second, the most appropriate seed measurements for each gene pair are selected with a grouping method. Third, the parameters of a meta-heuristic search model are estimated by maximizing the distances between distinct EC groups. The experimental results on molecular function, biological process and protein sequence indicate that our integrative measurement performs better than all the seed measurements (see details in [3]).

J. Peng and H. Li – Equal contributors.

© Springer International Publishing Switzerland 2015
R. Harrison et al. (Eds.): ISBRA 2015, LNBI 9096, pp. 431–432, 2015.
DOI: 10.1007/978-3-319-19048-8

Various web tools have been proposed to compute gene functional similarities using GO, including GossToWeb [1], FunSimMat [8] and G-SESAM [2]. Choosing the right measurement is difficult for users, but none of these tools provide a solution to it. In addition, most tools simply list the gene similarity values as the final output, neglecting the fact that appropriate data visualization is essential towards result interpretation and hypothesis testing. It is desirable to develop an easy-to-use web tool that allows researchers to conveniently measure gene functional similarities, and to visualize the functional interactions with an easy-to-use graphical interface. In this article, we present a novel web tool named *InteGO2*, which is available at http://mlg.hit.edu.cn:8089/.Comparing with the existing web tools, the major contributions of our work are:

- *InteGO2* supplies researchers an integrative approach that automatically choose and weigh appropriate gene functional similarity measurements for the input genes.
- *InteGO2* is an easy-to-use HTML5 based web interface to visualize gene functional associations.
- *InteGO2* can measure gene functional similarities using 98 types of gene IDs in 24 species.

References

1. Caniza, H., et al.: Gossto: a user-friendly stand-alone and web tool for calculating semantic similarities on the gene ontology. Bioinformatics btu144 (2014)
2. Du, Z., et al.: G-sesame: web tools for go-term-based gene similarity analysis and knowledge discovery. Nucleic Acids Res., W345–W349 (2009)
3. Peng, J., Li, H., Jiang, Q., Wang, Y., Chen, J.: An integrative approach for measuring semantic similarities using gene ontology. BMC Syst. Biol. 8(suppl. 5), S8 (2014)
4. Peng, J., Wang, Y., Chen, J.: Towards integrative gene functional similarity measurement. BMC Bioinformatics 15(suppl. 2), S5 (2014)
5. Pesquita, C., Faria, D., Falcao, A., Lord, P., Couto, F.: Semantic similarity in biomedical ontologies. PLoS Comput. Biol. 5(7), e1000443 (2009)
6. Resnik, P.: Semantic similarity in a taxonomy: An information-based measure and its application to problems of ambiguity in natural language. J. Artif Intell. Res. 11, 95–130 (1999)
7. Schlicker, A., Domingues, F., Rahnenfhrer, J., Lengauer, T.: A new measure for functional similarity of gene products based on gene ontology. BMC Bioinformatics 7, 302 (2006)
8. Schlicker, A., Albrecht, M.: Funsimmat update: new features for exploring functional similarity. Nucleic Acids Res. pp. D244–D248 (2009)
9. The Gene Ontology Consortium.: Gene ontology consortium: going forward. Nucleic Acids Res. 43(D1), D1049–D1056 (2015)
10. Wang, Z., Du, Z., Payattakool, R., Philip, Y., Chen, F.: A new method to measure the semantic similarity of go terms. Bioinformatics 23, 1274–1281 (2007)
11. Wu, X., Pang, E., Lin, K., Pei, Z.: Improving the measurement of semantic similarity between gene ontology terms and gene products: Insights from an edge-and ic-based hybrid method. PloS One 8, e66745 (2013)

Predicting Drug-Target Interactions for New Drugs via Strategies for Missing Interactions

Jian-Yu Shi[✉], Jia-Xin Li, and Hui-Meng Lu

School of Life Sciences, Northwestern Polytechnical University, Xi'an, China
jianyushi@nwpu.edu.cn, lijiaxin0932@mail.nwpu.edu.cn,
luhuimeng@nwpu.edu.cn

Identifying protein targets for novel drug candidates is a crucial step in drug discovery. However, testing drug candidates by turn in wet lab would require a huge amount of money and take a very long time. As one of dry-lab means, computational methods have shown their power when predicting novel drug–target interactions (DTI) by making use of both drug similarity and target similarity.

Compared with network inference-based methods and matrix factorization-based methods, supervised learning models (SLM) are able to cope with more scenarios of predicting DTIs, and to provide the better elucidation of why a drug interacts with a target as well. Thus, they have been gained a lot of concerns for predicting DTI in both drug discovery and drug repositioning.

Former approaches base on SLM can be roughly categorized into two groups: global model and local model. The global model holds the underlying assumption that all DTIs follow a common distribution in terms of feature or similarity [1]. Usually, it suffers the computation of large complexity due to the tensor product between drug similarity matrix and target similarity matrix. In contrast, the local model considers that all drugs linking to a concerned target follow a common distribution specific to the target [2]. It generally requires less computation than the global model. In common, two types of SLM treat known DTIs as positive instances and all unknown drug-target pairs (DTP) as negative instances respectively. However, SLM has no consideration that some of unknown DTPs are true or potential DTIs which were not explicitly labeled when the dataset was being built. They are called missing DTIs.

Deriving from missing DTIs hidden in unlabeled DTPs, several issues have not yet been addressed appropriately by SLM in the field of DTI prediction. First, simply regarded as negatives, missing DTIs give rise to a bad decision boundary of the trained classifier which predicts more positives as negatives. Secondly, they lead existing DTIs away from the popularly acceptable assumption of interactions: similar drugs tend to interact with similar targets. Thirdly, they may aggravate the inherent imbalance caused by few DTIs and many unknown DTPs.

In this work, we extended the local model of SLM to cope with the abovementioned issues. Focusing on the predicting scenario in which predicting

The work was supported by the Fundamental Research Funds for the Central Universities (Grant No. 3102015ZY081), and partially supported by NSFC (Grant No. 31402019) and the Natural Science Foundation of Shaanxi Province, China (Grant No. 2014JQ4140).

© Springer International Publishing Switzerland 2015
R. Harrison et al. (Eds.): ISBRA 2015, LNBI 9096, pp. 433–434, 2015.
DOI: 10.1007/978-3-319-19048-8

interactions between known targets and new drugs is anticipated [1], we proposed the approach, K Nearest Neighbors with strategies for missing DTI (KNNm), which includes a semi-supervised strategy (named Spy [3]), a clustering strategy (named Super-target[4]) and an instance-based classifier (named KNN [5]).

KNNm has the following advantages: (1) Spy can identify reliable non-DTIs among unknown DTP by means of investigating the behavior of DTIs in unknown DTP, so as to be able to train a less biased decision boundary by known DTIs and reliable non-DTIs. (2) Super-target enables missing DTIs in unknown DTPs as few as possible by grouping similar targets as well as their interacting drugs. Thus, the assumption of interactions can be matched better when determining how likely a drug interacts with a Super-target containing a group of similar targets. (3) KNN is particularly helpful to relax the inherent imbalance of few DTIs and many unknown DTPs, and has less computational complexity as well

To predict how likely a new drug interacts with a known target, KNNm perform Spy and Super-target strategies simultaneously and combine them together. First, using KNN in a traditional supervised way, we predicted how likely a new drug interacts with a Super-target which contains the concerned target. Secondly, we applied Spy to identify reliable non-DTIs, and predicted how likely the new drug interacts with the individual target in the Super-target directly by training KNN on both the known DTIs and the reliable non-DTIs. Moreover, considering the possibly bad case that drugs interacting with a Super-target are very different, we designed an adaptive rule to determine when to combine Super-target with Spy or not.

Based on four benchmark datasets [2], we adopted the five-fold cross validation (CV) in [6] to evaluate KNNm. The results show that KNNm is overall superior to two best existing approaches [1, 6] which also treat DTPs as unlabeled instances not simply as non-DTIs.

References

1. Pahikkala, T., Airola, A., Pietila, S., Shakyawar, S., Szwajda, A., Tang, J., Aittokallio, T.: Toward more realistic drug-target interaction predictions. Brief Bioinform. 16, 325–337 (2015)
2. Bleakley, K., Yamanishi, Y.: Supervised prediction of drug-target interactions using bipartite local models. Bioinformatics 25, 2397–2403 (2009)
3. Liu, B., Lee, W.S., Yu, P.S., Li, X.: Partially supervised classification of text documents. In: Proceedings of the Nineteenth International Conference on Machine Learning, pp. 387–394. Morgan Kaufmann Publishers Inc., San Francisco (2002)
4. Shi, J.Y., Yiu, S.M., Li, Y.M., Leung, H.C.M., Chin, F.Y.L.: Predicting Drug-Target Interaction for New Drugs Using Enhanced Similarity Measures and Super-Target Clustering. In: Proceedings of the IEEE International Conference on Bioinformatics and Biomedicine, pp. 45–50. IEEE Press, New York (2014)
5. Chen Zhang, M.L., Zhou, Z.H.: ML-KNN: A lazy learning approach to multi-label learning. Pattern Recognition 40, 2038–2048 (2007)
6. Chen, H., Zhang, Z.: A semi-supervised method for drug-target interaction prediction with consistency in networks. PLoS One. 8, e62975 (2013)

A Genome-Wide Drug Repositioning Approach toward Prostate Cancer Drug Discovery

Rong Xu[1] and QuanQiu Wang[2]

[1] Case Western Reserve University, Cleveland OH 44106, USA
[2] ThinTek, LLC, Palo Alto CA 94306, USA

Introduction. Prostate cancer (PC) is the most common cancer and the third leading cause of cancer death in men worldwide [1]. Despite its high incidence and mortality, the likelihood of a cure is low for late-stages of PC [2]. There is an unmet need for more effective agents for treating PC. Traditional drug development is expensive and time-consuming. Computation-based drug repositioning approaches that automatically search vast amounts of genetic, genomic, chemical, and phenotypic data for thousands of drugs and tens of thousands of diseases can greatly speed up the traditional drug discovery process [3]. Here we propose to develop a computation-based drug repositioning system, GenoPredict, that capitalizes on comprehensive disease genetic data generated and a unique large-scale drug treatment database that we recently constructed [4-6] in order to rapidly identify drug candidates for PC. Our study is based on the premise that the genetic overlap among diseases reflects pathophysiological overlap. Though the majority of such shared pathophysiological features remain unknown, treatment insights from one disease may be used to inform our knowledge of others and potentiate their treatments. In order to systematically reposition drug treatments from one disease to another, it is critical to have a comprehensive drug treatment knowledge base. In our recent studies, we constructed a comprehensive drug-disease treatment knowledge base (TreatKB) using computational techniques including natural language processing, text mining and data mining from multiple heterogeneous and complementary data resources, including FDA drug labels, the FDA Adverse Event Reporting System, clinical trial reports, and biomedical literature [4-6]. TreatKB contains 208,330 unique drug-disease treatment pairs, representing 2,484 drugs and 24,511 unique disease concepts.

Methods. The experiment framework consists of three steps: (1) we constructed a genetic disease networks (GDN) using disease-gene association data from the Catalog of Published Genome-Wide Association Studies (the GWAS catalog) from the US National Human Genome Research Institute (NHGRI) [7] On GDN, two diseases were connected if their associated genes overlaped. The edge weights were determined by the cosine similarity of disease-associated genes. GDN comprised of 882 disease nodes and 200,758 edges; (2) we applied the network-based ranking algorithms that we recently developed [8-9] to find diseases that are genetically related to PC. We tested the network construction and ranking algorithms by examining disease class distribution among ranked diseases; (3) we developed an approach to systematically reposition drugs from PC-related diseases to treat PC. We ranked drugs based on the number of PC-related diseases that they could treat as well as the ranking scores

© Springer International Publishing Switzerland 2015
R. Harrison et al. (Eds.): ISBRA 2015, LNBI 9096, pp. 435–437, 2015.
DOI: 10.1007/978-3-319-19048-8

of these diseases. We evaluated GenoPredict using 27 FDA-approved PC drugs. We compared GenoPredict to one of currently most comprehensive drug repositioning systems, PREDICT [10], in novel predictions using 172 PC drugs extracted from 172,888 clinical trials; and (4) to better understand top-ranked drug candidates, we examined their class distributions, i.e., the drug classes that most of the top candidates belong to. We analyzed genetic pathways targeted by top-ranked 124 candidates. Functions of enriched pathways might provide novel insights into common molecular mechanisms targeted by drug candidates.

Results. When evaluated in a *de-novo* prediction setting using 27 FDA-approved PC drugs. GenoPredict found 25 of 27 FDA-approved PC drugs and ranked them highly (recall: 0.925, mean ranking: 27.3%, median ranking: 15.6%). When compared to PREDICT, GenoPredict clearly dominated PREDICT in Precision-Recall (PR) curves across two evaluation datasets. GenoPredict achieved a mean average precision (MAP) of 0.447 when evaluated with 172 PC drugs extracted from 172,888 clinical trial reports, representing a 164.5% improvement as compared to a MAP of 0.169 for PREDICT. When evaluated with 72 PC drugs extracted from 43,811 ongoing clinical trial reports, GenoPredict achieved a MAP of 0.278, representing a 231.1% improvement as compared to a MAP of 0.084 for PREDICT.

Conclusions and Future Work. We developed a drug repositioning system, Geno-Predict, to exploit the genetic and treatment connections among a large number of diseases and applied it to identify drug candidates for PC. GenoPredict found 25 of 27 FDA-approved PC drugs and ranked them highly. Future works include testing GenoPredict on other common complex diseases to assess its generalizability. We expect that its performance will vary according to different diseases. This study focused on disease genetics-based drug repositioning. Additional invaluable data resources such as other disease-related data (i.e. disease phenotypic data or gene expression data) and drug-related data (i.e. drug side effects, drug chemical structure, and gene expression) can be incorporated into GenePredict to further improve its performance. However, integrating and reasoning over such complex biological data poses a significant challenge that bears future investigation.

References

1. Siegel, R., Naishadham, D., Jemal, A.: Cancer statistics, 2013. CA: A Cancer Journal for Clinicians 63(1), 11–30 (2013)
2. Trewartha, D., Carter, K.: Advances in prostate cancer treatment. Nature Reviews Drug Discovery 12(11), 823–824 (2013)
3. Hurle, M.R., Yang, L., Xie, Q., Rajpal, D.K., Sanseau, P., Agarwal, P.: Computational drug repositioning: From data to therapeutics. Clinical Pharmacology and Therapeutics (2013)
4. Xu, R., Wang, Q.: Large-scale extraction of drug-disease treatment pairs from biomedical literature for drug repurposing. BMC Bioinformatics 14(1), 181 (2013)
5. Xu, R.: Li, Li. and Wang, Q.: Towards building a disease-phenotype relation- ship knowledge base: large scale extraction of disease-manifestation relationship from literature. Bioinformatics (2103), doi: 10.1093/bioinformatics/btt359

6. Xu, R., Wang, Q.: Automatic signal prioritizing and filtering approaches in detecting post-marketing cardiovascular events associated with targeted cancer drugs from the FDA Adverse Event Reporting System (FAERS). Journal of Biomedical Informatics, 171–177 (2014)
7. Welter, D., MacArthur, J., Morales, J., Burdett, T., Hall, P., Junkins, H., Parkinson, H.: The NHGRI GWAS Catalog, a curated resource of SNP- trait associations. Nucleic acids research, 42(D1), D1001–D1006 (2014)
8. Chen, Y., Xu, R.: Network-based Gene Prediction for Plasmodium falci- parum Malaria Towards Genetics-based Drug Discovery. BMC Genomic (in press)
9. Xu, R., Wang, Q., Li, L.: Genome-wide systems analysis reveals strong link between colorectal cancer and trimethylamine N-oxide (TMAO), a gut microbial metabolite of dietary meat and fat. BMC Genomics (in press)
10. Gottlieb, A., Stein, G.Y., Ruppin, E., Sharan, R.: PREDICT: A method for inferring novel drug indications with application to personalized medicine. Molecular Systems Biology 7(1) (2011)

Clustering Analysis of Proteins from Microbial Genomes at Multiple Levels of Resolution

Leonid Zaslavsky[✉] and Tatiana Tatusova

National Center for Biotechnology Information, National Library of Medicine,
National Institutes of Health, Bethesda, MD 20854, USA
{zaslavsk,tatiana}@ncbi.nlm.nih.gov

Microbial genomes at NCBI represent a large collection containing almost 30,000 genomes from more than 5,000 species [5]. The quality and sampling density of the bacterial genome assemblies vary greatly: human pathogens are densely sampled while other bacteria are less represented. The variation in frequency of occurrences of different proteins in genome annotation is another factor contributing to the complexity of the analysis and presentation of the data. Redundancy in the results make them difficult to analyze and use, as the nearest-neighbor lists may often contain many nearly identical objects making it difficult or impossible to reflect more distant neighbor relationships. The complex data we work with requires the information to be organized, processed and shown at multiple levels of resolution, with appropriate levels of phylogenomic resolution and protein similarity and an adequate sampling strategy.

Our approach is to perform detailed clustering in the groups of closely-related genomes (species-level clades), and then cluster proteins globally, starting with *seed* clusters built from clustroids of conservative in-clade clusters of large clades and building other global clusters around them. Utilizing protein clusters built at different phylogenomic and sequence similarity levels, and links between them, allows us to explore the protein space for close and distant neighbor relationships important for automatic protein annotation and intelligent presentation of protein data.

In-clade Clusters. Genomic and proteomic structure of a densely-sampled group of related strains is usually described by the concept of pan-genome [3], [7]. Clusters within a clade are created using a combined approach that takes into account both sequence similarity and genome context. First, proteins are tentatively clustered by sequence similarity. Then local genome context and protein phylogeny are used to separate paralogs in the tentative clusters. This combined approach defines core and conservative clusters in a pan-genome more accurately than by sequence-based clustering alone. For computational efficiency, protein redundancy and near-redundancy is eliminated, with one representative sequence from each near-redundant group used [2]. In each in-clade cluster, a clustroid is selected to serve as a representative in global clustering. Using clustroids to represent all proteins of the in-clade clusters allows us, to some extent, to filter out the outliers.

The rights of this work are transferred to the extent transferable according to title 17 U.S.C. 105.

© US Government (outside the US) 2015
R. Harrison et al. (Eds.): ISBRA 2015, LNBI 9096, pp. 438–439, 2015.
DOI: 10.1007/978-3-319-19048-8

Global Clusters. In order to denoise the data and achieve robustness of the results as well as computational efficiency, global clustering is performed in three steps. First, seed clusters are built from the clustroids of conservative in-clade clusters. Then UCLUST [2] is used to select representatives from tight groups of proteins not represented by the seed cluster proteins. Finally, global clustering is performed and the selected UCLUST-representatives are either associated with seed clusters or form new global clusters. Building seed clusters from the stabilized and denoised data results in larger and more stable global clusters.

Implementation Details. Prokaryotic genomes from NCBI Refseq collection are organized in related groups (species-level clades), using robust distance between sets of universally conserved ribosomal proteins[1], [5]. Our clustering procedure is based on hierarchical clustering algorithm with additional restrictions on minimum alignment score and coverage (see section "Protein Clustering Procedure" in [6]). The available hardware that includes NCBI's Grid Engine-based computer farm and PanFS scalable storage system connected through a router, requires a coarse-grain parallelization, which is done in three stages: (1) The dataset is partitioned in disjoint sets using a parallel implementation based on a disjoint-set forest with union-by-rank heuristics [4]; (2) the data are redistributed according to the partitioning; (3) clustering in partitions is performed.

Conclusions. The computational infrastructure developed at NCBI provides a foundation for prokaryotic gene and genome analysis allowing easy access to pre-calculated genome groups (clades), protein clusters and pan-genome representation. It provides a solution to one of the most challenging problems in the data deluge - presenting the relevant data at different levels of details and eliminating data redundancy while keeping biologically interesting variations.

Acknowledgments. This research was supported by the Intramural Research Program of the NIH, National Library of Medicine.

References

1. Ciccarelli, F., Doerks, T., von Mering, C., Creevey, C., Snel, B., Bork, P.: Toward automatic reconstruction of a highly resolved tree of life. Science 311(5765), 1283–1287 (2006)
2. Edgar, R.: Search and clustering orders of magnitude faster than blast. Bioinformatics 26(19), 2460–2461 (2010)
3. Medini, D., Donati, C., Tettelin, H., Masignani, V., Rappuoli, R.: The microbial pangenome. Curr. Opin. Genet. 15(6), 589–594 (2005)
4. Tarjan, R.: Data structures and network algorithms. CBMS, vol. 44. Society for Industrial and Applied Mathematics, Philadelphia (1983)
5. Tatusova, T., Ciufo, S., Federhen, S., et al.: Update on refseq microbial genomes resources. Nucl. Acids Res. (database issue) (2015)
6. Tatusova, T., Zaslavsky, L., Fedorov, F., et al.: Protein clusters. In: The NCBI Handbook [Internet]. National Center for Biotechnology Information, 2nd edn., Bethesda, Maryland, USA (2013), http://www.ncbi.nlm.nih.gov/books/NBK242632
7. Tettelin, H., Masignani, V., Cieslewicz, M., et al.: Genome analysis of multiple pathogenic isolates of streptococcus agalactiae: implications for the microbial "pan-genome". Proc. Natl. Acad. Sci. USA 102(39), 13950–13955 (2005)

Systematic Analyses Reveal Regulatory Mechanisms of the Flexible Tails on Beta-catenin

Bi Zhao[1] and Bin Xue[2(✉)]

[1]Department of Molecular Medicine, College of Medicine, University of South Florida,
Tampa, FL 33620, USA
bizhao@mail.usf.edu
[2] Department of Cell Biology, Microbiology and Molecular Biology, School of Natural
Sciences and Mathematics, College of Arts and Sciences, University of South Florida,
Tampa, FL 33620, USA
binxue@usf.edu

Abstract. Beta-catenin has two major functions: coordinate cell-cell adhesion by interacting with cadherin; regulate gene expression through Wnt pathway. The armadillo domain in the central region of beta-catenin is the critical structural unit for its functions. Recently, the flexible tails at both N- and C-termini of beta-catenin were observed to regulate the functions of armadillo domain. However, the mechanisms are still elusive. In this study, we identified multiple functional motifs in the tail regions and analyzed the conserved hydrophobic sequential patterns on these regions, which act as interaction motifs between tails, armadillo domain, and other molecules. The interactome of beta-catenin is enriched of protein intrinsic disorder. The interactions between beta-catenin and its partners are modulated by post-translational modifications and the movement of tail regions. The "open" and "close" states of the tail regions determine the function of beta-catenin.

Keywords: Beta-Catenin · Intrinsic disorder · Tails · Functional motif · Conservation · Post-translational modification · Network · Pathway

Beta-catenin is composed of a central structured armadillo domain (ARM) and two flexible tails. The interactions between ARM of beta-catenin and its partners are critical for embryonic development, cell division, and maintenance of pluripotency. Disorganized expression of beta-catenin is associated with cancer development. Recent studies discovered that the above-mentioned interactions were mediated by both flexible tails of beta-catenin [1]. These discoveries extended our comprehension on the interaction patterns between beta-catenin and its partners, opened a new field on the functional roles of flexible tails. However, the detailed mechanisms are still largely unknown. In this study, we integrated multiple bioinformatics strategies and analyzed

© US Government (outside the US) 2015
R. Harrison et al. (Eds.): ISBRA 2015, LNBI 9096, pp. 440–441, 2015.
Doi: 10.1007/978-3-319-19048-8

beta-catenin on various aspects, including: intrinsic disorder, functional motifs, Evolutionary conservation, diseases-associated mutations, signaling pathways, and interaction networks. We also proposed a model through which the flexible tails regulate the function of beta-catenin.

Although being very flexible as shown by the results of disorder predictions, the tail regions of beta-catenin contain multiple hydrophobic segments that overlap with predicted binding motifs. In addition, the tail regions carry more than thirty database-retrieved functional motifs, which are categorized into four functional groups: regulation, interaction, recognition, and cleavage. Among these motifs, the function of GSK and SCF-TRCP1 motifs at N-terminal, and the function of PKA and PDZ motifs at C-terminal were experimentally validated [2-4]. Our further analysis demonstrated that these functional motifs were highly conserved across species. Many human-diseases-associated mutations were found on or near these functional motifs.

The interaction partners of beta-catenin are enriched of protein intrinsic disorder. Out of forty interaction partners, fourteen are completely disordered, seven accommodate large portion of disordered regions. Since protein interactions involving disordered regions are often characterized by high-specificity and low-affinity, they are extremely important in signaling and regulation. The interaction between beta-catenin and its partners on disordered regions contribute to related functional pathways.

In the two major functional pathways where beta-catenin is involved, the flexibility of both tails of beta-catenin and the post-translational modifications on ARM domain regulate the interaction between tail regions and ARM domain, as well as the interaction between beta-catenin and its partners. The phosphorylation on Tyr-654 was found crucial for many interactions of beta-catenin [5, 6]. Our studies demonstrate that the phosphorylation of Tyr-654 is also critical in modulating local polarity, packing of both tails on ARM domain, and regulating the interaction with other partners. The "close" and "open" status of both tails determines the function of beta-catenin.

References

1. Castano, J., et al.: Beta-catenin N- and C-terminal tails modulate the coordinated binding of adherens junction proteins to beta-catenin. The Journal of Biological Chemistry 277(35), 31541–31550 (2002)
2. Wu, G., et al.: Structure of a beta-TrCP1-Skp1-beta-catenin complex: destruction motif binding and lysine specificity of the SCF(beta-TrCP1) ubiquitin ligase. Mol. Cell. 11(6), 1445–1456 (2003)
3. Zhang, J., et al.: Structural basis of beta-catenin recognition by Tax-interacting protein-1. J. Mol. Biol 384(1), 255–263 (2008)
4. Taurin, S., et al.: Phosphorylation of beta-catenin by cyclic AMP-dependent protein kinase. The Journal of Biological Chemistry 281(15), 9971–9976 (2006)
5. Roura, S., et al.: Regulation of E-cadherin/Catenin association by tyrosine phosphorylation. The Journal of Biological Chemistry 274(51), 36734–36740 (1999)
6. Piedra, J., et al.: Regulation of beta-catenin structure and activity by tyrosine phosphorylation. The Journal of Biological Chemistry 276(23), 20436–20443 (2001)

GRASPx: Efficient Homolog-Search of Short-Peptide Metagenome Database Through Simultaneous Alignment and Assembly

Cuncong Zhong, Youngik Yang, and Shibu Yooseph[✉]

Informatics Department, J. Craig Venter Institute, La Jolla, CA 92037, USA
{czhong,yyang,syooseph}@jcvi.org

Abstract. To improve the reference-based homolog search of metagenomic sequencing reads, we have recently developed a program called GRASP based on a paradigm involving simultaneous search and assembly of the reads. GRASP shows substantially improved performances but is relatively slower than other homology search programs. In this abstract we present GRASPx, a computationally efficient improvement to GRASP. GRASPx achieves a 30X speedup compared to GRASP while maintaining its superior search performance.

1 Introduction

Metagenomic samples are routinely sequenced using high-throughput next-generation sequencing (NGS) technologies that generate short reads. These reads are subsequently aligned to reference sequences for the annotation of their functions. However, these short reads often represent only partial gene sequences (the corresponding protein sequence is referred to as a short peptide [1]) and it is challenging to accurately identify their homologs in reference databases. To address this issue, we developed a method called GRASP that identifies the homologs of a given reference protein sequence from a database of short peptide sequences [2]. GRASP has improved homology detection capability, primarily due to its ability to assemble the short peptide sequences in the database during the search step. GRASP has ~20% higher sensitivity than other homolog search programs (such as BLAST [3] and FASTM [4]). However, the overall computational efficiency of GRASP is adversely impacted by the assembly process; therefore, a substantial speedup is required for its applications on large data sets.

In this abstract we present GRASPx, a computational efficient improvement of GRASP. GRASPx is designed with the following major improvements. (1) *Prebuilt extension links*: Overlaps between the reads are resolved during the indexing step, substantially reducing the time spent on the search/assembly. (2) *Local assembly*: Each seed is allowed to be extended up to a pre-defined depth, therefore reducing the computation time wasted on extending non-homologous seeds. (3) *Query level parallelism*: Each thread is dedicated to the search/assembly of one query (instead of one seed as in GRASP), therefore minimizing the inter-thread communication. GRASPx executables are freely available at http://graspx.sourceforge.net/.

© Springer International Publishing Switzerland 2015
R. Harrison et al. (Eds.): ISBRA 2015, LNBI 9096, pp. 442–443, 2015.
DOI: 10.1007/978-3-319-19048-8

2   Results

GRASPx was benchmarked with GRASP [2], BLASTP, PSI-BLAST [3], and FASTM [4]. We constructed a simulated data set from 20 marine microbial genomes [1] using WGSIM (10X, 100bp pair-ended, 1% error rate). Short peptides were called using FragGeneScan [5], resulted in 6,273,043 short-peptide reads. 198 *Dehalococcoides sp. CBDB1* marker genes (Amphora2) were used as the queries. In this experiment, GRASPx shows ~30X speedup compared to GRASP; and GRASPx also shows the same performance as GRASP (see Fig. 1). Additional benchmark results on a real human saliva data set (SRS013942) generated by the Human Microbiome Project further confirm that GRASPx is capable of identifying many more homologs than the other tools with high accuracy (data not shown). In conclusion, GRASPx has a substantially improved computational efficiency over GRASP while keeping the same level of performance, enabling homolog search on large metagenomic data sets with superior sensitivity and specificity.

(A)   **(B)**

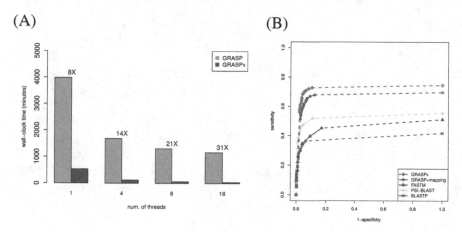

Fig. 1. (A) Total runtime of GRASPx and GRASP (numbers above bars indicate speedups). (B) ROC (Receiver Operating Characteristic) curve for performance comparison between GRASPx, GRASP+mapping, FASTM, PSI-BLAST (with 3 iterations), and BLASTP.

References

1. Yang, Y., Yooseph, S.: SPA: a short peptide assembler for metagenomic data. Nucleic Acids Res. 41(8), e91 (2013)
2. Zhong, C., Yang, Y., Yooseph, S.: GRASP: Guided Reference-based Assembly of Short Peptides. Nucleic Acids Res (2014)
3. Altschul, S.F., et al.: Gapped BLAST and PSI-BLAST: a new generation of protein database search programs. Nucleic Acids Res. 25(17), 3389–3402 (1997)
4. Goujon, M., et al.: A new bioinformatics analysis tools framework at EMBL-EBI. Nucleic Acids Res, 38(web server issue), W695-W699 (2010)
5. Rho, M., Tang, H., Ye, Y.: FragGeneScan: predicting genes in short and error-prone reads. Nucleic Acids Res. 38(20), e191 (2010)

Author Index

Alonso, Jose M.   138
Anannya, Tasmiah Tamzid   309
Ayada, Emi   12
Aygün, Ramazan S.   84
Azim, Md. Aashikur Rahman   24

Bansal, Mukul S.   187
Baudet, Christian   272
Bedford, John   417
Betkier, Arkadiusz   36
Bhattacharyya, Sourya   48
Biswas, Abhishek   60
Borges, Carlos Cristiano H.   247
Bowers, John C.   72

Cai, Zhipeng   96, 355
Capriles, Priscila V.S.Z.   247
Caragea, Doina   125, 331
Chauve, Cedric   260
Chen, Jin   431
Chen, Yang   413
Cheng, Dayou   355
Cheng, Micheal   211
Chernikov, Andrey N.   343
Chrétien, Stéphane   415
Collins, Jason C.   417
Conesa, Bastien   415

Dai, Cuihong   355
Delage-Mouroux, Régis   415
Descôtes, Françoise   415
Dias, Zanoni   272
Dinç, İmren   84
Du, Ding-Zhu   96

Erill, Ivan   420

Faria-Pinto, Priscila   247

Galvão, Gustavo Rodrigues   272
Górecki, Paweł   36
Greene, Lesley H.   417
Gu, Feng   366
Guo, Xuan   96, 366
Guyeux, Christophe   415

Hasegawa, Takanori   12, 422
He, Dan   108
He, Jing   1, 60
Heber, Steffen   138
Herndon, Nic   125, 331
Hon, Wk   211
Hoque, Mohammad Mozammel   309
Hoskins, William   402
Hou, Aiju   355
Hu, Qiwen   138
Huang, Yuanyuan   150
Huetz, Philippe   415

Iliopoulos, Costas S.   24
Imoto, Seiya   12

Jernigan, Robert   150
Ji, Hao   162
Jia, Kejue   150
Jiang, Qinghua   431
John, Rose Tharail   72
Jouvenot, Michèle   415
Juan, Liran   431

Kabir, Kazi Lutful   309
Karim, Rashid Saadman   309
Kawai, Yosuke   422
Kilic, Sefa   420
Kojima, Kaname   422
Kordi, Misagh   187

Lajoie, Gilles   223
Lajoie, Gilles A.   320
Lam, Tak-Wah   424, 429
Lam, Tw   211
Li, Dinghua   429
Li, Gaoshi   427
Li, Hongxiang   431
Li, Jia-Xin   433
Li, Jin   199
Li, Menglu   211
Li, Min   427
Li, Shuai Cheng   199
Li, Yaohang   162
Li, Yingrui   429

Lin, Yu   402
Liu, Binghang   429
Liu, Chi-Man   429
Liu, Xingwu   199
Liu, Yi   223
Liu, Yongzhuang   431
Lu, Hui-Meng   433
Luo, Ruibang   429

Ma, Bin   223, 320
Mäkinen, Veli   235
Merchante, Catharina   138
Mimori, Takahiro   422
Miyano, Satoru   12
Molloy, Kevin   175
Mukhopadhyay, Asish   284
Mukhopadhyay, Jayanta   48

Nagasaki, Masao   422
Nakhleh, Luay   378
Nariai, Naoki   422
Nasr, Kamal Al   1
Ng, Yen Kaow   199
Niida, Atsushi   12
Nunes, Vinicius Schmitz   247

Pan, Yi   96, 366, 390, 427
Panigrahi, Satish Chandra   284
Parada, Laxmi   108
Peng, Jiajie   431
Ponty, Yann   260
Pusey, Marc L.   84

Rahman, M. Sohel   24, 309
Rajaraman, Ashok   260
Rajasekaran, Sanguthevar   297
Ranjan, Desh   60
Ridwan, Iffatur   309
Roy, Kaushik   284

Saha, Subrata   297
Samiruzzaman, Mohammad   24
Shafin, Md. Kishwar   309
Shehu, Amarda   175
Shi, Jian-Yu   433
Stepanova, Anna N.   138

Streinu, Ileana   72
Sun, Weiping   223, 320
Szczęsny, Paweł   36

Tang, Cy   211
Tang, Jijun   402
Tangirala, Karthik   331
Tatusova, Tatiana   438
Ting, Hf   211
Ting, Hing-Fung   424, 429

Valenzuela, Daniel   235
Vasconcelos, Eveline Gomes   247

Wang, Jianxin   390, 427
Wang, QuanQiu   435
Wang, Yadong   431
Wang, Zhanyong   108
Weinberg, Seth H.   162
Wong, Thomas   211
Wu, Fang-Xiang   390
Wu, Zhijun   150

Xu, Dechang   355
Xu, Jing   343
Xu, Rong   413, 435
Xue, Bin   440

Yang, Youngik   442
Yang, Yubo   355
Ye, Yongtao   211, 424
Yiu, Siu-Ming   429
Yiu, Sm   211
Yooseph, Shibu   442
Yu, Ning   366
Yu, Yun   378

Zaslavsky, Leonid   438
Zhang, Jing   96
Zhang, Kaizhong   223, 320
Zhao, Bi   440
Zhao, Bihai   390
Zhong, Cuncong   442
Zhou, Jun   402
Zubair, Mohammad   60

Printed in the United States
By Bookmasters